Molecular Biology and Genetics of the Lepidoptera

CONTEMPORARY TOPICS in ENTOMOLOGY SERIES

THOMAS A. MILLER EDITOR

Insect Symbiosis
Edited by Kostas Bourtzis and Thomas A. Miller

Insect Sounds and Communication: Physiology, Behaviour, Ecology, and Evolution
Edited by Sakis Drosopoulos and Michael F. Claridge

Insect Symbiosis, Volume 2
Edited by Kostas Bourtzis and Thomas A. Miller

Insect Symbiosis, Volume 3
Edited by Kostas Bourtzis and Thomas A. Miller

Food Exploitation by Social Insects: Ecological, Behavioral, and Theoretical Approaches
Edited by Stefan Jarau and Michael Hrncir

Molecular Biology and Genetics of the Lepidoptera
Edited by Marian R. Goldsmith and František Marec

CONTEMPORARY TOPICS in ENTOMOLOGY SERIES

THOMAS A. MILLER EDITOR

Molecular Biology and Genetics of the Lepidoptera

Edited by

Marian R. Goldsmith

František Marec

CRC Press
Taylor & Francis Group
Boca Raton London New York

CRC Press is an imprint of the
Taylor & Francis Group, an **informa** business

CRC Press
Taylor & Francis Group
6000 Broken Sound Parkway NW, Suite 300
Boca Raton, FL 33487-2742

Printed in the United States of America on acid-free paper
10 9 8 7 6 5 4 3 2 1

International Standard Book Number: 978-1-4200-6014-0 (Hardback)

Library of Congress Cataloging-in-Publication Data

Molecular biology and genetics of the Lepidoptera / editors, Marian R. Goldsmith and Frantisek Marec.
p. cm. -- (Contemporary topics in entomology series)
Includes bibliographical references and index.
ISBN 978-1-4200-6014-0 (alk. paper)
1. Lepidoptera--Physiological genomics. 2. Lepidoptera--Molecular genetics. I. Goldsmith, Marian R. II. Marec, Frantisek. III. Title. IV. Series.

QL562.2.M635 2010
595.78--dc22 2009012588

Visit the Taylor & Francis Web site at
http://www.taylorandfrancis.com

and the CRC Press Web site at
http://www.crcpress.com

Contents

Preface

Moths and butterflies, insects of the order Lepidoptera, are among the most diverse and species-rich groups of organisms. (With more than 150,000 species, they comprise the second largest order of animals, after beetles, Coleoptera.) Their primarily phytophagous caterpillars represent important components of mainly terrestrial habitats all over the world. Many species have a significant impact on human society, whether negative as major pests of agriculture and forestry or beneficial as pollinators and food sources for other animals. A special case is the domesticated silkworm, *Bombyx mori*. For many years the rearing of silkworms for silk production, sericulture, was one of the most important industries in more than 30 countries, especially in Asia (China, Japan, Korea, Thailand), South Asia (India), Europe (France, Italy, Russia, Rumania, Bulgaria), and, in the twentieth century, South America (Brazil).

The Lepidoptera are probably the most widely studied group of invertebrates. Butterflies particularly, with their beautiful wing patterns, have attracted the attention of professional researchers and amateur entomologists. However, other than the domesticated silkworm, only a few lepidopteran species have attracted the deeper interest of geneticists. For example, the female heterogamety typical of all Lepidoptera had been first inferred at the beginning of the last century from sex-linked inheritance in the magpie moth, *Abraxas grossulariata*, by L. Doncaster and G.H. Raynor before the sex chromosomes were identified in several lepidopteran species by J. Seiler. The gypsy moth, *Lymantria dispar*, had become well known for the development of intersexuality from studies by R. Goldschmidt in the 1930s. Among early genetic models, we can certainly count the flour moth, *Ephestia kuehniella*, with a collection of eye-pigment mutants that were used in numerous studies on the ommochrome synthesis pathway by A. Kühn and colleagues and were used to develop the one-gene-one-enzyme hypothesis by a member of Kühn's group, E. Caspari. All these and other genetic studies until the late 1960s on formal genetics including chromosome numbers for several hundred species of moths and butterflies have been reviewed comprehensively in *Lepidoptera Genetics* by R. Robinson, published by Pergamon Press in 1971. This book also presents in detail the famous industrial melanism story, which began in 1848 with the discovery of the first *carbonaria* morph of the peppered moth (*Biston betularia*) in Manchester, England. In this book, considerable attention is paid to the evolution and genetics of Batesian mimicry in *Hypolimnas* and *Papilio* butterflies that mimic wing markings of poisonous milkweed butterflies of the genus *Danaus*. Today's textbook theory of Batesian mimicry is based on studies of a number of researchers (among others, J.W.Z. Brower, C.A. Clarke, and P.M. Sheppard) and was further elaborated by D. Charlesworth and B. Charlesworth in 1975.

In many ways this volume is a follow-up to *Molecular Model Systems in the Lepidoptera* (Goldsmith and Wilkins 1995), a collection of articles published in 1995 that highlighted research in selected current or emerging models with the aim of drawing the attention of the scientific community to their special qualities and experimental strengths. Adaptation of phytophagous lab-reared animals to artificial diet and relatively large size were major conveniences at the time that placed the silkworm in the forefront of basic research in countries where sericulture was deep-rooted and established the tobacco hornworm, *Manduca sexta*, as the premier model in the United States for many kinds of fundamental studies, though not genetics. Both insects remain central models in the lepidopteran pantheon, with *Bombyx* remaining the most advanced in molecular genetics and structural and functional genomics and *Manduca* still among the first in areas like innate immunity, olfaction and neurobiology, endocrinology, and biochemistry. With advances in molecular tools and genomic techniques, size and mass-rearing are no longer critical, so that the choice of lepidopteran model systems, as seen in the current volume, increasingly reflects their advantages for illuminating experimental questions peculiar to butterflies and moths. We have not attempted to be comprehensive, but again to provide a showcase for projects representing fundamental biological problems and practical ones arising from a pressing need to find new ways to control lepidopteran pests. We apologize for omitting important model systems and key areas of investigation, and trust that the research presented here will inspire readers to look further into the literature for additional fascinating and inspiring stories.

This volume begins with an overview of the current status of lepidopteran phylogenetics (Chapter 1), which is essential for placing what we know in an evolutionary context, with a focus on clades representing model systems. Subsequent chapters review the richness and diversity of genomic and post-genomic resources now available or under development for individual species or genera, including the silkworm (Chapter 2), the remarkable Müllerian mimics, the *Heliconius* butterflies (Chapter 6), and the major crop pest genus, *Helicoverpa* (Chapter 12). An important new theme that has emerged since the earlier volume is the role of sexual dimorphism on many aspects of lepidopteran biology, including chromosome structure, sex chromosome systems, and sex determination (Chapters 3 and 4), and mating behavior associated with butterfly vision (Chapter 7), circadian clocks (Chapter 8), and pheromone production and perception (Chapter 10). Similarly, in the past decade significant progress has been made in uncovering genes controlling the development of butterfly wings and wing patterns, which is gradually yielding information in an evolutionary context (Chapters 5 and 6). Knowledge of the diversity in structure and function of lepidopteran chemoreceptors in comparison with those of other major insect orders (Chapter 9) represents another recent breakthrough that has depended on the extensive genome sequencing project in *Bombyx*.

Turning to another dominant theme, many topics covered in this volume are connected directly or indirectly with the critical area of insect control. These include using Lepidoptera to explore the genetics and neurobiology of host range specificity (Chapter 11), to define mechanisms of insecticide resistance (Chapter 13), to augment already well-established fundamental knowledge of the innate insect immune response (Chapter 14), and to explore the interactions between polydnavirus-carrying parasitoids and their lepidopteran hosts (Chapter 17). Examples of practical applications in pest control include testing lepidopteran-derived antimicrobial peptides and virulence factors as potential therapeutic agents against human pathogens or to produce disease-resistant plants (Chapter 15), and delivering a variety of intrahemocoelic toxins by different means (Chapter 16). Encompassing the final theme of virus delivery systems and function, we end with the first reported case of map-based or positional cloning in a lepidopteran, the successful isolation of *nsd-2*, a gene conferring resistance to the densovirus *Bm*DNV-2, a serious pathogen in sericulture (Chapter 18). This, like so many of the projects described in this volume, represents a high point of molecular genetic research in Lepidoptera, using strategies, methods, and sequence information that were barely conceived in the mid-1990s. With the recent publication of "build 2" of the *Bombyx* genome (see Special Issue on the Silkworm Genome, *Insect Biochemistry and Molecular Biology*, 38(12), December 2008) and the prospects of many more lepidopteran genome sequencing projects to follow, we look forward to being able to report on even more rapid progress in the near future.

In closing we would like to express heartfelt thanks to Tom Miller, the editor for the Contemporary Topics in Entomology Series of CRC Press, who conceived of the new Lepidoptera book project and recruited us as editors; without his cheerful encouragement, constant cajoling, gentle arm twisting, creative ways of helping us keep up momentum, and faith in us it is doubtful we could have gotten the job done. We would also like to thank John Sulzycki, senior editor at CRC Press, for his infinite patience, and Pat Roberson, our Taylor & Francis Group production coordinator, for hers; her helpful and timely responses to our questions moved the project from conception to reality in ways she will probably never know. Our thanks are also due to Gail Renard for preparation of the proofs, and to Kelly Pennoyer, who performed endless tedious tasks to get the manuscript ready for the first stage of publication. And finally, grateful thanks to our many authors for their hard work, timely submissions, gracious responses to all of our comments and suggestions, and enthusiastic support of the project, despite busy schedules and occasional communication gaps on our end. Your contributions are invaluable.

Marian R. Goldsmith and František Marec
Kingston and České Budějovice

About the Editors

Marian R. Goldsmith, Ph.D., is professor and chair of the Department of Biological Sciences at the University of Rhode Island. Together with Fotis Kafatos, in 1988 she cofounded the International Workshop on the Molecular Biology and Genetics of the Lepidoptera, which is held every two to three years in Kolympari, Crete, bringing together lepidopteran researchers from around the world and serving as the only forum of its kind. Working primarily as a geneticist to bring molecular tools and techniques to bear on the domesticated silkworm as a model system, she has established strong collaborations with silkworm groups in Japan, France, and China, starting in her postdoctoral years when she first brought *Bombyx mori* to the laboratory of Fotis Kafatos at Harvard University in 1972 to study regulation and expression of chorion genes after a two-month stay in the laboratories of Toshio Ito at the Sericultural Experiment Station (now National Institute of Agrobiological Sciences [NIAS]) and Yataro Tazima at the National Institute of Genetics in Misima, sponsored by a National Science Foundation fellowship. This was followed by many sabbatical leaves and shorter stays during which she established collaborations with Bungo Sakaguchi and Hiroshi Doira at Kyushu University, Hideaki Maekawa at the National Institute of Health in Tokyo, Toshiki Tamura and Wajirou Hara at the National Institute of Sericultural and Entomological Science in Tsukuba, and Masahiko Kobayashi and Toru Shimada at Tokyo University, where she was a visiting professor in 1997–1998. At that time she began a long-standing collaboration with Kazuei Mita at NIAS helping to bring the silkworm genome project to fruition. More recently she served as a guest professor at the Shanghai Institute of Plant Physiology and Ecology in the laboratory of Yongping Huang and Xuexia Miao, developing tools for mapping and positional cloning of complex traits. She is a fellow of the American Association for the Advancement of Science, and in 2002 received the Sanshi-gaku-sho Annual Prize of the Japan Society of Sericultural Science for "Scientific Achievement Related to the Molecular Genetic Study on Chorion Gene Groups in *Bombyx mori*." She coedited *Molecular Model Systems in the Lepidoptera* (Cambridge University Press) with Adam Wilkins in 1995.

František Marec, CSc., is senior researcher at the Institute of Entomology, Biology Centre of the Academy of Sciences of the Czech Republic, and professor of molecular and cell biology and genetics at the Faculty of Science, University of South Bohemia, both in České Budějovice, Czech Republic. Since 1990 he has been a research fellow of the Alexander von Humboldt Foundation, Bonn, Germany. He has had a long-standing collaboration with Walther Traut at the Institute of Biology, Medical University of Lübeck, Germany, where he worked as a Humboldtian in 1991–1992 and 1998. At the end of the nineties he also established a strong collaboration with Ken Sahara at the Laboratory of Applied Molecular Entomology at the Hokkaido University, Sapporo, Japan. Like many geneticists, he started his career in *Drosophila*, being interested in mutation genetics, but soon become fascinated by Lepidoptera genetics. His early research interests included chemical and radiation mutagenesis targeted to the development of genetic control strategies against lepidopteran pests, chromosomal mechanisms of resistance of lepidopteran species to ionizing radiation, and meiotic chromosome pairing including formation of the synaptonemal complex, a unique nuclear structure mediating intimate association of two homologous chromosomes. Most of his early work was done in the flour moth *Ephestia kuehniella*, historically the second model of Lepidoptera genetics after *Bombyx mori*. More recent projects have focused on molecular composition of insect telomeres and phylogeny of insect telomeric DNA repeats, and on the study of molecular differentiation and evolution of sex chromosomes in Lepidoptera using advanced methods of molecular cytogenetics. In addition, his co-operation with the Insect Pest Control Section, Joint FAO/IAEA Division of Nuclear Techniques in Food and Agriculture, International Atomic Energy Agency, Vienna, Austria, resulted in a new approach for the construction of genetic sexing strains in lepidopteran pests. Based on this approach, he works on the development of transgenic genetic sexing strains in the codling moth, *Cydia pomonella*, aimed at generating male-only progeny for the control of this pest using sterile insect technique or inherited sterility strategies. This work is done in collaboration with Lisa G. Neven at the USDA—Agricultural Research Service, Yakima Agricultural Research Laboratory, Wapato, Washington.

Contributors

Hiroaki Abe
Department of Biological Production
Faculty of Agriculture
Tokyo University of Agriculture and Technology
Tokyo, Japan

Joaquin Baixeras
Cavanilles Institute of Biodiversity and Evolutionary Biology
University of Valencia
Valencia, Spain

Philip Batterham
Department of Genetics
Bio21 Institute
University of Melbourne
Parkville, Victoria, Australia

Simon W. Baxter
Department of Zoology
School of Biology
University of Cambridge
Cambridge, United Kingdom

Patrícia Beldade
Evolutionary Biology Group
Institute of Biology
Leiden University
Leiden, The Netherlands

Bryony C. Bonning
Department of Entomology
Iowa State University
Ames, Iowa

Adriana D. Briscoe
Department of Ecology and Evolutionary Biology
University of California, Irvine
Irvine, California

John Brown
Systematic Entomology Laboratory
USDA
National Museum of Natural History
Smithsonian Institution
Washington, D.C.

Nicola Chamberlain
School of Biological Sciences
University of Exeter in Cornwall
Penryn, United Kingdom

Derek Collinge
CSIRO Entomology
Canberra, Australia
and
School of Biochemistry and Molecular Biology
Australia National University
Canberra, Australia

Michael P. Cummings
Center for Bioinformatics and Computational Biology
University of Maryland
College Park, Maryland

Donald R. Davis
Department of Entomology
National Museum of Natural History
Smithsonian Institution
Washington, D.C.

Richard H. ffrench-Constant
School of Biological Sciences
University of Exeter in Cornwall
Penryn, United Kingdom

Tsuguru Fujii
Graduate School of Agriculture and Life Science
The University of Tokyo
Tokyo, Japan

Marian R. Goldsmith
Biological Sciences Department
University of Rhode Island
Kingston, Rhode Island

Karl Gordon
CSIRO Entomology
Canberra, Australia

Fred Gould
Department of Entomology and the Keck Center for Behavioral Biology
North Carolina State University
Raleigh, North Carolina

Astrid T. Groot
Max Planck Institute for Chemical Ecology
Department of Entomology
Jena, Germany

David G. Heckel
Department of Entomology
Max Planck Institute for Chemical Ecology
Jena, Germany

Keith R. Hopper
United States Department of Agriculture
Agricultural Research Service
Newark, Delaware

Chris D. Jiggins
Department of Zoology
School of Biology
University of Cambridge
Cambridge, United Kingdom

Keiko Kadono-Okuda
National Institute of Agrobiological Sciences
Tsukuba, Japan

Michael R. Kanost
Department of Biochemistry
Kansas State University
Manhattan, Kansas

Akito Y. Kawahara
Department of Entomology
University of Maryland
College Park, Maryland

František Marec
Biology Centre ASCR
Institute of Entomology
and
Faculty of Science
University of South Bohemia
České Budějovice, Czech Republic

Owen McMillan
Department of Genetics
North Carolina State University
Raleigh, North Carolina

Christine Merlin
Department of Neurobiology
University of Massachusetts Medical School
Worcester, Massachusetts

James B. Nardi
Department of Entomology
University of Illinois
Urbana, Illinois

Sara J. Oppenheim
Department of Entomology
North Carolina State University
Raleigh, North Carolina

Cynthia S. Parr
Human-Computer Interaction Lab
University of Maryland
College Park, Maryland

Jerome C. Regier
University of Maryland Biotechnology Institute
University of Maryland
College Park, Maryland

Steven M. Reppert
Department of Neurobiology
University of Massachusetts Medical School
Worcester, Massachusetts

Hugh M. Robertson
Department of Entomology
University of Illinois at Urbana-Champaign
Urbana, Illinois

Amanda D. Roe
Department of Entomology
University of Minnesota
St. Paul, Minnesota

Daniel Rubinoff
Department of Plant and Environmental Protection Sciences
University of Hawaii
Honolulu, Hawaii

Suzanne V. Saenko
Evolutionary Biology Group
Institute of Biology
Leiden University
Leiden, The Netherlands

Ken Sahara
Laboratory of Applied Molecular Entomology
Research Institute of Agriculture
Hokkaido University
Sapporo, Japan

Coby Schal
Department of Entomology and the Keck Center for Behavioral Biology
North Carolina State University
Raleigh, North Carolina

Nina Richtman Schmidt
Department of Entomology
Iowa State University
Ames, Iowa

Toru Shimada
Graduate School of Agriculture and Life Science
The University of Tokyo
Tokyo, Japan

Thomas J. Simonsen
Department of Biological Sciences
University of Alberta
Edmonton, Alberta, Canada

Marilou P. Sison-Mangus
Department of Ecology and Evolutionary Biology
University of California, Irvine
Irvine, California

Michael R. Strand
Department of Entomology
University of Georgia
Athens, Georgia

Wee Tek Tay
CSIRO Entomology
Canberra, Australia

Walther Traut
Universität Lübeck
Zentrum für Medizinische Strukturbiologie
Institut für Biologie
Lübeck, Germany

Gissella M. Vásquez
Department of Entomology and the Keck Center for Behavioral Biology
North Carolina State University
Raleigh, North Carolina

Andreas Vilcinskas
Institute of Phytopathology and Applied Zoology
Interdisciplinary Research Center
Justus-Liebig-University of Giessen
Giessen, Germany

Niklas Wahlberg
Department of Biology
University of Turku
Turku, Finland

Kevin W. Wanner
Department of Plant Sciences and Plant Pathology
Montana State University
Bozeman, Montana

Susan J. Weller
Department of Entomology
University of Minnesota
St. Paul, Minnesota

Adam Williams
Department of Genetics
Bio21 Institute
University of Melbourne
Parkville, Victoria, Australia

Andreas Zwick
Department of Entomology
University of Maryland
College Park, Maryland

1 Evolutionary Framework for Lepidoptera Model Systems

Amanda D. Roe, Susan J. Weller, Joaquin Baixeras, John Brown, Michael P. Cummings, Donald R. Davis, Akito Y. Kawahara, Cynthia S. Parr, Jerome C. Regier, Daniel Rubinoff, Thomas J. Simonsen, Niklas Wahlberg, and Andreas Zwick

CONTENTS

INTRODUCTION

Lepidoptera are among the most diverse and easily recognized organisms on the planet, with at least 150,000 described species (Kristensen and Skalski 1998). They are one of the four mega-diverse orders of holometabolous insects, together with Diptera (flies), Coleoptera (beetles), and Hymenoptera (wasps, bees, and ants). Butterflies alone are more numerous than birds, Class Aves, with approximately 18,000 species (Kristensen and Skalski 1998). Generally, Lepidoptera are characterized by the presence of scaled wings, elongate sucking mouthparts (proboscis), and complete (holometabolous) development where the larval stages are commonly referred to as "caterpillars." Historically, species of Lepidoptera have proven invaluable model systems in the fields of development, genetics, molecular biology, physiology, evolution, and ecology (e.g., Bates 1861; Müller 1879; Ford 1964; Ehrlich and Raven 1967; Kettlewell 1973). Interest in Lepidoptera species as model systems stems from a number of biological characteristics that render this group amenable for study (Bolker 1995). Lepidopterans are charismatic, due mainly to their striking variety of wing color patterns and larval morphologies, and they are avidly collected by professionals and amateurs alike

FIGURE 1.1 ***A color version of this figure follows page 176.*** Representatives of superfamilies containing model systems. A: Bombycoidea, *Anthela oressarcha* (A. Zwick); B: Bombycoidea, *Antheraea larissa* (A. Kawahara); C: Noctuoidea, *Trichoplusia ni* (M. Dreiling); D: Noctuoidea, *Tyria jacobaeae* (D. Dictchburn); E: Papilionoidea, *Bicyclus anynana* (A. Monteiro and W. Piel); F: Papilionoidea, *Heliconius erato* (K. Garwood); G: Pyraloidea, *Ostrinia nubilalis* (S. Nanz); H: Tortricoidea, *Cydia pomonella* (N. Schneider).

(Figure 1.1; Salmon 2000). Many common species have large larvae, which facilitated early studies in development and disease, and these larvae are often relatively easy to rear in the laboratory. The economic impact of Lepidoptera on human society also has contributed to the development of lepidopteran model systems. Silk moths (*Bombyx mori*), for example, are among the few insects considered "domesticated" by humans because of the long tradition of sericulture in Asia. As a primarily phytophagous clade, many lepidopterans are economically important as major pests of agriculture and forestry. Several species discussed in this book were developed initially as model systems to understand how Lepidoptera locate mates and host plants, with the goal of using this knowledge to manage pest populations.

In addition to practical applications, lepidopteran model systems have provided insights into basic research including wing pattern formation, neural development, and the interaction of developmental genes (e.g., *Bicyclus*, *Manduca*). Further, there are a number of lepidopteran genome projects (Mita et al. 2004; Xia et al. 2004; Jiggins et al. 2005). Currently, four Genome Projects are listed on GenBank (http://www.ncbi.nlm.nih.gov, accessed April 16, 2008), including *B. mori* (Bombycidae), *Bicyclus anynana* (Nymphalidae), *Melitaea cinxia* (Nymphalidae), and *Spodoptera frugiperda* (Noctuidae). Comparatively, lepidopterans have relatively large genomes: *B. mori* is estimated at 475 Mbp, *Manduca sexta* at 500 Mbp, *Heliothis virescens* at 400 Mbp, and *Heliconius* at 292 Mbp (J.S. Johnston, unpublished; Goldsmith, Shimada, and Abe 2005; Jiggins et al. 2005). These genomes can be up to ~2.5 times larger than the previously described genome for *Drosophila melanogaster* (175 Mbp), and up to ~1.6 times larger than either *Apis mellifera* (236 Mbp) or *Anopheles gambiae* (280 Mbp) (Goldsmith, Shimada, and Abe 2005; Honeybee Genome Sequencing Consortium 2006). Further insights into lepidopteran genomes are provided in Chapters 2 and 6 and references therein.

In this chapter, we review how initial model system choice affects subsequent generalizations and the role of phylogenetic studies in placing model systems into a broader evolutionary context. We then review known phylogenetic relationships within superfamilies that contain multiple model systems and phylogenetic placement of these model systems. Our current knowledge of relationships among these superfamilies is being challenged by recent molecular studies, and dating the newly proposed divergences is complicated by an incomplete and poorly identified fossil record. Finally, we examine new global initiatives in lepidopteran phylogenetics that hold promise to connect a historically fragmented community. These global initiatives promise to foster a new age of lepidopteran systematics research.

Phylogenetics and Model Systems

Model systems allow researchers to focus resources and effort on examining fundamental biological questions in detail. Although this focused study is essential, the true power of model systems lies in the subsequent ability to extrapolate these details across larger groups of organisms (Kellogg and Shaffer 1993; Bolker 1995). To generalize these results, comparative studies are essential and require that model systems be placed into their evolutionary context. An evolutionary framework or phylogeny can be inferred using a number of analytical approaches (Swofford et al. 1996; Holder and Lewis 2003). Phylogenetic analyses have used a range of heritable, independent characters, such as molecular (e.g., DNA, RNA, and amino acid sequences), phenotypic (e.g., morphological structures, allozymes), and developmental traits (e.g., ontogenetic stages or pathways) to infer evolutionary relationships among organisms.

Without phylogenies, knowledge gained from model systems would remain in isolation. We would be unable to generalize among silk moths (e.g., *B. mori*), European corn borer (*Ostrinia nubilalis*), and fruit fly (*Drosophila*). By understanding evolutionary relationships among organisms, hypotheses concerning the origin of key innovations (i.e., character evolution) can be generated throughout a group (Mabee 2000; Collins et al. 2005). Phylogenetics provides the means to reconstruct ancestral character states and can provide insight into character polarity and homology (Mabee 2000; Felsenstein 2004). For example, the phylogeny of Arctiidae shows that the key innovation of pyrrolizidine alkaloid sequestration is ancestral to Arctiinae, the largest subfamily. Reconstruction of subsequent losses and gains within the subfamily then can be examined on a finer scale (e.g., Weller, Jacobson, and Conner 1999; DaCosta et al. 2006).

Phylogenetic reconstruction also allows the correlation of gene expression to morphological character expression, which may be used to hypothesize the role genes play on morphological character evolution. Long-standing morphological arguments concerning evolution of head segmentation and brains in arthropods are resolved in large part through studies on *Hox* gene expression (e.g., Cook et al. 2001). Thus, when placed in a phylogenetic context, research on model systems can provide important insights into previously intractable questions concerning morphological evolution.

In addition to providing insight into the evolution of genetic, developmental, and morphological traits, phylogenetics also can be used to identify potential bias in the use of model systems (Bolker 1995). The characteristics of model systems (e.g., ease of culture, body size, and economic importance) can inadvertently influence conclusions drawn from other studies conducted on the system. If not placed into a proper phylogenetic context, model systems may result in misleading inferences about the traits of the larger group. By placing model systems into an evolutionary framework with other models and nonmodels, potential sources of bias can be identified and erroneous generalizations avoided.

Overview of Lepidopteran Phylogeny

Currently, Lepidoptera are arranged into 126 families and 46 superfamilies (Kristensen and Skalski 1998). The current estimate of evolutionary relationships among superfamilies (summarized in Figure 1.2) is a patchwork of variably resolved phylogenies. The basal lepidopteran relationships are relatively well established based on morphology (Kristensen 1984; Davis 1986; Nielsen and Kristensen 1996; Kristensen and Skalski 1998) and confirmed with molecular data (Wiegmann et al. 2000; Wiegmann, Regier, and Mitter 2002). These early-diverging lineages or "non-ditrysians" (named groups below node 7) contain only a fraction of the total lepidopteran species diversity and contain very few model systems (Figure 1.2), and therefore are not discussed further. In contrast, the clade Ditrysia (node 7) includes approximately 99 percent of the described species, but among-superfamily relationships are poorly understood. This clade, however, contains all current model systems. Members in this lineage are grouped by a number of shared derived morphological traits (termed "synapomorphies") including specialized female genitalia (Kristensen and Skalski 1998). A review by Minet (1991) proposed several nested higher-groupings within Ditrysia based on the morphology of all life stages and hypothesized ground plans for superfamilies. We follow recent convention in provisionally adopting Minet's arrangement for presentation purposes, although this arrangement has not been tested by phylogenetic analysis until recently (see Current Research and Future Directions).

We provide a phylogenetic overview of those superfamilies that possess multiple model systems. We comment, when possible, on whether the distribution of model systems adequately captures the phylogenetic diversity of the superfamily and whether results of these studies can be extrapolated confidently to other members. We do not treat superfamilies with single model systems even if they have an extensive literature because phylogenetically, it is a sample size of one. We present our discussion in order of their phylogenetic placement in Figure 1.2, working from the tree base toward the crown.

Overview of Selected Ditrysian Superfamilies

Within Ditrysia, multiple model systems are found in six of the thirty-three superfamilies: Bombycoidea (e.g., silkworm moths), Noctuoidea (e.g., tiger, gypsy, cutworm moths), Papilionoidea and Hesperioidea (butterflies and skippers), Pyraloidea (snout moths), and Tortricoidea (leaf rollers) (Figure 1.1). Tortricoidea, along with several basal lineages of Lepidoptera, are sometimes referred to as "microlepidoptera" because of their small size. However, the term is confusing because

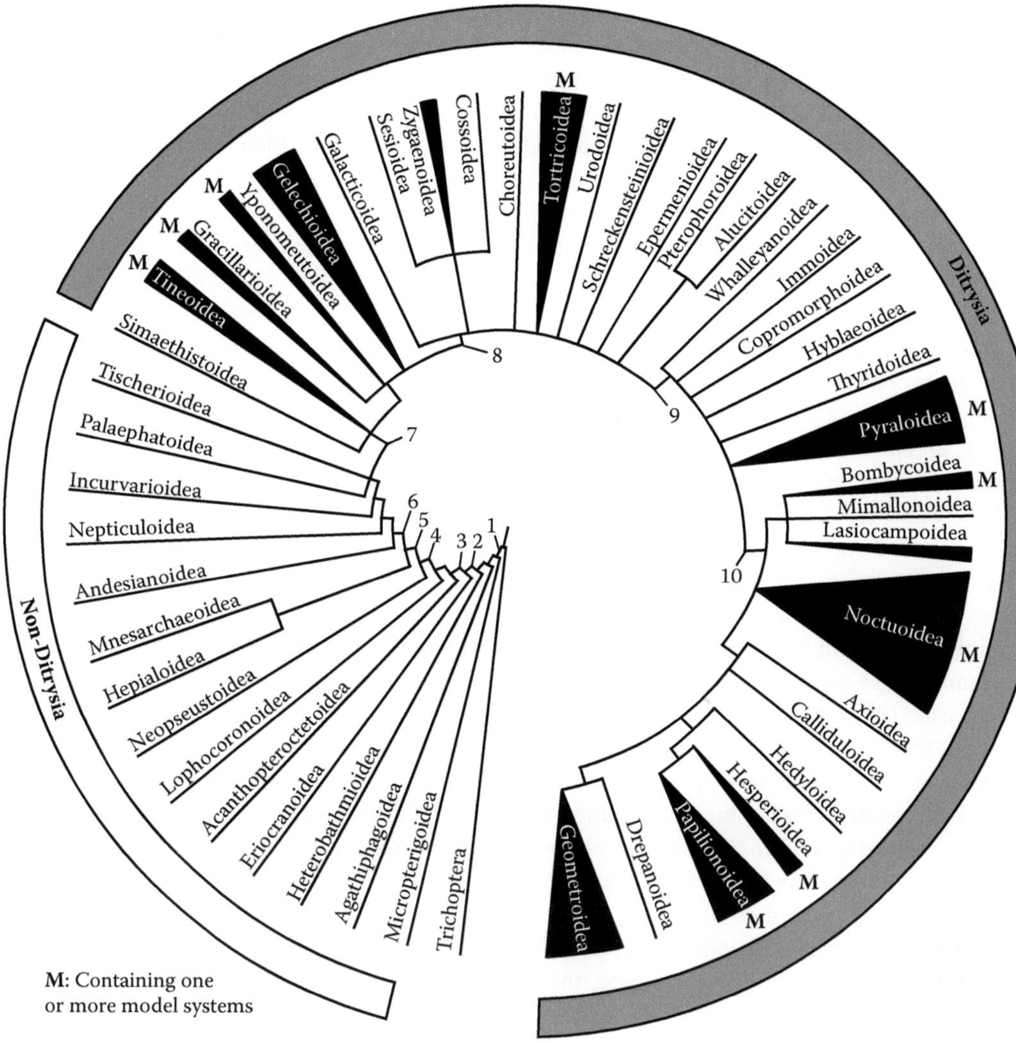

FIGURE 1.2 Current hypothesized phylogeny of Lepidoptera superfamilies, based on morphological characters, including Trichoptera, sister group to Lepidoptera (adapted from Figure 2.2 in Kristensen and Skalski 1998). Nodes representing higher-level classifications are labeled, for example, macrolepidoptera; 1 = Lepidoptera, 2 = Glossata, 3 = Coelolepida, 4 = Myoglossata, 5 = Neolepidoptera, 6 = Heteroneura, 7 = Ditrysia (shown above), 8 = Apoditrysia, 9 = Obtectomera, 10 = Macrolepidoptera. Superfamilies containing model systems are shown. Width of branches represents approximate proportions of currently described species.

small body size occurs in "macrolepidoptera" (e.g., Micronoctuidae; Fibiger and Lafontaine 2005) and large moths can occur in lineages of microlepidoptera (e.g., Limacodidae, slug caterpillars; Hepialidae, ghost moths). Despite the concentration of lepidopteran model systems in Ditrysia, a number of superfamilies are not represented by model systems (Figure 1.2). Three diverse superfamilies that lack developed model systems are Geometroidea (e.g., inchworms), Gelechioidea (e.g., case bearers, concealer moths, and twirler moths), and Zygaenoidea (e.g., burnets and slug caterpillars), and these superfamilies should be important targets for identifying future model systems.

Tortricoidea

Tortricoidea, comprised of the single family Tortricidae, are second only to Gelechioidea in terms of species richness among major microlepidopteran lineages, with about 9,100 described species

(Brown 2005). The monophyly of Tortricidae is supported by a variety of characters of the adults, larvae, and pupae. Although several groups within Tortricidae have been considered distinct families by one or more authors over the last century, it is now generally accepted that these groups represent subordinate taxa within the family (Horak 1998). The family is currently comprised of three subfamilies, Tortricinae, Chlidanotinae, and Olethreutinae, into which twenty-two tribes and 957 genera are arranged (Horak 1998; Brown 2005). Olethreutinae and Chlidanotinae are supported by convincing morphological synapomorphies, but Tortricinae is almost certainly para- or polyphyletic. Tortricid species occur on all continents except Antarctica; greatest species richness is attained in the New World tropics, where a large percent of the fauna remains to be described (Horak 1998; Brown 2005).

The common name "leaf rollers" has been applied to the family, owing to the prevalent larval habit of shelter building by folding or rolling leaves of the food plant, but tortricid larvae employ a wide range of feeding strategies, including gall inducing, stem and root boring, fruit boring, seed predating, and flower feeding. Additionally, a very few are predators or occur as inquilines. Among other characters, many tortricid adults can be recognized by a somewhat bell-shaped silhouette with the wings held rooflike when at rest.

Many tortricids are important pests of agricultural, forest, and ornamental plants—164 genera and 687 species have been recorded worldwide as economically important (Zhang 1994). Among the most thoroughly studied tortricids is spruce budworm (*Choristoneura fumiferana,* Tortricinae: Archipini), which is an important forest pest in North America, specializing on species of Pinaceae. This organism has been investigated from the perspectives of morphology (e.g., Walters, Albert, and Zacharuk 1998), pheromone chemistry (e.g., Delisle, Picimbon, and Simard 1999), host plant preferences (e.g., Albert 1991, 2003), physiology (e.g., Hock, Albert, and Sandoval 2007), behavior (e.g., Wallace, Albert, and McNeil 2004), and parasitoids and pathogens (e.g., Quayle et al. 2003). Owing to the relatively unambiguous relationship among *Choristoneura*, *Archips*, *Argyrotaenia*, *Pandemis*, *Adoxophyes*, *Clepsis*, and several other genera in Archipini, many of the morphological, biological, and ecological features of spruce budworm can be extrapolated to a single, relatively large clade with a degree of confidence.

The codling moth (*Cydia pomonella*, Olethreutinae: Grapholitini) represents another tortricid model system. Described from Europe, the species is virtually cosmopolitan today. Although primarily a pest of cultivated apples and pears (*Malus* spp. and *Pyrus* spp.; Rosaceae), this species' documented host range includes plants in six different families of dicotyledons: Fagaceae, Rutaceae, Rosaceae, Moraceae, Juglandaceae, and Proteaceae. Countless scientific studies over the past thirty years have focused on characterizing its sex pheromones, antennal receptors, pheromone production glands, and mating behaviors with the goal of disrupting reproduction (e.g., Ahmad and Al-Gharbawi 1986; Arn 1991; McDonough et al. 1993; Backman 1997; El-Sayed et al. 1999; Addison 2005; Trematerra and Sciarretta 2005). The quarantine significance of this species exerts considerable pressure on international trade agreements in regard to specific agricultural commodities (e.g., Wearing et al. 2001).

Other species of Tortricidae that have received substantial attention include Oriental fruit moth (*Grapholita molesta*, Olethreutinae: Grapholitini), which, like the codling moth, has become nearly cosmopolitan as a pest of stone and pome fruits; red-banded leafroller (*Argyrotaenia velutinana*, Tortricinae: Archipini), a highly polyphagous pest of fruit trees in North America; false codling moth (*Thaumatotibia leucotreta*, Olethreutinae: Grapholitini), an important pest of species of Solanaceae (*Capsicum* and *Solanum*) and Rutaceae (*Citrus*) in Africa; and obliquebanded leaf roller (*Choristoneura rosaceana*, Tortricinae: Archipini), another broadly polyphagous North American leaf roller. Despite the economic importance of species in Tortricidae, we lack a phylogeny for the family, and we cannot comment on the current phylogenetic distribution of these model systems, which are concentrated in only two tribes, Archipini and Grapholitini.

Pyraloidea

Pyraloidea, or snout moths, are a large superfamily with approximately sixteen thousand described species (Heppner 1991; Munroe and Solis 1998) and at least as many remaining to be described. A number of morphological characters support the monophyly of Pyraloidea, including the presence of paired tympanal organs (membranous hearing structures) on the second abdominal segment, a basally scaled proboscis (when present), and a characteristic wing venation pattern (Munroe and Solis 1998; Nuss 2006).

Larval habits in Pyraloidea are highly diverse. Most larvae are concealed feeders, feeding primarily on plant tissue either internally or in webbed foliage, silk, or frass (Neunzig 1987; Munroe and Solis 1998). Larvae of some species are predatory or parasitic in nests of Hymenoptera or on scale insects (e.g., Neunzig 1997), while others scavenge on nonliving plant material (e.g., stored products). Many larvae of Acentropinae (Crambidae) have developed aquatic lifestyles, feeding on submerged plants (e.g., water veneer, *Acentria ephemerella*, Crambidae: Acentropinae). Many pyraloid larvae are also important economic pests. European corn borer (*Ostrinia nubilalis*, Crambidae: Pyraustinae) and rice stem borers (*Chilo* spp. and *Scirpophaga* spp., Crambidae: Crambinae and Schoenobiinae, respectively) damage a variety of field crops, and Indian meal moth (*Plodia interpunctella*, Pyralidae: Phycitinae), Mediterranean flour moth (*Ephestia kuehniella*, Pyralidae: Phycitinae), and almond moth (*Cadra cautella*, Pyralidae: Phycitinae) are important dried-product pests (Neunzig 1987). Wax moths (*Galleria mellonella* and *Achroia grisella*, Pyralidae: Galleriinae) damage the nests and hives of bees (Neunzig 1987). Conversely, some pyralids have been used for biological control of invasive plants, although doing so carries risks. *Cactoblastis cactorum* (Pyralidae: Phycitinae), introduced for control of prickly pear cactus in Australia and elsewhere (Common 1990; Zimmermann, Moran, and Hoffmann 2000), recently has become an invasive pest of native cacti in North America (Solis, Hight, and Gordon 2004).

Currently, Pyraloidea is divided into two families, Pyralidae and Crambidae, based on a number of morphological characters (Minet 1983; Munroe and Solis 1998; Goater, Nuss, and Speidel 2005). Pyralidae contains five subfamilies, and Crambidae, sixteen to seventeen (Munroe and Solis 1998; Solis and Maes 2002; Goater, Nuss, and Speidel 2005). As is evident in the composite tree (Figure 1.3), existing morphology-based hypotheses provide little resolution of relationships among subfamilies (Solis and Mitter 1992; Solis and Maes 2002). However, ongoing molecular studies show strong promise for sorting out pyraloid relationships.

At least five pyraloid species, all pests, have become major model systems (Figure 1.3), including European corn borer (*O. nubilalis*) and rice stem borer (*Chilo suppressalis*, Crambidae: Crambinae). These two model systems have been used to study the genetics of pheromone synthesis and neural biology (Roelofs and Rooney 2003; Jurenka 2004), physiology (Hodkova and Hodek 2004; Srinivasan, Giri, and Gupta 2006), and development of insecticide resistance (Coates, Hellmich, and Lewis 2006). Greater wax moth (*Galleria mellonella*) has been used extensively to study the genetics and physiology of immune response and to model human disease pathogens (see Chapter 15). Indian meal moth (*Plodia interpunctella*) and the Mediterranean meal moth (*Ephestia kuehniella*) have been used to study, among other topics, lepidopteran gut physiology, pheromone detection, silk biosynthesis and structure, and the development of insecticide resistance (Beckemeyer and Shirk 2004; Srinivasan, Giri, and Gupta 2006; Siaussat et al. 2007). Historically, *E. kuehniella* was an early genetic model system for the study of pigment biosynthesis and wing pattern development, even prior to the development of *Drosophila melanogaster* as a standard laboratory model (Robinson 1971; Leibenguth 1986).

Although presumably not selected for this purpose, the pyraloid model systems are dispersed in an almost ideal fashion across the phylogeny, representing a diversity of subfamilies in both families. This distribution should maximize the confidence with which conclusions from the model systems collectively can be extrapolated to the rest of the superfamily, although these conclusions could be biased from being based only on pest species.

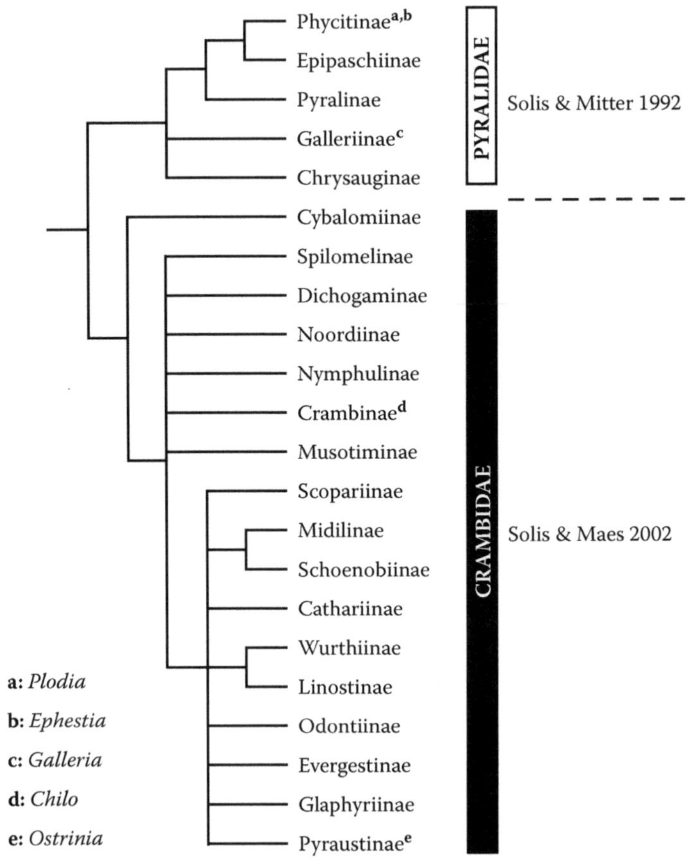

FIGURE 1.3 A composite tree illustrating the inferred relationships among the subfamilies of Pyraloidea based on adult morphology, following Solis and Mitter (1992; Crambidae) and Solis and Maes (2002; Pyralidae). Superscripts indicate placement of model systems.

Papilionoidea and Hesperioidea

The Papilionoidea (true butterflies) and Hesperioidea (skippers) are undoubtedly among the better known groups of insects, both among scientists and the general public (Grimaldi and Engel 2005). Their diurnal habits, aesthetically appealing appearance, and ease of study place them among the most collected and observed insects. With approximately eighteen thousand species described, the alpha taxonomy of Papilionoidea is fairly well studied compared with moths, and the total number of species may be close to its actual diversity. Also, the natural history of many species has been documented. Butterflies have been model organisms in numerous areas of biological sciences, including conservation biology, ecology, physiology, evolution, evolutionary developmental biology (evo-devo), and molecular biology. Several volumes attest to this diversity of scientific interest (Vane-Wright and Ackery 1984; Nijhout 1991; Boggs, Watt, and Ehrlich 2003; Ehrlich and Hanski 2004).

The evolutionary relationships of the major lineages have been the subject of numerous studies starting with the work of Reuter (1896). The seminal paper by Ehrlich (1958) set the stage for future work based on rigorous analyses of character sets to infer relationships of the major lineages in the group. Papilionoidea and Hesperioidea have long been considered sister taxa, mainly based on their diurnal habits, but six potential morphological characters also unite them (de Jong, Vane-Wright, and Ackery 1996). Recent molecular data support the hypothesis that the two superfamilies are sister groups (Wahlberg et al. 2005).

Hesperioidea contains just Hesperiidae, the skippers, with about four thousand species. Papilionoidea contains the true butterflies, which traditionally have been placed in four or five families (e.g., Ehrlich 1958; Kristensen 1976; de Jong, Vane-Wright, and Ackery 1996; Ackery, de Jong, and Vane-Wright 1998; Wahlberg et al. 2005). Papilionidae and Pieridae are well defined, and their circumscriptions have remained stable ever since they were proposed in the early 1800s. Recent work within these families has resolved relationships of subfamilies and other subordinate groups (Caterino et al. 2001; Braby, Vila, and Pierce 2006; Nazari, Zakharov, and Sperling 2007). Riodinidae, in contrast, has been treated as either an independent family (e.g., Eliot 1973; Lamas 2004; Wahlberg et al. 2005) or a subfamily of Lycaenidae (e.g., Ehrlich 1958; Kristensen 1976; de Jong, Vane-Wright, and Ackery 1996; Ackery, de Jong, and Vane-Wright 1998). Recent molecular data suggest that riodinids form an independent lineage sister to Lycaenidae, and thus they should be considered a separate family (Wahlberg et al. 2005).

Nymphalidae has been a source of much confusion, with some authors dividing it into nine different families including Danaidae and Ithomiidae (e.g., Smart 1975). However, many of these "families" place as lineages nested within Nymphalidae (Ehrlich 1958; de Jong, Vane-Wright, and Ackery 1996; Brower 2000; Wahlberg, Weingartner, and Nylin 2003; Freitas and Brown 2004), although the only morphological synapomorphy for the family is three longitudinal ridges on the antennae (Kristensen 1976; Ackery, de Jong, and Vane-Wright 1998). A recent molecular study (Wahlberg et al. 2005) confirms that these lineages form a monophyletic group, supporting their inclusion in Nymphalidae as subfamilies or tribes.

Phylogenetic relationships of the five papilionoid families (i.e., Papilionidae, Pieridae, Riodinidae, Lycaenidae, and Nymphalidae) and the family Hesperidae have been studied using morphological and molecular data (Ehrlich 1958; Kristensen 1976; de Jong, Vane-Wright, and Ackery 1996; Weller, Pashley, and Martin 1996; Wahlberg et al. 2005). Our current understanding of relationships is shown in Figure 1.4, although the position of Pieridae is unstable. In some analyses, Pieridae places as sister to Papilionidae, and more data are needed to resolve this issue.

Several butterfly species have been used as model systems in molecular biology and genetics. *Papilio* species (Papilionidae: Papilioninae) have been the focus of chemical ecology and speciation studies (e.g., Scriber, Tsubaki, and Lederhouse 1995). In Nymphalidae, checkerspot butterflies (Nymphalinae) figure prominently in studies of population biology (e.g., Ehrlich and Hanski 2004). *Heliconius* species (Heliconiinae) have been central to studies of mimicry, speciation, and the genetics of mimetic systems (e.g., Mallet, McMillan, and Jiggins 1998; see Chapter 6 for genetics of color pattern in *Heliconius*). Another important nymphalid model system is *Bicyclus anynana* (Satyrinae), a species central to studies of wing pattern formation (e.g., Beldade and Brakefield 2002; see Chapter 5 for evolution and genetics of eyespots in *Bicyclus*).

Bombycoidea

The cosmopolitan Bombycoidea *sensu lato* ("bombycoid complex" *sensu* Minet 1994) comprises approximately five thousand described species of medium- to very large–sized moths (Figure 1.1A, B) in 650 genera and twelve families: Anthelidae, Apatelodidae, Bombycidae, Brahmaeidae (Lemoniidae), Carthaeidae, Endromidae, Eupterotidae, Lasiocampidae, Mimallonidae, Mirinidae, Saturniidae, and Sphingidae. Monophyly of Bombycoidea is supported by one thoracic (Minet 1991) and one forewing vein synapomorphy (shared derived trait) (A. Zwick, unpublished) and potentially by modification in the larval proleg cuticle (Hasenfuss 1999). The shortage of synapomorphies stems largely from a shortened adult life span, and consequently a reduction of structures commonly used for inferring phylogenetic relationships (e.g., mouthparts and wing coupling mechanisms; Minet 1991, 1994). Therefore, the relationships among bombycoid families are generally poorly understood (Figure 1.5).

Within Bombycoidea, only three families—Bombycidae *s. str.*, Sphingidae, and Saturniidae—contain species that are widely used as model systems. These three families probably form a monophyletic group as indicated by two independent molecular studies (Regier et al. 2008;

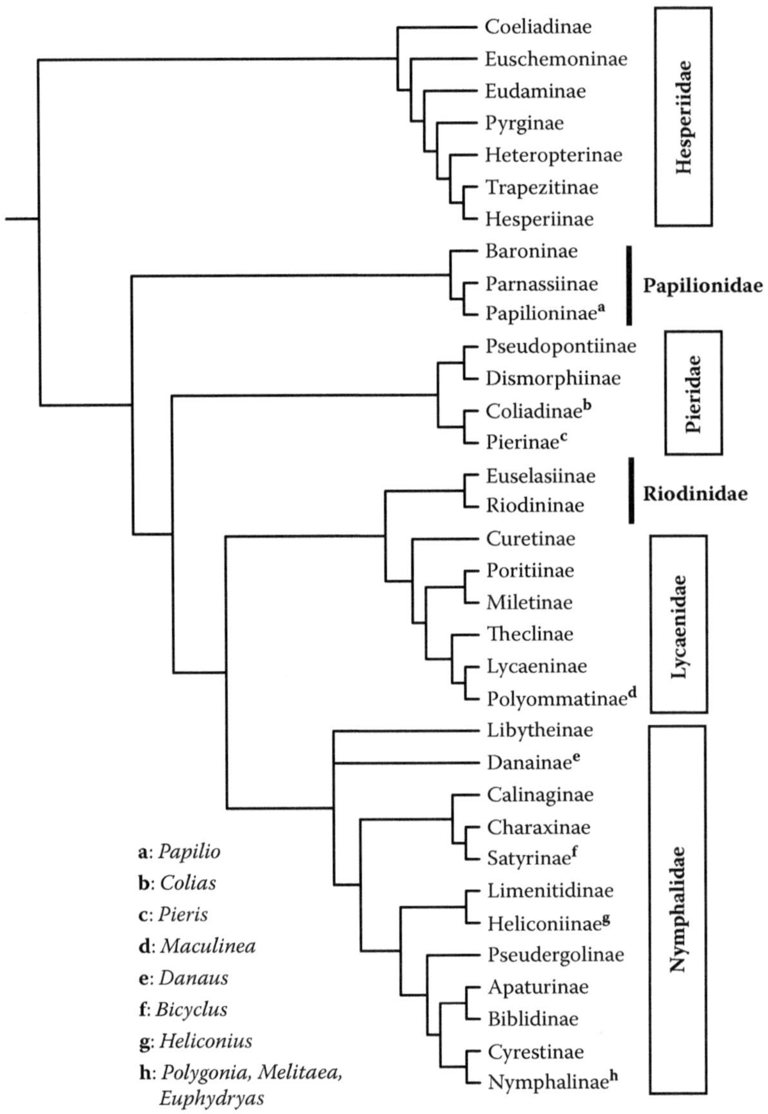

FIGURE 1.4 A composite tree showing the inferred relationships of butterfly families and subfamilies. Family relationships taken from Wahlberg et al. (2005), subfamily relationships within families taken from various sources: Hesperiidae (Warren 2006), Papilionidae (Caterino et al. 2001), Pieridae (Braby, Vila, and Pierce 2006), Riodinidae and Lycaenidae (Wahlberg et al. 2005), and Nymphalidae (Wahlberg and Wheat 2008).

Zwick 2008), both of which partially contradict morphology-based studies of Brock (1971) and Minet (1991, 1994).

Bombycidae are best known for the domestic silkworm *B. mori* (Bombycinae), the most prominent model system in Lepidoptera. Recent molecular studies (Regier et al. 2008; Zwick 2008; Figure 1.5) demonstrate that the family (*sensu* Minet 1994) is unnatural (e.g., polyphyletic) and should be restricted to the nominate subfamily Bombycinae. Despite a wealth of knowledge for *B. mori*, virtually nothing is known about the biology of most other bombycid species. The family is in urgent need of a comprehensive taxonomic revision, and phylogenetic hypotheses are lacking for its 350 described species that are currently placed in forty genera. As a model system, the domestic

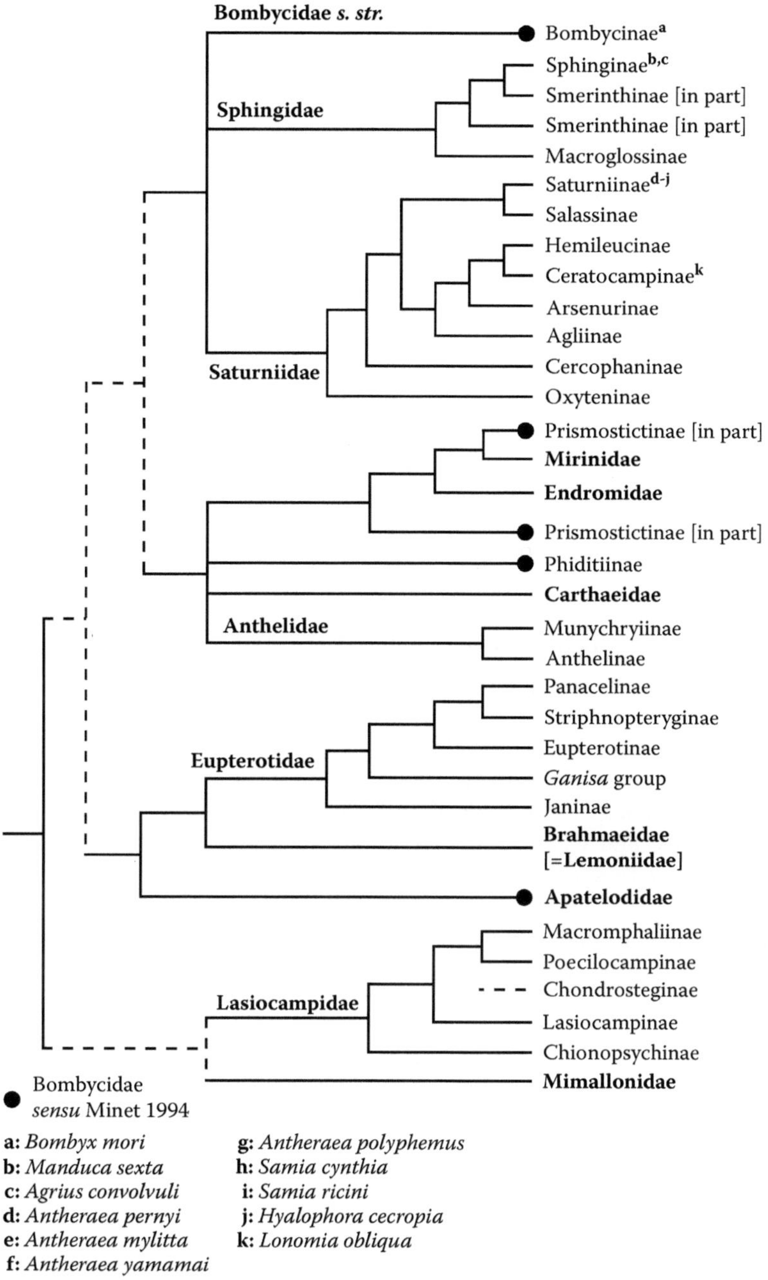

FIGURE 1.5 Composite tree based on the molecular studies of Kawahara et al. (2009), Regier et al. (2001, 2002, 2008), and Zwick (2008). Bombycidae *sensu* Minet 1994 are marked with black circles.

silkworm has been the object of extensive basic, biotechnological, and sericultural research (see Goldsmith, Shimada, and Abe 2005; for a review on molecular biology and genetics, see Chapters 2 and 4). A huge quantity of molecular data is publicly available, including two genome drafts (Mita et al. 2004; Xia et al. 2004); bacterial artificial chromosome (BAC) libraries; expressed sequence tags (ESTs); and molecular linkage, genetic, and physical maps (Goldsmith, Shimada, and Abe 2005). Some genetic work has focused on *Bombyx mandarina*, the species thought to represent the wild ancestor of *B. mori* (Arunkumar, Metta, and Nagaraju 2006). However, this wealth of genetic information on *B. mori* and *B. mandarina* contrasts starkly with the nearly complete lack of data for other bombycids.

Sphingidae, another family with model species, is comprised of three subfamilies: Sphinginae, Smerinthinae, and Macroglossinae. Recent molecular studies (Regier et al. 2001; Kawahara et al. 2009) reveal evolutionary relationships that partly contradict current subfamily and tribal classification but concur with morphological interpretations of sphingid relationships (e.g., Rothschild and Jordan 1903; Nakamura 1976; Kitching and Cadiou 2000; Kitching 2002, 2003). The taxonomy, immature stages, and biology of many sphingid species are thoroughly studied, although there are still many species that are poorly understood (Kitching and Cadiou 2000).

A well-known model organism is the tobacco hornworm, *Manduca sexta* (Sphinginae: Sphingini). This species has been used in a broad range of research on biochemistry, physiology, morphology, and nutritional ecology (Slansky 1993; Willis, Wilkins, and Goldsmith 1995). Its universality as a model system is reflected by the almost thirty-eight hundred GenBank accessions (including ESTs) and the construction of two BAC libraries (Sahara et al. 2007). An emerging model sphingid species is the cosmopolitan sweet potato hornworm *Agrius convolvuli* (Sphinginae: Acherontiini), which has been used in numerous physiological and immunological studies, some of which include molecular genetics.

Saturniidae represent the third well-known family of bombycoid moths. Current hypotheses of its higher phylogeny (Figure 1.5) are largely based on molecular studies (Regier et al. 2002, 2008; Zwick 2008), with Michener (1952) providing the only morphology-based hypothesis for relationships within the entire family. In the molecular studies, relationships among the eight subfamilies are fully resolved and statistically well supported. Phylogenetic hypotheses for other levels of divergence are limited (e.g., Friedlander et al. 1998; Rubinoff and Sperling 2002, 2004; Regier et al. 2005). As with Sphingidae, the taxonomy of Saturniidae historically has been studied extensively on a global scale, and information on immatures and life history is available for a large number of species.

Nine saturniid species are used as molecular model systems, representing only two of the eight subfamilies, Saturniinae and Ceratocampinae. Four of these models are congeners: *Antheraea pernyi*, *A. mylitta*, *A. yamamai*, and *A. polyphemus*. *Antheraea* collectively exhibits a Holarctic distribution, and the genus includes approximately seventy described species and numerous subspecies (Paukstadt, Brosch, and Paukstadt 2000). Regier et al. (2005) present a phylogenetic hypothesis for sixteen of seventy species, including four model systems. Their study examines the evolution of morphology and development of chorionic aeropyle crowns on the molecular phylogeny. *Antheraea* species have been used in sericulture and, like *B. mori* and *M. sexta*, for a wide range of fundamental research (see Goldsmith and Wilkins 1995). *Antheraea pernyi* has been an important model for studying the molecular mechanisms of the circadian clock (see Chapter 8 and references therein). However, compared to *B. mori* and *M. sexta*, the number of distinct GenBank accessions for these four *Antheraea* species is negligible, except for *A. mylitta* (i.e., approximately four thousand) and *A. yamamai* (i.e., approximately seven hundred), each of which is represented by numerous ESTs.

Three additional model systems occur in Attacini, the sister tribe to Saturniini (Figure 1.5): *Samia cynthia*, *S. ricini*, and *Hyalophora cecropia*. The taxonomy of the Asian genus *Samia* (*Phylosamia*; nineteen species) is particularly complex and confusing due to countless synonyms and inconsistent use of names. Fortunately, Peigler and Naumann (2003) recently published a comprehensive revision of the genus. In nontaxonomic literature, the species "*Samia ricini*" has been problematic and inconsistently treated as a valid species, as a subspecies of *S. cynthia* or as a form of *S. cynthia*. However, it is possible that none of these treatments is correct; *S. ricini* now is thought to be a domesticated form of *S. canningi* and unrelated to wild *S. cynthia* (Peigler and Naumann 2003). Consequently, the identity of *Samia* species in past studies has to be viewed with caution, which provides a compelling case for routine deposition of voucher specimens even in studies of model systems. Similarly, the taxonomy of the Nearctic *Hyalophora*—containing three species and numerous subspecies—continues to be controversial. All taxa readily interbreed in captivity, and natural hybrids occur in some contact zones (Tuskes, Tuttle, and Collins 1996). The breadth of research areas for the model systems in *Samia* and *Hyalophora* is similar but not quite as large as for *Antheraea* (see Goldsmith and Wilkins 1995). The number of distinct GenBank accessions for these species is smaller than for *Antheraea*; however, an international initiative to database ESTs for these genera is under way (www.cdfd.org.in/wildsilkbase/home.php), which promises to greatly increase our knowledge of their genomes.

Unlike other saturniid models, the South American *Lonomia obliqua* (Ceratocampinae), and to a lesser extent *L. achelous*, are used specifically in research on its highly poisonous larvae and the anticoagulating properties of their poison (Veiga et al. 2005). Despite this limited research scope, *L. obliqua* has the second largest number of saturniid GenBank accessions, which includes one of the few EST libraries available for Bombycoidea. *Lonomia* contains more than a dozen described species based on revisions by Lemaire (1972, 2002).

The remaining bombycoid families do not contain molecular model systems, and most are relatively poorly studied, economically insignificant, and low in species numbers. Lasiocampidae (e.g., tent caterpillars) is an important exception, with approximately fifteen hundred species in 150 genera and several significant pest species in several genera (e.g., *Malacosoma*, *Dendrolimus*, and *Trabala*). The monophyly of Lasiocampidae is strongly supported by a combination of molecular and morphological data (Zwick 2008). However, phylogenetic relationships within the family are poorly studied (Regier et al. 2001, 2008; Zwick 2008).

Based on our current, limited knowledge, the three bombycoid families containing molecular model systems seem to be more closely related to each other than to any other family (Figure 1.5) (Regier et al. 2001, 2008; Zwick 2008). To maximize the utility and value of the extraordinarily large quantity of existing data for molecular model systems in Bombycoidea, future phylogenetic studies are needed at many taxonomic levels. A model system needs to be developed in Lasiocampidae or another early diverging lineage to increase the phylogenetic breadth of current model systems in Bombycoidea. Currently, generalizations about Bombycoidea are limited by the phylogenetic placement of its model systems.

Noctuoidea

Noctuoidea is the largest superfamily of Lepidoptera—approximately seventy thousand described species in over seventy-two hundred genera (Kitching and Rawlins 1998). The superfamily includes some of our best-known lepidopterans: Arctiidae (ermines, footman, and tiger moths), Lymantriidae (gypsy and tussock moths), Noctuidae (cutworms, deltoids, owlets, and underwings), and Notodontidae (prominents and processionary moths). Noctuoidea is defined by the possession of a thoracic tympanum (hearing organ) and associated abdominal structures in the adults, and the presence of two microdorsal setae on the larval metathorax (Hinton 1946)—except oenosandrid larvae, which have only one (Miller 1991; Kitching and Rawlins 1998).

Noctuoidea include a number of ecologically important species, including some of our most damaging forest and agricultural pests (Kitching and Rawlins 1998). Others, particularly arctiids, have been the focus of studies on chemical ecology and mating behavior (reviewed in Conner and Weller 2004). Well-known species in North America include gypsy moth (*Lymantria dispar*, Lymantridae: Orgyiinae), corn earworm (*Helicoverpa zea*, Noctuidae: Heliothinae), tobacco budworm (*H. armigera*), cabbage looper (*Trichoplusia ni*, Noctuidae: Plusiinae), and fall armyworm (*Spodoptera frugiperda*, Noctuidae: Xyleninae). Caterpillars of some species possess urticating (stinging), deciduous setae that can be hazardous to humans (e.g., *L. dispar*, *Thaumetopoeia processionaria*, Notodontidae: Thaumetopoeinae; review Kitching and Rawlins 1998).

In this superfamily, basal lineages Oenosandridae, Notodontidae, and Doidae lack model systems (Miller 1991; Kitching and Rawlins 1998). Rather, model systems are concentrated in three large families, Noctuidae (about fifty thousand species; e.g., *Helicoverpa*, *Heliothis*, *Spodoptera*), Arctiidae (about eleven thousand species; e.g., *Utetheisa ornatrix*, *Arctia caja*, *Creatonotus gangis*), and Lymantriidae (about five hundred species; e.g., *L. dispar*). These families form a clade with a handful of problematic lineages recognized as either subfamilies or families, depending on the authority consulted (Kitching and Rawlins 1998). These families and problematic lineages have been arranged variously based on easily observed traits such as their hindwing venation, leg spination, and aspects of their larval morphology (Kitching and Rawlins 1998).

The classification of noctuoids (Figure 1.6), particularly Noctuidae, has recently undergone major shifts and rearrangements. Traditionally, Arctiidae was placed as sister to Lymantriidae; and four taxa—Aganainae (Hypsidae), Nolinae, Hermiinae, and Pantheinae—were treated as subfamilies of Arctiidae, subfamilies of Noctuidae, or separate families allied to Arctiidae (review Kitching and Rawlins 1998; Jacobson and Weller 2002; Fibiger and Lafontaine 2005). Molecular studies starting over a decade ago have suggested consistently that the remaining "noctuid" lineage divide into two main lineages or clades (Figure 1.6), superficially diagnosed by their hindwing venation. One of these, the "trifines," contains noctuids with trifine hindwing venation (i.e., M_3, Cu1A, Cu1B associated; vertical bars; Mitchell, Mitter, and Regier 2006). The other clade consists of most noctuid subfamilies with quadrifine hindwing venation and Lymantriidae and Arctiidae (Weller et al. 1994; Mitchell et al. 1997; Mitchell, Mitter, and Regier 2000). In Figure 1.6A, all taxa except Lymantriidae and Arctiidae have at one time or another been included in Noctuidae. Thus, Noctuidae in the traditional sense is not monophyletic.

Numerous modifications to classification ensued (review Lafontaine and Fibiger 2006). Most dramatically, beginning in June 2005, three landmark publications presented detailed phylogenies and completely rewrote the classification of Noctuoidea three times (Fibiger and Lafontaine 2005; Mitchell, Mitter, and Regier 2006; Lafontaine and Fibiger 2006). These publications each have their own limitations and strengths. The molecular study (Mitchell, Mitter, and Regier 2006) has incomplete sampling of more obscure lineages but provides a rigorous data analysis of two nuclear genes encoding proteins, *elongation factor 1-alpha* (*EF-1α*) and *dopa decarboxylase* (*Ddc*; Figure 1.6A), with very strong statistical support for the clades contradicted by the morphological hypotheses. The morphological conclusions (Fibiger and Lafontaine 2005; Lafontaine and Fibiger 2006) are not based on formal phylogenetic data analyses but provide an authoritative review of morphological ground plans (Figure 1.6B). Thus, discordance among results of these studies cannot be resolved without further investigation. All studies demonstrated the nonmonophyly of the traditional Noctuidae (Figure 1.6).

Model systems in Noctuoidea are concentrated in the "Noctuidae" (Figure 1.6A), which contain the majority of agricultural pests. These models span phylogenetic diversity within this clade, from *T. ni* (Plusiinae) to *Spodoptera* (Noctuinae; shaded, Figure 1.6). Recent phylogenetic studies are helping to clarify the relationship in some of these subfamilies (e.g., Heliothinae, Fang et al. 1997; Cho et al. 2008) and genera (e.g., *Spodoptera*; Pogue 2002). The remaining model systems occur only in Arctiidae and Lymantriidae (Figure 1.6B; Arctiinae and Lymantriinae, respectively; Lafontaine and Fibiger 2006). None occur in Notodontidae, Doidae, or Oenosandridae, earlier-diverging lineages. This spotty distribution of model systems undermines the confidence with which observations can be generalized across the superfamily.

CURRENT RESEARCH AND FUTURE DIRECTIONS

Molecular Ditrysian Phylogenies and the Role of Fossils

All the preceding discussions rest upon the phylogeny presented in Figure 1.2. While this phylogeny serves as a useful working hypothesis, the proposed superfamily relationships are highly provisional within Ditrysia. Relationships among ditrysian superfamilies are rarely examined and have never been subject to explicit phylogenetic analysis. Only recently has research focused on understanding relationships within a few superfamilies (e.g., Bombycoidea, Papilionoidea, Pyraloidea, and Noctuoidea). Thus, a comprehensive phylogenetic framework for Lepidoptera is not yet achieved.

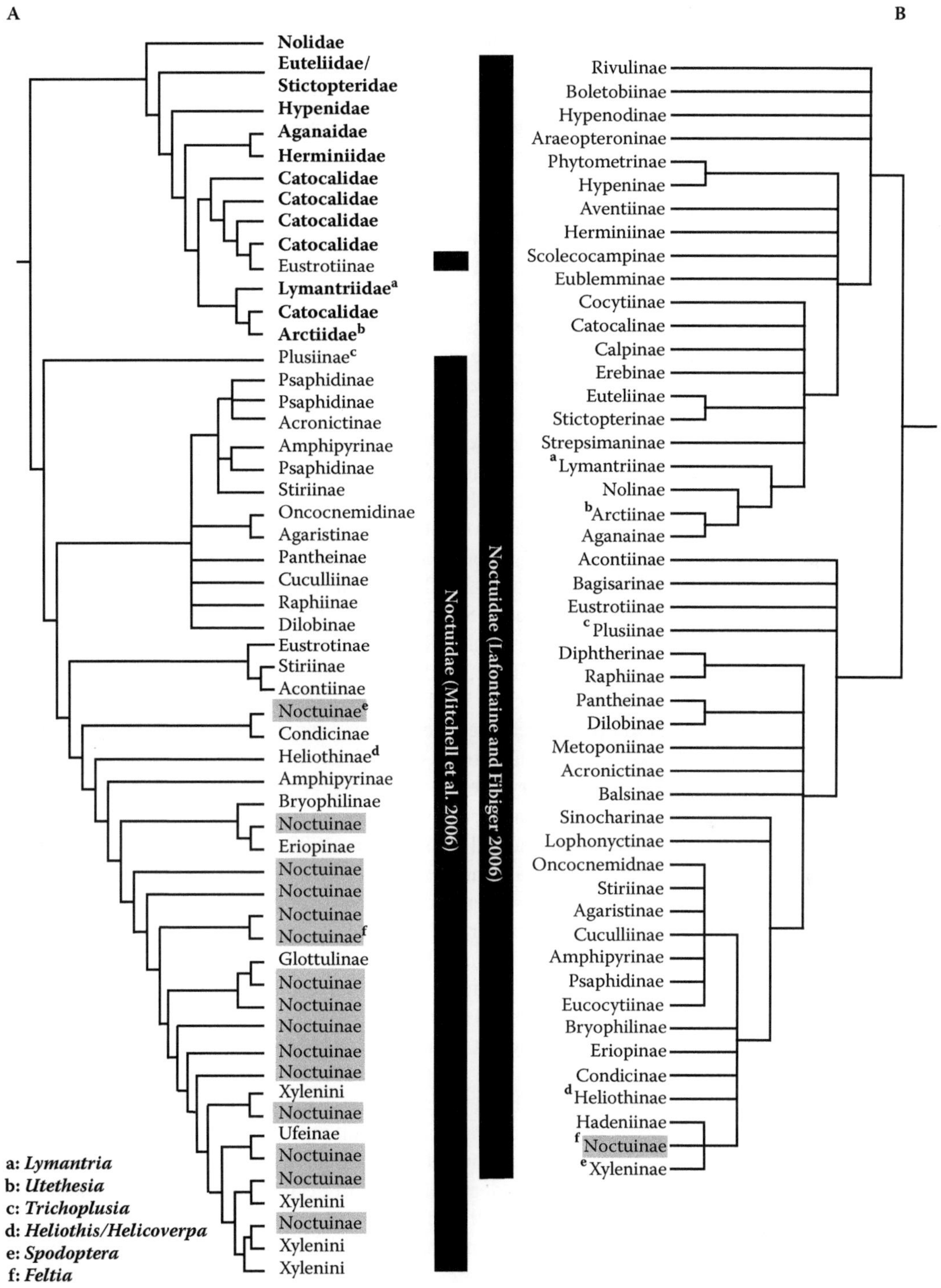

FIGURE 1.6 Hypothesized relationships in the quadrafid forewing clade of Noctuoidea. no bar = quadrifine hindwing; vertical bar = trifine hindwing subfamilies; superscripts indicate placement of model systems. A. Maximum parsimony molecular tree adapted from Mitchell, Mitter, and Regier (2006) based on *EF-1α* and *Ddc* sequences (Figures 2 and 3 within). Clades reduced to represent family and subfamily relationships. B. Alternative classification and relationships proposed by Lafontaine and Fibiger (2006), with Micronoctuidae not shown due to uncertain placement.

Fortunately, lepidopteran phylogenetic research is currently focused on understanding superfamily relationships within Ditrysia. Two ongoing, complementary initiatives began in 2006. The first, Assembling the Lepidopteran Tree of Life (LepAToL, based in the United States), is examining superfamily relationships across all lepidopteran families using nuclear gene sequences (up to twenty-six loci) from several hundred species. The second, Systematics of Ditrysian Lepidoptera (based in Finland), is examining a comparable number of species for eight nuclear and mitochondrial loci (a mostly nonoverlapping set of genes with LepAToL) and two hundred morphological characters. Progress updates for both projects are posted regularly at http://www.leptree.net.

An initial study of superfamily relationships by the LepAToL group were posted online as of December 2007. This study included 123 species (twenty-six superfamilies) concentrated in the clade Obtectomera (node 9, Figure 1.2). These were sequenced for about 6.7 kb from five nuclear genes encoding for proteins, including dopa decarboxylase, period, wingless, enolase, and CAD (carbamoyl-phosphate synthetase 2, aspartate transcarbamylase, and dihydroorotase).

The main results of this study can be summarized as follows: Most superfamilies were recovered as monophyletic, and when several representatives were included, relationships within superfamilies were resolved and well supported (e.g., Tortricoidea, Pyraloidea, Noctuoidea). On a broad scale, the analysis recovered several clades depicted in Figure 1.2; however, there were two significant exceptions. First, butterflies (Papilionoidea) and allies (Hesperioidea, Hedyloidea) fall outside of the macrolepidoptera clade, and these superfamilies are consistently grouped more closely with one or more microlepidopteran superfamilies. Second, Pyraloidea consistently form a sister relationship to the remaining macrolepidoptera, a novel association. Although individual nodes subtending multiple superfamilies were not strongly supported, tests (Shimodaira 2002) comparing the overall fit of the molecular data to prior phylogenetic hypotheses significantly rejected the monophyly of macrolepidoptera as previously defined (Figure 1.2). Thus, findings to date suggest that a vastly improved phylogenetic framework for Lepidoptera at many levels will be available in the near future.

Well-resolved phylogenies at multiple levels are essential for comparative studies of Lepidoptera, but a full understanding of the evolution of many traits of interest will also require knowledge of the absolute timing of reconstructed evolutionary events. The most reliable source of such knowledge is, in most cases, the fossil record. Unfortunately, the lepidopteran fossil record is sparser than that of any other major insect order (Labandeira and Sepkoski Jr. 1993), particularly in compression fossils, that is, those formed in rocks (Grimaldi and Engel 2005). Grimaldi and Engel (2005) suggest that lepidopterans do not sink easily in water because their scales interfere with wetting, and consequently they are rarely buried in underwater sediments.

On a broad scale, fossil evidence on the timing of lepidopteran evolution can be summarized as follows: The earliest lepidopteran fossils date to the lower Jurassic, about 190 million years ago (mya; Whalley 1986; Grimaldi and Engel 2005). The oldest fossils that can be assigned to any extant group are found in Lebanese amber of the lower Cretaceous (125 mya), and include *Parasabatinca aftimacrai* (Whalley 1978), which is assignable to the oldest extant family, Micropterigidae. A fossil larva, also from Lebanese amber, has been placed in the clade Glossata (Figure 1.2, node 2; Grimaldi 1996, 1999), the clade in which the characteristic adult proboscis first appears. A probable first record of Ditrysia occurs in the mid-Cretaceous (97 mya) in the form of fossil leaf mines attributed to the extant Phyllocnistinae (Gracillariidae) (Labandeira et al. 1994), although caution has been urged in interpreting such fossils (Grimaldi and Engel 2005). Although additional ditrysian fossils are lacking from the Cretaceous, the occurrence of Paleocene fossils of macrolepidoptera (56 mya) suggests that many additional ditrysian lineages arose during the Cretaceous, which ended 65.5 mya (Grimaldi and Engel 2005). Many extant ditrysian families first appear in Eocene Baltic amber (44 mya), and many others are first found in the Eocene-Oligocene Florissant shales (38 mya; reviewed in Grimaldi and Engel 2005).

To integrate fossils with current phylogenetic studies, two challenges must be addressed. First, fossils have to be placed in current molecular phylogenies. Minimally, fossil assignments to superfamily (or more precise identification to family or genus) need to be based on diagnostic morphological features (synapomorphies). Currently, most existing identifications are based solely on subjective impressions of similarity to extant species, and most fossils have never been examined by the relevant expert (de Jong 2007). The second problem is methodological. Molecular evolution often departs from rate constancy; that is, mutations do not accumulate in a "clock-like" fashion (e.g., Kumar 2005). To allow more accurate dating of nodes based on molecular analyses, development of more sophisticated analytical approaches is under way to account for uneven mutation rates (e.g., Welch and Bromham 2005; Rutschmann 2006). In the best scenario, future analyses will include morphological matrices as well as molecular data, so that fossil placement is testable despite fragmentary remains. Thus, somewhat ironically, the full promise of molecular systematics for revealing the evolutionary history of Lepidoptera cannot be realized until morphological studies are as comprehensive in species' sampling as molecular analyses.

Virtual Community Building in Lepidopteran Systematics

Until the last couple decades, the work of placing lepidopteran model systems into an evolutionary context fell largely to systematists specializing in morphology. Morphologists historically worked in relative isolation with publications authored typically by one or occasionally by two researchers. The typically slow pace of research has been frustrating to others who rely on their phylogenies to place their model systems. In contrast, many researchers working on model systems publish in large collaborations (Cronin and Franks 2006). One historical barrier to community building has been the slow dissemination of taxonomic studies. Systematic works often were published as monographs by museums or regional journals, available only in the largest libraries. Only recent work is likely to be available digitally, usually beyond the financial reach of many in the global community. Historically, researchers maintained a network of personal contacts to keep abreast of advances in the field, which led to exchanges of reprints and specimens, as well as occasional visits, supplemented by conferences that facilitated scientific exchanges and growth and maintenance of social networks. Consequently, much work was conducted in isolation or small communities of researchers. In Lepidoptera, as in other insect orders, the situation led to an idiosyncratic development of morphological terminologies, often highly specialized for a particular group (e.g., genitalic terms for butterflies). Isolated by fragmented lexicons, researchers interested in studying multiple groups of Lepidoptera had to learn new terms specific to a particular family or superfamily. This situation has impeded rapid progress in lepidopteran systematics.

To remediate these barriers and foster a large, global, collaborative virtual community, several Web site initiatives have been undertaken. Some of these are taxon based, focusing on a particular superfamily; for example, the GlobIZ Pyraloidea Database (http://www.pyraloidea.org), geometroid Web site Forum Herbulot (http://www.herbulot.de), the Nymphalidae Systematics Group (http://www.nymphalidae.utu.fi), Tortricid.net (http://www.tortricidae.com), and the gelechioid work group (Gelechioidea Framework, http://www.msstate.edu/org/mississippientmuseum/Lepidoptera/GelechioideaFramework.htm).

Some Web-based projects attempt to link all taxonomists or all lepidopterists in a virtual community. The European Distributed Institute of Taxonomy (EDIT; http://www.e-taxonomy.eu) aims to unite European taxonomic research. EDIT seeks to coordinate and digitally reorient the European contribution to the global taxonomic effort, particularly with the Global Taxonomy Initiative. Lepidoptera is a priority target group for EDIT, and along with Lepsys, a consortium for European Lepidoptera Systematists (http://www.lepsys.eu), coordinates lepidopteran research in Europe.

LepTree (http://www.leptree.net), like EDIT, supports a virtual community where both molecular and morphology-based projects are supported. To maximize openness and interactions, this Web site was created with the open source content management system Drupal (http://www.drupal.org) to take advantage of its discussion and collaborative authoring tools. This site is further customized to store biological and social data in an open source triple store (http://www.openrdf.org), and the semantically rich OWL (Web Ontology Language) and RDF (Resource Description Framework) formats for data storage and sharing (e.g., Mabee et al. 2007). As an example of these technologies, subcommunities work together to build, cross-reference, and widely share glossaries of morphological terms that are illustrated with annotated images. Community data will be available automatically to portals such as the Encyclopedia of Life (http://www.eol.org) and to digital taxonomists everywhere.

These global Web site initiatives (EDIT, Lepsys, LepTree) represent an intermediate state between casual discussions and data-rich collaborations of "big science" (e.g., "collaboratories" *sensu* Arzberger and Finholt 2002). They facilitate an exchange of knowledge, terminology, and protocols, thereby fostering a larger, more global community of practice (Preece 2000). Web sites with such tools can establish virtual communities that encourage collaboration and training over large distances, in addition to disseminating results quickly and easily. Thus, the global community of lepidopteran systematists is evolving to work together efficiently to answer a broad spectrum of comparative and evolutionary questions.

Summary and Future Model Systems

Model systems are scattered throughout Ditrysia, and these provide insight into a wide range of biological processes. By reviewing the phylogenetic placement of these model systems, underrepresentation of several major lepidopteran groups becomes obvious, most notably the non-ditrysians as well as Geometroidea and Zygaenoidea. These gaps will need to be addressed with the choice and examination of new models in these lineages.

Evolutionary frameworks provide guidance to choosing new model systems or organisms for comparative analyses. Several authors have outlined factors to consider when developing new model systems (Mabee 2000; Santini and Stellwag 2002; Collins et al. 2005; Jenner and Willis 2007), and careful selection of comparative taxa will maximize mechanistic and comparative results. First, a model should be chosen to help further a research theme, identify plesiomorphic (ancestral or primitive) character states, or seek to resolve the origins of novel traits. For example, understanding the evolution and development of lepidopteran wing patterns (Ramos and Monteiro 2007; reviewed in Chapters 5 and 6), genome size (Gregory and Hebert 2003; Goldsmith, Shimada, and Abe 2005), or pheromone synthesis (reviewed in Chapter 10) will require judicious choice of models throughout the lepidopteran tree, as well as among members of Trichoptera. Second, model systems should be practical to obtain and culture in a laboratory setting. For example, *B. mori* proved amenable to culture, having been domesticated several thousand years ago for silk production, leading to its prominence in scientific studies. On the other hand, the larval diets and adult breeding behavior are not known for many lepidopterans, particularly the basal lineages that currently lack model systems (Figure 1.2). We need to advance our understanding of the basic biology of these lineages before we can successfully culture these taxa and develop them as viable model systems. Finally, phylogenetic sampling and relationships within and between lineages containing model systems must be examined to ensure appropriate breadth of taxon sampling relative to the trait being examined. Current European and North American initiatives will provide more robust phylogenetic frameworks for accurately assessing relationships of model and nonmodel taxa and for choosing new model systems for future exploration.

ACKNOWLEDGMENTS

We thank the LepTree community for conversations and suggestions that improved this chapter, particularly the detailed and insightful comments and recommendations from M. Horak and C. Mitter. We thank our anonymous SEL reviewers for their helpful comments. This research was supported by National Science Foundation Grant No. 0531639 (Mitter, Cummings, Parr, Regier, Roe, Weller), NSF DEB-0212910 (Regier), Minnesota Experiment Station Grant No. MN-17-022 (Roe, Weller), Carlsberg Foundation (Simonsen), University of Valencia grant UV-AE-20070204

(Baixeras), and a grant from the National Science and Engineering Research Council of Canada (Sperling, Simonsen).

REFERENCES

Ackery, P.R., R. de Jong, and R.I. Vane-Wright. 1998. The butterflies: Hedyloidea, Hesperoidea and Papilionoidea. In *Lepidoptera, moths and butterflies. 1 Evolution, systematics and biogeography*, ed. N.P. Kristensen, 263–300. *Handbook of Zoology*, Vol. 4, Part 35, *Arthropoda: Insecta*. Berlin and New York: Walter de Gruyter.

Addison, M.F. 2005. Suppression of codling moth *Cydia pomonella* L. (Lepidoptera: Tortricidae) populations in South African apple and pear orchards using sterile insect release. *Acta Hort.* 671:255–57.

Ahmad, T.R., and Z.A. Al-Gharbawi. 1986. Effects of pheromone trap design and placement on catches of codling moth *Cydia pomonella* males. *J. Appl. Entomol.* 102:52–57.

Albert, P.J. 1991. A review of some host-plant chemicals affecting the feeding and oviposition behaviours of the eastern spruce budworm, *Choristoneura fumiferana* Clem. (Lepidoptera: Tortricidae). *Mem. Entomol. Soc. Can.* 159:13–18.

Albert, P.J. 2003. Electrophysiological responses to sucrose from a gustatory sensillum of the larval maxillary palp of the spruce budworm, *Choristoneura fumiferana* (Clem.) (Lepidoptera: Tortricidae). *J. Insect Physiol.* 49:733–38.

Arn, H. 1991. Sex pheromones. In *Tortricid pests: Their biology, natural enemies and control*, ed. L.P.S. Van der Gees and H.H. Evenhuis, 187–207. Amsterdam: Elsevier.

Arunkumar, K.P., M. Metta, and J. Nagaraju. 2006. Molecular phylogeny of silkmoths reveals the origin of domesticated silkmoth, *Bombyx mori* from Chinese *Bombyx mandarina* and paternal inheritance of *Antheraea proylei* mitochondrial DNA. *Mol. Phylogenet. Evol.* 40:419–27.

Arzberger, P., and T. Finholt. 2002. *Data and collaboratories in the biomedical community*. Ballston,VA: Nat. Inst. Health.

Backman, A.-C. 1997. Pheromones released by codling moth females and mating disruption dispensers. *IOBC West Palearc. Reg. Sec. Bull.* 20:175–80.

Bates, H.W. 1861. Contributions to an insect fauna of the Amazon valley. Lepidoptera: Heliconidae. *Trans. Linn. Soc.* 23:495–566.

Beckemeyer, E.F., and P.D. Shirk. 2004. Development of the larval ovary in the moth, *Plodia interpunctella*. *J. Insect Physiol.* 50:1045–51.

Beldade, P., and P.M. Brakefield. 2002. The genetics and evo-devo of butterfly wing patterns. *Nat. Rev. Genet.* 3:442–52.

Boggs, C.L., W.B. Watt, and P.R. Ehrlich. 2003. *Butterflies: Ecology and evolution taking flight*. Chicago: Univ. Chicago Press.

Bolker, J.A. 1995. Model systems in developmental biology. *Bioessays* 17:451–55.

Braby, M.F., R. Vila, and N.E. Pierce. 2006. Molecular phylogeny and systematics of the Pieridae (Lepidoptera: Papilionoidea): higher classification and biogeography. *Zool. J. Linn. Soc.* 147:239–75.

Brock, J.P. 1971. A contribution towards an understanding of the morphology and phylogeny of the ditrysian Lepidoptera. *J. Nat. Hist.* 5:29–102.

Brower, A.V.Z. 2000. Phylogenetic relationships among the Nymphalidae (Lepidoptera), inferred from partial sequences of the *wingless* gene. *Proc. Biol. Sci.* 267:1201–11.

Brown, J.W. 2005. *World catalogue of insects, Volume 5: Tortricidae (Lepidoptera)*. Stenstrup, Denmark: Apollo Books.

Caterino, M.S., R.D. Reed, M.M. Kuo, and F.A.H. Sperling. 2001. A partitioned likelihood analysis of swallowtail butterfly phylogeny (Lepidoptera: Papilionidae). *Syst. Biol.* 50:106–27.

Cho, S., A. Mitchell, C. Mitter, J.C. Regier, M. Matthews, and R. Robertson. 2008. Molecular phylogenetics of heliothine moths (Lepidoptera: Noctuidae: Heliothinae), with comments on the evolution of host range and pest status. *Syst. Entomol.* 33:581–94.

Coates, B.S., R.L. Hellmich, and L.C. Lewis. 2006. Sequence variation in trypsin- and chymotrypsin-like cDNAs from the midgut of *Ostrinia nubilalis*: Methods for allelic differentiation of candidate *Baccillus thuringiensis* resistance genes. *Insect Mol. Biol.* 15:13–24.

Collins, A.G., P. Cartwright, C.S. McFadden, and B. Schierwater. 2005. Phylogenetic context and basal metazoan model systems. *Integr. Comp. Biol.* 45:585–94.

Common, I.F.B. 1990. *Moths of Australia*. Carlton: Melbourne Univ. Press.

Conner, W.E., and S.J. Weller. 2004. A quest for alkaloids. The curious relationships between tiger moths and plants containing pyrrolizidine alkaloids. In *Advances in insect chemical ecology*, ed. R.T. Cardé and J. Millar, 248–82. Cambridge: Cambridge Univ. Press.

Cook, C., M.L. Smith, M.J. Telford, A. Bastianello, and M. Akam. 2001. *Hox* genes and the phylogeny of the arthropods. *Curr. Biol.* 11:759–63.

Cronin, B., and S. Franks. 2006. Trading cultures: Resource mobilization and service rendering in the life sciences as revealed in the journal article's paratext. *J. Am. Soc. Inf. Sci. Technol.* 57:1909–18.

DaCosta, M.A., P. Larson, J.P. Donahue, and S.J. Weller. 2006. Phylogeny of milkweed tussocks (Arctiidae: Arctiinae, Phaegopterini) and its implications for evolution of ultrasound communication. *Ann. Entomol. Soc. Am.* 99:723–42.

Davis, D.R. 1986. A new family of monotrysian moths from austral South America (Lepidoptera: Palaephatidae), with a phylogenetic review of the Monotrysia. *Smithson. Cont. Zool.* 434:1–202.

de Jong, R. 2007. Estimating time and space in the evolution of the Lepidoptera. *Tijdschr. Entomol.* 150:319–46.

de Jong, R., R.I. Vane-Wright, and P.R. Ackery. 1996. The higher classification of butterflies (Lepidoptera): Problems and prospects. *Entomol. Scand.* 27:65–101.

Delisle, J., J.-F. Picimbon, and J. Simard. 1999. Physiological control of pheromone production in *Choristoneura fumiferana* and *C. rosaceana*. *Arch. Insect Biochem. Physiol.* 42:253–68.

Ehrlich, P.R. 1958. The comparative morphology, phylogeny and higher classification of the butterflies (Lepidoptera: Papilionoidea). *Univ. Kan. Sci. Bull.* 39:305–70.

Ehrlich, P.R., and I. Hanski. 2004. *On the wings of checkerspots: A model system for population biology.* Oxford: Oxford Univ. Press.

Ehrlich, P.R., and P.H. Raven. 1967. Butterflies and plants: A study in coevolution. *Evolution* 18:586–608.

El-Sayed, A., M. Bengtsson, S. Rauscher, J. Lofqvist, and P. Witzgall. 1999. Multicomponent sex pheromone in codling moth (Lepidoptera: Tortricidae). *Environ. Entomol.* 28:775–79.

Eliot, J.N. 1973. The higher classification of the Lycaenidae (Lepidoptera): A tentative arrangement. *Bull. Br. Mus. Nat. Hist. Entomol.* 28:373–505.

Fang, Q.Q., C. Soowon, J.C. Regier, et al. 1997. A new nuclear gene for insect phylogenetics: Dopa decarboxylase is informative of relationships within Heliothinae (Lepidoptera: Noctuidae). *Syst. Biol.* 46:269–83.

Felsenstein, J. 2004. *Inferring phylogenies.* Sunderland, MA: Sinauer Associates.

Fibiger, M., and J.D. Lafontaine. 2005. A review of the higher classification of the Noctuoidea (Lepidoptera) with special reference to the Holarctic fauna. *Esperiana* 11:7–92.

Ford, E.B. 1964. *Ecological genetics.* London: Chapman & Hall.

Freitas, A.V.L., and K.S.J. Brown. 2004. Phylogeny of the Nymphalidae (Lepidoptera: Papilionoidea). *Syst. Biol.* 53:363–83.

Friedlander, T.P., K.R. Horst, J.C. Regier, C. Mitter, R.S. Peigler, and Q.Q. Fang. 1998. Two nuclear genes yield concordant relationships within Attacini (Lepidoptera: Saturniidae). *Mol. Phylogenet. Evol.* 9:131–40.

Goater, B., M. Nuss, and W. Speidel. 2005. Pyraloidea I (Crambidae: Acentropinae, Evergestinae, Heliothelinae, Schoenobiinae, Scopariinae). In *Microlepidoptera of Europe 4,* ed. P. Huemer and O. Karsholt, 1–304. Stenstrup, Denmark: Apollo Books.

Goldsmith, M.R., T. Shimada, and H. Abe. 2005. The genetics and genomics of the silkworm, *Bombyx mori. Annu. Rev. Entomol.* 50:71–100.

Goldsmith, M.R., and A.S. Wilkins. 1995. *Molecular model systems in the Lepidoptera.* New York: Cambridge Univ. Press.

Gregory, T.R., and P.D.N. Hebert. 2003. Genome size and variation in lepidopteran insects. *Can. J. Zool.* 81:1399–1405.

Grimaldi, D.A. 1996. *Amber: Window to the past.* New York: Abrams/Am. Mus. Nat. Hist.

Grimaldi, D.A. 1999. The co-radiations of pollinating insects and angiosperms in the Cretaceous. *Ann. Mo. Bot. Gard.* 86:373–406.

Grimaldi, D.A., and M.S. Engel. 2005. *Evolution of the insects.* New York: Cambridge Univ. Press.

Hasenfuss, I. 1999. The adhesive devices in larvae of Lepidoptera (Insecta, Pterygota). *Zoomorphology* 119:143–62.

Heppner, J.B. 1991. Faunal regions and the diversity of Lepidoptera. *Trop. Lepid.* 2, Suppl. 1:1–85.

Hinton, H.E. 1946. On the homology and nomenclature of the setae of lepidopterous larvae, with some notes on the phylogeny of the Lepidoptera. *Trans. Entomol. Soc. Lond.* 97:1–37.

Hock, V., P.J. Albert, and M. Sandoval. 2007. Physiological differences between two sugar-sensitive neurons in the galea and the maxillary palp of the spruce budworm larva *Choristoneura fumiferana* (Clem.) (Lepidoptera: Tortricidae). *J. Insect Physiol.* 53:59–66.

Hodkova, M., and I. Hodek. 2004. Photoperiod, diapause, and cold-hardiness. *Eur. J. Entomol.* 101:445–58.

Holder, M., and P.O. Lewis. 2003. Phylogeny estimation: traditional and Bayesian approaches. *Nat. Rev. Genet.* 4:275–84.

Honeybee Genome Sequencing Consortium. 2006. Insights into social insects from the genome of the honeybee *Apis mellifera*. *Nature* 443:931–49.

Horak, M. 1998. The Tortricoidea. In *Lepidoptera, moths and butterflies. 1 Evolution, systematics and biogeography*, ed. N.P. Kristensen, 199–215. *Handbook of Zoology*, Vol. 4, Part 35, *Arthropoda: Insecta*. Berlin and New York: Walter de Gruyter.

Jacobson, N.L., and S.J. Weller. 2002. *A cladistic study of the tiger moth family Arctiidae (Noctuoidea) based on larval and adult morphology*. Thomas Say Monograph Series, Entomol. Soc. Amer.

Jenner, R.A., and M.A. Willis. 2007. The choice of model organisms in evo-devo. *Nat. Rev. Genet.* 8:311–19.

Jiggins, C.D., J. Mavarez, M. Beltrán, J.S. Johnston, and E. Bermingham. 2005. A genetic linkage map of the mimetic butterfly *Heliconius melpomene*. *Genetics* 171:557–70.

Jurenka, R. 2004. Insect pheromone biosynthesis. *Top. Curr. Chem.* 239:97–132.

Kawahara, A.Y., A.A. Mignault, J.C. Regier, I. J. Kitching, and C. Mitter. 2009. Phylogeny and biogeography of hawkmoths (Lepidoptera: Sphingidae): Evidence from five nuclear genes./PLoS ONE/ 4:e5719.

Kellogg, E.A., and H.B. Shaffer. 1993. Model organisms in evolutionary studies. *Syst. Biol.* 42:409–14.

Kettlewell, H.B.D. 1973. *The evolution of melanism, the study of a recurring necessity; with special reference to industrial melanism in Lepidoptera*. Oxford: Clarendon Press.

Kitching, I.J. 2002. The phylogenetic relationships of Morgan's Sphinx, *Xanthopan morganii* (Walker), the tribe Acherontiini, and allied long-tongued hawkmoths (Lepidoptera: Sphingidae, Sphinginae). *Zool. J. Linn. Soc.* 135:471–527.

Kitching, I.J. 2003. Phylogeny of the death's head hawkmoths, *Acherontia* [Laspeyres], and related genera (Lepidoptera: Sphingidae: Sphinginae: Acherontiini). *Syst. Entomol.* 28:71–88.

Kitching, I.J., and J.-M. Cadiou. 2000. *Hawkmoths of the world: An annotated and illustrated revisionary checklist (Lepidoptera: Sphingidae)*. Ithaca: Nat. Hist. Mus. London & Cornell Univ. Press.

Kitching, I.J., and J.E. Rawlins. 1998. The Noctuoidea. In *Lepidoptera, moths and butterflies. 1. Evolution, systematics, and biogeography*, ed. N. P. Kristensen, 351–401. *Handbook of Zoologie*, Vol. 4, Part 35, *Arthropoda: Insecta*. Berlin and New York: Walter de Gruyter.

Kristensen, N.P. 1976. Remarks on the family-level phylogeny of butterflies (Insecta, Lepidoptera, Rhopalocera). *Z. Zool. Syst. Evol.* 14:25–33.

Kristensen, N.P. 1984. Studies on the morphology and systematics of primitive Lepidoptera. *Steenstrupia* 10:141–91.

Kristensen, N.P., and A.W. Skalski. 1998. Phylogeny and palaeontology. In *Lepidoptera, moths and butterflies. 1. Evolution, systematics, and biogeography*, ed. N.P. Kristensen, 7–25. *Handbook of Zoologie*, Vol. 4, Part 35, *Arthropoda: Insecta*. Berlin and New York: Walter de Gruyter.

Kumar, S. 2005. Molecular clocks: Four decades of evolution. *Nat. Rev. Genet.* 6:654–62.

Labandeira, C.C., D.L. Dilcher, D.R. Davis, and D.L. Wagner. 1994. Ninety-seven million years of angiosperm-insect association: Palaeobiological insights into the meaning of coevolution. *Proc. Natl. Acad. Sci. U.S.A.* 9:12278–82.

Labandeira, C.C., and J.J. Sepkoski, Jr. 1993. Insect diversity in the fossil record. *Science* 261:310–15.

Lafontaine, J.D., and M. Fibiger. 2006. Revised higher classification of the Noctuoidea (Lepidoptera). *Can. Entomol.* 138:610–35.

Lamas, G. 2004. *Atlas of Neotropical Lepidoptera. Checklist: Part 4A, Hesperioidea – Papilionoidea*. Gainesville: Scientific Publishers.

Leibenguth, F. 1986. Genetics of the flour moth, *Ephestia kühniella*. *Agric. Zool. Rev.* 1:39–72.

Lemaire, C. 1972. Révision du genre *Lonomia* (Walker) [Lep. Attacidae]. *Ann. Soc. Entomol. Fr.* 8:767–861.

Lemaire, C. 2002. *The Saturniidae of America (= Attacidae). Hemileucinae*. Keltern: Goecke & Evers.

Mabee, P.M. 2000. Development data and phylogenetic systematics: Evolution of the vertebrate limb. *Amer. Zool.* 40:789–800.

Mabee, P.M., M. Ashburner, Q. Cronk, et al. 2007. Phenotype ontologies: the bridge between genomics and evolution. *Trends Ecol. Evol.* 22:345–50.

Mallet, J., W.O. McMillan, and C.D. Jiggins. 1998. Mimicry and warning color at the boundary between races and species. In *Endless forms: Species and speciation*, ed. D.J. Howard and S.H. Berlocher, 390–403. New York: Oxford Univ. Press.

McDonough, L.M., H.G. Davis, P.S. Chapman, and C.L. Smithhisler. 1993. Response of male codling moths (*Cydia pomonella*) to components of conspecific female sex pheromone glands in flight tunnel tests. *J. Chem. Ecol.* 19:1737–48.

Michener, C.D. 1952. The Saturniidae (Lepidoptera) of the Western hemisphere. *Bull. Am. Mus. Nat. Hist.* 98:335–502.

Miller, J.S. 1991. Cladistics and classification of the Notodontidae (Lepidoptera: Noctuoidae) based on larval and adult morphology. *Bull. Am. Mus. Nat. Hist.* 204:1–230.

Minet, J. 1983. Etude morphologique et phylogénétique des organs tympaniques des Pyraloidea. 1 – généralitiés et homologies. (Lep. Glossata). *Ann. Soc. Entomol. Fr.* 19:175–207.

Minet, J. 1991. Tentative reconstruction of the ditrysian phylogeny (Lepidoptera: Glossata). *Entomol. Scand.* 22:69–95.

Minet, J. 1994. The Bombycoidea: Phylogeny and higher classification (Lepidoptera: Glossata). *Entomol. Scand.* 25:63–88.

Mita, K., M. Kasahara, S. Sasaki, et al. 2004. The genome sequence of silkworm, *Bombyx mori*. *DNA Res.* 11:27–35.

Mitchell, A., S. Cho, J.C. Regier, C. Mitter, R.W. Poole, and M. Matthews. 1997. Phylogenetic utility of elongation factor-1α in Noctuoidea (Insecta: Lepidoptera): The limits of synonymous substitution. *Mol. Biol. Evol.* 14:381–90.

Mitchell, A., C. Mitter, and J.C. Regier. 2000. More taxa or more characters revisited: Combining data from nuclear protein-encoding genes for phylogenetic analyses of Noctuoidea (Insecta: Lepidoptera). *Syst. Biol.* 49:202–24.

Mitchell, A., C. Mitter, and J.C. Regier. 2006. Systematics and evolution of the cutworm moths (Lepidoptera: Noctuidae): Evidence from two protein-coding nuclear genes. *Syst. Entomol.* 31:21–46.

Müller, F. 1879. *Ituna* and *Thyridia*; a remarkable case of mimicry in butterflies. *Trans. Entomol. Soc. Lond.* 20–29.

Munroe, E., and M.A. Solis. 1998. The Pyraloidea. In *Lepidoptera, moths and butterflies. 1 Evolution, systematics and biogeography*, ed. N.P. Kristensen, 233–56. *Handbook of Zoology*, Vol. 4, Part 35, *Arthropoda: Insecta*. Berlin and New York: Walter de Gruyter.

Nakamura, M. 1976. An inference on the phylogeny of Sphingidae in relation to habits and the structures of their immature stages. *Yugatô* 63:19–28.

Nazari, V., E.V. Zakharov, and F.A.H. Sperling. 2007. Phylogeny, historical biogeography, and taxonomic ranking of Parnassiinae (Lepidoptera, Papilionidae) based on morphology and seven genes. *Mol. Phylogenet. Evol.* 42:131–56.

Neunzig, H.H. 1987. Pyralidae (Pyraloidea). In *Immature insects*, ed. F.W. Stehr, 462–94. Dubuque: Kendall Hunt Publishing Co.

Neunzig, H.H. 1997. Pyraloidea, Pyralidae (part). In *The moths of America north of Mexico*, ed. R.B. Dominick, D.C. Ferguson, J.G. Franclemont, R.W. Hodges, and E.G. Munroe, fascicle 15.4:1–157. Washington, DC: The Wedge Entomological Research Foundation.

Nielsen, E.S., and N.P. Kristensen. 1996. The Australian moth family Lophocoronidae and the basal phylogeny of the Lepidoptera (Glossata). *Invertebr. Taxon.* 10:1199–1302.

Nijhout, H.F. 1991. *The development and evolution of butterfly wing patterns*. Washington, DC: Smithsonian Institution Press.

Nuss, M. 2006. Global information system of Pyraloidea (GlobIZ). http://www.pyraloidea.org (accessed July 31, 2007).

Paukstadt, U., U. Brosch, and L.H. Paukstadt. 2000. Preliminary checklist of the names of the worldwide genus *Antheraea* Hübner, 1819 ("1816") (Lepidoptera: Saturniidae). Part I. *Galathea. Ber. Kreis. Nürnber. Entomol. e. V.*, Suppl. 9:1–59.

Peigler, R.S., and S. Naumann. 2003. *A revision of the silkmoth genus* Samia. San Antonio: Univ. Incarnate Word.

Pogue, M., 2002. *A world revision of the genus Spodoptera Guenée (Lepidoptera: Noctuidae)*. Washington, DC: Mem. Am. Entomol. Soc. no. 43.

Preece, J. 2000. *Online communities: Designing usability, supporting sociability*. Chichester: John Wiley & Sons.

Quayle, D., J. Régnière, N. Cappuccino, and A. Dupont. 2003. Forest composition, host-population density, and parasitism of spruce budworm *Choristoneura fumiferana* eggs by *Trichogramma minutum*. *Entomol. Exp. Appl.* 107:215–27.

Ramos, D.M., and A. Monteiro. 2007. Transgenic approaches to study wing color pattern development in Lepidoptera. *Mol. Biosyst.* 3:530–35.

Regier, J.C., C.P. Cook, C. Mitter, and A. Hussey. 2008. A phylogenetic study of the "bombycoid complex" (Lepidoptera) using five protein-coding nuclear genes, with comments on the problem of macrolepidopteran phylogeny. *Syst. Entomol.* 33:175–89.

Regier, J.C., C. Mitter, T.P. Friedlander, and R.S. Peigler. 2001. Phylogenetic relationships in Sphingidae (Insecta: Lepidoptera): Initial evidence from two nuclear genes. *Mol. Phylogenet. Evol.* 20:311–25.

Regier, J.C., C. Mitter, R.S. Peigler, and T.P. Friedlander. 2002. Monophyly, composition, and relationships within Saturniinae (Lepidoptera: Saturniidae): Evidence from two nuclear genes. *Insect Syst. Evol.* 33:9–21.

Regier, J.C., U. Paukstadt, L.H. Paukstadt, C. Mitter, and RS. Peigler. 2005. Phylogenetics of eggshell morphogenesis in *Antheraea* (Lepidoptera: Saturniidae): Unique origin and repeated reduction of the aeropyle crown. *Syst. Biol.* 54:254–67.

Reuter, E. 1896. Über die Palpen der Rhopaloceren. *Acta Soc. Sci. Fennicae* 22:1–578.

Robinson, R. 1971. *Lepidoptera genetics*. Oxford: Pergamon Press.

Roelofs, W.L., and A.P. Rooney. 2003. Molecular genetics and evolution of pheromone biosynthesis in Lepidoptera. *Proc. Natl. Acad. Sci. U.S.A.* 100:9179–84.

Rothschild, W.V., and K. Jordan. 1903. A revision of the lepidopterous family Sphingidae. *Novit. Zool.* 9 (Suppl.):1–972.

Rubinoff, D., and F.A.H. Sperling. 2002. Evolution of ecological traits and wing morphology in *Hemileuca* (Saturniidae) based on a two-gene phylogeny. *Mol. Phylogenet. Evol.* 25:70–86.

Rubinoff, D., and F.A.H. Sperling. 2004. Mitochondrial DNA sequence, morphology and ecology yield contrasting conservation implications for two threatened buckmoths (*Hemileuca*: Saturniidae). *Biol. Conserv.* 118:341–51.

Rutschmann, F. 2006. Molecular dating of phylogenetic trees: A brief review of current methods that estimate divergence times. *Divers. Distrib.* 12:35–48.

Sahara, K., A. Yoshido, F. Marec, et al. 2007. Conserved synteny of genes between chromosome 15 of *Bombyx mori* and a chromosome of *Manduca sexta* shown by five-color BAC-FISH. *Genome* 50:1061–65.

Salmon, M.A. 2000. *The Aurelian legacy: British butterflies and their collectors*. Berkeley: Univ. California Press.

Santini, F., and E.J. Stellwag. 2002. Phylogeny, fossils and model systems in the study of evolutionary developmental biology. *Mol. Phylogenet. Evol.* 24:379–83.

Scriber, J.M., Y. Tsubaki, and R.C. Lederhouse. 1995. *Swallowtail butterflies: Their ecology and evolutionary biology*. Gainesville: Scientific Publishers.

Shimodaira, H. 2002. An approximately unbiased test of phylogenetic tree selection. *Syst. Biol.* 51:492–508.

Siaussat, D., F. Bozzolan, P. Porcheron, and S. Debernard. 2007. Identification of steroid hormone signaling pathway in insect cell differentiation. *Cell. Mol. Life Sci.* 64:365–76.

Slansky Jr., F., 1993. Nutritional ecology: The fundamental quest for nutrients. In *Caterpillars: Ecological and evolutionary constraints on foraging*, ed. N.E. Stamp and T.M. Casey, 29–91. New York: Chapman and Hall.

Smart, P. 1975. *The illustrated encyclopedia of the butterfly world*. London: Salamander Books.

Solis, M.A, S.D. Hight, and D.R. Gordon. 2004. Alert: Tracking the cactus moths as it flies and eats it way westward in the U.S. *News Lepid. Soc.* 46:3–4.

Solis, M.A., and K.V.N. Maes. 2002. Preliminary phylogenetic analysis of the subfamilies of Crambidae (Pyraloidea Lepidoptera). *Belg. J. Entomol.* 4:53–95.

Solis, M.A., and C. Mitter. 1992. Review and preliminary phylogenetic analysis of the subfamilies of the Pyralidae (sensu stricto) (Lepidoptera: Pyraloidea). *Syst. Entomol.* 17:79–90.

Srinivasan, A., A.P. Giri, and V.S. Gupta. 2006. Structural and functional diversities in Lepidoptera serine proteases. *Cell. Mol. Biol. Lett.* 11:132–54.

Swofford, D.L., G.J. Olsen, P.J. Waddell, and D.M. Hillis. 1996. Phylogenetic inference. In *Molecular Systematics*, 2nd edition, ed. D.M. Hillis, C. Moritz, and B.K. Mable, 407–514. Sunderland: Sinauer Associates.

Trematerra, P., and A. Sciarretta. 2005. Activity of the kairomone ethyl (E,Z)-2,4-decadienoate in the monitoring of *Cydia pomonella* (L.) during the second annual flight. *Redia* (Italy) 88:57–67.

Tuskes, P.M., J.P. Tuttle, and M.M. Collins. 1996. *The wild silk moths of North America. A natural history of the Saturniidae of the United States and Canada*. Ithaca: Cornell Univ. Press.

Vane-Wright, R.I., and P.R. Ackery. 1984. *The Biology of Butterflies*. Princeton: Princeton Univ. Press.

Veiga, A.B.G., J.M.C. Ribeiro, J.A. Guimarães, and I.M.B. Francischetti. 2005. A catalog for the transcripts from the venomous structures of the caterpillar *Lonomia obliqua*: Identification of the proteins potentially involved in the coagulation disorder and hemorrhagic syndrome. *Gene* 355:11–27.

Wahlberg, N., M.F. Braby, A.V. Brower, et al. 2005. Synergistic effects of combining morphological and molecular data in resolving the phylogeny of butterflies and skippers. *Proc. Biol. Sci.* 272:1577–86.

Wahlberg, N., E. Weingartner, and S. Nylin. 2003. Towards a better understanding of the higher systematics of Nymphalidae (Lepidoptera: Papilionoidea). *Mol. Phylogenet. Evol.* 28:473–84.

Wahlberg, N., and C.W. Wheat. 2008. Genomic outposts serve the phylogenomic pioneers: Designing novel nuclear markers for genomic DNA extractions of Lepidoptera. *Syst. Biol.* 57:231–42.

Wallace, E.K., P.J. Albert, and J.N. McNeil. 2004. Oviposition behavior of the eastern spruce budworm *Choristoneura fumiferana* (Clemens) (Lepidoptera: Tortricidae). *J. Insect Behav.* 17:145–54.

Walters, B.D., P.J. Albert, and R.Y. Zacharuk. 1998. Morphology and ultrastructure of chemosensilla on the proboscis of the adult spruce budworm *Choristoneura fumiferana* (Clem.) (Lepidoptera: Tortricidae). *Can. J. Zool.* 76:466–79.

Warren, A. 2006. The higher classification of the Hesperiidae (Lepidoptera: Hesperioidea). PhD diss., Oregon State Univ.

Wearing, C.H., J. Hansen, C. Whyte, C.E. Miller, and J. Brown. 2001. The potential for spread of codling moth via commercial sweet cherry: A critical review and risk assessment. *Crop Prot.* 20:465–88.

Welch, J.J., and L. Bromham, 2005. Molecular dating when rates vary. *Trends Ecol. Evol.* 20:320–27.

Weller, S.J., N.L. Jacobson, and W.E. Conner. 1999. The evolution of chemical defences and mating systems in tiger moths (Lepidoptera: Arctiidae). *Biol. J. Linn. Soc.* 68:557–78.

Weller, S.J., D.P. Pashley, and J.A. Martin. 1996. Reassessment of butterfly family relationships using independent genes and morphology. *Ann. Entomol. Soc. Am.* 89:184–92.

Weller, S.J., D.P. Pashley, J.A. Martin, and J.L. Constable. 1994. Phylogeny of noctuoid moths and the utility of combining independent nuclear and mitochondrial genes. *Syst. Biol.* 43:194–211.

Whalley, P.E.S. 1978. New taxa of fossil and recent Micropterigidae with a discussion of their evolution and a comment on the origin of the Lepidoptera. *Ann. Tvl. Mus.* 31:71–86.

Whalley, P.E.S. 1986. A review of the current fossil evidence of Lepidoptera in the Mesozoic. *Biol. J. Linn. Soc.* 28:253–71.

Wiegmann, B.M., C. Mitter, J.C. Regier, T.P. Friedlander, D.M. Wagner, and E.S. Nielsen. 2000. Nuclear genes resolve Mesozoic-aged divergences in the insect order Lepidoptera. *Mol. Phylogenet. Evol.* 15:242–59.

Wiegmann, B.M., J.C. Regier, and C. Mitter. 2002. Combined molecular and morphological evidence on phylogeny of the earliest lepidopteran lineages. *Zool. Scripta* 31:67–81.

Willis, J.H., A.S. Wilkins, and M.R. Goldsmith. 1995. A brief history of Lepidoptera as model systems. In *Molecular model systems in the Lepidoptera*, ed. M.R. Goldsmith and A.S. Wilkins, 1–20. Cambridge: Cambridge Univ. Press.

Xia, Q., Z. Zhou, C. Lu, et al. 2004. A draft sequence for the genome of the domesticated silkworm (*Bombyx mori*). *Science* 306:1937–40.

Zhang, B.-C. 1994. *Index of economically important Lepidoptera.* Oxon: CAB International.

Zimmermann, H.G., V.C. Moran, and J.H. Hoffmann. 2000. The renowned cactus moth, *Cactoblastis cactorum*: Its natural history and threat to native *Opuntia* floras in Mexico and United States of America. *Divers. Distrib.* 6:259–69.

Zwick, A. 2008. Molecular phylogeny of Anthelidae and other bombycoid taxa (Lepidoptera: Bombycoidea). *Syst. Entomol.* 33:190–209.

2 Recent Progress in Silkworm Genetics and Genomics

Marian R. Goldsmith

CONTENTS

INTRODUCTION: THE SILKWORM MODEL

The domesticated silkworm, *Bombyx mori*, has long served as a model for lepidopteran biology. Its relatively short, predictable life cycle, large size, and adaptation to mass rearing and laboratory culture, coupled with the high degree of fundamental knowledge needed for successful practice of sericulture, explains in part the silkworm's use as a representative for Lepidoptera in basic research. It became an object of genetic studies early in the twentieth century (for accounts of the history of silkworm genetics and sericulture, see Eickbush 1995; Yasukochi, Fujii, and Goldsmith 2008), leading to the collection of many spontaneous mutations found in the course of mass rearing or introgressed from its nearest wild relative, *Bombyx mandarina*, which is present in mulberry fields in China, Japan, and Korea, and can still form partially fertile hybrids with its domesticated cousin. Major germ plasm collections in China, Japan, Korea, and India maintain hundreds of "geographic" or local strains, and genetically improved varieties that, like any agricultural breeding stock, are classified according to their economic characters, such as rearing properties and robustness, cocoon and silk qualities, fertility, and fecundity (Sohn 2003; Table 2.1). These centers also maintain various morphological, biochemical, and behavioral mutations affecting all metamorphic stages (about 430 mutations in Japan (Fujii 1998; Table 2.1) and more than 600 mutant strains in China (Lu, Dai, and Xiang 2003). The silkworm was used for irradiation studies shortly after the first report of using X-rays to mutate *Drosophila* by Müller in 1927 (reviewed in Tazima 1964). This early work led to the induction and selection of many morphological and behavioral mutations and chromosome aberration stocks. Among the latter are sex-linked translocations involving visible mutations affecting egg, larva, or cocoon, which were transferred from autosomes to the female W chromosome, and have been used to breed autosexing stocks for sericulture (Tazima 1964; Nagaraju 1996; Fujii et al. 2006). These aberrations have also been useful for fine structure mapping of the affected autosomes (Fujiwara and Maekawa 1994) and the W chromosome itself, which carries the putative female sex-determining gene, *Fem* (see Chapter 4 for a discussion of W-translocations and their application in studies of sex determination).

In the 1990s the silkworm continued to serve as the lepidopteran model for molecular genetics and genomics studies, owing to the properties that led to its earlier use for classical or Mendelian genetics, namely, plentiful well-differentiated, stable laboratory stocks and a large body of knowledge about its fundamental biology. These and other technical features, such as the ability to provide starting material with low DNA polymorphism by intensive inbreeding, together with its central historical role as an insect model, led *B. mori* to become the first lepidopteran used for large-scale sequencing of cDNAs (Mita et al. 2003) and, ultimately, a full-scale genome project (International Silkworm Genome Consortium 2008).

In this chapter I will focus on recent developments in the molecular genetics and genomics of the silkworm; reviews of earlier work may be found in Nagaraju and Goldsmith (2002) and Goldsmith, Shimada, and Abe (2005); an overview of recent developments in silkworm genomics may be found in Zhou, Yang, and Zhong (2008).

TABLE 2.1
Silkworm Stocks and Mutant Resources

Database	Contents	Host	Web Site
SilkwormBase	Silkworm strains and mutant resources; information on mutations	Institute of Genetic Resources Center, Kyushu University	http://www.shigen.nig.ac.jp/silkwormbase/ViewStrainGroup.do
Silkworm Genome Database	Descriptions and photographs of mutations	University of Tokyo Department of Agricultural and Environmental Biology	http://www.ab.a.u-tokyo.ac.jp/bioresource/shimada/database.html
Silkworm "Aburako" (translucent) mutation database	Descriptions, images, and linkage assignments of translucent mutations	National Institute of Agrobiological Sciences, Tsukuba, Japan	http://cse.nias.affrc.go.jp/natuo/en/aburako_top_en.htm
Database of Sericultural Genebank	Silkworm strain resources (453 stocks; mutant and commercial races); mulberry stocks; insect cell lines	National Institute of Agrobiological Sciences, Kobuchizawa, Japan	http://ss.nises.affrc.go.jp/nises/db-eng.html
Silkworm Germplasm	Practical breeding strains (430 stocks); economic characters and morphological traits	Central Sericultural Germplasm Resources Centre, Central Silk Board, Tamil Nadu, India	http://www.silkgermplasm.com/

BOMBYX CYTOGENETICS

The haploid chromosome number in *B. mori* is 28, close to the modal number of 31 characteristic of Lepidoptera (De Prins and Saitoh 2003), with a WZ:ZZ female:male sex chromosome constitution (see Chapter 3 for a discussion on the sex chromosome system in Lepidoptera; see Chapter 4 for a discussion of sex chromosomes and sex determination in *B. mori*). The silkworm karyotype was first differentiated cytogenetically in pachytene oocytes by W. Traut in the mid-1970s (Traut 1976). Although this provided a look at the relative sizes of silkworm chromosomes and some chromosome regions were differentially stained, the WZ bivalent was not evident. Further, because, silkworm chromosomes lack a distinct primary constriction (the centromere) and have a nearly holokinetic structure typical of Lepidoptera (see Chapter 3), it was possible to identify only six chromosomes by distinctive chromomere patterns.

Silkworm cytogenetics progressed rapidly since the 1990s in parallel with technological developments driven by the availability of radioactive and, more recently, fluorescent nucleotide derivatives that can be used for direct labeling of DNA probes for chromosomal in situ hybridization. The earliest study reported in silkworm was with a radioactively labeled probe for BMC1, an approximately 5-kb member of the long interspersed nuclear element one (LINE-1) family, which was estimated to occur in about thirty-five hundred copies in the genome and was seen to be widely dispersed chromosomally (Ogura et al. 1994). Using the more precise and sensitive technique of fluorescence in situ hybridization (FISH), Sahara and colleagues identified telomeres containing the canonical insect sequence $(TTAGG)_n$ on silkworm chromosome ends (Sahara, Marec, and Traut 1999). These probes were composed of long stretches of repeats, which were needed to detect localized signals on the relatively small lepidopteran chromosomes, and had a predictable pattern and location (doublets on the bivalent chromosome ends), making them good subjects for working out the basic methodology. Traut et al. (1999) adapted comparative genomic hybridization (CGH), the technique originally devised to detect chromosomal abnormalities in cancer tissues, for the study of molecular differentiation of sex chromosomes. With the help of CGH they were the first to identify the wild-type W chromosome in female silkworms and show that this female-determining chromosome is largely composed of ubiquitous DNA repeats (the principle of CGH is described in Chapter 3).

The breakthrough for visualizing single-copy sequences was the construction of large fragment BAC libraries (Wu et al. 1999) from which clones with genes of interest could be isolated and labeled directly with fluorochromes. This led to the easy visual identification of the W in the silkworm. In a pioneering study, K. Sahara and colleagues were able to light up the whole W chromosome using BACs isolated with primers derived from non-LTR retrotransposons known to be linked to the W (Sahara et al. 2003). This study also provided the first direct evidence for the kinds of repeated elements comprising the W (see Chapter 4 for additional details on the structure of the silkworm W), foreshadowing the analytic power of the BAC-FISH technique for cytogenetic investigation in Lepidoptera. In a further refinement, use of two fluorescent labels allowed Yasukochi and colleagues to image the 5′ and 3′ ends of a large BAC contig containing the *BmHox* gene cluster on linkage group (LG) 6 containing the *Bmzen*, *Dfd*, *Scr*, *Antp*, *Ubx*, *abd-A*, and *Abd-B* genes, which they estimated to be 2–3 Mb long (Yasukochi et al. 2004). By adding a third probe for the distally linked gene, *labial*, labeled with the same fluorochrome as that on the proximal end of the *Hox* cluster, they were able to show the location of this single-copy gene directly, about one-tenth chromosome length away, consistent with the genetic linkage map. This landmark study led the way to defining a complete *Bombyx* karyotype using two-color FISH in which BACs carrying previously mapped markers were anchored to each LG (Yoshido et al. 2005), and localization of individual genes with BAC-FISH (e.g., Niimi et al. 2006). Since the technique was established as many as five colors have been used to compare the location of conserved markers between LG 15 of *B. mori* and the corresponding LG of its close relative, the sphingid *Manduca sexta* (Sahara et al. 2007), showing the power of this method not only for direct linkage mapping in Lepidoptera but also for comparative genomics.

"CLASSICAL" LINKAGE MAPS

The "classical" linkage maps consist of about 240 loci marked by morphological, behavioral, and biochemical mutations (Fujii 1998). Using morphological mutations with similar phenotypes as reference loci to mark a particular linkage group can pose problems; recently, LGs 26 and 28, which were defined by two independent yellow larval mutations, *Sel* (Sepialumazine) and *Xan* (Xanthous), were shown to be on the same linkage group using sequence-tagged sites (STS; Yasukochi et al. 2005) and likely allelic or very closely linked using microsatellites or simple sequence repeats (SSRs; Miao et al. 2007). This suggests that although easily scored visible mutations will continue to be used to detect contamination in stock maintenance and for rapid screening for recombinants where they flank a chromosome region of interest, the ready availability of many types of molecular markers that can be scored quickly and precisely by the polymerase chain reaction (PCR), single nucleotide polymorphisms (SNPs), or high-throughput sequencing techniques will render visibly marked stocks obsolete for many applications, and their greatest value will remain as sources of altered genes for functional analysis.

MOLECULAR LINKAGE MAPS

Molecular linkage maps are the foundation of modern genetics. They provide a reference framework for many applications: determining chromosomal locations of spontaneous and induced mutations, including genes with a Mendelian type of inheritance (i.e., with a major effect) and complex or quantitative traits with relatively small additive or nonadditive effects; providing scaffolds to link contiguous components of large-scale physical maps and whole genome shotgun (WGS) sequence data; enabling positional or map-based cloning of genes known only by their phenotypic effects; and revealing global patterns genome organization and evolution. Genetic markers can be anonymous, such as random amplification of polymorphic DNA sequences (RAPDs) or amplified fragment length polymorphisms (AFLPs), which are limited to a single laboratory strain or field population, or derived from species-specific single-copy chromosomal sites or genes via restriction fragment length polymorphisms (RFLPs), cleaved amplified polymorphic sequences (CAPS), STS, SSRs, and SNPs. All of these types of markers are now in use in *B. mori*, and significant progress has been made in the construction of linkage maps for silkworm since preliminary ones were published more than thirteen years ago using RAPDs (Promboon et al. 1995) and RFLPs detected with cloned cDNAs whose sequences were unknown at the time (Shi, Heckel, and Goldsmith 1995). Details of early linkage maps have been reviewed (Nagaraju and Goldsmith 2002; Goldsmith, Shimada, and Abe 2005; Zhou, Yang, and Zhong 2008). In the following section, I describe the salient features of current linkage maps, emphasizing strategic decisions for their uses as resources for genetics of silkworm and other lepidopteran species.

Reference Strains

The silkworm can tolerate a relatively high level of inbreeding, likely the result of its long years of domestication. This contrasts sharply with laboratory strains of many "wild" lepidopteran species, which must be replenished regularly by outcrossing to field-caught insects to avoid population crashes, presumably caused by inbreeding depression, that is, homozygosis of deleterious recessive mutations. This property has made the silkworm an especially good model for molecular genetic and genomic studies, as it has been possible to use a few well-characterized, widely disseminated, stable inbred strains as common references, allowing sharing of resources and ready comparison of results. Hence, two strains—one called p50 (or Daizo) in Japanese stock centers and Dazao in Chinese stock centers, and the other, Chinese 108 (C108), a relatively old genetically improved strain—have been used for most linkage mapping projects. p50/Dazao, which originated in China and is considered a geographic strain, has also been used for construction of many cDNA libraries used for EST projects (Mita et al. 2003), for several BAC libraries used in large-scale physical mapping projects (Yamamoto et al. 2006, 2008; Suetsugu et al. 2007), for chromosome walking (Koike et al. 2003), for BAC-FISH (Yoshido et al. 2005; Sahara et al. 2007), and for the main WGS projects (Mita et al. 2004; Xia et al. 2004). Although some polymorphism remains in these strains (Mills and Goldsmith 2000; Cheng et al. 2004), as might be expected, they are homozygous enough to enable consistent scoring of such variable sites as RAPDs and RFLPs in stocks of the same strains maintained in different laboratories (M.R. Goldsmith, personal observations; Miao et al. 2005), to facilitate construction of BAC contigs (Suetsugu et al. 2007; Yamamoto et al. 2008), and to allow alignment and assembly of WGS sequences from independent projects (Mita et al. 2004; Xia et al. 2004). This relative homozygosity was a major reason for choosing the silkworm as the initial lepidopteran model for whole genome sequencing (see below).

Bombyx mandarina and Domestication of *B. mori*

Distinct populations of *B. mandarina* exist, differentiated by karyotype. Those in China (ChBm) have a haploid chromosome number of n = 28, the same as *B. mori*, whereas those found in Japan (JaBm) have a haploid chromosome number of n = 27. These differences led in part to the prevailing notion that *B. mori* was domesticated in China, although whether this occurred once or twice in different parts of the country is still under discussion (Li et al. 2005).

The karyotype difference is likely the result of a fusion between *B. mori* chromosomes 14 and 27 or 28 (Banno et al. 2004). Populations of *B. mandarina* with 27 or 28 chromosomes can be found in different regions of South Korea, suggesting they were separated by the formation of the Korean peninsula, on the order of two mya (Kawanishi et al. 2008). Application of BAC-FISH using probes derived from *B. mori* indicates that the W chromosomes of the two species have retained a high degree of similarity, both largely composed of the same types of repeated sequences; however, use of a BAC that gives a distinctive hybridization pattern suggests the possibility that some rearrangements have occurred (Yoshido et al. 2007; for composition of lepidopteran W chromosomes see Chapter 3). Further detailed cytogenetic analysis may uncover additional large-scale karyotypic changes that have occurred between populations of *B. mandarina* or during its domestication, providing landmarks for tracing the evolution of different subgroups.

The complete mitochondrial DNA sequences of representative *B. mandarina* strains are available. Comparisons of single (Li et al. 2005) or multiple (Arunkumar, Metta, and Nagaraju 2006; Pan et al. 2008) mitochondrial genes and regions with those of *B. mori* also show that the domesticated silkworm is more closely related to ChBm, consistent with the cytogenetic evidence and historical records, although divergence estimates based on various sequences vary from one to two mya (Pan et al. 2008) to as much as seven mya (Yukuhiro et al. 2002). Evidence from variants of nuclear 5.8S-28S ribosomal DNA (rDNA) units suggests that ChBm and JaBm diverged about three mya, in line with the estimate of their geographic separation (Maekawa et al. 1988). Heterozygosity in this rDNA locus is present among individuals from Chinese, Japanese, and Korean sources, a sign that the strains continue to hybridize in nature and with *B. mori*. Hence, investigations of nuclear genes potentially associated with domestication must account for the parental origin of a given chromosome that could have been introduced from either species and one must interpret the data with this caveat in mind (Kawanishi et al. 2008). Phylogenetic analysis of mariner-like elements (MLEs) isolated from geographically diverse populations of *B. mandarina* is consistent with a Chinese origin for *B. mori* (Kawanishi et al. 2007); interestingly, many MLEs have full-length open reading frames and features consistent with the possibility that they are still active (Kawanishi et al. 2008).

In contrast with larvae and adults of *B. mandarina*, which are both darkly pigmented and well-camouflaged, larvae of the domesticated insect are plain and white or with limited markings on the second ("eyespot"), fifth ("crescent"), and eighth ("star") segments, and moths are generally beige with faint bands on the wings. Many of the pigment pattern genes have been introgressed from

B. mandarina into this "null" background of *B. mori* and mapped as Mendelian traits that affect such characters as larval markings, wing pattern, and cocoon color (Fujii 1998). Other more subtle traits that differ between the species, like behavioral differences (e.g., dispersed vs. clumped egg-laying patterns, twig mimicry) and immune response, could be studied as quantitative traits after crossing with *B. mori* (see below).

Domestication of *B. mori* from *B. mandarina* has long been of interest, especially as *B. mori* is the only insect completely dependent on humans for survival. Behavioral, morphological, physiological, and biochemical differences between *B. mandarina* and *B. mori* could be exploited to inform an understanding of the impact of directional selection and insect biology in general. What underlying genetic changes were imposed to improve the silkworm's qualities for agriculture—that is, high survival under mass rearing conditions; high production/food ratio; high yield of silk with fine textile properties; lack of flight and other behavioral changes guaranteeing easy mating, high fertility, and fecundity; long-term egg diapause; plump, edible pupae—and which were simply hitchhikers, that is, unselected genes with no evident advantage for the silkworm's commercial use? Answers to these long-standing questions in sericulture seem to be close at hand with new genomic data and tools available, but will require a thoughtful choice of characters and gene systems to yield new insights for understanding the mechanisms that underlie speciation and the processes of artificial and natural selection.

Mapping Strategies

Most silkworm linkage maps have been constructed in two steps, first taking advantage of lack of crossing over in females (Maeda 1939) by establishing linkage groups with a female informative F_1 backcross to a male recurrent parent and genotyping a small population of ten to fifteen offspring. This is followed by scoring a much larger population, on the order of one hundred to two hundred individuals, using a male informative F_1 backcross to a female recurrent parent in order to calculate a recombination frequency at the level of a few percent of crossing over (see detailed examples of this approach in Chapter 6 for *Heliconius* butterflies and Chapter 18 for *B. mori*).

Using this strategy, Miao et al. (2005) developed a recombination map composed of more than five hundred SSR markers anchored with approximately eighty CAPS markers corresponding to genes whose sequences were registered in the National Center for Biotechnology Information (NCBI); LGs and assignments were verified by an independent cross using stocks from another laboratory that were genotyped for both RFLPs and reference SSRs. Similarly, Kadono-Okuda et al. (2002) and Nguu et al. (2005) produced independent RFLP maps with strains RF02 and RF50, selected to maximize polymorphism (W. Hara, pers. comm.). Integration of independently constructed maps requires genotyping with the same markers; this becomes relatively easy once reference loci are established for each LG. Use of a single F_2 cross is also a viable approach, with the advantage of reducing the number of strains to maintain and rear in each generation, but with the disadvantage of having to sort out the nonrecombinant female-derived chromosome for each linkage group in the progeny. This adds to the complexity of the analysis, especially for dominant characters such as RAPDs and AFLPs, which produce two separate linkage maps for each contributing parent that must be integrated using common, reference, codominant markers (Yasukochi 1998; see Chapter 6 for detailed discussion of this strategy). Nevertheless, projects based on RFLPs (Shi, Heckel, and Goldsmith 1995) and RAPDs (Promboon et al. 1995; Yasukochi 1998) successfully used F_2 crosses of p50 and C108 (Promboon et al. 1995; Yasukochi 1998) to establish preliminary maps. One of these yielded a primary reference RAPD map of more than one thousand markers (Yasukochi 1998) to which 523 BAC contigs carrying 427 STS representing cloned genes and ESTs were subsequently anchored (Yasukochi et al. 2006).

The *tour de force* in silkworm genetics using the two-stage mapping strategy involved the assignment of 1,688 nuclear genes to linkage groups by probing Southern blots of subpopulations of only fifteen individuals from a female informative backcross (Yamamoto et al. 2008). This was accompanied by construction of a recombination map using a male-informative backcross comprised initially of more than 500 SNPs (Yamamoto et al. 2006) and later expanded to 1,755 SNPs (Yamamoto et al. 2008) derived from BAC-end sequences (BESs). The BESs were anchored to BAC contigs assembled by a combination of restriction fragment fingerprinting and high-density replica BAC filter hybridization with ESTs (Yamamoto et al. 2008). This integrated genetic-physical map provides a robust scaffold for assembling the WGS (see below).

Map Characteristics

To date, several independent reference maps have been established: low- and medium-density maps with about two hundred markers using RAPDs (Promboon et al. 1995) and RFLPs (Hara et al. 2002; Nguu et al. 2005); medium-density maps with about five hundred markers using AFLPs (Tan et al. 2001), SSRs (Miao et al. 2005) or SNPs (Yamamoto et al. 2006); and high-density maps with more than a thousand markers using RAPDs (Yasukochi 1998; Yasukochi et al. 2006) or SNPs (Yamamoto et al. 2008; see also Yasukochi, Fujii, and Goldsmith 2008). The RFLP maps are based on cloned cDNAs; hence, they provide common reference points for integrating these projects and for comparison with other species where conserved sequences/genes are involved. Despite the use of common reference stocks for most of these projects, independently constructed maps have not yet been integrated; although this can now be done to some extent using assembled WGS data (see below), to improve resolution and resolve ambiguities a concerted effort to integrate all of these maps is needed, especially to fill in gaps left by using different marker systems. Consequently, map lengths based on different types of markers have not yet converged.

In addition to genome-wide linkage maps, molecular maps have been developed for the Z-chromosome using a combination of RAPDs, SSRs, and fluorescent intersimple sequence repeats (FISSRs; Nagaraja et al. 2005), and SSRs (Miao et al. 2008). This is an especially important target given the hypothesized evolutionary role of sex linkage in speciation (Prowell 1998; Qvarnström and Bailey 2009). Once populated with genes, the silkworm Z-map will provide references for uncovering and comparing sex-linked traits in other Lepidoptera.

Integration of Classical and Molecular Linkage Maps

Integration of the molecular maps with the classical morphological/biochemical mutant maps is an important strategic goal to facilitate positional cloning of genes with unique mutations available in no other lepidopteran. Some offer a promise of illuminating fundamental problems such as embryonic and larval development (e.g., *ki*, kidney-shaped egg, a maternal lethal that forms only ectodermal derivatives, as well as numerous larval lethals that survive only until the first molt), mutations affecting overall larval shape (e.g., *gn*, gooseneck, slender body, and constricted between segments; *e*, elongated, intersegmental membrane stretched between fourth and fifth segments; *nb*, narrow breast, short thorax, and stout abdomen), larval markings (e.g., the famous *p*-locus with at least fifteen diverse alleles), wing development and patterning (for a description of silkworm wing mutants see Chapter 6), and larval cuticle (e.g., *K*, knobbed, exaggerated protrusions associated with characteristic "crescent" and "star" markings), and more than 35 strains with independent translucent skin mutations affecting the formation and deposition of nitrogen waste products (e.g., *od*, sex-linked translucent Nagaraja et al. 2005; w^{3oe}, white egg, Kômoto et al. 2009). Determining the underlying basis of other mutations such as food preference, diapause, moltinism, and resistance to fungi and viruses (for virus resistance, see Chapter 18) promises to open up new avenues of research for control of lepidopteran pests. For a current list of silkworm mutations and their linkage map assignments, see SilkwormBase (Table 2.1).

Thus far, most efforts to integrate the classical and molecular maps have been directed at associating one or two standard, easily scored morphological mutations with a specific molecular LG (e.g., Miao et al. 2005; Yasukochi et al. 2005). Given the high density of molecular markers and anchor loci now available on genetic and physical maps, plus the ease of generating anonymous markers such as AFLPs, RAPDs, and SNPs in stocks that have not been characterized on the DNA level, a global map integration effort is urgently needed.

Molecular Analysis of Silkworm Mutations

To date, only a small number of silkworm mutations have been investigated on the DNA level. Included are the oily or translucent larval mutations, *oq* (Yasukochi, Kanda, and Tamura 1998; Kômoto 2002) and *og* (Kômoto et al. 2003), which fail to deposit uric acid crystals in the epidermis because of defects in the genes encoding xanthine dehydrogenase and molybdenum cofactor sulfurase, respectively; two white egg mutations—*w-1* (white egg 1, Quan et al. 2002), which encodes kynurenine-3-monooxygenase, and *w-3* (white egg 3, Abraham et al. 2000; Kômoto et al. 2009), an adenosine triphosphate-binding cassette (ABC) transporter which is an ortholog to the *Drosophila* white gene (reviewed in Goldsmith, Shimada, and Abe 2005)—and *Y* (yellow hemolymph and cocoon color), which encodes a carotenoid-binding protein (Tabunoki et al. 2004; Tsuchida et al. 2004; Sakudoh et al. 2007). These mutations have been cloned by shrewd guesswork using a correlation of their approximate map location with what was basically a candidate gene approach, followed by molecular evidence for disrupted function of the putative mutant, such as a deletion (Sakudoh et al. 2007) or premature stop codon (Kômoto et al. 2009).

The availability of large EST collections, BAC libraries, and an integrated physical-genetic map has made it possible to isolate mutant genes and their corresponding wild-type forms using map-based or positional-cloning strategies. This involves fine-structure mapping of a mutation to narrow the chromosome region of interest, followed by sequencing and functional assays, ideally, mutant rescue by germ line transformation, to confirm the target gene (for discussions of positional cloning strategies, see Chapters 6 and 18). Three mutations were recently identified in this way. The first was *nsd-2* (densovirus-2 resistance), a recessive mutation that renders silkworms refractory to a form of densovirus, which is a major pathogen for sericulture and proved to encode a defective twelve-pass membrane protein, the putative *Bm*dnv-2 receptor (see Chapter 18 for a detailed description of the cloning and analysis of *nsd-2*). The second silkworm gene identified by positional cloning was the wingless mutation *fl* (*flugellos*, lacking pupal and adult wings). Interestingly, initial studies of RNA expression patterns in cultured wing discs after treatment with ecdysteroids (Matsuoka and Fujiwara 2000) and in differential display (Matsunaga and Fujiwara 2002) uncovered several over- or underexpressed candidate genes. It required the more direct approach of positional cloning coupled with the molecular analysis of four *fl* alleles to identify *fringe* (*fng*), the actual *fl* locus, which encodes a glycosyltransferase involved in *Notch* signaling, which is essential for wing disc morphogenesis (Sato et al. 2008). The third was *Ek*^{p}, a mutation mapping to the homeotic *Hox* gene cluster that perturbs development of larval prolegs. It was found by detailed comparison of wild-type and mutant sequences after first developing a fine-structure genetic map using more than two thousand progeny from nine segregating backcross populations, and then identifying the mutation in a single BAC covering the region (Xiang et al. 2008).

Reports localizing mutations between molecular markers in preparation for map-based cloning are beginning to appear. Such projects aim at using the target genes for strain improvement (e.g., *pph*, polyphagous, a spontaneous mutation that induces polyphagy in normally monophagic silkworm larvae and hence feeds well on artificial diet; Mase et al. 2007), for marker-assisted selection (e.g., *sch*, sex-linked chocolate, used for male-specific selection; Miao et al. 2008), or for other practical applications in sericulture (e.g., *sli*, short lifespan in imago, in which moths die within a few days after eclosion; Li et al. 2008; see a discussion of genes that confer resistance to virus attack and positional cloning of *nsd-2* in Chapter 18). Other work in progress promises to answer fundamental questions about gene function such as *C*, outer-layer yellow cocoon, which affects absorption of carotenoids from the hemolymph to the silk gland (Tabunoki et al. 2004) and was recently localized with SSRs (Zhao et al. 2008). Though compared with colorful butterflies and many decorative moth species, the light beige, largely unpigmented adult *B. mori* is a drab relative; nevertheless, a few body and wing pattern mutations (e.g., *Bm*, black moth; *Ws*, wild wingspot; *Wm*, wild melanism; and *wb*, white banded wing), likely introgressed from *B. mandarina*, are potential targets to shed light on the origin of wing pattern mimicry and melanism, key topics in lepidopteran biology (for discussions of butterfly wing patterns, see Chapters 5 and 6).

Quantitative Trait Loci (QTL)

A second class of mutations that holds promise for silkworm-lepidopteran genetics/genomics is quantitative or complex traits that are well-segregated in the wild-type strains used for practical breeding in sericulture. For example, stocks vary in their seasonal productivity and survival, so that certain hybrids are best reared in spring and others in late summer or fall, impacted by variation in rearing conditions such as temperature, humidity, and quality of mulberry. Determining the underlying genetic basis of these characters may be relevant not only for marker-assisted selection in sericulture, but also for defining new targets for control of lepidopteran pests. Although teasing out the genetic basis of traits under the control of many genes with small effects may prove difficult, preliminary reports of QTL mapping using intersimple sequence repeats (Chatterjee and Mohandas 2003) and AFLPs (Lu et al. 2004) have appeared. Further, some practical breeding characters have been associated with Mendelizing genes, making them good targets for positional cloning. A prime example is *Lm*, late maturity, a Z-linked gene that affects larval stage duration, total cocoon (including pupa), and cocoon shell (silk) weight. Traits affecting fertility, fecundity, and other aspects of reproduction may also prove to be valuable resources for silkworm breeding. In addition, they may provide candidate genes to test hypotheses of speciation and reproductive isolation formulated using other lepidopteran species (e.g., Sperling 1994; Prowell 1998).

Conserved Synteny between *Bombyx* and Other Lepidopterans

Hints of synteny in Lepidoptera—the presence of genes of common ancestry (orthologs) on the same linkage group—first arose in studies of sex- or Z-linked traits, notably allozymes and phenotypic characters affecting reproductive isolation among various taxa (Sperling 1994). The establishment of a well-developed molecular genetic reference map in *B. mori* made it possible to test synteny on any well-mapped autosome. The first lepidopterans in which synteny was tested using this strategy were the nymphalid *Heliconius melpomene* (Yasukochi et al. 2006; Pringle et al. 2007) and the sphingid *M. sexta* (Sahara et al. 2007). Using recombination mapping (Pringle et al. 2007) and BAC-FISH (Yasukochi et al. 2006), a common set of reference markers for known genes was found to reside in a single chromosome in all three species, with conserved gene order. More detailed genetic analysis and physical mapping of additional linkage groups in *H. melpomene* and *H. erato* relative to *B. mori* and to each other has enabled a prediction of chromosome equivalence among the species (Papa et al. 2008). This study has also revealed considerable microsynteny, showing the presence of well-conserved gene clusters with potential functional significance (Papa et al. 2008). A recent karyotype-wide comparison of another nymphalid, *Bicyclus anynana*, with *B. mori* using a map based on >460 orthologs detected significant chromosome conservation and synteny blocks along with many large and small translocations and inversions (Beldade et al. 2009). The demonstration of synteny among such widely diverged taxa offers the promise that silkworm reference chromosome maps and other genomic resources will facilitate future identification of candidate genes and *de novo* gene discovery in poorly mapped species. For further discussion of synteny, see Chapter 6.

GENOMIC RESOURCES

Extensive genomics resources are now available for silkworm, including, as noted earlier, large EST collections (about 238,000 sequences in GenBank, October 2008), BAC libraries, the integrated physical-genetic map, and a draft genome sequence. Searchable online databases are provided for some of these resources (Table 2.2); in addition, microarray and proteomics databases are being established, making *Bombyx* a leading model for insects together with *Drosophila,* mosquitoes, *Tribolium castaneum,* and *Apis melifera*. Following is a brief overview of these silkworm resources.

EST Collections

The first large EST collection, comprising about 35,000 sequences, was made by randomly sequencing approximately 1,000 ESTs from each of 36 normalized cDNA libraries constructed from different tissues and developmental stages with the aim of identifying a large number of distinct silkworm gene products (now increased to 48 libraries; see SilkBase, Table 2.2). Approximately 11,000 independent sequences were reported (Mita et al. 2003), a yield of one cluster or contig per 2.7 ESTs, suggesting this is a reasonable strategy for gene discovery. The second large EST project, which aimed at finding SNPs for linkage mapping and gene association tests, generated 12,980 contigs from 73,235 high-quality sequences derived from 12 tissue- or stage-specific libraries (Cheng et al. 2004). Although the yield of independent gene products was lower (about one independent cluster per 5.6 ESTs), as expected, the project showed this to be an effective strategy for identifying SNPs and small insertions or deletions, while significantly expanding the EST database. The overall yield of SNPs was low (101 SNPs and 27 insertions/deletions, or about 1.4 per 1,000 base pairs), verifying that the inbred strain chosen for the WGS project, Dazao (p50), is relatively homozygous at the level of coding regions. Later projects have focused on sequencing sets of ESTs from a single source such as wing disc (Kawasaki et al. 2004), posterior silk gland (Zhong et al. 2005), epidermis (Okamoto et al. 2008), fat body (Cheng et al. 2006), and whole animals (Oh et al. 2006) at one or several developmental stages, in order to characterize gene expression patterns (see below) and capture new or rare sequences involved in a process of interest. Contigs assembled from the EST collections (Zhang et al. 2007b) and sequencing of full-length cDNAs, which is now under way (K. Mita and T. Shimada, pers. comm.; see Full-Length cDNA Database, Table 2.2), will be used to provide a more complete resource for annotating the silkworm genome sequence.

WildSilkbase, a searchable database containing ESTs and full-length cDNAs (Table 2.2) has been constructed with sequences from three wild silk moths that produce commercially viable silk, namely, the Indian golden silk moth, *Antheraea assama*; the Indian tassar silk moth, *A. mylitta*; and the eri silk moth, *Samia cynthia ricini* (Arunkumar et al. 2008). The number of ESTs for these species currently registered in GenBank has increased significantly from the original publication of about 57,000 sequences (4,000 contigs and 10,000 singletons) to 94,500 sequences (about 35,200 for *A. assama*; 39,000 for *A. mylitta*; and 20,300 for *S. c. ricini*; October 2008); they comprise important resources for finding genes of economic interest as well as for fundamental and evolutionary studies.

BAC Libraries

Four BAC libraries are now available for *B. mori*, three constructed from mixed samples of male and female posterior silk glands from strain p50, and one from strain C108 using partial digests with one of three restriction enzymes, *Eco*RI (Koike et al. 2003), *Hind*III (Wu et al. 1999), or *Bam*HI (Yamamoto et al. 2006). Paired-end BESs from these libraries were used for constructing BAC contigs, assembling the WGS, and, as noted earlier, to find SNPs and ESTs for the construction of an integrated physical-genetic map (Yamamoto et al. 2006, 2008). The value of using BACs digested with different restriction enzymes became apparent upon detailed comparison of the BESs from the *Eco*RI and *Bam*HI libraries. Based on 95,000 BESs (55 Mb) representing more than 10 percent of the silkworm genome, Suetsugu et al. (2007) found that long interspersed nuclear elements (LINEs) were more abundant in the BESs from the *Bam*HI (16.40 percent total length nucleotides sequenced) than the *Eco*RI library (5.14 percent); further, the relative content of shorter transposable elements (TEs) was reversed (2.07 percent in *Bam*HI BESs vs. 5.40 percent in *Eco*RI BESs). The overall guanosine–cytosine (GC) content of BESs derived from the two digests also differed

TABLE 2.2
Genomics and Functional Genomics Resources

Database	Contents	Host	Web Site
SilkDB	Silkworm genome database	The Institute of Sericulture and Systems Biology, Southwest University, Chongqing, China	http://silkworm.genomics.org.cn/
KAIKObase	Silkworm genome database	National Institute of Agrobiological Sciences, Tsukuba, Japan	http://sgp.dna.affrc.go.jp/KAIKObase/
SilkBase	ESTs from *Bombyx mori*, *Samia cynthia ricini*	University of Tokyo, Laboratory of Insect Genetics and BioSciences	http://morus.ab.a.u-tokyo.ac.jp/cgi-bin/index.cgi
Full-Length cDNA Database	ESTs from *Bombyx mori* full-length cDNA libraries	University of Tokyo, Laboratory of Insect Genetics and Biosciences	http://papilio.ab.a.u-tokyo.ac.jp/Bombyx_EST/
ButterflyBase	Genomics database for Lepidoptera	Max Planck Society, Jena, Germany; Consortium for Comparative Genomics of Lepidoptera	http://butterflybase.ice.mpg.de/
WildSilkBase	ESTs, *Antheraea assama*, *A. mylitta*, *Samia cynthia ricini*	Center for DNA Fingerprinting and Diagnostics, Hyderabad, India	http://www.cdfd.org.in/wildsilkbase/home.php
SilkSatDB	Silkworm microsatellites	Center for DNA Fingerprinting and Diagnostics, Hyderabad, India	http://210.212.212.7:9999/PHP/SILKSAT/index.php
BmMDB: *Bombyx mori* Microarray Database	Microarray profiles In process: Web site established; one set of published data currently available	The Institute of Sericulture and Systems Biology, Southwest University, Chongqing, China	http://silkworm.swu.edu.cn/microarray/
Silkworm Proteome Database	Two-dimensional gel electrophoresis patterns and spot identification	Center for Genetic Resource Information, National Institute of Genetics, Misima, Japan	http://www.shigen.nig.ac.jp/ISPD/index.jsp
KAIKO 2DDB	Silkworm proteome database: two-dimensional gel electrophoresis patterns and spot identification	National Institute of Agrobiological Sciences, Tsukuba, Japan	http://kaiko2ddb.dna.affrc.go.jp/cgi-bin/search_2DDB.cgi
Bombyx Trap Database	Enhancer trap, GAL4/UAS, and other transgenic lines	Transgenic Silkworm Research Center, National Institute of Agrobiological Sciences, Tsukuba, Japan	http://sgp.dna.affrc.go.jp/ETDB/

substantially (40.30 percent for *Bam*HI vs. 37.45 percent for *Eco*RI), suggested to result from differences in the GC content of the enzyme target sites. Analysis of clustered BESs from the two libraries by BLASTx (Basic Local Alignment Search Tool) revealed that *Bam*HI BESs corresponded more frequently to protein coding sequences (28.2 percent of *Bam*HI-clusters vs. 20.2 percent of *Eco*RI-derived clusters), consistent with the relative overall GC content of silkworm protein coding versus intergenic DNA.

These observations illustrate the importance of using complementary digests to obtain a complete representation of the genome, especially in species for which a WGS is not yet available. The BACs from these projects have also been used extensively for cloning and detailed analysis of gene clusters such as *Hox* and neighboring genes (Yasukochi et al. 2004) and the *Broad-Complex* (Ijiro et al. 2004), positional cloning (see above and Chapter 18), physical mapping of genes by BAC-FISH (Yoshido et al. 2005; Sahara et al. 2007), and detailed analysis of chromosome structure (for example, the W; Sahara et al. 2003; Yoshido et al. 2007; see also Chapters 3 and 4).

Draft WGS Sequence

A complete genome sequence has become the *sine qua non* of modern molecular and evolutionary biology. Whereas deep sequencing of ESTs from cDNA libraries can discover a high proportion of expressed gene products, many important ones, such as the messenger RNAs (mRNAs) encoding chemosensory proteins, small nuclear RNAs (snRNAs), and microRNAs (miRNAs), are too rare to detect in cDNA libraries without considerable effort and are much more efficiently cataloged by genome-wide surveys using computational methods designed explicitly for their unique characteristics. Although structural features such as introns and exons may be characterized by sequencing selected BAC clones, the availability of a completely sequenced genome opens up the possibility of finding many other important or unsuspected genomic features, such as unexpressed pseudogenes, closely and distantly related family members, gene clusters, and duplicated chromosome segments, which require a global analysis. Having a reference genome for a model species like the silkworm makes gene discovery and even genome assembly and annotation much easier for related species, whose genomes can then be sequenced to a lower coverage while still providing a high yield of information. Finally, having a complete genome sequence for a representative lepidopteran immediately opens up an opportunity for comparative genomics, providing clues to the evolutionary history of different clades and answers to such questions as which genes and features are unique to the silkworm or shared with other Lepidoptera or other taxonomic groups of insects, with other arthropods, or with all Metazoa.

Two draft sequences of the silkworm genome using a shotgun approach were published in 2004, one at a coverage of approximately six-fold (Xia et al. 2004) and the other at three-fold (Mita et al. 2004). The low coverage of these initial projects left the genome in relatively small pieces: the average size of contigs (overlapping stretches of sequenced DNA) was estimated at 12.5 kb and 17.9 kb for the six-fold and three-fold projects, respectively; scaffolds (contigs shown to be physically connected) averaged 26.9 kb (six-fold) and 7.8 kb (three-fold). The use of p50/Dazao for the two projects will facilitate merging of the data to increase genome coverage and enable a more complete assembly, now under way, and will include sequencing of additional fosmids and full-length cDNAs, and alignment with the integrated SNP-EST-genetic-physical map (International Silkworm Genome Consortium 2008). Searchable databases of the extended scaffolds, predicted gene models, aligned ESTs, and other features are available in online databases (Table 2.2; Wang et al. 2005).

Despite its fragmentation, the extent of the genome sequenced or covered in the two projects appears to be relatively high. By BLAST search and sequence alignment, an estimated 91 percent of known genes and available ESTs were found in the six-fold WGS (Xia et al. 2004). Similarly, 97 percent of the genome was estimated to be organized into scaffolds in the three-fold WGS; further, alignment of a set of completely sequenced and assembled BAC clones with the WGS data from this project yielded an estimated 78–87 percent coverage (Mita et al. 2004). Contigs lengths and scaffolds promise to be much longer in the joint assembly, which will have eight- to nine-fold coverage; the integration of these data with the additional sequence information and the integrated physical-genetic maps will provide a more robust platform for accurate gene annotation as well.

Note: Chinese and Japanese groups recently reported combining their WGS *Bombyx* genome sequences with additional fosmid and BAC clone sequence data for a new assembly totaling 8.5X genome coverage (432 Mb) (International Silkworm Genome Consortium 2008). It shows significantly improved continuity with half or more of the assembled sequences in contigs or scaffolds of at least 3.7 Mb (maximum length 14.3 Mb) of which 87 percent are anchored to the 28 chromosomes by the SNP map. See Couble, Xia and Mita (2008) for related articles on genome organization and evolution and function of silkworm genes and gene families.

Genome Features

Haploid genome size estimates ranged from 428.7 Mb for the total length of sequenced contigs and singletons in the six-fold project (Xia et al. 2004) to 514 Mb for maximum scaffold lengths in the three-fold project (Mita et al. 2004). These data are in reasonable agreement with the early standard values of 0.53 pg per 1C DNA derived from DNA-DNA reannealing (Cot) kinetics (Gage 1974) and 0.52 pg per 1C DNA by cytophotometric measurements of stained sperm and hemocyte nuclei (Rasch 1974). Note that errors have crept into the conversion value for these data, which have been taken as equivalent to 530 Mb but are closer to 518 Mb using 978×10^6 bp/pg DNA (Doležel et al. 2003) for calculating base pair equivalence. Recent measurements with the more precise technique of flow cytometry using propidium iodide stained mitotic chromosomes from adult brain with *D. melanogaster* as a standard yielded a somewhat lower overall value of 450–493 Mb (J.S. Johnston, pers. comm.), closer to that obtained in the six-fold sequencing project, and suggesting that the reference value will likely be reduced further after integration of the two WGS projects (International Silkworm Genome Consortium 2008).

Using an *ab initio* gene-finder algorithm with corrections for potential errors, repeated sequences or transposable elements (TEs), and pseudogenes, the six-fold project predicted that the silkworm genome contains 18,510 genes with a mean size about 2.3 times longer than those of the fruit fly, based on aligned reciprocal best matches (Xia et al. 2004). Only slightly more exons were found per gene (a mean ratio of 1.15) than in *Drosophila*, suggesting that the size difference results from larger introns in silkworm genes, attributable at least in part to an increase in the number of inserted TEs (Xia et al. 2004). Cot curve measurements indicate that repeated elements comprise up to 45 percent of the silkworm genome (Gage 1974), consistent with a report of more than 180,000 copies of various repeat families in the initial three-fold WGS (Mita et al. 2004). Prominent among these are the superabundant short interspersed repetitive nuclear element (SINE) *Bm1* (121,000 copies), the LINEs or non-LTR retrotransposons *BMC1* (37,000 copies), *Bm5886* (28,000 copies), and *HOPE Bm2* (3,380 copies), three types of Class II *mariner* transposons (about 10,000 copies), and various LTR retrotransposons (about 400 copies). Except for *Bm1*, *BMC1* and LINEs, the majority are truncated at the five prime end (Mita et al. 2004). A concerted search for LTR retrotransposons in the WGS data detected twenty-nine families comprising nearly 12 percent of the genome, many of which were previously unknown but, with few exceptions, present in relatively small numbers (about 20–700; Xu et al. 2005). Most fall into three major groups: *Gypsy*-like, *Copia*-like, and the *Pao-Bel* group, which together comprised nearly 12 percent of the genome, excluding the W chromosome, which is composed almost entirely of retrotransposable elements (for molecular characterization of the W chromosome see Chapter 4; for reviews of silkworm TE composition and structure see Eickbush 1995; Abe et al. 2005; Goldsmith, Shimada, and Abe 2005).

Whereas most elements are widely distributed chromosomally, two other families of non-LTR retrotransposons, present in about 1,000 copies per haploid genome, are associated with TTAGG telomeric repeats, called TRAS1 and SART1 (Fujiwara et al. 2005). They are actively transcribed;

furthermore, evidence suggests that they transpose into the telomeric repeats by several mechanisms and are required for telomere maintenance (Fujiwara et al. 2005).

Another group of Class II elements of interest are *piggyBac* elements. The *piggyBac* transposable element was discovered jumping from a cell line of the cabbage looper, *Trichoplusia ni*, into a baculovirus (Fraser et al. 1995), and subsequently engineered into versatile, widely used vectors for germ line transformation of insects (e.g., Sarkar et al. 2006), including *B. mori* (see below). Ninety-eight *piggyBac*-like elements (PBLEs) were identified in the silkworm genome, most of which are truncated; however, transcription was observed in cDNA from a variety of tissues and developmental stages, and five elements with ORFs (open reading frames) encoding apparent full-length transposase proteins were found, suggesting that some may be active (Xu et al. 2006). *PBLE*s have since been found in other Lepidoptera (Wang et. al. 2006; Sun et al. 2008).

One of the most elusive classes of potentially repeated sequences in the genome of the silkworm—or any lepidopteran—are the specialized DNA elements associated with the specific kinetochore structure of holokinetic chromosomes, formerly called "diffuse kinetochores." It will be interesting to find out whether they are numerous and widely distributed, as has been speculated (Ogura et al. 1994), and therefore constitute a class of repeated sequence, or correspond to any of the ninety or so homologues of "holocentric protein" elements of other species with holokinetic chromosomes found in the six-fold draft genome sequence (Xia et al. 2004). For a discussion of lepidopteran chromosome structure see Chapter 3.

The first fruits of genome-wide surveys of the current silkworm genome include catalogs of genes involved in embryonic development, identified by homology to known *Drosophila* genes (Xia et al. 2004), as well as rare transcribed sequences difficult to find in EST databases such as chemoreceptors (Zhou et al. 2006; Gong et al. 2007; Wanner et al. 2007; for discussions of lepidopteran chemoreceptors, see also Chapters 9, 10, and 11), snRNAs (Sierra-Montes et al. 2005; Smail et al. 2006), and miRNAs (Tong, Jin, and Zhang 2006; He et al. 2008; Yu et al. 2008). With the new genome assembly and more advanced computational methods to identify members of diverged gene families, many more are expected in the near future.

FUNCTIONAL GENOMICS

Many post-genomic tools have been developed for silkworm with the aim of determining the role or function(s) of genes and genome sequences, called *functional genomics.* These include methods to examine gene expression patterns of selected tissues and developmental stages (e.g., EST profiling, microarrays, and proteomics), to produce transgenic insects, and to disrupt gene function by, for example, RNA interference (RNAi). Following are highlights of recent work in these areas; early studies are summarized in Goldsmith, Shimada, and Abe (2005). See also Zhou, Yang, and Zhong (2008).

Transcription and Protein Profiling

Transcriptional profiling is carried out to reveal changing patterns of gene expression, often to document a metamorphic event, a response to an experimental treatment, or the impact of a pathogen or mutation. Thus far, most gene expression studies in silkworm are based on sequencing ESTs from cDNA libraries constructed from tissues and developmental stages of interest. This approach can give a rough estimate of expression based on the frequency of finding a particular mRNA, which can be verified by more direct and sensitive measures of transcription levels using Northern blots, RT-PCR, or microarrays. EST profiling has been carried out to measure changing gene activity in metamorphosis for early embryos (Hong et al. 2006), embryonic versus larval stages (Oh et al. 2006), fat body (Cheng et al. 2006), silk gland (Zhong et al. 2005), wing disc (Kawasaki et al. 2004), and larval epidermis (Okamoto et al. 2008).

Although it is possible to obtain a rough estimate of transcriptional activity in a particular tissue or developmental stage by surveying ESTs, microarrays provide a more sensitive and comprehensive snapshot of gene expression. Before the WGS was available, microarrays were developed for *Bombyx* using a few thousand ESTs collected from individual tissues such as differentiating wing disc (Kawasaki et al. 2004; Ote et al. 2004), or particular developmental stages such as germ band embryos (Hong et al. 2006); these projects were able to catalog a set of up- or downregulated genes for further characterization (e.g., Ote et al. 2005), but were not comprehensive. Subsequently, a genome-wide microarray containing about twenty-three thousand oligonucleotide probes was designed from known genes and those predicted from the draft WGS and used to survey expression profiles of ten tissues and organs from mid-fifth instar larvae (Xia et al. 2007). This study will serve as a baseline for documenting changes in gene expression throughout larval life. Equally important, it is a resource for annotating the silkworm genome, in part to enable inference of common function by comparing expression profiles of a particular sequence or subset of sequences in different tissues, at other developmental stages, and in other insects. A microarray representing sixteen thousand genes designed from more than fifty thousand ESTs was used to detect differences in the response of cultured hemocyte-like cells upon exposure to bacteria versus putative biochemical elicitors (Ha Lee et al. 2007). Although very few sequences showed changes in expression levels, their patterns were distinctly different, illustrating the power of this technique to uncover coordinate expression patterns and the involvement of unrecognized genes in a process.

Use of proteomic assays, namely, two-dimensional electrophoresis coupled with MS/MS or Maldi-TOF (matrix-assisted laser desorption/ionization—time of flight) to identify peptides, is also starting to accelerate. Although identifying some proteins has been possible using EST databases and a comparative approach, the need for a well-annotated lepidopteran genome undoubtedly slowed progress in this area. As with early transcriptome data, preliminary proteomic profiling has focused mostly on larval tissues such as midgut (Kajiwara et al. 2005; Yao et al. 2008), fat body (Kajiwara et al. 2006), hemolymph (Li et al. 2006), skeletal muscle at larval-pupal metamorphosis (Zhang et al. 2007a), and middle and posterior silk gland (Hou et al. 2007; Liu et al. 2008). Silkworm proteome databases are now available online (Table 2.2), providing useful public resources. A description of proteomics techniques and additional applications in silkworm are reviewed in Zhou, Yang, and Zhong (2008).

Transgenics

Producing Transgenic Silkworms

The technology for making transgenic silkworms was first reported in 2000 (Tamura et al. 2000) and is now widely available; early efforts to develop efficient gene delivery systems and expression vectors are reviewed in Goldsmith, Shimada, and Abe (2005). Injection into preblastoderm embryos remains the most common method of gene delivery. A variety of vectors has been developed based on *piggyBac*, as noted earlier, usually with a *BmactinA3* promoter or a generic eye- and neuron-specific *3x3P* promoter driving an enhanced green fluorescent protein (EGFP) or *Discosoma* red fluorescent protein (DsRed) reporter. Conditional expression can be driven by a standard *hsp70* promoter from *Drosophila* (Yamamoto et al. 2004; Dai et al. 2007), and a GAL4/UAS system has been developed for targeting gene expression in specific tissues with other promoters (Imamura et al. 2003; Tan et al. 2005; Uchino et al. 2006). Successful use of the transposase and other components derived from another transposable element, *Minos*, first found in *D. hydei* (Franz and Savakis 1991), was recently reported in silkworm (Uchino et al. 2007). Although transformation levels with *Minos* were lower than with *piggyBac*-based vectors, it provided an alternative source of transposase to make possible the development of an enhancer trap system, opening up an exciting and powerful approach for uncovering tissue- and stage-specific regulatory elements (Uchino et al. 2008). Recently, vectors based on the *B. mori* LINEs SART1, which integrates preferentially into telomeres, and R1, which integrates into 28S ribosomal DNA, were engineered using the baculovirus AcNPV, and infected into fifth instar larvae as a strategy to produce high-yield recombinant proteins while avoiding disruption of functioning genes (Kawashima et al. 2007). A new database has been established for results from transgenic experiments (*Bombyx* trap database; Table 2.2).

Applications

Genetic engineering of the silkworm has been used as a tool to study gene function and regulation, to produce high-yield, value-added products that can be harvested from cocoons or hemolymph, and for strain improvement, notably to increase resistance to Bm nucleopolyhedrosis virus, although complete resistance has not yet been achieved (Isobe et al. 2004; Kanginakudru et al. 2007). Examples of the first two applications are given below, without attempting to be comprehensive.

The use of transgenic silkworms to study gene function has taken several approaches, including introduction of wild-type genes to confirm candidate genes by mutant rescue (e.g., *BmY*, Sakudoh et al. 2007; *Nd-sD*, Inoue et al. 2005; and *Bm-nsd-2*, Ito et al. 2008), ectopic expression of wild-type genes to test their role in development and differentiation (e.g., *Juvenile Hormone Esterase*, Tan et al. 2005; and *B. mori doublesex, Bmdsx*, Suzuki et al. 2005), teasing apart promoter (*fibroin-H*, Shimizu et al. 2007) and RNA splicing elements (*Bmdsx*, Funaguma et al. 2005), and gene knockdown or silencing using RNAi (see next section for examples). With vectors incorporating additional transcriptional control elements, perhaps discovered with the new enhancer trap system, the usefulness of this transgenic system for fundamental studies of silkworm and potentially other lepidopteran gene function seems assured.

Using the silkworm as a "bioreactor," gene-specific promoters have been used to drive expression in different compartments of the silk gland, for example, *fibroin-L* (Inoue et al. 2005; Adachi et al. 2006; Hino, Tomita, and Yoshizato 2006), *fibrohexamerin* (Royer et al. 2005), and *fibroin-H* (Kojima et al. 2007; Kurihara et al. 2007) in the posterior silk gland, and *sericin* (Ogawa et al. 2007) in the middle silk gland. A "naked pupa" mutant, *Nd-sD*, which lacks the fibroin-L component of the silk fiber and thus cannot secrete a filament, can be used to increase the yield of recombinant protein in the cocoon (Inoue et al. 2005; Yanagisawa et al. 2007); further, introduction of a transcriptional enhancer from a *B. mori* baculovirus immediate-early 1 (Tomita et al. 2007) gene or translational enhancer from the 5′ UTR (untranslated region) of a polyhedrin gene (Iizuka et al. 2008) can help achieve high levels of expression. Value-added products made by silkworms include human beta fibroblast growth factor (Hino, Tomita, and Yoshizato 2006), bioactive silk films for use in cell culture incorporating collagen (Adachi et al. 2006; Yanagisawa et al. 2007) or fibronectin (Yanagisawa et al. 2007), and model pharmaceuticals like feline interferon (Kurihara et al. 2007). A refinement of this system was the achievement of collagen proline hydroxylation by engineering silkworms to express an enzyme subunit normally present at levels too low to yield the necessary post-translational modification for biological activity (Adachi et al. 2006).

RNAi

RNAi is a versatile approach used in many organisms to probe gene function and confirm the identity of candidate genes in positional cloning studies. It works by selectively cleaving mRNA through the action of an endogenous nucleolytic RNA-induced silencing complex (RISC); the latter is activated by exposure to a catalytic double-stranded RNA (dsRNA) corresponding to the target mRNA, which can be introduced into the organism by various routes (for more information on the mechanism of RNAi, see Chapter 12). The first successful RNAi experiments in silkworm involved injection of dsRNA for *Bmwh3*, the ortholog of the *Drosophila white* gene, into preblastoderm eggs, yielding white egg and translucent embryo phenotypes (Quan, Kanda, and Tamura 2002). Successful downregulation of genes by injecting dsRNA into fifth instar larvae or pupae has since been achieved for a variety of functions, including programmed cell death (*Bm cathepsin D*, Gui et al. 2006), the immune response (*Bombyx lysozyme-like protein I*, Gandhe, Janardhan, and Nagaraju

2007), larval-pupal metamorphosis (*Bmftz*, Cheng et al. 2008; *cathepsin B*, Wang et al. 2008), wing expansion (Huang et al. 2007), and, in one of the most comprehensive studies published so far, a set of factors implicated in pheromone production (Matsumoto et al. 2007). In addition, transgenic constructs expressing an inverted repeat of the target RNA under control of a constitutive (*period*, Sandrelli et al. 2007; *Bmrelish*, Tanaka et al. 2007) or heat shock (*eclosion hormone*, Dai et al. 2007) promoter have been developed; the latter will allow fine-tuning of knockout conditions at specific developmental stages. RNAi has also been used to transfect cultured silkworm cells to examine the role of an RNA methyltransferase in programmed cell death (Nie et al. 2008) and, interestingly, to implicate *BmAGO2*, a central component of the RISC complex, in repressing double-stranded break repair (Tsukioka et al. 2006). This method seems to be working well in the silkworm, although few reports are published in other lepidopteran species, suggesting that some technical difficulties or biological barriers remain to be solved before it can be used routinely in other taxa.

CONCLUSION: WHERE NEXT ON THE SILK ROAD?

It is evident that we have reached a new era on the long march down the silk road: the role of the silkworm, long the leading model for Lepidoptera, has been significantly enhanced by the development of new genetic and genomic resources that will make possible more rapid progress in other species in many areas. The new molecular cytogenetics, especially BAC-FISH, is a critical companion to physical and fine-structure linkage mapping and makes possible investigations of chromosome and gene organization in species without much genetics. The new integrated molecular genetic maps, apparent widespread synteny, and, of course, the complete genome sequence will accelerate gene discovery in all Lepidoptera and provide tools for investigating many important developmental functions and interesting evolutionary questions. Among these is to identify as many genes as possible on the sex chromosome Z, which has been hypothesized to play such a leading role in speciation. Identification of genes involved in the sex-determining pathway including the *Fem* factor would help to clarify the mysterious role of the W chromosome in the female genome. Similarly, there remains a need to apply these tools, along with those of functional genomics, to investigate the large collection of silkworm mutants, which are a unique resource for lepidopteran biology. Another remaining frontier is to explore more fully the genetic basis of the variation present in geographic races, as well as in *B. mandarina*. Undoubtedly some of the complex traits that are segregated in inbred lines will turn out to have their origins in the silkworm's wild ancestor, and some undoubtedly will be found to have arisen *de novo* under the strong selection pressure of domestication. Both are of interest and may illuminate evolutionary questions for other Lepidoptera as well.

REFERENCES

Abe, H., K. Mita, Y. Yasukochi, T. Oshiki, and T. Shimada. 2005. Retrotransposable elements on the W chromosome of the silkworm, *Bombyx mori*. *Cytogenet. Genome Res.* 110:144–51.

Abraham, E.G., H. Sezutsu, T. Kanda, T. Sugasaki, T. Shimada, and T. Tamura. 2000. Identification and characterisation of a silkworm ABC transporter gene homologous to *Drosophila* white. *Mol. Gen. Genet.* 264:11–19.

Adachi, T., M. Tomita, K. Shimizu, S. Ogawa, and K. Yoshizato. 2006. Generation of hybrid transgenic silkworms that express *Bombyx mori* prolyl-hydroxylase alpha-subunits and human collagens in posterior silk glands: Production of cocoons that contained collagens with hydroxylated proline residues. *J.Biotechnol.* 126:205–19.

Arunkumar, K.P., M. Metta, and J. Nagaraju. 2006. Molecular phylogeny of silkmoths reveals the origin of domesticated silkmoth, *Bombyx mori* from Chinese *Bombyx mandarina* and paternal inheritance of *Antheraea proylei* mitochondrial DNA. *Mol. Phylogenet. Evol.* 40:419–27.

Arunkumar, K.P., A. Tomar, T. Daimon, T. Shimada, and J. Nagaraju. 2008. WildSilkbase: An EST database of wild silkmoths. *BMC Genomics* 9:338.

Banno, Y., T. Nakamura, E. Nagashima, H. Fujii, and H. Doira. 2004. M chromosome of the wild silkworm, *Bombyx mandarina* (n = 27), corresponds to two chromosomes in the domesticated silkworm, *Bombyx mori* (n = 28). *Genome* 47:96–101.

Beldade, P., S.V. Saenko, N. Pul, and A.D. Long. 2009. A gene-based linkage map for *Bicyclus anynana* butterflies allows for a comprehensive analysis of synteny with the lepidopteran reference genome. *PLoS Genet.*5:e1000366.

Chatterjee, S.N., and T.P. Mohandas. 2003. Identification of ISSR markers associated with productivity traits in silkworm, *Bombyx mori* L. *Genome* 46:438–47.

Cheng, D., Q.Y. Xia, J. Duan, et al. 2008. Nuclear receptors in *Bombyx mori*: Insights into genomic structure and developmental expression. *Insect Biochem. Mol. Biol.* 38:1130–37.

Cheng, D.J., Q.Y. Xia, P. Zhao, et al. 2006. EST-based profiling and comparison of gene expression in the silkworm fat body during metamorphosis. *Arch. Insect Biochem. Physiol.* 61:10–23.

Cheng, T.C., Q.Y. Xia, J.F. Qian, et al. 2004. Mining single nucleotide polymorphisms from EST data of silkworm, *Bombyx mori*, inbred strain Dazao. *Insect Biochem. Mol. Biol.* 34:523–30.

Couble, P., K. Mita, and Q. Xia. 2008. Editorial: Silkworm genome. *Insect Biochem. Mol. Biol.* 38:1035.

Dai, H., R. Jiang, J. Wang, et al. 2007. Development of a heat shock inducible and inheritable RNAi system in silkworm. *Biomol. Eng.* 24:625–30.

De Prins, J., and K. Saitoh. 2003. Karyology and sex determination. In *Lepidoptera, moths and butterflies: Morphology, physiology, and development*, ed. N. P. Kristensen, 449–68. Berlin: Walter de Gruyter.

Doležel, J., J. Bartoš, H. Voglmayr, and J. Greilhuber. 2003. Nuclear DNA content and genome size of trout and human. *Cytometry A* 51:127–28.

Eickbush, T.H. 1995. Mobile elements in lepidopteran genomes. In *Molecular model systems in the Lepidoptera*, ed. M.R. Goldsmith and A.S. Wilkins, 77–105. New York: Cambridge University Press.

Franz, G., and C. Savakis. 1991. *Minos*, a new transposable element from *Drosophila hydei*, is a member of the Tc1-like family of transposons. *Nucleic Acids Res.* 19:6646.

Fraser, M.J., L. Cary, K. Boonvisudhi, and H.G. Wang. 1995. Assay for movement of lepidopteran transposon IFP2 in insect cells using a baculovirus genome as a target DNA. *Virology* 211:397–407.

Fujii, H. 1998. *Genetical stocks and mutations of Bombyx mori: Important genetic resources.* Fukuoka, Japan: Kyushu University.

Fujii, T., N. Tanaka, T. Yokoyama, et al. 2006. The female-killing chromosome of the silkworm, *Bombyx mori*, was generated by translocation between the Z and W chromosomes. *Genetica* 127:253–65.

Fujiwara, H., and H. Maekawa. 1994. RFLP analysis of chromosomal fragments in genetic mosaic strains of *Bombyx mori*. *Chromosoma* 103:468–74.

Fujiwara, H., M. Osanai, T. Matsumoto, and K.K. Kojima. 2005. Telomere-specific non-LTR retrotransposons and telomere maintenance in the silkworm, *Bombyx mori*. *Chromosome Res.* 13:455–67.

Funaguma, S., M.G. Suzuki, T. Tamura, and T. Shimada. 2005. The *Bmdsx* transgene including trimmed introns is sex-specifically spliced in tissues of the silkworm, *Bombyx mori*. *J. Insect Sci.* 5:17.

Gage, L.P. 1974. The *Bombyx mori* genome: Analysis by DNA reassociation kinetics. *Chromosoma* 45:27–42.

Gandhe, A.S., G. Janardhan, and J. Nagaraju. 2007. Immune upregulation of novel antibacterial proteins from silkmoths (Lepidoptera) that resemble lysozymes but lack muramidase activity. *Insect Biochem. Mol. Biol.* 37:655–66.

Goldsmith, M.R., T. Shimada, and H. Abe. 2005. The genetics and genomics of the silkworm, *Bombyx mori*. *Annu. Rev. Entomol.* 50:71–100.

Gong, D.P., H.J. Zhang, P. Zhao, Y. Lin, Q.Y. Xia, and Z.H. Xiang. 2007. Identification and expression pattern of the chemosensory protein gene family in the silkworm, *Bombyx mori*. *Insect Biochem. Mol. Biol.* 37:266–77.

Gui, Z.Z., K.S. Lee, B.Y. Kim, et al. 2006. Functional role of aspartic proteinase cathepsin D in insect metamorphosis. *BMC Dev. Biol.* 6:49.

Ha Lee, J., I. Hee Lee, H. Noda, K. Mita, and K. Taniai. 2007. Verification of elicitor efficacy of lipopolysaccharides and peptidoglycans on antibacterial peptide gene expression in *Bombyx mori*. *Insect Biochem. Mol. Biol.* 37:1338–47.

Hara, W., E. Kosegawa, K. Mase, and K. Kadono-Okuda. 2002. Improvement of linkage analysis in the silkworm, *Bombyx mori*, by using cDNA clones' RFLP. *J. Sericol. Sci. Jpn.* 71:95–100.

He, P.A., Z. Nie, J. Chen, et al. 2008. Identification and characteristics of microRNAs from *Bombyx mori*. *BMC Genomics* 9:248.

Hino, R., M. Tomita, and K. Yoshizato. 2006. The generation of germline transgenic silkworms for the production of biologically active recombinant fusion proteins of fibroin and human basic fibroblast growth factor. *Biomaterials* 27:5715–24.

Hong, S.M., S.K. Nho, N.S. Kim, J.S. Lee, and S.W. Kang. 2006. Gene expression profiling in the silkworm, *Bombyx mori,* during early embryonic development. *Zool. Sci.* 23:517–28.

Hou, Y., Q. Xia, P. Zhao, et al. 2007. Studies on middle and posterior silk glands of silkworm (*Bombyx mori*) using two-dimensional electrophoresis and mass spectrometry. *Insect Biochem. Mol. Biol.* 37:486–96.

Huang, J., Y. Zhang, M. Li, et al. 2007. RNA interference-mediated silencing of the *bursicon* gene induces defects in wing expansion of silkworm. *FEBS Lett.* 581:697–701.

Iizuka, M., M. Tomita, K. Shimizu, Y. Kikuchi, and K. Yoshizato. 2008. Translational enhancement of recombinant protein synthesis in transgenic silkworms by a 5'-untranslated region of polyhedrin gene of *Bombyx mori* nucleopolyhedrovirus. *J. Biosci. Bioeng.* 105:595–603.

Ijiro, T., H. Urakawa, Y. Yasukochi, M. Takeda, and Y. Fujiwara. 2004. cDNA cloning, gene structure, and expression of Broad-Complex (BR-C) genes in the silkworm, *Bombyx mori. Insect Biochem. Mol. Biol.* 34:963–69.

Imamura, M., J. Nakai, S. Inoue, G.X. Quan, T. Kanda, and T. Tamura. 2003. Targeted gene expression using the GAL4/UAS system in the silkworm *Bombyx mori. Genetics* 165:1329–40.

Inoue, S., T. Kanda, M. Imamura, et al. 2005. A fibroin secretion-deficient silkworm mutant, *Nd-sD*, provides an efficient system for producing recombinant proteins. *Insect Biochem. Mol. Biol.* 35:51–59.

The International Silkworm Genome Consortium. 2008. The genome of a lepidopteran model insect, the silkworm, *Bombyx mori. Insect Biochem. Mol. Biol.* 38:1036–45.

Isobe, R., K. Kojima, T. Matsuyama, et al. 2004. Use of RNAi technology to confer enhanced resistance to BmNPV on transgenic silkworms. *Arch. Virol.* 149:1931–40.

Ito, K., K. Kidokoro, H. Sezutsu, et al. 2008. Deletion of a gene encoding an amino acid transporter in the midgut membrane causes resistance to a *Bombyx* parvo-like virus. *Proc. Natl. Acad. Sci. U.S.A.* 105:7523–27.

Kadono-Okuda, K., E. Kosegawa, K. Mase, and W. Hara. 2002. Linkage analysis of maternal EST cDNA clones covering all twenty-eight chromosomes in the silkworm, *Bombyx mori. Insect Mol. Biol.* 11:443–51.

Kajiwara, H., Y. Ito, A. Imamaki, M. Nakamura, K. Mita, and M. Ishizaka. 2005. Protein profile of silkworm midgut of fifth-instar day-3 larvae. *J. Electrophoresis* 49:61–69.

Kajiwara, H., Y. Itou, A. Imamaki, M. Nakamura, K. Mita, and M. Ishizaka. 2006. Proteomic analysis of silkworm fat body. *J. Insect Biotech. Sericol.* 75:47–56.

Kanginakudru, S., C. Royer, S.V. Edupalli, et al. 2007. Targeting *ie-1* gene by RNAi induces baculoviral resistance in lepidopteran cell lines and in transgenic silkworms. *Insect Mol. Biol.* 16:635–44.

Kawanishi, Y., R. Takaishi, Y. Banno, et al. 2007. Sequence comparison of *Mariner*-like elements among the populations of *Bombyx mandarina* inhabiting China, Korea and Japan. *J. Insect Biotech. Sericol.* 76:79–87.

Kawanishi, Y., R. Takaishi, M. Morimoto, et al. 2008. A novel maT-type transposable element, BmamaT1, in *Bombyx mandarina*, homologous to the *B. mori* mariner-like element Bmmar6. *J. Insect Biotech. Sericol.* 77:45–52.

Kawasaki, H., M. Ote, K. Okano, T. Shimada, Q. Guo-Xing, and K. Mita. 2004. Change in the expressed gene patterns of the wing disc during the metamorphosis of *Bombyx mori. Gene* 343:133–42.

Kawashima, T., M. Osanai, R. Futahashi, T. Kojima, and H. Fujiwara. 2007. A novel target-specific gene delivery system combining baculovirus and sequence-specific long interspersed nuclear elements. *Virus Res.* 127:49–60.

Koike, Y., K. Mita, M.G. Suzuki, et al. 2003. Genomic sequence of a 320-kb segment of the Z chromosome of *Bombyx mori* containing a *kettin* ortholog. *Mol. Genet. Genomics* 269:137–49.

Kojima, K., Y. Kuwana, H. Sezutsu, et al. 2007. A new method for the modification of fibroin heavy chain protein in the transgenic silkworm. *Biosci. Biotechnol. Biochem.* 71:2943–51.

Kômoto, N. 2002. A deleted portion of one of the two xanthine dehydrogenase genes causes translucent larval skin in the oq mutant of the silkworm (*Bombyx mori*). *Insect Biochem. Mol. Biol.* 32:591–97.

Kômoto, N., G.X. Quan, H. Sezutsu, and T. Tamura. 2009. A single-base deletion in an ABC transporter gene causes white eyes, white eggs, and translucent larval skin in the silkworm *w-3(oe)* mutant. *Insect Biochem. Mol. Biol.* 39:152–56.

Kômoto, N., H. Sezutsu, K. Yukuhiro, Y. Banno, and H. Fujii. 2003. Mutations of the silkworm molybdenum cofactor sulfurase gene, *og*, cause translucent larval skin. *Insect Biochem. Mol. Biol.* 33:417–27.

Kurihara, H., H. Sezutsu, T. Tamura, and K. Yamada. 2007. Production of an active feline interferon in the cocoon of transgenic silkworms using the fibroin H-chain expression system. *Biochem. Biophys. Res. Commun.* 355:976–80.

Li, A., Q. Zhao, S. Tang, Z. Zhang, S. Pan, and G. Shen. 2005. Molecular phylogeny of the domesticated silkworm, *Bombyx mori*, based on the sequences of mitochondrial cytochrome b genes. *J. Genet.* 84:137–42.

Li, O., W. Hara, T. Yokoyama, and O. Ninagi. 2008. Gene mapping of *short lifespan in imago* by cDNA linkage in the silkworm, *Bombyx mori*. *Sanshi-Konchu Biotec* 77:47–52 (in Japanese).

Li, X., X. Wu, W. Yue, J. Liu, G. Li, and Y. Miao. 2006. Proteomic analysis of the silkworm (*Bombyx mori*) hemolymph during developmental stage. *J. Proteome Res.* 5:2809–14.

Liu, W., F. Yang, S. Jia, X. Miao, and Y. Huang. 2008. Cloning and characterization of Bmrunt from the silkworm *Bombyx mori* during embryonic development. *Arch. Insect Biochem. Physiol.* 69:47–59.

Lu, C., F.Y. Dai, Z.H. Xiang. 2003. Studies on the mutation strains of the *Bombyx mori* gene bank. *Scientia Agricultura Sinica* 36:968–75. (In Chinese)

Lu, C., B. Li, A. Zhao, and Z. Xiang. 2004. QTL mapping of economically important traits in silkworm (*Bombyx mori*). *Sci. China C Life Sci.* 47:477–84.

Maeda, T. 1939. Chiasma studies in the silkworm *Bombyx mori*. *Jpn. J. Genet.* 15:118–27.

Maekawa, H., N. Takada, K. Mikitani, et al. 1988. Nucleolus organizers in the wild silkworm *Bombyx mandarina* and the domesticated silkworm *Bombyx mori*. *Chromosoma* 96:263–69.

Mase, K., T. Iizuka, T. Yamamoto, E. Okada, and W. Hara. 2007. Genetic mapping of a food preference gene in the silkworm, *Bombyx mori*, using restriction fragment length polymorphisms (RFLPs). *Genes Genet. Syst.* 82:249–56.

Matsumoto, S., J.J. Hull, A. Ohnishi, K. Moto, and A. Fonagy. 2007. Molecular mechanisms underlying sex pheromone production in the silkmoth, *Bombyx mori*: Characterization of the molecular components involved in bombykol biosynthesis. *J. Insect Physiol.* 53:752–59.

Matsunaga, T.M., and H. Fujiwara. 2002. Identification and characterization of genes abnormally expressed in wing-deficient mutant (flugellos) of the silkworm, *Bombyx mori*. *Insect Biochem. Mol. Biol.* 32:691–99.

Matsuoka, T., and H. Fujiwara. 2000. Expression of ecdysteroid-regulated genes is reduced specifically in the wing discs of the wing-deficient mutant (fl) of *Bombyx mori*. *Dev. Genes Evol.* 210:120–28.

Miao, X., M. Li, F. Dai, C. Lu, M.R. Goldsmith, and Y. Huang. 2007. Linkage analysis of the visible mutations *Sel* and *Xan* of *Bombyx mori* (Lepidoptera: Bombycidae) using SSR markers. *Eur J. Entomol.* 107:647–52.

Miao, X.X., W.H. Li, M.W. Li, Y.P. Zhao, X.R. Guo, and Y.P. Huang. 2008. Inheritance and linkage analysis of co-dominant SSR markers on the Z chromosome of the silkworm (*Bombyx mori* L.). *Genet. Res.* 90:151–56.

Miao, X.X., S.J. Xub, M.H. Li, et al. 2005. Simple sequence repeat-based consensus linkage map of *Bombyx mori*. *Proc. Natl. Acad. Sci. U.S.A.* 102:16303–08.

Mills, D.R., and M.R. Goldsmith. 2000. Characterization of early follicular cDNA library suggests evidence for genetic polymorphisms in the inbred strain C108 of *Bombyx mori*. *Genes Genet. Syst.* 75:105–13.

Mita, K., M. Kasahara, S. Sasaki, et al. 2004. The genome sequence of silkworm, *Bombyx mori*. *DNA Res.* 11:27–35.

Mita, K., M. Morimyo, K. Okano, et al. 2003. The construction of an EST database for *Bombyx mori* and its application. *Proc. Natl. Acad. Sci. U.S.A.* 100:14121–26.

Nagaraja, G.M., G. Mahesh, V. Satish, M. Madhu, M. Muthulakshmi, and J. Nagaraju. 2005. Genetic mapping of Z chromosome and identification of W chromosome-specific markers in the silkworm, *Bombyx mori*. *Heredity* 95:148–57.

Nagaraju, J. 1996. Sex determination and sex limited traits in the silkworm, *Bombyx mori*; their application in sericulture. *Indian J. Sericol.* 35:83–89.

Nagaraju, J., and M.R. Goldsmith. 2002. Silkworm genomics—progress and prospects. *Curr. Sci.* 83:415–25.

Nguu, E.K., K. Kadono-Okuda, K. Mase, E. Kosegawa, and W. Hara. 2005. Molecular linkage map for the silkworm, *Bombyx mori*, based on restriction fragment length polymorphisms of cDNA clones. *J. Insect Biotechnol. Sericol.* 74:5–13.

Nie, Z., R. Zhou, J. Chen, et al. 2008. Subcellular localization and RNA interference of an RNA methyltransferase gene from silkworm, *Bombyx mori*. *Comp. Funct. Genomics* 2008:571023.

Niimi, T., K. Sahara, H. Oshima, Y. Yasukochi, K. Ikeo, and W. Traut. 2006. Molecular cloning and chromosomal localization of the *Bombyx Sex-lethal* gene. *Genome* 49:263–68.

Ogawa, S., M. Tomita, K. Shimizu, and K. Yoshizato. 2007. Generation of a transgenic silkworm that secretes recombinant proteins in the sericin layer of cocoon: Production of recombinant human serum albumin. *J. Biotechnol.* 128:531–44.

Ogura, T., K. Okano, K. Tsuchida, et al. 1994. A defective non-LTR retrotransposon is dispersed throughout the genome of the silkworm, *Bombyx mori*. *Chromosoma* 103:311–23.

Oh, J.H., Y.J. Jeon, S.Y. Jeong, et al. 2006. Gene expression profiling between embryonic and larval stages of the silkworm, *Bombyx mori*. *Biochem. Biophys. Res. Commun.* 343:864–72.

Okamoto, S., R. Futahashi, T. Kojima, K. Mita, and H. Fujiwara. 2008. A catalogue of epidermal genes: Genes expressed in the epidermis during larval molt of the silkworm *Bombyx mori*. *BMC Genomics* 9:396.

Ote, M., K. Mita, H. Kawasaki, M. Kobayashi, and T. Shimada. 2005. Characteristics of two genes encoding proteins with an ADAM-type metalloprotease domain, which are induced during the molting periods in *Bombyx mori*. *Arch. Insect Biochem. Physiol.* 59:91–98.

Ote, M., K. Mita, H. Kawasaki, et al. 2004. Microarray analysis of gene expression profiles in wing discs of *Bombyx mori* during pupal ecdysis. *Insect Biochem. Mol. Biol.* 34:775–84.

Pan, M., Q. Yu, Y. Xia, et al. 2008. Characterization of mitochondrial genome of Chinese wild mulberry silkworm, *Bombyx mandarina* (Lepidoptera: Bombycidae). *Sci. China C Life Sci.* 51:693–701.

Papa, R., C.M. Morrison, J.R. Walters, et al. 2008. Highly conserved gene order and numerous novel repetitive elements in genomic regions linked to wing pattern variation in *Heliconius* butterflies. *BMC Genomics* 9:345.

Pringle, E.G., S.W. Baxter, C.L. Webster, A. Papanicolaou, S.F. Lee, and C.D. Jiggins. 2007. Synteny and chromosome evolution in the lepidoptera: Evidence from mapping in *Heliconius melpomene*. *Genetics* 177:417–26.

Promboon, A., T. Shimada, H. Fujiwara, and M. Kobayashi. 1995. Linkage map of random amplified DNAs (RAPDs) in the silkworm, *Bombyx mori*. *Genet. Res.* 66:1–7.

Prowell, D.P. 1998. Sex linkage and speciation in Lepidoptera. In *Endless forms: Species and speciation*, ed. D.J. Howard and S.H. Berlocher, 309–19. New York: Oxford University Press.

Quan, G.X., T. Kanda, and T. Tamura. 2002. Induction of the *white egg 3* mutant phenotype by injection of the double-stranded RNA of the silkworm *white* gene. *Insect Mol. Biol.* 11:217–22.

Quan, G.X., I. Kim, N. Kômoto, et al. 2002. Characterization of the kynurenine 3-monooxygenase gene corresponding to the *white egg 1* mutant in the silkworm *Bombyx mori*. *Mol. Genet. Genomics* 267:1–9.

Qvarnström, A., and R.I. Bailey. 2009. Speciation through evolution of sex-linked genes. *Heredity* 102:4–15.

Rasch, E.M. 1974. The DNA content of sperm and hemocyte nuclei of the silkworm, *Bombyx mori* L. *Chromosoma* 45:1–26.

Royer, C., A. Jalabert, M. Da Rocha, et al. 2005. Biosynthesis and cocoon-export of a recombinant globular protein in transgenic silkworms. *Transgenic Res.* 14:463–72.

Sahara, K., F. Marec, and W. Traut. 1999. TTAGG telomeric repeats in chromosomes of some insects and other arthropods. *Chromosome Res.* 7:449–60.

Sahara, K., A. Yoshido, N. Kawamura, et al. 2003. W-derived BAC probes as a new tool for identification of the W chromosome and its aberrations in *Bombyx mori*. *Chromosoma* 112:48–55.

Sahara, K., A. Yoshido, F. Marec, et al. 2007. Conserved synteny of genes between chromosome 15 of *Bombyx mori* and a chromosome of *Manduca sexta* shown by five-color BAC-FISH. *Genome* 50:1061–65.

Sakudoh, T., H. Sezutsu, T. Nakashima, et al. 2007. Carotenoid silk coloration is controlled by a carotenoid-binding protein, a product of the *Yellow* blood gene. *Proc. Natl. Acad. Sci. U.S.A.* 104:8941–46.

Sandrelli, F., S. Cappellozza, C. Benna, et al. 2007. Phenotypic effects induced by knock-down of the *period* clock gene in *Bombyx mori*. *Genet. Res.* 89:73–84.

Sarkar, A., A. Atapattu, E.J. Belikoff, et al. 2006. Insulated *piggyBac* vectors for insect transgenesis. *BMC Biotechnol.* 6:27.

Sato, K., T.M. Matsunaga, R. Futahashi, et al. 2008. Positional cloning of a *Bombyx* wingless locus *flugellos* (*fl*) reveals a crucial role for fringe that is specific for wing morphogenesis. *Genetics* 179:875–85.

Shi, J., D.G. Heckel, and M.R. Goldsmith. 1995. A genetic linkage map for the domesticated silkworm, *Bombyx mori*, based on restriction fragment length polymorphisms. *Genet. Res.* 66:109–26.

Shimizu, K., S. Ogawa, R. Hino, T. Adachi, M. Tomita, and K. Yoshizato. 2007. Structure and function of 5′-flanking regions of *Bombyx mori* fibroin heavy chain gene: Identification of a novel transcription enhancing element with a homeodomain protein-binding motif. *Insect Biochem. Mol. Biol.* 37:713–25.

Sierra-Montes, J.M., S. Pereira-Simon, S.S. Smail, and R.J. Herrera. 2005. The silk moth *Bombyx mori* U1 and U2 snRNA variants are differentially expressed. *Gene* 352:127–36.

Smail, S.S., K. Ayesh, J.M. Sierra-Montes, and R.J. Herrera. 2006. U6 snRNA variants isolated from the posterior silk gland of the silk moth *Bombyx mori*. *Insect Biochem. Mol. Biol.* 36:454–65.

Sohn, K-W. 2003. Conservation status of sericulture germplasm resources in the world—II. Conservation status of silkworm (*Bombyx mori*) genetic resources in the world. Papers contributed to expert consultation on promotion of global exchange of sericulture germplasm, Bangkok, Thailand; September 2002. http://www.fao.org/DOCREP/005/AD108E/AD108E00.HTM#Contents.

Sperling, F. 1994. Sex-linked genes and species differences in Lepidoptera. *Can. Entomol.* 126:807–18.

Suetsugu, Y., H. Minami, M. Shimomura, et al. 2007. End-sequencing and characterization of silkworm (*Bombyx mori*) bacterial artificial chromosome libraries. *BMC Genomics* 8:314.

Sun, Z.C., M. Wu, T.A. Miller, and Z.J. Han. 2008. *piggyBac*-like elements in cotton bollworm, *Helicoverpa armigera* (Hubner). *Insect Mol. Biol.* 17:9–18.

Suzuki, M.G., S. Funaguma, T. Kanda, T. Tamura, and T. Shimada. 2005. Role of the male BmDSX protein in the sexual differentiation of *Bombyx mori*. *Evol. Dev.* 7:58–68.

Tabunoki, H., S. Higurashi, O. Ninagi, et al. 2004. A carotenoid-binding protein (CBP) plays a crucial role in cocoon pigmentation of silkworm (*Bombyx mori*) larvae. *FEBS Lett.* 567:175–78.

Tamura, T., C. Thibert, C. Royer, et al. 2000. Germline transformation of the silkworm *Bombyx mori* L. using a *piggyBac* transposon-derived vector. *Nat. Biotechnol.* 18:81–84.

Tan, A., H. Tanaka, T. Tamura, and T. Shiotsuki. 2005. Precocious metamorphosis in transgenic silkworms overexpressing juvenile hormone esterase. *Proc. Natl. Acad. Sci. U.S.A.* 102:11751–56.

Tan, Y.D., C. Wan, Y. Zhu, C. Lu, Z. Xiang, and H.W. Deng. 2001. An amplified fragment length polymorphism map of the silkworm. *Genetics* 157:1277–84.

Tanaka, H., H. Matsuki, S. Furukawa, et al. 2007. Identification and functional analysis of *Relish* homologs in the silkworm, *Bombyx mori*. *Biochim. Biophys. Acta* 1769:559–68.

Tazima, Y. 1964. *The genetics of the silkworm*. London: Academic Press.

Tomita, M., R. Hino, S. Ogawa, et al. 2007. A germline transgenic silkworm that secretes recombinant proteins in the sericin layer of cocoon. *Transgenic Res.* 16:449–65.

Tong, C.Z., Y.F. Jin, and Y.Z. Zhang. 2006. Computational prediction of microRNA genes in silkworm genome. *J. Zhejiang Univ. Sci. B* 7:806–16.

Traut, W. 1976. Pachytene mapping in the female silkworm, *Bombyx mori* L. (Lepidoptera). *Chromosoma* 58:275–84.

Traut, W., K. Sahara, T.D. Otto, and F. Marec. 1999. Molecular differentiation of sex chromosomes probed by comparative genomic hybridization. *Chromosoma* 108:173–80.

Tsuchida, K., C. Katagiri, Y. Tanaka, et al. 2004. The basis for colorless hemolymph and cocoons in the *Y*-gene recessive *Bombyx mori* mutants: A defect in the cellular uptake of carotenoids. *J. Insect Physiol.* 50:975–83.

Tsukioka, H., M. Takahashi, H. Mon, et al. 2006. Role of the silkworm *argonaute2* homolog gene in double-strand break repair of extrachromosomal DNA. *Nucleic Acids Res.* 34:1092–101.

Uchino, K., M. Imamura, H. Sezutsu, et al. 2006. Evaluating promoter sequences for trapping an enhancer activity in the silkworm *Bombyx mori*. *J. Insect Biotech. Sericol.* 75:89–97.

Uchino, K., M. Imamura, K. Shimizu, T. Kanda, and T. Tamura. 2007. Germ line transformation of the silkworm, *Bombyx mori*, using the transposable element Minos. *Mol. Genet. Genomics* 277:213–20.

Uchino, K., H. Sezutsu, M. Imamura, et al. 2008. Construction of a *piggyBac*-based enhancer trap system for the analysis of gene function in silkworm *Bombyx mori*. *Insect Biochem. Mol. Biol.* 38:1165–73.

Wang, G.H., C. Liu, Q.Y. Xia, Y.F. Zha, J. Chen, and L. Jiang. 2008. Cathepsin B protease is required for metamorphism in silkworm, *Bombyx mori*. *Insect Sci.* 15:201–08.

Wang, J., X. Ren, T. Miller, and Y. Park. 2006. *piggyBac*-like elements in the tobacco budworm, *Heliothis virescens* (Fabricius). *Insect Mol. Biol.* 15: 435–43.

Wang, J., Q. Xia, X. He, et al. 2005. SilkDB: a knowledgebase for silkworm biology and genomics. *Nucleic Acids Res.* 33:D399–402.

Wanner, K.W., A.R. Anderson, S.C. Trowell, D.A. Theilmann, H.M. Robertson, and R.D. Newcomb. 2007. Female-biased expression of odourant receptor genes in the adult antennae of the silkworm, *Bombyx mori*. *Insect Mol. Biol.* 16:107–19.

Wu, C., S. Asakawa, N. Shimizu, S. Kawasaki, and Y. Yasukochi. 1999. Construction and characterization of bacterial artificial chromosome libraries from the silkworm, *Bombyx mori*. *Mol. Gen. Genet.* 261:698–706.

Xia, Q., D. Cheng, J. Duan, et al. 2007. Microarray-based gene expression profiles in multiple tissues of the domesticated silkworm, *Bombyx mori*. *Genome Biol.* 8:R162.

Xia, Q., Z. Zhou, C. Lu, et al. 2004. A draft sequence for the genome of the domesticated silkworm (*Bombyx mori*). *Science* 306:1937–40.

Xiang, H., M. Li, F. Yang, et al. 2008. Fine mapping of *E(kp)-1*, a locus associated with silkworm (*Bombyx mori*) proleg development. *Heredity* 100:533–40.

Xu, H.F., Q.Y. Xia, C. Liu, et al. 2006. Identification and characterization of *piggyBac*-like elements in the genome of domesticated silkworm, *Bombyx mori*. *Mol. Genet. Genomics* 276:31–40.

Xu, J.-S., Q.-Y. Xia, J. Li, G.-Q. Pan, and Z.-Y. Zhou. 2005. Survey of long terminal repeat retrotransposons of domesticated silkworm (*Bombyx mori*). *Insect Biochem. Mol. Biol.* 35:921–29.

Yamamoto, K., J. Narukawa, K. Kadono-Okuda, et al. 2006. Construction of a single nucleotide polymorphism linkage map for the silkworm, *Bombyx mori*, based on bacterial artificial chromosome end sequences. *Genetics* 173:151–61.

Yamamoto, K., J. Nohata, K. Kadono-Okuda, et al. 2008. A BAC-based integrated linkage map of the silkworm *Bombyx mori*. *Genome Biol.* 9:R21.

Yamamoto, M., M. Yamao, H. Nishiyama, et al. 2004. New and highly efficient method for silkworm transgenesis using *Autographa californica* nucleopolyhedrovirus and *piggyBac* transposable elements. *Biotechnol. Bioeng.* 88:849–53.

Yanagisawa, S., Z. Zhu, I. Kobayashi, et al. 2007. Improving cell-adhesive properties of recombinant *Bombyx mori* silk by incorporation of collagen or fibronectin derived peptides produced by transgenic silkworms. *Biomacromolecules* 8:3487–92.

Yao, H.P., X.W. Xiang, L. Chen, et al. 2008. Identification of the proteome of the midgut of silkworm, *Bombyx mori* L. by multidimensional liquid chromatography LTQ-Orbitrap mass spectrometry. *Biosci. Rep.* doi:10.1042/BSR20080144.

Yasukochi, Y. 1998. A dense genetic map of the silkworm, *Bombyx mori*, covering all chromosomes based on 1018 molecular markers. *Genetics* 150:1513–25.

Yasukochi, Y., L.A. Ashakumary, K. Baba, A. Yoshido, and K. Sahara. 2006. A second-generation integrated map of the silkworm reveals synteny and conserved gene order between lepidopteran insects. *Genetics* 173:1319–28.

Yasukochi, Y., L.A. Ashakumary, C. Wu, et al. 2004. Organization of the *Hox* gene cluster of the silkworm, *Bombyx mori*: A split of the *Hox* cluster in a non-*Drosophila* insect. *Dev. Genes Evol.* 214:606–14.

Yasukochi, Y., Y. Banno, K. Yamamoto, M.R. Goldsmith, and H. Fujii. 2005. Integration of molecular and classical linkage groups of the silkworm, *Bombyx mori* (n = 28). *Genome* 48:626–29.

Yasukochi, Y., H. Fujii, and M.R. Goldsmith. 2008. Silkworm. In *Genome mapping and genomics in arthropods*, ed. C. Kole and W. Hunter, 43–57. Berlin: Springer-Verlag.

Yasukochi, Y., T. Kanda, and T. Tamura. 1998. Cloning of two *Bombyx* homologues of the *Drosophila* rosy gene and their relationship to larval translucent skin colour mutants. *Genet. Res.* 71:11–19.

Yoshido, A., H. Bando, Y. Yasukochi, and K. Sahara. 2005. The *Bombyx mori* karyotype and the assignment of linkage groups. *Genetics* 170:675–85.

Yoshido, A., Y. Yasukochi, F. Marec, H. Abe, and K. Sahara. 2007. FISH analysis of the W chromosome in *Bombyx mandarina* and several other species of Lepidoptera by means of *B. mori* W-BAC probes. *J. Insect Biotech. Sericol.* 76:1–7.

Yu, X., Q. Zhou, S.C. Li, et al. 2008. The silkworm (*Bombyx mori*) microRNAs and their expressions in multiple developmental stages. *PLoS ONE* 3:e2997.

Yukuhiro, K., H. Sezutsu, M. Itoh, K. Shimizu, and Y. Banno. 2002. Significant levels of sequence divergence and gene rearrangements have occurred between the mitochondrial genomes of the wild mulberry silkmoth, *Bombyx mandarina*, and its close relative, the domesticated silkmoth, *Bombyx mori*. *Mol. Biol. Evol.* 19:1385–89.

Zhao, Y.-P., M.-W. Li, A.-Y. Xu, et al. 2008. SSR based linkage and mapping analysis of *C*, a yellow cocoon gene in the silkworm, *Bombyx mori*. *Insect Sci.* 15:399–404.

Zhang, P., Y. Aso, H. Jikuya, et al. 2007a. Proteomic profiling of the silkworm skeletal muscle proteins during larval-pupal metamorphosis. *J. Proteome Res.* 6:2295–303.

Zhang, Y.Z., J. Chen, Z.M. Nie, et al. 2007b. Expression of open reading frames in silkworm pupal cDNA library. *Appl. Biochem. Biotechnol.* 136:327–43.

Zhong, B., Y. Yu, Y. Xu, et al. 2005. Analysis of ESTs and gene expression patterns of the posterior silkgland in the fifth instar larvae of silkworm, *Bombyx mori* L. *Sci. China C Life Sci.* 48:25–33.

Zhou, J.J., Y. Kan, J. Antoniw, J.A. Pickett, and L.M. Field. 2006. Genome and EST analyses and expression of a gene family with putative functions in insect chemoreception. *Chem. Senses* 31:453–65.

Zhou, Z., H. Yang, and B. Zhong. 2008. From genome to proteome: Great progress in the domesticated silkworm (*Bombyx mori* L.). *Acta Biochim. Biophys. Sin. (Shanghai)* 40:601–11.

3 Rise and Fall of the W Chromosome in Lepidoptera

František Marec, Ken Sahara, and Walther Traut

CONTENTS

INTRODUCTION

Among several peculiarities of Lepidoptera cytogenetics, the WZ sex chromosome system is the most conspicuous. Whereas most animal and some dioecious plant species have an XY or X0 system, relatively few share the WZ or Z0 system with Lepidoptera, the most prominent ones being birds, snakes, and caddis flies (Trichoptera). Sex chromosomes and sex determination in Lepidoptera have been reviewed in a recent article (Traut, Sahara, and Marec 2007). We focus here on the fate of the W chromosome. Since most extant lepidopteran species possess a W chromosome, it may come as a surprise that the Lepidoptera clade started evolution with a Z0 system, and the W chromosome appeared much later in its history. Thereafter, as we will show, the evolution of the W chromosome followed the rules already learned from Y chromosomes: It decayed, fused with other chromosomes, and sometimes even got lost again. Hence, the W chromosome exhibits the full evolutionary life cycle of a typical sex chromosome in a single clade.

For a better understanding of the chromosome system, we briefly summarize basic peculiarities of lepidopteran cytogenetics and new techniques of chromosome identification and then discuss sex chromosomes, specifically the W chromosome, in more detail. We review data on the composition of W chromosomes, their molecular differentiation and evolution, and their role in sex determination.

CYTOGENETIC PECULIARITIES OTHER THAN SEX CHROMOSOMES

CHROMOSOME NUMBERS

Chromosomes in Lepidoptera are usually small, numerous, and uniform in shape. Most species are reported to have haploid numbers close to 30 (Robinson 1971). The modal, and possibly also ancestral, chromosome number for Lepidoptera appears to be n = 31, and a typical karyotype shows a gradual decrease in chromosome size (reviewed by De Prins and Saitoh 2003). In certain branches of the lepidopteran phylogenetic tree, species with low-number karyotypes evolved independently by multiple chromosome fusions; accordingly, their chromosomes are relatively large. Notable examples are moths of the genus *Orgyia* (Lymantriidae): *O. antiqua* with n = 14 and *O. thyellina* with n = 11 (Traut and Clarke 1997); the large white butterfly, *Pieris brassicae* (Pieridae), with n = 15 (Doncaster 1912); and the silk moth *Samia cynthia* (Saturniidae) with variable n = 12–14 depending on the population studied (Yoshido, Marec, and Sahara 2005). Species with reduced chromosome numbers are particularly common in Papilionoidea (Brown et al. 2007), where the smallest number of n = 5 is also found in two species: a neotropical butterfly, *Hypothyris thea* (Nymphalidae; Brown, von Schoultz, and Suomalainen 2004), and the Arizona giant-skipper, *Agathymus aryxna* (Hesperiidae; De Prins and Saitoh 2003). On the other hand, karyological studies have identified species with high chromosome numbers that are most probably the result of chromosome fission. For example, the Chinese oak silk moth, *Antheraea pernyi* (Saturniidae), has n = 49, whereas the closely related Japanese oak silk moth, *A. yamamai*, has n = 31 chromosomes, the modal number for Lepidoptera (Kawaguchi 1934). A typical example of multiple chromosome fission is the *Agrodiaetus dolus* species complex (Lycaenidae), which includes species with haploid chromosome numbers as high as n = 90 to n = 125 (Lukhtanov, Vila, and Kandul 2006). Indeed, the genus *Agrodiaetus* shows exceptional diversity, with haploid chromosome numbers ranging from n = 10 to n = 134 (Lukhtanov and Dantchenko 2002; Lukhtanov et al. 2005). Another lycaenid butterfly, the Atlas blue, *Polyommatus atlantica*, is known as the species with the highest chromosome number, n = 223, not only of Lepidoptera but of all animals (de Lesse 1970).

CHROMOSOME STRUCTURE

Lepidoptera and their sister group, Trichoptera, are regarded as organisms with holokinetic chromosomes. In contrast to the typical monocentric chromosomes of, for example, Diptera and Coleoptera, holokinetic chromosomes lack a distinct primary constriction (the centromere) and, as a result, sister chromatids separate by parallel disjunction during mitotic metaphase (Suomalainen 1966; Murakami and Imai 1974). Proper disjunction is ensured by a large kinetochore plate, to which the spindle microtubules are attached. In fact, the chromosomes are not properly "holokinetic," but the plate covers a significant part of the chromatid length (Gassner and Klemetson 1974; Wolf 1996).

Lepidopteran species are known to be rather resistant to ionizing radiation. Although the mechanisms underlying this phenomenon are not fully understood, a significant role in the high radioresistance has been attributed to the specific chromosome structure (reviewed by Carpenter, Bloem, and Marec 2005). The large kinetochore plates ensure that most radiation-induced breaks do not lead to the loss of chromosome fragments as is typical in species with monocentric chromosomes. Instead, the fragments may persist for a number of mitotic cell divisions and can even be transmitted through germ cells to the next generation (Rathjens 1974; Marec et al. 2001). The extended kinetochore plates reduce the risk of lethality otherwise caused by the formation of dicentric chromosomes and acentric fragments (Tothová and Marec 2001). These characteristics, resulting from the near-holokinetic structure of chromosomes, are responsible for a unique feature of Lepidoptera—the so-called inherited sterility (Carpenter, Bloem, and Marec 2005)—and they presumably facilitated the conservation of chromosome fusion and fission products in the karyotype evolution of some Lepidoptera groups, such as those mentioned previously.

Absence of Crossing Over and Chiasmata in Female Meiosis

Crossing over and chiasma formation do not take place in meiotic prophase I in lepidopteran females, whereas in males the chromosomes display a normal sequence of meiotic events with typical chiasmata in diplotene and diakinesis (e.g., Traut 1977; Nokkala 1987). In this respect, Lepidoptera present the reverse situation of *Drosophila* and other higher Diptera, where meiotic recombination is absent in males. In both cases, though, achiasmatic meiosis is linked to the heterogametic sex.

In contrast to *Drosophila* males, however, early meiotic prophase I in lepidopteran females is conventional and resembles that of lepidopteran males up to the pachytene stage, when the chromosomes pair and form regular bivalents. The process is mediated by the proteinaceous synaptonemal complexes (SCs), which are absent in *Drosophila* males (reviewed by Marec 1996). The only noticeable difference between lepidopteran male and female SCs in that stage is the presence of recombination nodules in male SCs and their absence in female SCs. Recombination nodules are multienzyme nanomachines that mediate recombination. Female meiosis proceeds without meiotic recombination; consequently, no chiasmata are formed, and hence, no diplotene or diakinesis stage can be recognized. The SCs, which remain associated with the bivalents in a modified form until metaphase I, are transformed into "elimination chromatin," which glues homologues together, and stays in the equatorial plane during anaphase I, while chromosomes segregate. Elimination chromatin is considered a substitute for crossing over and chiasma formation in order to ensure regular disjunction of homologous chromosomes (Rasmussen 1977).

Polyploidy in Somatic Tissues

Most somatic tissues are polyploid in larvae and adults of Lepidoptera. The levels of polyploidization cover a huge span. Diploid nuclei are found in imaginal disks of larvae, neural ganglia, and hemolymph (e.g., Traut and Scholz 1978; F. Marec, unpublished). In imaginal disks, tetraploid cells can also be seen regularly. At least some of them are dividing mitotically, and hence, the tetraploid status is easily determined from the chromosome number. Above the tetraploid level, endopolyploidization is the rule, so that the level of polyploidy has to be determined cytophotometrically. For example, we find intermediate levels of polyploidization in intestinal cells, high levels of several hundred C in Malpighian tubules, and still higher levels in silk glands and mandibular glands (Buntrock, Marec, and Traut, unpublished). The record of polyploidization is set by the silk glands of *Bombyx mori* with at least 4×10^5 C (Gage 1974). Nuclei of high-level polyploid cells often have an unusual form; they are branched, not more or less spherical as in ordinary cells (e.g., Traut and Scholz 1978).

CHROMOSOME IDENTIFICATION

The small size and uniform morphology of lepidopteran chromosomes, especially the absence of localized centromeres as morphological landmarks together with the lack of convenient banding techniques, inhibited identification of individual chromosomes by standard cytogenetic methods. For a long time this situation restricted their study to counting chromosome numbers. The conditions for identification improved when meiotic chromosomes of the pachytene stage were used for so-called pachytene mapping (Traut 1976). In this stage, the chromosomes form a haploid number of bivalents that are much longer than mitotic chromosomes and, in addition, display a specific chromomere pattern. Pachytene mapping allowed the identification of WZ bivalents in several species and helped to identify a few autosomal bivalents (Traut 1976; Schulz and Traut 1979). But the problem of general chromosome identification remained unsolved.

All modern techniques take advantage of the favorable morphology of the pachytene stage. Recently, probes prepared from bacterial artificial chromosomes (BACs) were successfully used with fluorescence in situ hyridization (FISH) to pachytene complements to establish the complete

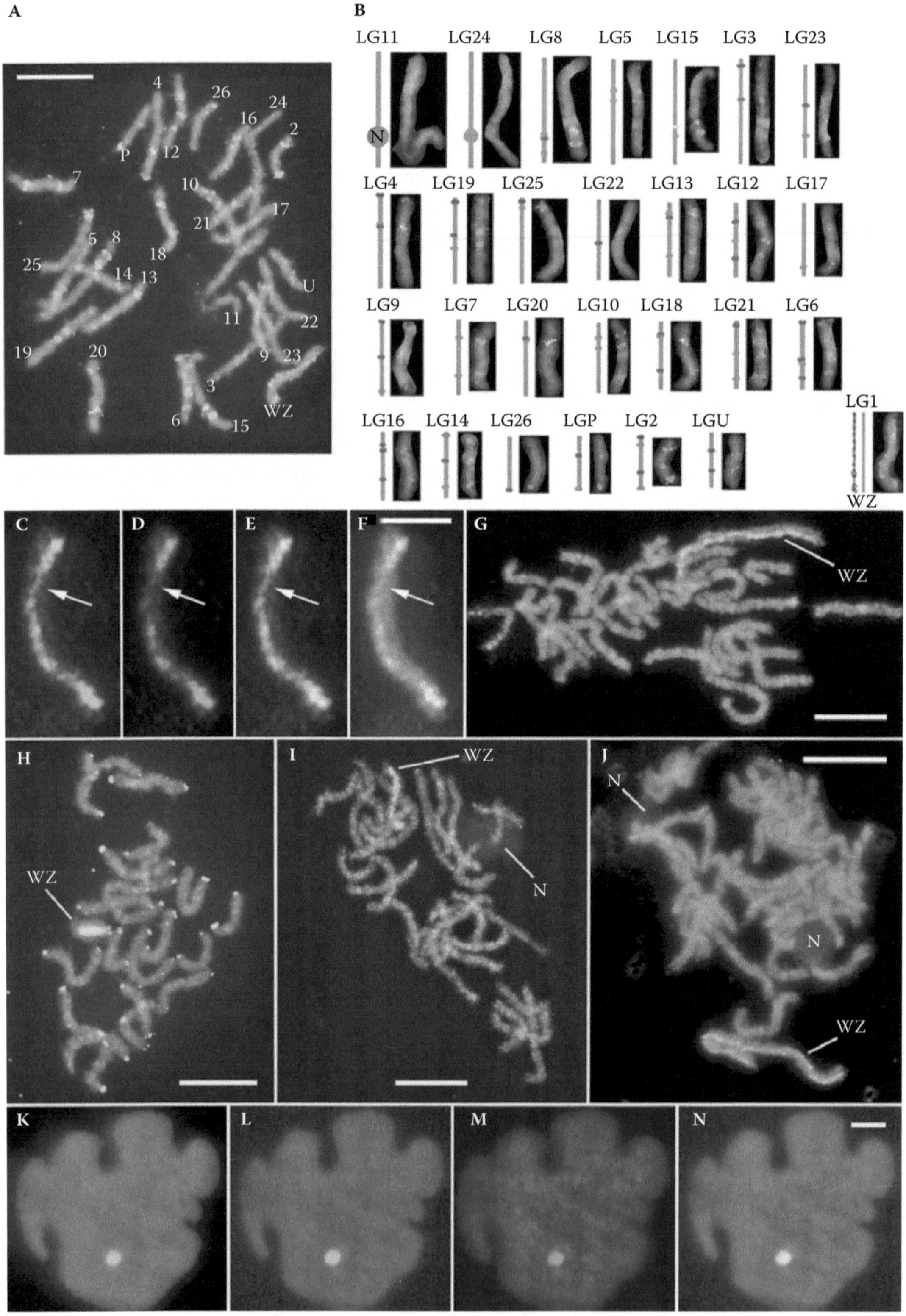
A
B
LG11
LG24
LG8
LG5
LG15
LG3
LG23
N
LG4
LG19
LG25
LG22
LG13
LG12
LG17
LG9
LG7
LG20
LG10
LG18
LG21
LG6
LG16
LG14
LG26
LGP
LG2
LGU
LG1
WZ
C
D
E
F
G
WZ
H
WZ
I
WZ
N
J
N
N
WZ
K
L
M
N

karyotype of the silkworm *B. mori* (Figure 3.1a,b; Yoshido et al. 2005). The BAC-FISH technique provided specific recognition and anchor points and, hence, not only identified individual chromosomes but also proved to be a powerful tool for direct physical mapping of genes (Yasukochi et al. 2004; Niimi et al. 2006). Conserved synteny of genes in *B. mori* and other lepidopteran species has been studied using this technique (Yasukochi et al. 2006; Sahara et al. 2007). BAC-FISH is presently the state of the art for karyotyping and physical gene mapping in Lepidoptera.

Pachytene mapping alone is sufficient to recognize well-differentiated sex chromosome pairs in some species. Although autosomes and the Z chromosome display distinctive chromomere-interchromomere patterns, the W chromosome often forms an evenly stained thread over part or all of its length (e.g., Schulz and Traut 1979; Traut and Marec 1997). When this method fails, several molecular techniques are now available to identify the W chromosome unambiguously: comparative genomic hybridization (CGH), genomic in situ hybridization (GISH), W-BAC-FISH, and W-chromosome painting.

CGH takes its advantage from cohybridization of two differently labeled probes, in this case whole genomic DNAs from females and males. CGH discriminates the W chromosome by color and/or signal intensity, that is, by binding of female-specific probe components and hybridization of accumulated repetitive elements, respectively (Traut et al. 1999; Traut, Eickhoff, and Schorch 2001; Sahara et al. 2003a; Vítková et al. 2007). Comparison of signals from hybridizing male and female DNA with CGH, therefore, not only differentiates sex chromosomes but also supplies some additional information about sequence composition (Figure 3.1c–f).

GISH is based on hybridization of a labeled probe (here, whole genomic DNA of the heterogametic sex) in the presence of an excess of unlabeled competitor DNA (here, genomic DNA from the homogametic sex). GISH visualizes the W chromosome by strong binding of fluorescently labeled

FIGURE 3.1 ***A color version of this figure follows page 176.*** Fluorescence *in situ* hybridization (FISH) images of pachytene bivalents (**A–J**) and a polyploid interphase nucleus (**K–N**) from Lepidoptera females. Chromosomes and the polyloid nucleus were counterstained with DAPI (blue). (**A–B**) BAC-FISH in *Bombyx mori*; (**A**) an oocyte pachytene nucleus hybridized with a probe cocktail composed of 62 BAC probes, each labeled with Cy-3 (red) and/or fluorescein (green); (**B**) BAC-FISH karyotype assembled from 28 bivalents of the same nucleus; the color code of the respective BAC signals is diagnostic for discrimination of individual bivalents. The bivalents are labeled here with the numbers and letters (P and U) of the corresponding linkage groups (LGs); WZ (=LG1), sex chromosome bivalent. (**C–F**) WZ bivalent of *Galleria mellonella* stained with CGH; (**C**) hybridization signals of the Alexa Fluor 488-labeled (green) female genomic probe, bound to the whole W chromosome; (**D**) hybridization signals of the Cy3-labeled (red) male genomic probe; (**E**) merged image of both probes; (**F**) merged image of both probes including counterstaining (blue); note that the W chromosome is highlighted evenly with both probes, indicating the presence of ubiquitous repetitive sequences, except for an interstitial segment (arrow) identified only by the female probe, indicating a region of W-specific DNA. (**G**) a pachytene oocyte nucleus of *Cydia pomonella* (n = 28) stained with GISH using Cy3-labeled female genomic DNA (red); note the highlighted W chromosome in the WZ bivalent. (**H**) a pachytene oocyte nucleus of *Artaxa subflava* (n = 22; A. Yoshido and K. Sahara, unpublished) stained with GISH using Cy3-labeled female genomic DNA (red) and the telomeres visualized by a fluorescein-labeled $(TTAGG)_n$ probe (green); GISH identified the small W chromosome, wrapped by the much longer Z chromosome. (**I**) a pachytene oocyte nucleus of *B. mori* stained with W-BAC-FISH; note the highlighted W chromosome of the WZ bivalent identified with a W-derived BAC probe, the FluorX-labeled 5H4C clone (green); a single nucleolus (N) is associated with an autosome bivalent. (**J**) a pachytene complement of *Cadra cautella* (n = 30) subjected to FISH using a SpectrumOrange-labeled W-painting probe; note the strong binding of the probe to the W chromosome of the WZ bivalent and two nucleoli (N), each associated with a different autosome bivalent. (**K–N**) a polyploid interphase nucleus of a Malpighian tubule cell in *Plodia interpunctella*, stained with CGH, showing a conspicuous sex chromatin body highlighted with DAPI and with both female and male genomic DNA probes; (**k**) DAPI counterstaining (blue); (**l**) Alexa Fluor 488-labeled (green) female probe; (**M**) Cy3-labeled (red) male probe; (**N**) merged image of both probes; note stronger binding of the female probe to the sex chromatin in comparison with the male probe, indicating a significant fraction of the W-specific DNA. **A**, **B**, and **H** courtesy of Atsuo Yoshido (Sapporo); **C–F**, **J**, and **K–N** courtesy of Magda Vítková (České Budějovice). Bar = 10 μm.

female genomic DNA, while binding to autosomes and the Z chromosome is suppressed by the competitor DNA (Figure 3.1g; Sahara et al. 2003b; Mediouni et al. 2004; Fuková, Nguyen, and Marec 2005; Yoshido, Yamada, and Sahara 2006). A combination of FISH with the insect telomeric probe $(TTAGG)_n$ and GISH proved to be particularly useful in resolving the sex chromosome constitution in the unusual sex chromosome pair of the oriental tussock moth *Artaxa subflava* (Figure 3.1h), and in pachytene complements of species with multiple sex chromosomes (Yoshido, Marec, and Sahara 2005).

Sahara et al. (2003b) made use of the distribution of repetitive elements throughout the W chromosome in a refinement of whole genome techniques. They performed FISH with a single W-derived BAC as a probe (W-BAC-FISH), which sufficed to label the whole W chromosome of *B. mori* (Figure 3.1i), although it was derived from a relatively small portion of the W-DNA molecule (about 170 kb). This demonstrated a molecular homogeneity of the W chromosome (for sequence composition of the *B. mori* W chromosome, see Chapter 4).

Recently, a new, universally applicable approach has been developed to visualize and analyze lepidopteran W chromosomes. It is based on laser microdissection of W-chromatin bodies from highly polyploid nuclei, for example, those of Malpighian tubules. Since W-chromatin bodies are composed of many copies of the W chromosome (see below), they provide material for easy sampling of large numbers of W-chromosomes. After amplification of the DNA by degenerate oligonucleotide-primed polymerase chain reaction (DOP-PCR) and labeling with a fluorescent dye, the probe can be used directly in FISH to paint the W chromosome (Figure 3.1j; Fuková et al. 2007; Vítková et al. 2007).

SEX CHROMOSOME SYSTEMS

Species with XY or X0 sex chromosomes generate female- and male-determining sperm ("male heterogamety"), whereas WZ and Z0 species produce female- and male-determining eggs ("female heterogamety"). In Lepidoptera, females are the heterogametic sex. This is known from numerous cases of sex-linked, more specifically Z-linked, inheritance and from cytogenetic evidence.

Simple sex chromosome systems with female heterogamety are referred to as WZ/ZZ and Z/ZZ (alternatively designated WZ and Z0), depending on the presence or absence of a W chromosome. WZ/ZZ is the most common type among Lepidoptera (Figure 3.2a). The model species *B. mori* and experimental species like *Ephestia kuehniella* and *Galleria mellonella* have it, as do well-known pest species like *Cydia pomonella*, *Lymantria dispar*, *Ectomyelois ceratoniae*, *Cadra cautella*, and *Plodia interpunctella* (Traut and Marec 1997; Mediouni et al. 2004; Fuková, Nguyen, and Marec 2005; Vítková et al. 2007). The Z/ZZ system is less frequent (Figure 3.2b). It is found not only in basal lepidopteran groups, for example, Micropterigidae (Traut and Marec 1997), but also sporadically in more advanced groups, for example, in the noctuid moth *Orthosia gracilis* (Traut and Mosbacher 1968) and the saturniid moth *Samia cynthia ricini*, the Eri silkworm (Yoshido, Marec, and Sahara 2005).

Multiple sex chromosome systems have also been described. A system with two different W chromosomes (W_1W_2Z/ZZ; Figure 3.2c,d) was found in a pyralid moth, *Witlesia murana*; in a tortricid moth, *Bactra lacteana* (Suomalainen 1969); in the ruby tiger moth *Phragmatobia fuliginosa* (Traut and Marec 1997); and in one of two chromosomal races of the tussock moth, *O. thyellina* (Yoshido, Marec, and Sahara 2005). A system with two different Z chromosomes ($WZ_1Z_2/Z_1Z_1Z_2Z_2$) was reported for six species of the small ermine moth genus *Yponomeuta* (Figure 3.2f; Nilsson, Löfstedt, and Dävring 1988). A similar type was found in the lasiocampid moth *Trabala vishnu* (Figure 3.2e; Rishi, Sahni, and Rishi 1999) and in the Shinju silkworm, *S. cynthia* subsp. indet. (Yoshido, Marec, and Sahara 2005).

Multiple sex chromosome systems may have originated from sex chromosome fusions or fissions. Because of their holokinetic organization, fission events are not as unlikely in Lepidoptera as they are in other groups of organisms. Candidates for an origin by fission are the W_1 and W_2 chromosomes of *W. murana* and *B. lacteana*. The resulting sex chromosome system is of the W_1W_2Z/ZZ type (Figure 3.2c). In most cases, however, the fusion of an autosome with a sex chromosome appears to

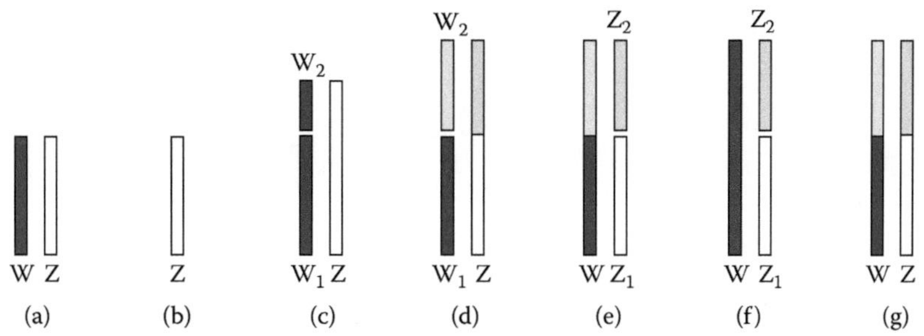

FIGURE 3.2 Schemes of sex chromosome complements in Lepidoptera females. W and Z, sex chromosomes; black bars represent W-chromosome heterochromatin; white bars, the original Z chromosome; and gray bars, autosomes or chromosome segments of autosomal origin. (**a**) standard WZ, as in *Ephestia kuehniella*, *Cydia pomonella*, and *Bombyx mori*. (**b**) Z0, as in *Micropterix calthella* and *Samia cynthia ricini*. (**c**) W_1W_2Z system, consisting of the original Z chromosome and two heterochromatic W chromosomes, the longer W_1 and shorter W_2, indicating a fission of the original W chromosome, as in *Witlesia murana* and *Bactra lacteana*. (**d**) W_1W_2Z system, consisting of the original W (W_1), a neo-Z chromosome that arose by a fusion of the original Z with an autosome, and a W_2 chromosome that is a homologue of the fused autosome, as in *Phragmatobia fuliginosa*. (**e**) WZ_1Z_2 consisting of the original Z (Z_1), a neo-W chromosome that arose by a relatively recent fusion of the original W with an autosome (note that the autosomal part is not yet heterochromatic), and a Z_2 chromosome that is a homologue of the fused autosome, as in *Trabala vishnu*. (**f**): WZ_1Z_2 similar to (**e**) but with a fully heterochromatinized W chromosome, indicating an old fusion event between the original W and an autosome, as in *Yponomeuta cagnagellus* and *Y. padellus*. (**g**) WZ consisting of neo-W and neo-Z chromosomes that arose by fusion of the original W and Z with one homologue each of an autosome pair, as in *Orgyia antiqua* and *Samia cynthia walkeri*.

be the more likely source of multiple sex chromosomes. Fusion of an autosome with the Z chromosome gives rise to a neo-Z chromosome (alternatively designated Z^A or A^Z) with two pairing partners, the free autosome and the original W, both behaving like W chromosomes in meiosis. The resulting sex chromosome system has again to be designated W_1W_2Z/ZZ formally (Figure 3.2d). Fusion of an autosome with the W chromosome brings about a neo-W chromosome (alternatively designated W^A or A^W), which has two pairing partners in meiosis, the free autosome and the original Z chromosome, both behaving like Z chromosomes. Thus a $WZ_1Z_2/Z_1Z_1Z_2Z_2$ system is created (Figure 3.2e,f). But such a system may have also resulted from a fission of the original Z chromosome. An investigation of the ancestral state and of homology between chromosomes is required to decide between different interpretations of the origin of a multiple sex chromosome system in Lepidoptera.

HETEROCHROMATINIZATION OF THE W

In some species such as *E. kuehniella*, *L. dispar*, and *C. cautella*, the W chromosome can be easily identified at the pachytene stage with 4′, 6-diamino-2-phenylindole (DAPI) (Figure 3.3a,b) or even by conventional staining with Giemsa or orcein (Schulz and Traut 1979; Traut and Marec 1997; Vítková et al. 2007). It lacks the typical chromomere-interchromomere pattern found in autosomes and the Z chromosome and forms instead a thread of extended segments of apparently thicker or more condensed chromatin. CGH and GISH highlight this thread (see Figure 3.1c-f,g). In species like *B. mori*, *G. mellonella*, and *E. ceratoniae*, where the W is difficult to identify by conventional microscopy, it is well differentiated by CGH or GISH (Traut et al. 1999; Mediouni et al. 2004; Vítková et al. 2007). The highlighting is thought to be due to a considerable accumulation of repetitive sequences such as retrotransposons in the W (for examples of retrotransposons in the W chromosome of *B. mori*, see Chapter 4). We consider these chromosomes as heterochromatic. The heterochromatic nature is even more apparent in interphase nuclei where the W forms a dense chromatin body, the sex chromatin or W chromatin (see next section).

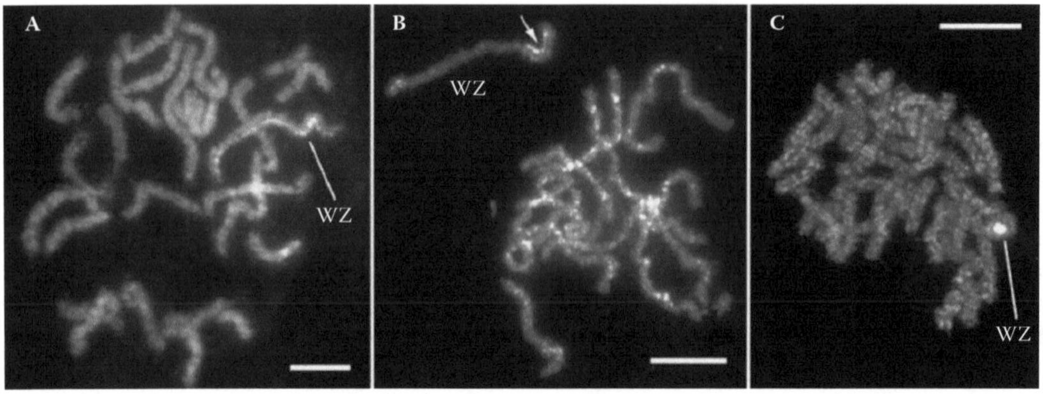

FIGURE 3.3 Fluorescent images of oocyte pachytenes, stained with DAPI (**a**, **b**) or YOYO-1 (**c**). (**a**) *Cydia pomonella* (n = 28); note the DAPI-highlighted heterochromatic thread of the W chromosome in the WZ bivalent. (**b**) *Orgyia antiqua* (n = 14); in this species, the WZ bivalent arose by fusion with 1 or 2 autosome bivalents; note the ancestral part of the W chromosome consisting of DAPI-highlighted heterochromatin (arrow). (**c**) *Bicyclus anynana* (n = 28); note the YOYO-1 highlighted tiny W chromosome which is embraced by the long Z chromosome. **c** courtesy of Arjen E. Van't Hof (Liverpool). Bar = 10 μm.

The heterochromatic W and the euchromatic Z chromosome, although obviously nonhomologous to a large extent, pair in meiosis and remain so until metaphase I (Weith and Traut 1986; Marec et al. 2001). Their pairing is, however, delayed in comparison with autosome bivalents. In *E. kuehniella*, length differences between W and Z are being adjusted during the pairing process so that the sex chromosomes are fully synapsed in late pachytene (Marec and Traut 1994). Even multiple sex chromosomes form a single multivalent of apparently adjusted length in meiosis (e.g., Seiler 1914; Traut and Marec 1997; Yoshido, Marec, and Sahara 2005). Exceptions in adjusting length differences during pairing are rare. For example, in *A. subflava*, the short W chromosome is wrapped by the much longer Z chromosome in the pachytene stage (see Figure 3.1h). Similarly, the tiny W chromosome of *Bicyclus anynana* forms a spherical body encircled by the long Z chromosome (Figure 3.3c; A.E. Van't Hof et al. 2008).

In some species, the W chromosome is only partly heterochromatic; the rest of the chromosome is apparently euchromatic (Traut and Marec 1997). This has been investigated most closely in the vapourer moth, *O. antiqua* (Lymantriidae). *O. antiqua* has a reduced chromosome number, n = 14, which indicates multiple chromosome fusions in the evolutionary history of the species (Traut and Clarke 1997). In the pachytene stage, W and Z form a long bivalent, in which a part of the W chromosome is heterochromatic, while the remaining part displays a homologous chromomere pattern in the W and the Z chromosomes (Traut and Marec 1997). Only the heterochromatic part is highlighted by CGH and GISH (Traut, Eickhoff, and Schorch 2001; Yoshido, Marec, and Sahara 2005). The W-Z pair is probably a neo-W, neo-Z pair that arose by fusion of a pair of autosomes with the ancestral W (now the heterochromatic segment) and the ancestral Z chromosome (Figure 3.2g; Yoshido, Marec, and Sahara 2005), since a close relative, *O. recens*, with a chromosome number of n = 30 still has a small and completely heterochromatic W chromosome (Robinson 1971; Yoshido, Yamada, and Sahara 2006). A similar origin has been inferred for the partly heterochromatic W chromosome of *S. cynthia walkeri* (Saturniidae) with 2n = 26 (Yoshido, Marec, and Sahara 2005).

W CHROMATIN

W chromosomes are responsible for a peculiarity of the lepidopteran interphase nucleus: Like mammals, Lepidoptera possess female-specific so-called sex chromatin. Most species display one or more heterochromatin bodies in female somatic interphase cells but not in male cells. This female-specific sex chromatin is derived from the W chromosome and therefore designated W chromatin (reviewed by Traut and Marec 1996). The mammalian sex chromatin, in contrast, is derived from one of the two X chromosomes present in somatic tissues of mammalian females.

W chromatin presents itself as an intensely staining heterochromatic body. The W chromatin body is especially conspicuous in highly polyploid tissues of Lepidoptera such as the Malpighian tubules and silk glands (Figure 3.1k–n). In species with a totally heterochromatic W, such as in *E. kuehniella*, it remains a single but increasingly larger body during the polyploidization process in most tissues. Only nurse cells and follicle cells have multiple W chromatin bodies. Species with a partly heterochromatic W, such as *O. antiqua*, or with an artificially induced W chromosome–autosome fusion, have multiple sex chromatin bodies in all polyploid tissues (Traut, Weith, and Traut 1986; Marec and Traut 1994; Traut and Marec 1997). In *E. kuehniella*, radiation-induced translocations of a Z-chromosome segment onto the W chromosome, T(W;Z) translocations, showed either malformed or fragmented sex chromatin, depending on the size of translocated segment (Marec and Traut 1994). Both the malformed and multiple sex chromatin bodies are thought to be due to the opposing tendencies of the heterochromatic W chromosomes to stick together and of euchromatic chromosomes to disperse in the nuclear space of the polyploid nucleus (Traut, Weith, and Traut 1986; Marec and Traut 1994).

RISE OF THE W CHROMOSOME IN THE LEPIDOPTERA CLADE

The closely related caddis flies, Trichoptera, share female heterogamety with Lepidoptera. This property, therefore, evolved in the common phylogenetic ancestor even before Trichoptera and Lepidoptera split, that is, more than 190 mya (Kristensen and Skalski 1999; Grimaldi and Engel 2005). Trichoptera, though, have Z/ZZ sex chromosome systems in all species investigated so far (Klingstedt 1931; Marec and Novák 1998; Lukhtanov 2000). A closer inspection of the distribution of Z/ZZ and WZ/ZZ systems in Lepidoptera shows that the Z/ZZ system is a common property of basal lineages of Lepidoptera, while it occurs only sporadically in more advanced groups (Figure 3.4). This indicates that Lepidoptera as a separate clade started evolution with a Z/ZZ sex chromosome mechanism, and the W chromosome was a later acquisition. The cladogram in Figure 3.4 reveals the W chromosome as a common trait of "advanced" Lepidoptera, the Ditrysia, which include 98 percent of the extant species of Lepidoptera and their sister clade, trumpet leaf-miner moths, Tischeriina (Traut and Marec 1996; Lukhtanov 2000). Hence, the W chromosome came into being in the common ancestor of Ditrysia and Tischeriina. In terms of time, this was between 180–190 mya, the earliest evidence of Trichoptera and Lepidoptera, and 97 mya, the earliest fossil record of a ditrysian species (Kristensen and Skalski 1999; reviewed in Grimaldi and Engel 2005), probably closer to the latter date because Adeliidae and Incruvariidae did not yet have the W chromosome (Lukhtanov 2000).

Occasionally and independently in different groups of advanced Lepidoptera the W chromosome was lost again. Those species returned secondarily to the Z/ZZ system. In other species again, the WZ/ZZ system was replaced by multiple sex chromosome systems, W_1W_2Z/ZZ and $WZ_1Z_2/Z_1Z_1Z_2Z_2$ (Figure 3.4).

There are different hypotheses on the source from which the W evolved in the common ancestor of Tischeriina and Ditrysia. A fusion of the Z chromosome with an autosome in the ancestral Z/ZZ system may have been the origin of the new WZ/ZZ system. According to this hypothesis, the new Z chromosome of advanced Lepidoptera consists partly of the ancestral Z and partly of the former autosome. The free autosome was the origin of the W chromosome (Traut and Marec 1996). Another hypothesis proposes a supernumerary chromosome (B chromosome) as the source of the W chromosome since the chromosome numbers in ditrysian and nonditrysian moth families do not readily suggest the loss of an autosome (Lukhtanov 2000). The problem will be solved when sufficient molecular data from Z chromosomes become available to compare the genetic content of Z chromosomes from ditrysian and basal nonditrysian species.

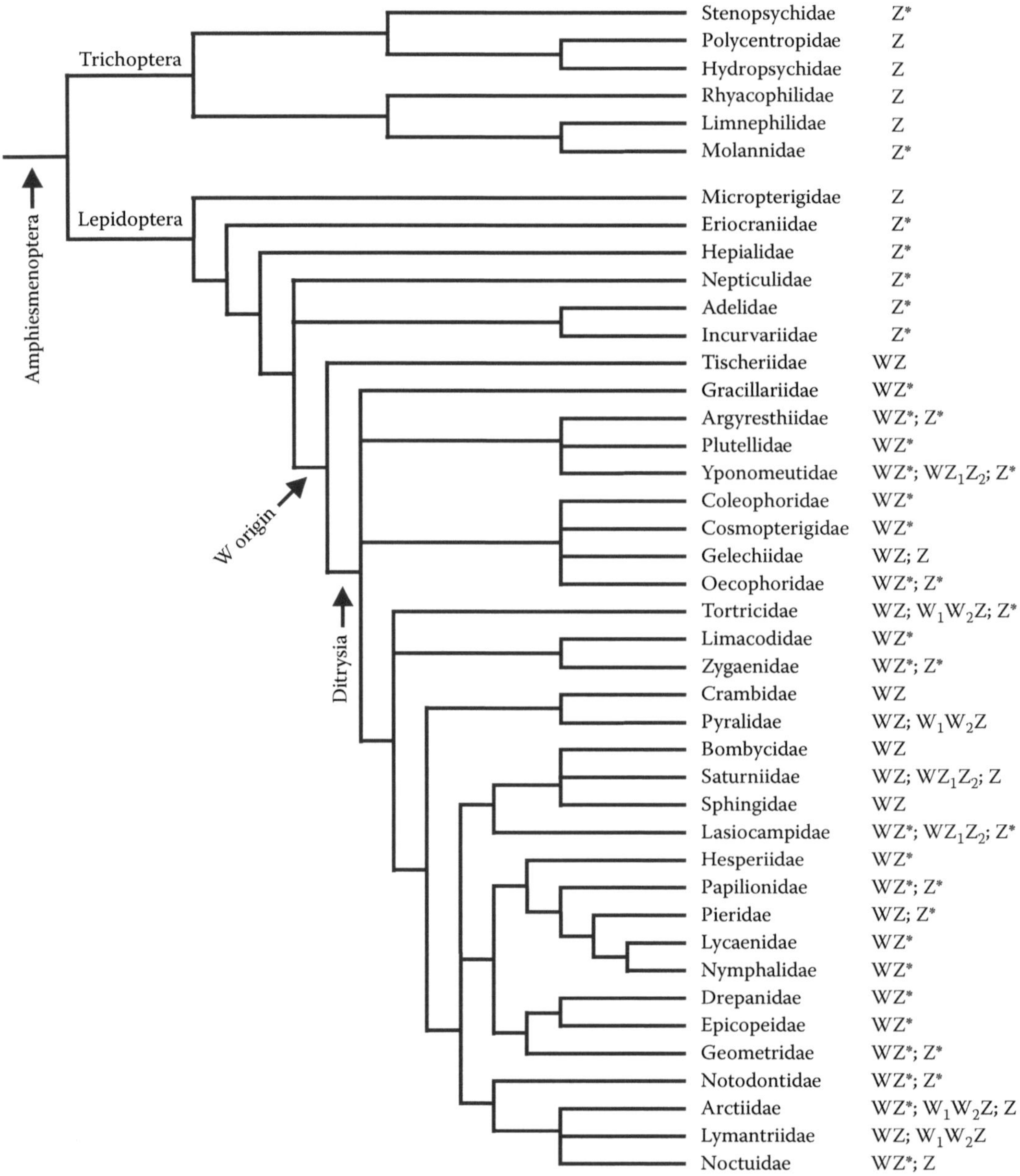

FIGURE 3.4 Phylogeny of lepidopteran sex chromosome systems. The right end of each branch lists the female sex-chromosome complements found in the respective family. Sex chromosome systems deduced from the absence or presence of the W chromatin are marked with an asterisk (*). Sex chromosome and W chromatin data from Traut and Marec (1996) and references therein, Marec and Novák (1998), Lukhtanov (2000), and Traut, Sahara, and Marec (2007) and references therein. The cladogram is based on Morse (1997) for Trichoptera and on Kristensen and Skalski (1999) for Lepidoptera.

FUNCTION IN SEX DETERMINATION

The pair of sex chromosomes, W and Z, plays an obvious role in sex determination. It produces the primary signal to start alternatively the female or male pathway of development in an embryo. But this does not tell whether the primary sex-determining genes are located on the W chromosome and/or on the Z chromosome. Cytogenetically speaking, it is either the presence/absence of the W chromosome or the number of Z chromosomes that determines female or male development.

Considering the origin of the present-day sex chromosome systems in Lepidoptera from an ancestral Z/ZZ mechanism, we expect male-promoting genes on the Z chromosomes functioning with a Z-counting mechanism as the basis of sex determination in Lepidoptera. The expectation is certainly met in species with Z/ZZ sex chromosome systems, not only in those of the basal lineages but also in those of the more advanced Lepidoptera that have lost their W chromosome. However, neither the nature of the male-promoting gene (or genes) nor its precise location has been revealed yet in any lepidopteran species.

In *B. mori*, however, presence or absence of the W chromosome determines sex. This is known from various polyploids and aneuploids, where the presence of a single W chromosome causes female development, independent of the number of Z chromosomes (Tazima 1964). The putative feminizing gene (*Fem*) has not been isolated yet nor precisely mapped to a specific location on the W chromosome. However, a series of translocations and deletions involving the W chromosome define segments of the W that do not contain *Fem* (Abe et al. 2005b; for details, see Chapter 4). Apart from this clear case, the role of the W chromosome in sex determination is uncertain because it has never been rigorously tested in polyploid and aneuploid combinations in species other than *B. mori*.

The primary sex-chromosomal signal triggers the activity of a cascade of subsequent sex-determining genes. The cascade is well known from *Drosophila*. It includes action of the *Sxl*, *tra-tra2*, and *dsx* genes, besides the primary signal. But the cascade is not fully conserved in other insects, not even in nondrosophilid flies (reviewed in Schütt and Nöthiger 2002). In *B. mori*, the *Sxl*, *tra2*, and *dsx* genes are conserved. Only the homologue of *dsx* (*Bmdsx*), however, plays a role in sex determination, while *Sxl* and probably *tra2* do not (Ohbayashi, Suzuki, and Shimada 2002; Niimi et al. 2006; Fujii and Shimada 2007; sex determination in *B. mori* is discussed in Chapter 4). Sex-determining genes in Lepidoptera species other than *B. mori* have not been investigated.

GENETIC EROSION OF THE W

Besides *Fem* in *B. mori*, few genes have been considered candidates for linkage to the W chromosome in any lepidopteran species. In *B. mori*, a gene causing giant eggs, *Esd*, was supposed to be associated with the W chromosome (Kawamura 1988, 1990), but its location has not been confirmed yet. In *E. kuehniella*, atypical transmission of two sex chromosome rearrangements in the female line only suggested the presence of a male-killing factor in the W chromosome (Marec et al. 2001). In *Papilio glaucus*, the dark morph trait was considered W-linked, but doubts have been published (Scriber, Hagen, and Lederhouse 1996; Andolfatto, Scriber, and Charlesworth 2003).

The classical way of searching for heritable traits and assigning them to specific LGs has exposed only these few putative W-linked genes. Molecular searches for genes have not been successful yet. In the Chinese oak silk moth, *A. pernyi*, the *period* gene is Z-linked, but several copies of the gene have also been detected in the W chromosome, two of them with possible functions. One, designated *perW*, encodes a truncated protein; the other produces an antisense RNA (Gotter, Levine, and Reppert 1999).

In *B. mori*, W-derived RAPDs (randomly amplified polymorphic DNAs), Lambda clones, and BACs have been isolated. They turned out to be predominantly composed of long terminal repeat (LTR) and non-LTR retrotransposons, indicating that these are the main structural components of the *B. mori* W chromosome (Abe et al. 2005a; molecular composition of the *B. mori* W chromosome is discussed in Chapter 4). Using a direct approach, Fuková et al. (2007) isolated W chromosomal

DNA by microdissection of W chromatin from the codling moth, *C. pomonella*. Besides retrotransposons, they detected noncoding sequences. Only few of the sequences were single-copy and W-specific; the majority were multicopy sequences whose predominant locations were either the autosomes, the W chromosome, or even the Z chromosome.

Although not satisfactorily complete enough to draw conclusions, the above studies show that the lepidopteran W chromosome is at least very poor in genes. Instead it has accumulated repetitive sequences. Evidently the W is eroded genetically, compared with its proposed autosomal origin. It should be dispensable if it plays no role in sex determination (see discussion above) and lacks other essential functions. Obviously this was true for species with a Z/ZZ sex chromosome system among the advanced groups of Lepidoptera (see Figure 3.4): They have indeed lost their W chromosome, thus demonstrating it had become unnecessary. But in the majority of advanced Lepidoptera (Ditrysia), the W chromosome has been retained. Hence, we suspect it contains a gene or genes for indispensable functions in these species.

The basic characteristics of W chromosomes thus are similar to those of Y chromosomes. Like many Y chromosomes, the lepidopteran W chromosome is partly or wholly heterochromatic, extremely rich in repetitive sequences, and almost devoid of genes. And like the Y in present-day XX/X systems, the W chromosome has occasionally been lost in the course of evolution. Constant heterozygosity and the inability to recombine with its chromosomal counterpart are considered the causes of this kind of genetic erosion. The mechanisms for the process, Muller's ratchet, background selection, and selective sweeps have been the subject of several articles and reviews (e.g., Charlesworth, Charlesworth, and Marais 2005). We see here that the whole evolutionary life cycle of a typical sex chromosome, from its origin through structural and molecular differentiation, rearrangements with autosomes, and eventual loss, can be studied in the insect order Lepidoptera and bring new insights into basic mechanisms underlying the evolution of sex chromosomes and sex determination.

ACKNOWLEDGMENTS

F.M. was supported by grant 206/06/1860 of the Grant Agency of the Czech Republic (Prague) and from the Entomology Institute project Z50070508. K.S. acknowledges support from the Program for Promotion of Basic Research Activities for Innovative Biosciences (PROBRAIN) and from grant 18380037 of Japan Society for the Promotion of Science (JSPS).

REFERENCES

Abe, H., K. Mita, T. Oshiki, and T. Shimada. 2005a. Retrotransposable elements on the W chromosome of the silkworm, *Bombyx mori*. *Cytogenet. Genome Res.* 110:144–51.

Abe, H., M. Seki, F. Ohbayashi, et al. 2005b. Partial deletions of the W chromosome due to reciprocal translocation in the silkworm *Bombyx mori*. *Insect Mol. Biol.* 14:339–52.

Andolfatto, P., J.M. Scriber, and B. Charlesworth. 2003. No association between mitochondrial DNA haplotypes and a female-limited mimicry phenotype in *Papilio glaucus*. *Evolution* 57:305–16.

Brown, K.S.J., A.V.L. Freitas, N. Wahlberg, B. von Schoultz, A.O. Saura, and A. Saura. 2007. Chromosomal evolution in the South American Nymphalidae. *Hereditas* 144:137–48.

Brown, K.S.J., B. von Schoultz, and E. Suomalainen. 2004. Chromosome evolution in Neotropical Danainae and Ithomiinae (Lepidoptera). *Hereditas* 141:216–36.

Carpenter, J.E., S. Bloem, and F. Marec. 2005. Inherited sterility in insects. In *Principles and practice in area-wide integrated pest management*, ed. V.A. Dyck, J. Hendrichs, and A.S. Robinson, 115–46. Dordrecht: Springer.

Charlesworth, D., B. Charlesworth, and G. Marais. 2005. Steps in the evolution of heteromorphic sex chromosomes. *Heredity* 95:118–28.

de Lesse, H. 1970. Les nombres de chromosomes dans le groupe de *Lysandra argester* et leur incidence sur la taxonomie. *Bull. Soc. Entomol. Fr.* 75:64–68.

De Prins, J., and K. Saitoh. 2003. Karyology and sex determination. In *Lepidoptera, moths and butterflies: Morphology, physiology, and development*, ed. N.P. Kristensen, 449–68. Berlin: Walter de Gruyter.

Doncaster, L. 1912. The chromosomes in the oogenesis and spermatogenesis of *Pieris brassicae*, and in the oogenesis of *Abraxas grossulariata*. *J. Genet.* 2:189–200.

Fujii, T., and T. Shimada. 2007. Sex determination in the silkworm, *Bombyx mori*: A female determinant on the W chromosome and the sex-determining gene cascade. *Sem. Cell Dev. Biol.* 18:379–88.

Fuková, I., P. Nguyen, and F. Marec. 2005. Codling moth cytogenetics: Karyotype, chromosomal location of rDNA, and molecular differentiation of sex chromosomes. *Genome* 48:1083–92.

Fuková, I., W. Traut, M. Vítková, P. Nguyen, S. Kubíčková, and F. Marec. 2007. Probing the W chromosome of the codling moth, *Cydia pomonella*, with sequences from microdissected sex chromatin. *Chromosoma* 116:135–45.

Gage, L.P. 1974. Polyploidization of the silk gland. *J. Mol. Biol.* 86:97–108.

Gassner, G., and D.J. Klemetson. 1974. A transmission electron microscope examination of hemipteran and lepidopteran gonial centromeres. *Can. J. Genet. Cytol.* 16:457–64.

Gotter, A.L., J.D. Levine, and S.M. Reppert. 1999. Sex-linked *period* genes in the silkmoth, *Antheraea pernyi*: Implications for circadian clock regulation and the evolution of sex chromosomes. *Neuron* 24:953–65.

Grimaldi, D.A., and M.S. Engel. 2005. *Evolution of the insects*. New York: Cambridge Univ. Press.

Kawaguchi, E. 1934. Zytologische Untersuchungen am Seidenspinner und seinen Verwandten. II. Spermatogenesese bei *Antheraea yamamai* Guérin, *Antheraea pernyi* Guérin, und ihrem Bastard. *Jpn. J. Genet.* 10:135–51.

Kawamura, N. 1988. The egg size determining gene, *Esd*, is a unique morphological marker on the W chromosome of *Bombyx mori*. *Genetica* 76:195–201.

Kawamura, N. 1990. Is the egg size determining gene, *Esd*, on the W chromosome identical with the sex-linked giant egg gene, *Ge*, in the silkworm? *Genetica* 81:205–10.

Klingstedt, H. 1931. Digametie bei Weibchen der Trichoptere *Limnophilus decipiens* Kol. *Acta Zool. Fennica* 10:1–69.

Kristensen, N.P., and A.W. Skalski. 1999. Phylogeny and palaeontology. In *Lepidoptera, moths and butterflies. 1. Evolution, systematics, and biogeography*, ed. N.P. Kristensen, 7–25. *Handbook of Zoologie*, Vol. 4, Part 35, *Arthropoda: Insecta*. Berlin and New York: Walter de Gruyter.

Lukhtanov, V.A. 2000. Sex chromatin and sex chromosome systems in non-ditrysian Lepidoptera (Insecta). *J. Zool. Syst. Evol. Res.* 38:73–79.

Lukhtanov, V.A., and A.D. Dantchenko. 2002. Principles of the highly ordered arrangement of metaphase I bivalents in spermatocytes of *Agrodiaetus* (Insecta, Lepidoptera). *Chromosome Res.* 10:5–20.

Lukhtanov, V.A., N.P. Kandul, J.B. Plotkin, A.V. Dantchenko, D. Haig, and N.E. Pierce. 2005. Reinforcement of pre-zygotic isolation and karyotype evolution in *Agrodiaetus* butterflies. *Nature* 436:385–89.

Lukhtanov, V.A., R. Vila, and N.P. Kandul. 2006. Rearrangement of the *Agrodiaetus dolus* species group (Lepidoptera, Lycaenidae) using a new cytological approach and molecular data. *Insect Syst. Evol.* 37:325–34.

Marec, F. 1996. Synaptonemal complexes in insects. *Int. J. Insect Morphol. Embryol.* 25:205–33.

Marec, F., and K. Novák. 1998. Absence of sex chromatin correponds with a sex-chromosome univalent in females of Trichoptera. *Eur. J. Entomol.* 95:197–209.

Marec, F., A. Tothová, K. Sahara, and W. Traut. 2001. Meiotic pairing of sex chromosome fragments and its relation to atypical transmission of a sex-linked marker in *Ephestia kuehniella* (Insecta: Lepidoptera). *Heredity* 87:659–71.

Marec, F., and W. Traut. 1994. Sex chromosome pairing and sex chromatin bodies in W-Z translocation strains of *Ephestia kuehniella* (Lepidoptera). *Genome* 37:426–35.

Mediouni, J., I. Fuková, R. Frydrychová, M.H. Dhouibi, and F. Marec. 2004. Karyotype, sex chromatin and sex chromosome differentiation in the carob moth, *Ectomyelois ceratoniae* (Lepidoptera: Pyralidae). *Caryologia* 57:184–94.

Morse, J.C. 1997. Phylogeny of Trichoptera. *Annu. Rev. Entomol.* 42:427–50.

Murakami, A., and H.T. Imai. 1974. Cytological evidence for holocentric chromosomes of the silkworms, *Bombyx mori* and *B. mandarina* (Bombycidae, Lepidoptera). *Chromosoma* 47:167–78.

Niimi, T., K. Sahara, H. Oshima, Y. Yasukochi, K. Ikeo, and W. Traut. 2006. Molecular cloning and chromosomal localization of the *Bombyx Sex-lethal* gene. *Genome* 49:263–68.

Nilsson, N.-O., C. Löfstedt, and L. Dävring. 1988. Unusual sex chromosome inheritance in six species of small ermine moths (*Yponomeuta*, Yponomeutidae, Lepidoptera). *Hereditas* 108:259–65.

Nokkala, S. 1987. Cytological characteristics of chromosome behaviour during female meiosis in *Sphinx ligustri* L. (Sphingidae, Lepidoptera). *Hereditas* 106:169–79.

Ohbayashi, F., M.G. Suzuki, and T. Shimada. 2002. Sex determination in *Bombyx mori*. *Curr. Sci.* 83:466–71.

Rasmussen, S.W. 1977. The transformation of the synaptonemal complex into the "elimination chromatin" in *Bombyx mori* oocytes. *Chromosoma* 60:205–21.

Rathjens, B. 1974. Zur funktion des W-chromatins bei *Ephestia kuehniella* (Lepidoptera): Isolierung und charakterisierung von W-chromatin-mutanten. *Chromosoma* 47:21–44.

Rishi, S., G. Sahni, and K.K. Rishi. 1999. Inheritance of unusual sex chromosome evidenced by $AA^{W}Z$ sex trivalent in *Trabala vishnu* (Lasiocampidae, Lepidoptera). *Cytobios* 100:85–94.

Robinson, R. 1971. *Lepidoptera genetics*. Oxford: Pergamon Press.

Sahara, K., F. Marec, U. Eickhoff, and W. Traut. 2003a. Moth sex chromatin probed by comparative genomic hybridization (CGH). *Genome* 46:339–42.

Sahara, K., A. Yoshido, N. Kawamura, et al. 2003b. W-derived BAC probes as a new tool for identification of the W chromosome and its aberrations in *Bombyx mori*. *Chromosoma* 112:48–55.

Sahara, K., A. Yoshido, F. Marec, et al. 2007. Conserved synteny of genes between chromosome 15 of *Bombyx mori* and a chromosome of *Manduca sexta* shown by five-color BAC-FISH. *Genome* 50:1061–65.

Schütt, C., and R. Nöthiger. 2002. Structure, function and evolution of sex-determining systems in dipteran insects. *Development* 127:667–77.

Schulz, H.-J., and W. Traut. 1979. The pachytene complement of the wildtype and a chromosome mutant strain of the flour moth, *Ephestia kuehniella* (Lepidoptera). *Genetica* 50:61–66.

Scriber, J.M., R.H. Hagen, and R.C. Lederhouse. 1996. Genetics of mimicry in the tiger swallowtail butterflies, *Papilio glaucus* and *P. canadensis* (Lepidoptera: Papilionidae). *Evolution* 50:222–36.

Seiler, J. 1914. Das Verhalten der Geschlechtschromosomen bei Lepidopteren. Nebst einem Beitrag zur Kenntnis der Eireifung, Samenreifung und Befruchtung. *Arch. Zellforsch.* 13:159–269.

Suomalainen, E. 1966. Achiasmatische Oogenese bei Trichopteren. *Chromosoma* 18:201–07.

Suomalainen, E. 1969. On the sex chromosome trivalent in some Lepidoptera females. *Chromosoma* 28:298–308.

Tazima, Y. 1964. *The genetics of the silkworm.* London: Academic Press.

Tothová, A., and F. Marec. 2001. Chromosomal principle of radiation-induced F_1 sterility in *Ephestia kuehniella* (Lepidoptera: Pyralidae). *Genome* 44:172–84.

Traut, W. 1976. Pachytene mapping in the female silkworm *Bombyx mori* L. (Lepidoptera). *Chromosoma* 58:275–84.

Traut, W. 1977. A study of recombination, formation of chiasmata and synaptonemal complexes in female and male meiosis of *Ephestia kuehniella* (Lepidoptera). *Genetica* 47:135–42.

Traut, W., and C.A. Clarke. 1997. Karyotype evolution by chromosome fusion in the moth genus *Orgyia*. *Hereditas* 126:77–84.

Traut, W., and F. Marec. 1996. Sex chromatin in Lepidoptera. *Q. Rev. Biol.* 71:239–56.

Traut, W., and F. Marec. 1997. Sex chromosome differentiation in some species of Lepidoptera (Insecta). *Chromosome Res.* 5:283–91.

Traut, W., and C. Mosbacher. 1968. Geschlechtschromatin bei Lepidopteren. *Chromosoma* 25:343–56.

Traut, W., and D. Scholz. 1978. Structure, replication and transcriptional activity of the sex-specific heterochromatin in a moth. *Exp. Cell Res.* 113:85–94.

Traut, W., U. Eickhoff, and J.-C. Schorch. 2001. Identification and analysis of sex chromosomes by comparative genomic hybridization (CGH). *Methods Cell Sci.* 23:157–63.

Traut, W., K. Sahara, and F. Marec. 2007. Sex chromosomes and sex determination in Lepidoptera. *Sex. Dev.* 1:332–46.

Traut, W., K. Sahara, T.D. Otto, and F. Marec. 1999. Molecular differentiation of sex chromosomes probed by comparative genomic hybridization. *Chromosoma* 108:173–80.

Traut, W., A. Weith, G. Traut. 1986. Structural mutants of the W chromosome in *Ephestia* (Insecta, Lepidoptera). *Genetica* 70:69–79.

Van't Hof, A.E., F. Marec, I.J. Saccheri, P.M. Brakefield, B.J. Zwaan. 2008. Cytogenetic characterization and AFLP-based genetic linkage mapping for the butterfly *Bicyclus anynana*, covering all 28 karyotyped chromosomes. *PLoS ONE* 3:e3882.

Vítková, M., I. Fuková, S. Kubíčková, and F. Marec. 2007. Molecular divergence of the W chromosomes in pyralid moths (Lepidoptera). *Chromosome Res.* 15:917–30.

Weith, A., and W. Traut. 1986. Synaptic adjustment, non-homologous pairing, and non-pairing of homologous segments in sex chromosome mutants of *Ephestia kuehniella* (Insecta, Lepidoptera). *Chromosoma* 94:125–31.

Wolf, K.W. 1996. The structure of condensed chromosomes in mitosis and meiosis of insects. *Int. J. Insect Morphol. Embryol.* 25:37–62.

Yasukochi, Y., L.A. Ashakumary, K. Baba, A. Yoshido, and K. Sahara. 2006. A second-generation integrated map of the silkworm reveals synteny and conserved gene order between lepidopteran insects. *Genetics* 173:1319–28.

Yasukochi, Y., L.A. Ashakumary, C.C. Wu, et al. 2004. Organization of the *Hox* gene cluster of the silkworm, *Bombyx mori*: A split of the *Hox* cluster in a non-*Drosophila* insect. *Dev. Genes Evol.* 214:606–14.

Yoshido, A., H. Bando, Y. Yasukochi, and K. Sahara. 2005. The *Bombyx mori* karyotype and the assignment of linkage groups. *Genetics* 170:675–85.

Yoshido, A., F. Marec, and K. Sahara. 2005. Resolution of sex chromosome constitution by genomic in situ hybridization and fluorescence in situ hybridization with $(TTAGG)_n$ telomeric probe in some species of Lepidoptera. *Chromosoma* 114:193–202.

Yoshido, A., Y. Yamada, and K. Sahara. 2006. The W chromosome detection in several lepidopteran species by genomic in situ hybridization (GISH). *J. Insect Biotech. Sericol.* 75:147–51.

4 Sex Chromosomes and Sex Determination in *Bombyx mori*

Hiroaki Abe, Tsuguru Fujii, and Toru Shimada

CONTENTS

INTRODUCTION

All insects are gonochoristic, and their sex is usually genetically determined. The genetic and molecular mechanism as to how sex is determined has been discussed in detail for *Drosophila* (Schütt and Nöthiger 2000), and recently, sex-determining genes have been identified in the honeybee *Apis mellifera* (Hasselmann et al. 2008).

Many researchers have long since made superfluous efforts to solve the mechanism of sex determination in the silkworm *Bombyx mori*. Sex strongly affects the efficiency of cocoon production, and therefore, an artificial sex control method is desirable. However, the mechanism of sex determination remains unclear. The most upstream factor for sex determination is located on the female-specific W chromosome, which determines the female sex autonomously acting through several genes on autosomes in somatic tissues. In this chapter, we first review the history of research on the W chromosome. Later, we introduce the function of Z chromosome–linked genes as well as autosomal sex-determining genes and discuss briefly the sex-determination pathway in germ line cells.

TRANSLOCATIONS BETWEEN THE W CHROMOSOME AND AUTOSOMES

The silkworm *B. mori* is an economically important insect for silk production. Study on sex determination in this insect has been associated with the requirements of the sericultural industry. In 1906, Kametaro Toyama, one of the founders of silkworm genetics, recognized the positive effect of heterosis in hybrids between Japanese and Thai strains (Toyama 1906; cited in Tazima 1964). Based on Toyama's suggestions, the Japanese government decided to enforce rearing of hybrid silkworms in sericultural farms and began distributing hybrid eggs in 1914. Since then, almost all silkworm strains reared in sericultural farms in Japan have been F_1 hybrids. To produce F_1 hybrids, males and females must be separated at or before the pupal stage; therefore, it becomes important to determine the sex of the silkworm for egg producers. Sex determination was usually carried out by professional "discriminators" through inspection of Ishiwata's germinal discs in the female and Herald's gland in the male on the postventral surface of the final instar larva. However, this procedure needs skillful specialists and is impractical for mass rearing. Therefore, a different and more convenient method of sex discrimination was desired.

Tanaka (1916) revealed that the sex of *B. mori* is determined by sex chromosomes using sex-linked (Z-linked) genes. Females are WZ (equivalent to XY) and males are ZZ (XX; the sex chromosome system with the heterogametic female sex and its variation in Lepidoptera is reviewed in Chapter 3). However, it was not clear which of the two sex chromosomes, Z or W, plays the principal role in sex determination. Subsequently, through experiments using polyploidy states, Hashimoto (1933) found that the W chromosome has a strong positive female-determining gene that directs sex determination irrespective of the number of autosomes or Z chromosomes present. Polyploid individuals, for example, WZZ and WWZZ, are phenotypically normal females. Therefore, Hashimoto (1933) postulated that femaleness is determined by the presence of a single W chromosome; the existence and function of the W chromosome was almost certain. However, at that time, there was no way to prove the existence of the W chromosome genetically because, despite the fact that some morphological gene loci had been found on the Z chromosome, no genes for morphological traits had been found on the W chromosome. Moreover, recombination (crossing over) is restricted to males in *B. mori* (Maeda 1939), and there was no cytological method to identify the W chromosome.

Yataro Tazima believed, "If any one of the autosomes which carry dominant genes for noticeable traits is translocated to the W chromosome artificially, discriminating the sex of the silkworm becomes very easy" (Tazima 2001). As a result, the effort to establish translocations between the W chromosome and autosomes began.

In the course of investigating chromosomal aberrations induced by X-ray irradiation, Y. Tazima had obtained a strain in which the two copies of chromosome 2 were fused (marked by $+^p$ and p^{Sa} for normal marking and sable marking, respectively). In 1941, Tazima discovered a translocation between the W chromosome and the compound form of chromosome 2 ($+^p$ p^{Sa}). In a single batch isolated in this study, all females displayed $+^p$ p^{Sa} traits and all males were *p* (plain). There were no $+^p$ p^{Sa} males. Believing that the compound chromosome ($+^p$ p^{Sa}) in this batch had fused to the W chromosome, Y. Tazima tested and verified his belief by crossing this female with a *p* male and observing the segregation pattern in the next generation. He named the strain having the W chromosome–autosome translocation as the W-PSa strain (technically called T(W;2)p^{Sa} $+^p$; Tazima 1941, 1944). As a result, Y. Tazima verified Hashimoto's view and concluded that femaleness is determined by the presence or absence of the W chromosome in *B. mori*.

Encouraged by Tazima's successful experimentation, several researchers undertook the challenge of inducing new translocations between the W chromosome and an autosome carrying dominant alleles of genes with noticeable traits (egg color, larval marking, and cocoon or blood color). As discussed below, several W-translocated strains, for example, T(W;10)+$^{w\text{-}2}$, T(W;2)p^{Sa}, T(W;2)p^{B}, and T(W;2)*Y*, were obtained. The main focus of the analysis and modification of the W chromosome was to enable practical use of the resulting strains. Therefore, molecular biological studies on the W chromosome were not conducted for many years. However, the W-translocated strains have unexpectedly proven to be very useful for molecular analysis of the W chromosome.

A study using a genetic mosaic strain (Goldschmidt and Katsuki 1931) is an example of fruitful analyses of sex determination. This mosaic strain has been used to study topics such as developmental biology and pathology (Mine et al. 1983; Shimada, Ebinuma, and Kobayashi 1986; Abe, Kobayashi, and Watanabe 1990). Using the *mo* gene with T(W;2)p^{Sa}, T(W;2)p^{B}, or T(W;10)+$^{w\text{-}2}$, sex mosaics that contain distinct boundaries between female and male tissues can be obtained at the embryonic, larval, and adult stages. Morphological and biochemical studies using these sex mosaics strongly suggest the absence of diffusible substances such as sex hormones in *B. mori* (Mine et al. 1983; Fujii and Shimada 2007).

An interesting feature of sex chromosomes in *B. mori* is that there are many functional genes on the Z chromosome (Koike et al. 2003), whereas the W chromosome is devoid of functional genes except for the putative *Fem* gene, which determines femaleness. Another W chromosome–specific property is that it does not recombine with the Z chromosome or autosomes. Without recombination, the W chromosome remains static, undergoing no changes during transmission from mother to daughter. Therefore, the structure of the W chromosome should be unusually stable.

MOLECULAR AND GENOMIC ANALYSIS OF THE W CHROMOSOME

Studies on the W chromosome focused on its modification to enable practical use of the resulting strains. Molecular biological studies on the W chromosome were not conducted for many years even after it became easy to analyze genomic DNA. Genome studies of *B. mori* were initiated in the 1990s. Promboon et al. (1995) constructed a linkage map of random amplified polymorphic DNAs (RAPDs), and Yasukochi (1998) established a denser genetic map based on 1018 molecular markers. However, no W-specific RAPD markers were obtained in these two linkage studies. To initiate molecular biological analysis of the W chromosome, Abe et al. (1995) attempted to obtain RAPD markers on the W chromosome using a large set of arbitrary 10-mer primers and obtained twelve female-specific RAPD markers (W-Kabuki, W-Kamikaze, W-Samurai, BMC1-Kabuki, W-Rikishi, W-Musashi, W-Sasuke, W-Sakura, W-Yukemuri-L, W-Yukemuri-S, W-Bonsai, and W-Mikan; Abe et al. 1998a; 2005b). Next, they obtained two lambda phage clones containing the W-Kabuki and W-Samurai RAPD sequences. Surprisingly, the DNA sequences of W-specific RAPD markers and phage clones comprised a nested structure of many retrotransposable elements (Abe et al. 1998a, 1998b, 2005b; Ohbayashi et al. 1998). From these results, it was predicted that the W chromosome is composed entirely of repetitive sequences. In addition, it was expected that, owing to the WZ sex chromosome pair, if female genomic DNA was used for WGS analysis, the assembly of sequence data would become very complicated and difficult. Therefore, in both Japan and China, the genomic DNA of the male (ZZ) was used for WGS sequencing (Mita et al. 2004; Xia et al. 2004). For this reason, the W chromosome has become a frontier in genome studies in *B. mori*.

The deduced amino acid sequences of eleven of the twelve W-specific RAPD markers show similarity to amino acid sequences of many retrotransposable elements from various organisms (Abe et al. 1998a; 2005b). Moreover, almost all amino acid sequences of W-specific RAPD markers contain boundaries of two or three retrotransposable elements (Abe et al. 1998a; 2005b).

COMPARISON OF ABERRANT W CHROMOSOMES USING MOLECULAR MARKERS

T(W;3)*Ze* Chromosome (Sex-Limited Zebra)

Several researchers undertook the challenge of inducing new translocations between the W chromosome and an autosome carrying dominant alleles of genes with visible traits. Hashimoto (1948) isolated a translocation of a chromosome 3 fragment bearing the *Ze* (*Zebra*) marker to the W chromosome after X-ray irradiation. In the resulting strain with the T(W;3)*Ze* chromosome, female larvae had zebra markings, whereas male larvae had a whitish skin. The W chromosome region of T(W;3)*Ze* chromosome lacked two of the twelve W-specific RAPD markers (W-Mikan and W-Samurai; Abe et al. 2005b; Figure 4.1).

T(W;10)$+^{w\text{-}2}$ Chromosome (Sex-Limited Black Egg)

The abundance of egg color (serosa color) is a unique feature of the silkworm. Egg color results from pigments produced in cells of the serosa, a single-layer membrane that covers the embryo. The cells of the serosa are derived from cleavage nuclei so that the genotype of serosa cells is identical to that of the embryo. Normal eggs have dark brown (black) pigmentation, whereas one of the mutants, *w-2* (*white* 2 gene), has yellowish-white pigmentation. Tazima, Harada, and Ohta (1951) isolated a translocation of a chromosome 10 fragment bearing the $+^{w\text{-}2}$ gene to the W chromosome (T(W;10)$+^{w\text{-}2}$) after X-irradiation. In this strain, white eggs are males of the Z/Z, *w-2/w-2* genotype, and black eggs are females of the Z/T(W;10)$+^{w\text{-}2}$, *w-2/w-2* genotype. The W chromosome region of the T(W;10)$+^{w\text{-}2}$ chromosome lacked one (W-Mikan) of the twelve W-specific RAPD markers (Abe et al. 2005b; Figure 4.2).

T(W;2)*Y* Chromosome (Sex-Limited Yellow Cocoon)

Many cocoon color mutants have been maintained in Japan. A white (colorless) cocoon is conventionally considered to be normal since almost all silkworms reared in Japan for industrial purposes have white cocoons. However, during the long history of sericulture, strains with other cocoon colors have been observed and maintained, including yellow, golden yellow, pinkish, and green. Yellow and pinkish colors are attributed to the presence of carotenoids, carotenes, and xanthophylls derived from mulberry leaves. Green color is due to flavonoids, and yellow-blooded larvae are generally also yellow cocoon spinners (Doira 1978). *Y* is an allele of the *yellow blood* gene (*Y*, 2–25.6). In larvae with the *Y* allele, the hemolymph is deep yellow because carotenoids from mulberry pass through the digestive organs.

To establish a sex-limited yellow cocoon strain, Kimura, Harada, and Aoki (1971) used γ-rays or X-rays to irradiate 707 female pupae of a strain with a normal W chromosome and with the *Y* allele, over the course of nine years (from 1961 to 1969). In these experiments, they isolated a translocation of a chromosome 2 fragment bearing the *Y* allele to the W chromosome, and obtained a sex-limited yellow cocoon strain (T(W;2)*Y*) in which females have yellow cocoons (Z/T(W;2)*Y*, $+^{Y}/+^{Y}$) and males have white cocoons (Z/Z, $+^{Y}/+^{Y}$). During subsequent breeding processes aimed at the commercial use of this strain, however, several physiological defects (less healthy, a lighter cocoon shell, and proctocele larvae) were recognized in females containing the T(W;2)*Y* chromosome (Niino et al. 1987). These defects were thought to be caused by the extra portion of the chromosome 2 fragment of T(W;2)*Y*. To eliminate them, Niino et al. (1988) irradiated this strain with γ-rays and isolated a derivative in which non-*Y* portions of the translocated chromosome 2 fragment were deleted. Thus, the T(W;2)*Y* chromosome had again been modified by irradiation, and it seems likely that several silkworm strains containing differently modified T(W;2)*Y* chromosomes were distributed to researchers.

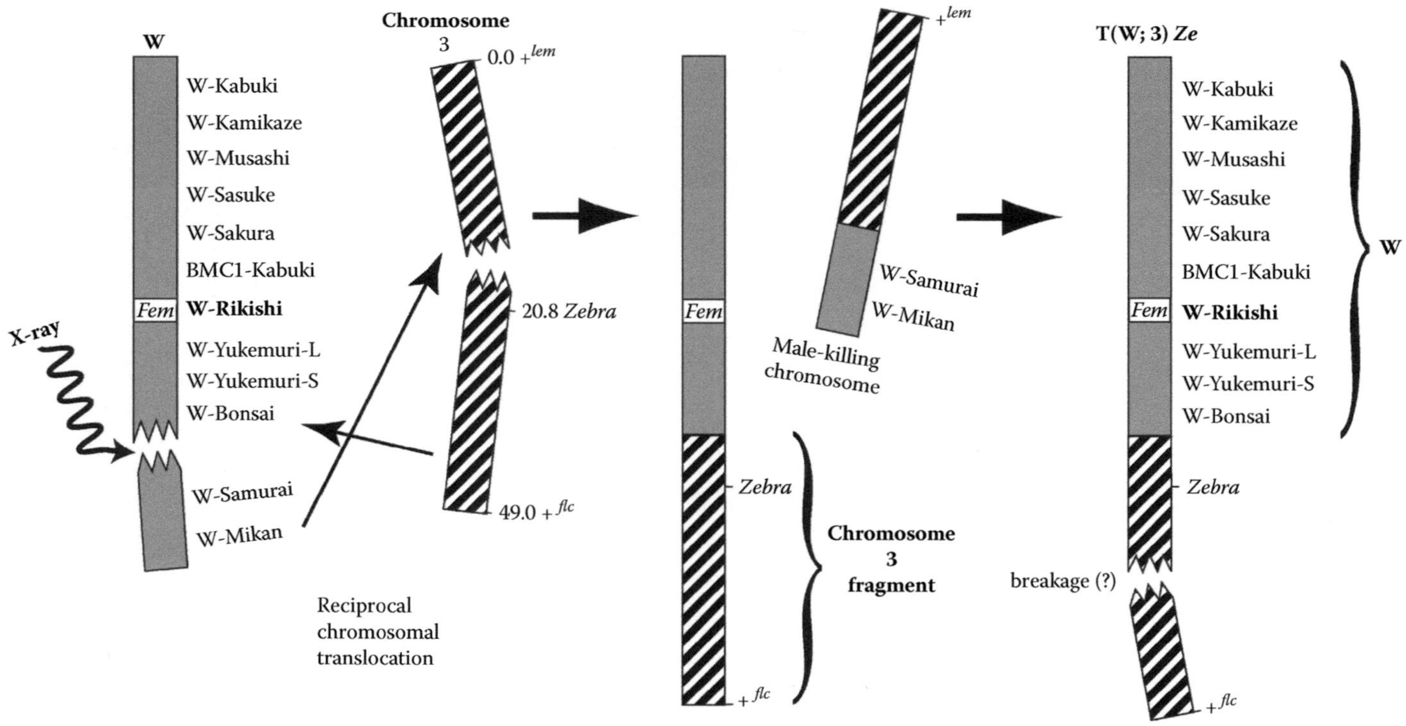

FIGURE 4.1 Schematic representation of the X-ray–induced reciprocal translocation between the W chromosome and chromosome 3 (*Zebra*), resulting in a male-killing chromosome and the T(W;3)*Ze* chromosome of the sex-limited Zebra-W strain. The male-killing chromosome was tolerated by Z/T(W;3) Ze females but caused the death of Z/Z male embryos. It was suggested that the chromosome carries a translocated fragment of the W chromosome, which contains a factor incompatible with male development (Hashimoto 1953). Unfortunately, the male-killing chromosome was lost before a detailed analysis could be performed (Y. Tazima, pers. comm.).

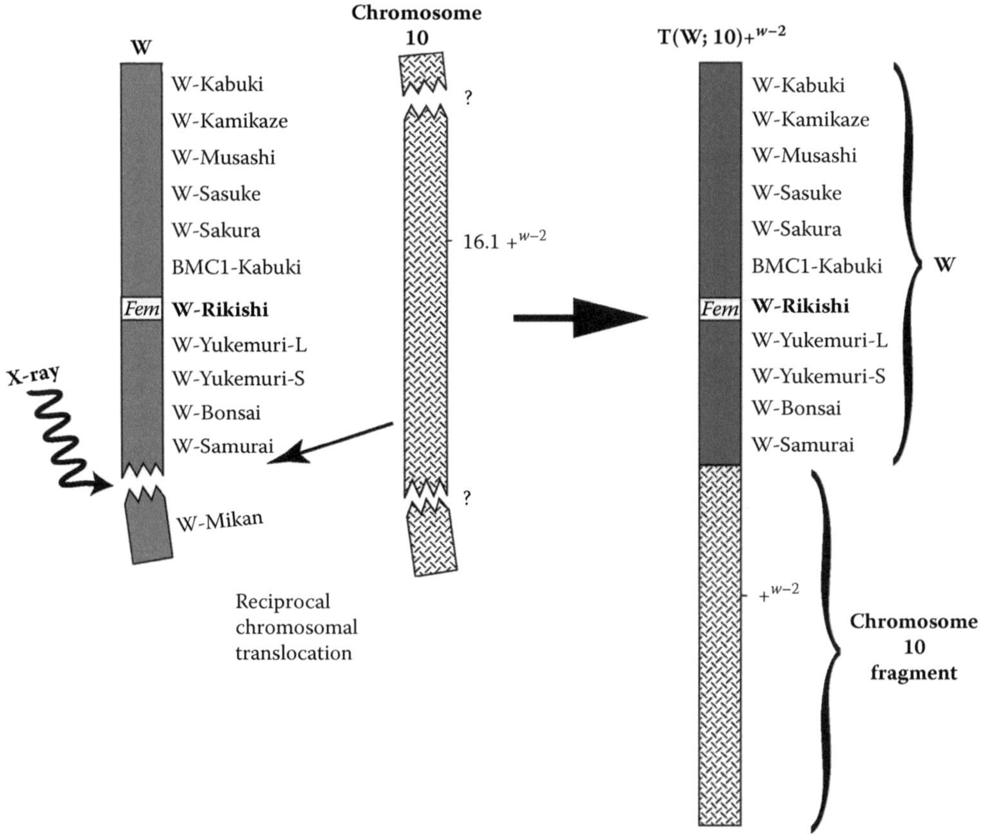

FIGURE 4.2 Schematic representation of the X-ray-induced reciprocal translocation between the W ($+^{w\text{-}2}$) chromosome and chromosome 10, resulting in the T(W;10)$+^{w\text{-}2}$ chromosome of the sex-limited Black-egg-W strain.

Strains each containing one of the three T(W;2)*Y* chromosomes—T(W;2)*Y*-Abe, T(W;2)*Y*-Chu, and T(W;2)*Y*-Ban—were maintained by three independent research groups. Abe et al. (2008) determined the presence or absence of W-specific RAPD markers on these three T(W;2)*Y* chromosomes. T(W;2)*Y*-Chu contained six of the twelve W-specific RAPD markers (W-Rikishi, W-Yukemuri-L, W-Yukemuri-S, W-Bonsai, W-Samurai, and W-Mikan). However, both T(W;2)*Y*-Abe and T(W;2)*Y*-Ban contained only one of the twelve W-specific RAPD markers (W-Rikishi; Figure 4.3).

p^{Sa} $+^{p}$ W $+^{od}$ and Df(p^{Sa} $+^{p}$ W $+^{od}$)*Fem* (Female-Killing) Chromosomes

Tazima (1948) obtained a complex translocation, p^{Sa} $+^{p}$ W $+^{od}$, which was presumably the product of a connection between the short fragment of the Z chromosome that includes the $+^{od}$ gene (*od*; distinct translucent, 1(Z chromosome)–49.6) and the p^{Sa} $+^{p}$ W chromosome. In a series of dissociation experiments using the p^{Sa} $+^{p}$ W $+^{od}$ chromosome, he found an exceptional individual that was phenotypically male and marked with $+^{p}$, p^{Sa}, and $+^{od}$ among the progeny of an X-ray–irradiated Z/p^{Sa} $+^{p}$ W $+^{od}$ female. Tazima (1952) presumed that this exceptional male had a mutant W chromosome that was generated by deletion of the female-determining region of the p^{Sa} $+^{p}$ W $+^{od}$ chromosome. For convenience, the mutant W chromosome was designated Df(p^{Sa} $+^{p}$ W $+^{od}$) *Fem* by Fujii et al. (2006), who then analyzed the genetic behavior of the Df(p^{Sa} $+^{p}$ W $+^{od}$) *Fem* chromosome

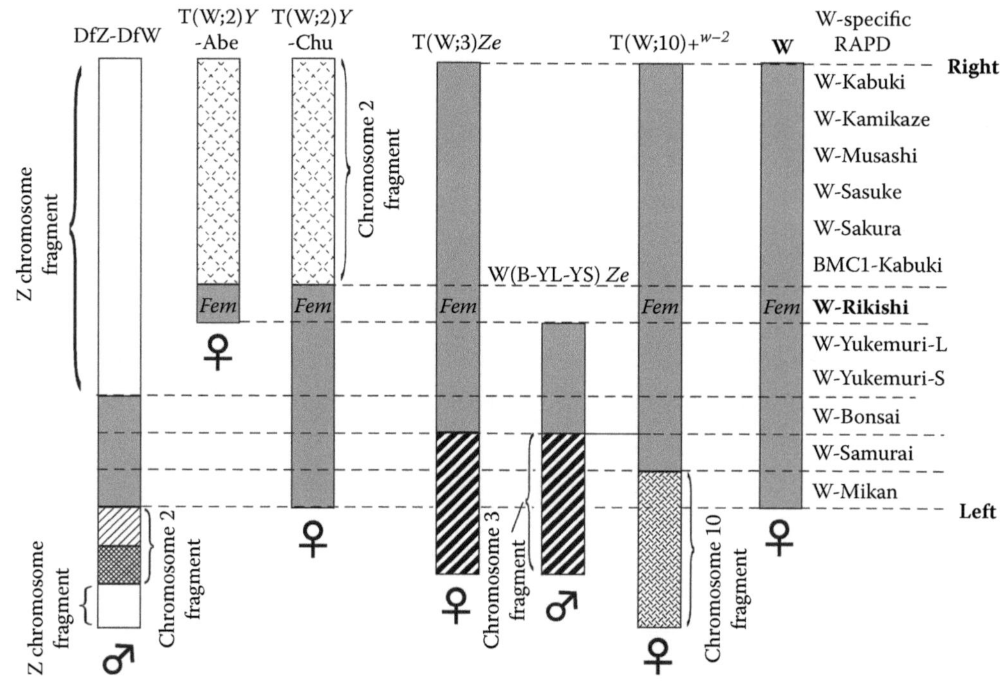

FIGURE 4.7 High-resolution mapping of W-specific RAPD markers and the putative *Fem* gene on the W chromosome using W chromosome variants. The order of six (W-Kabuki, W-Kamikaze, W-Musashi, W-Sasuke, W-Sakura, and BMC1-Kabuki) and two (W-Yukemuri-L and W-Yukemuri-S) W-specific RAPD markers could not be determined. Mapping of the other markers is shown. For details on T(W;3)*Ze*, T(W;10)+$^{w-2}$, and DfZ-DfW chromosomes, see Abe et al. (2005b) and Fujii et al. (2006).

two other *B. mori* BAC libraries were constructed using genomic DNA of the p50 and C108 strains (Wu et al. 1999). We used the W-specific RAPD marker sequence to identify W-specific BAC clones and subjected them to shotgun sequencing. However, due to the presence of many repetitive DNA elements, particularly the non-LTR retrotransposons BMC1 (Abe et al. 1998b) and *Kendo* (Abe et al. 2002), the assembly of the shotgun clone sequence proved difficult and did not result in a single contig. Nevertheless, subsections of the clone sequences could be assembled into contigs and have been analyzed in detail (Abe et al. 2005a). As shown in Figure 4.8, the structural features of the W chromosome of *B. mori* are quite different from those typical of other chromosomes. Many LTR and non-LTR retrotransposons, retroposons (Bm1; Adams et al. 1986), DNA transposons, and their derivatives have accumulated on the W chromosome. Combining the DNA sequence data of W-specific RAPD markers and BAC clones (Abe et al. 1998a, 2005a, 2005b) with the results of Abe et al. (2008), the W chromosome is thought to be a huge, functionless graveyard except for the region containing the putative *Fem* gene.

STRUCTURAL ANALYSES OF ABERRANT Z CHROMOSOMES USING PCR MARKERS

In 2004, a Japanese group and a Chinese group published their *Bombyx* WGS sequence data independently (Mita et al. 2004; Xia et al. 2004). Through the integration of both WGS data, five scaffolds composed of 20.4-Mb sequences were assigned to the Z chromosome (International Silkworm

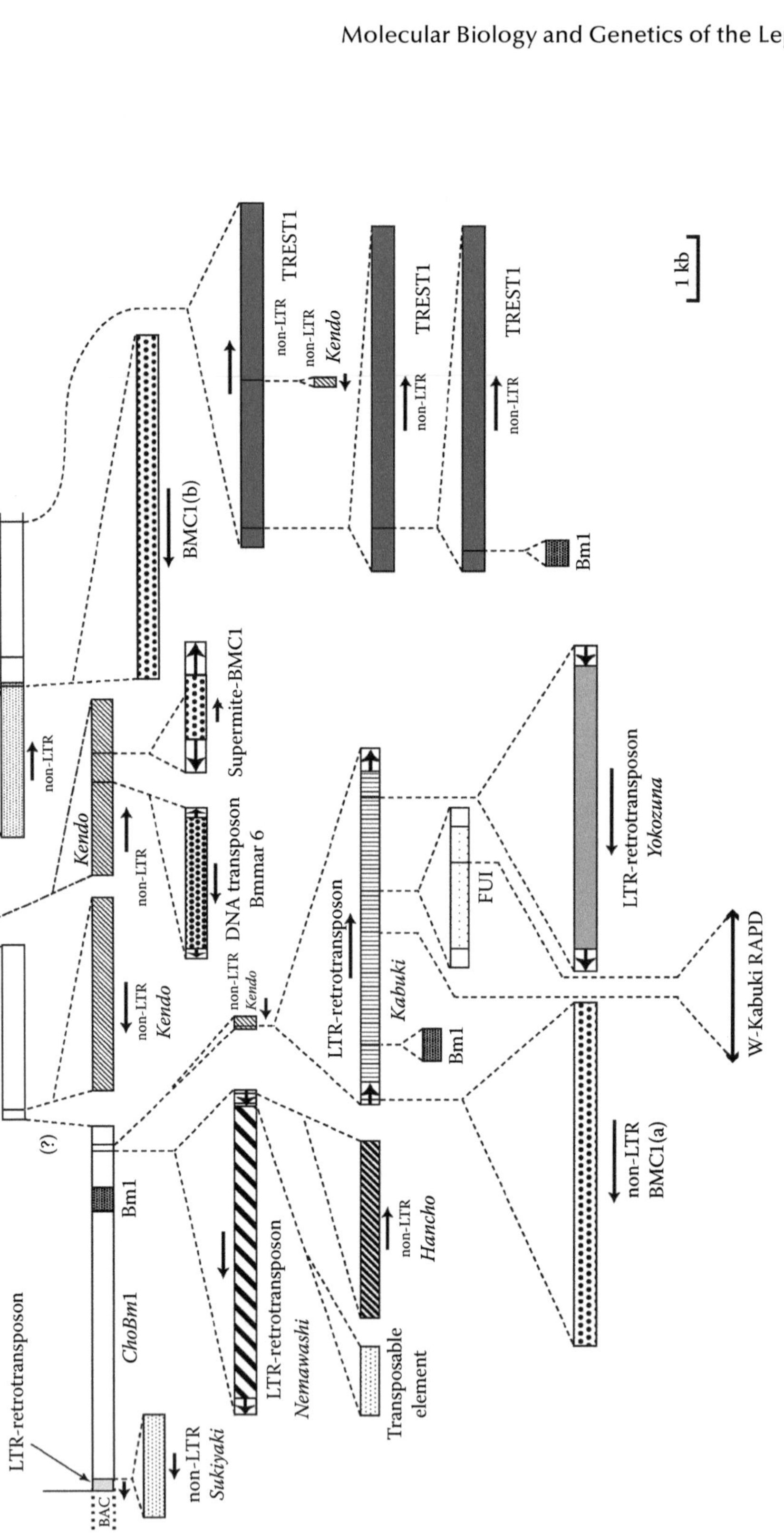

FIGURE 4.8 The presence of transposable elements on the W chromosome sequence. Nested transposable elements were found in the bacterial artificial chromosome clone that includes the W-Kabuki RAPD marker sequence. This figure is based on recent DNA sequence information. See Abe et al. (2005a) for details.

Genome Consortium 2008). These sequence data (build2) are very useful for the structural analysis of the aberrant Z chromosome.

As mentioned above, during the experiments to generate W chromosome aberrations, we obtained two Z chromosome fragments, Z_1 and Z_2. As a result, it was possible to analyze the region around the breakage points of Z_1 and Z_2. Therefore, we simultaneously analyzed Z^{Vg} and $+^{od}\, p^{Sa}\, +^{p}$ W chromosomes, and one of the previously obtained Z chromosome aberrations (Fujii et al. 2007).

Z_1 and Z_2 ($Ze^{W}Z_2$) Chromosomes

Earlier, we described the isolation of X-ray–induced deleted Z chromosomes designated Z_1 and $Z_2(Ze^{W}Z_2)$, which were generated by the breakage events between the *sch* (1–21.5) and $+^{od}$ (1–49.6) loci (Figure 4.6). Subsequently, we obtained embryonic lethal eggs with Z_1/W. Through PCR analysis using many markers for the Z chromosome based on the build2 sequence data, we found that the Z_1 chromosome lacks a 6-Mb portion including the proximal end (Fujii et al. 2007).

We analyzed the genetic behavior of the Z_1 and $Ze^{W}Z_2$ fragments during meiosis in males (Z_1 $Ze^{W}Z_2$/Z) and females (Z_1 $Ze^{W}Z_2$/W). The Z_1 fragment and Z or W chromosomes segregated properly, but nondisjunction was observed between the $Ze^{W}Z_2$ fragment and either the Z or W chromosome. Moreover, the 2A:Z/Z_1 males that were the products of nondisjunction between $Ze^{W}Z_2$ and W had observable phenotypic defects. Namely, Z/Z_1 male moths could walk and copulate with females but they could not flap their wings. To analyze why the Z/Z_1 males could not flap their wings, we looked at their indirect flight muscles, which we found were damaged or deteriorated (Fujii et al. 2007). This flapless phenotype and deterioration of flight muscles were not caused by Z_1, because Z/Z_1 $Ze^{W}Z_2$ males could flap their wings vigorously. This result indicates that a muscle-related, dose-sensitive gene is located on the deleted region of the Z_1 chromosome (Fujii et al. 2007).

Z^{Vg} Chromosome

Vg is an X-ray–induced dominant gene that results in the vestigial phenotype of the adult wings. Individuals of $+^{Vg}$ *od*/*Vg* always exhibit the *od* phenotype because the Z chromosome containing the *Vg* gene (Z^{Vg} chromosome) lacks the *od* locus (Tazima 1944). The fact that Z/Z_1 males cannot flap their wings but Z/Z^{Vg} males can indicates that the deleted region of the Z^{Vg} chromosome is smaller than that of the Z_1 chromosome. Through PCR analysis, we found that the Z^{Vg} chromosome lacks an interstitial region of 1.5 Mb. Moreover, we succeeded in obtaining a fragment containing the breakpoint junction (Fujii et al. 2008). Sequencing of the fragment enabled us to determine precisely the proximal and distal breakpoints. *Vg* is a dominant mutation. Therefore, we thought that a gene related to wing development would be disrupted by an interstitial deletion. However, no hypothetical genes of the build2 sequence data span the proximal and distal breakpoints. On the other hand, we found that a hypothetical gene located nearest to the distal breakpoint is a homologue of *apterous* (*ap*), a *Drosophila* LIM-HD family member that plays a critical role in the development of wings. *ap* expression is restricted to the dorsal compartment of the wing disc and is required for dorsal cell identity (Cohen et al. 1992; Diaz-Benjumea and Cohen 1993). Ectopic expression of *ap* causes severe wing defects (Rincón-Limas et al. 2000). It is probable that the expression of the *ap* homologue of *B. mori* is disrupted by the 1.5 Mb interstitial deletion located upstream of this gene.

SYNTENY OF THE Z CHROMOSOME AMONG LEPIDOPTERAN SPECIES

Yasukochi et al. (2006) revealed that both the *triosephosphate isomerase* and *ap* homologue genes are located on Z chromosomes not only in *Bombyx* but also in a butterfly, *Heliconius melpomene*, suggesting a chromosome-wide synteny among macrolepidopteran conserved species.

It is known that the Z chromosome controls voltinism (diapause) and photoperiodism (Morohoshi 1980), and the sex-linked *Lm* (*Late maturity*) locus is strongly associated with voltinism. That diapause is controlled by the Z chromosome has been reported in other lepidopteran species including *Samia cynthia* (Saito 1994) and *Antheraea pernyi* (Shimada, Yamauchi, and Kobayashi 1988). Interestingly, three genes for biological clock, *BmClk* (orthologous to *Drosophila clock*), *Bmper* (*period*), and *BmCyc* (*cycle*) are also located on the Z chromosome (http://sgp.dna.affrc.go.jp/KAIKObase/; T. Shimada et al., unpublished). Only one component of the central clock system, *Bmtim* (*timeless*), is on an autosome. In *A. pernyi*, the *period* gene is also located on the Z chromosome but has multiple incomplete copies on the W chromosome. The Z-linkage of the *period* gene may explain the sexual difference of the emergence timing and behaviors in adults as discussed by Gotter, Levine, and Reppert (1999). The present genome assembly may further explain sex-dependent functions of the biological clock in lepidopteran insects (for the detailed mechanism of circadian timing in Lepidoptera, see Chapter 8).

DOSAGE COMPENSATION

Animals in which the male is heterogametic (XY) or hemizygous (XO), such as *Drosophila* and *C. elegans*, have a dosage compensation mechanism to equalize the amount of transcripts on the X chromosome in male (single X) and female (two Xs) cells (Straub and Becker 2007). It is known that all mutations that affect dosage compensation are lethal in *Drosophila*, indicating the importance of dosage compensation (Bhadra et al. 2005). On the other hand, dosage compensation of the Z chromosome has not been observed in *B. mori*. For example, it was found that there is approximately twice as much mRNA from the Z-linked genes *T15.180a* and *Bmkettin* in males than in females (Suzuki, Shimada, and Kobayashi 1998, 1999). Furthermore, Koike et al. (2003) quantified mRNAs of thirteen Z-linked genes near the *Bmkettin* locus and found that ten of them were more abundantly expressed in males than in females. These results indicate the absence of general dosage compensation, though there may be a small number of genes that are dosage compensated.

In *Drosophila*, at least five protein-coding genes (*mof*, *mle*, *msl-1*, *msl-2*, and *msl-3*) are needed (Dahlsveen et al. 2006; Gilfillan et al. 2006). Screening of *Bombyx* EST and genomic sequences revealed putative orthologs of *mof*, *mle*, and *msl-3*. The *mof* gene encodes a histone acetyltransferase required for chromatin remodeling in *Drosophila*, and *mle* encodes a helicase family protein. In contrast to *mof*, *mle*, and *msl-3* genes, *msl-1* and *msl-2* appear to be absent from the *Bombyx* genome. The *Drosophila msl-2* mRNA precursor undergoes female-specific splicing with the help of the TRA (Transformer) protein and is functional only in males. *Drosophila* Msl proteins form a complex known as the MSL complex with two noncoding RNAs, roX1 and roX2. The fact that *Bombyx* lacks the homologues of *msl-1*, *msl-2*, *roX1*, and *roX2* would explain the lack of chromosome-wide dosage compensation of Z-linked genes in *Bombyx*.

Why does *Bombyx* lack a mechanism for chromosome-wide dosage compensation? The Z chromosome of *Bombyx* carries approximately eight hundred genes. It is known that the Z chromosome has several genes which control functions important for masculinization. A typical case is the gene encoding the receptor for the female sex pheromone bombycol, which is located on the Z chromosome (Sakurai et al. 2004) and expressed only in the antennae of adult males. As discussed by Koike et al. (2003), the Z chromosome carries many genes for components of indirect flight muscles, including *Bmkettin* and *Bmtitin*. Since male moths are much more active in flapping and moving, it is reasonable that the muscle protein genes are expressed in a male-biased manner.

ROLE OF *BMDSX*, HOMOLOGUE TO *DROSOPHILA DOUBLESEX*, IN SEX DETERMINATION

As described above, sex determination in *Bombyx* depends primarily on the presence or absence of the W chromosome. The W chromosome initiates a cascade of events that leads to the female mode of development both in somatic and germ line cells. However, until recent years, we did not know the downstream mechanism in the sex-determining cascade. One of the downstream genes, *B. mori doublesex* (*Bmdsx*), was found to play an important role in sex determination. *Bmdsx* is an ortholog of *Drosophila doublesex* (*dsx*), a key gene acting at the downstream end of the *Drosophila* sex differentiation cascade (Saccone et al. 2002).

In *Drosophila*, the primary signal in sex determination is the X/A ratio, the numeric balance of the X chromosome and autosomes. The number of X chromosomes controls the expression of *Sex lethal* (*Sxl*), the master gene for sex determination in early embryogenesis. In later embryonic and postembryonic stages, *Sxl* is regulated by a male-specific splicing inhibition depending on the SXL protein. As a result, the SXL protein is synthesized only in females. *Drosophila* SXL regulates the splicing of *Sxl* and *transformer* (*tra*). The functional TRA protein, which is produced only in females, leads to female-specific splicing of *dsx*, whereas in males the absence of functional TRA generates the male-specific (default) splicing of *dsx*. The female and male DSX proteins that result from sex-specific splicing regulate transcription of the genes required for female and male differentiation, respectively (Yamamoto et al. 1998). In the honeybee, *Apis mellifera*, the sex-determining region of a chromosome contains *complementary sex determination* (*csd*) and another gene, both of which encode TRA-like SR-type proteins (Hasselmann et al. 2008). This suggests that *dsx* in the honeybee is also regulated by TRA-like splicing activator(s), although the interaction between the SR-type proteins and the honeybee *dsx* has not yet been reported.

In *Bombyx*, *Bmdsx* is a single-copy gene on chromosome 25 which is regulated by sex-specific splicing similar to *dsx* in *Drosophila*. Different mRNA forms of *Bmdsx* are expressed in both females and males in this species (Ohbayashi et al. 2001). Using HeLa cell extracts, Suzuki et al. (2001) showed that the default mode of splicing of *Bmdsx* produces the female form (Figure 4.9). This indicates that the factor controlling sex-specific splicing of *Bmdsx* is not a splicing activator like TRA but some splicing repressor. Clearly the mechanism for regulating the splicing of *dsx* genes in *Drosophila* and *Bombyx* is different.

BmDSX protein contains a zinc-fingerlike DNA-binding domain. To understand the function of *Bmdsx*, we need to know the targets of the BmDSX protein. Suzuki et al. (2003) found that transgenic male silkworms carrying the female-type *Bmdsx* mRNA express a small amount of vitellogenin mRNA, which is not detectable in normal males. In contrast, in transgenic female silkworms that express the male-type *Bmdsx* mRNA, expression of vitellogenin mRNA is relatively repressed, expression of pheromone-binding protein mRNA is increased, and abnormal morphology of several genital organs can be observed (Suzuki et al. 2003, 2005). These results strongly suggest that *Bmdsx* affects not only the *vitellogenin* gene but also the genes that encode pheromone-binding proteins. Suzuki et al. (2003) found an element in the promoter region of the *vitellogenin* gene which is a putative binding site of the BmDSX protein based on experimental data that show a molecular interaction between the element and the female and male forms of BmDSX proteins. They also observed morphological abnormalities in the genital organs of *Bmdsx* transgenic moths, indicating that *Bmdsx* controls genes for genital morphogenesis. Additional genome-wide surveys of BmDSX targets will be required to understand fully the function of *Bmdsx*.

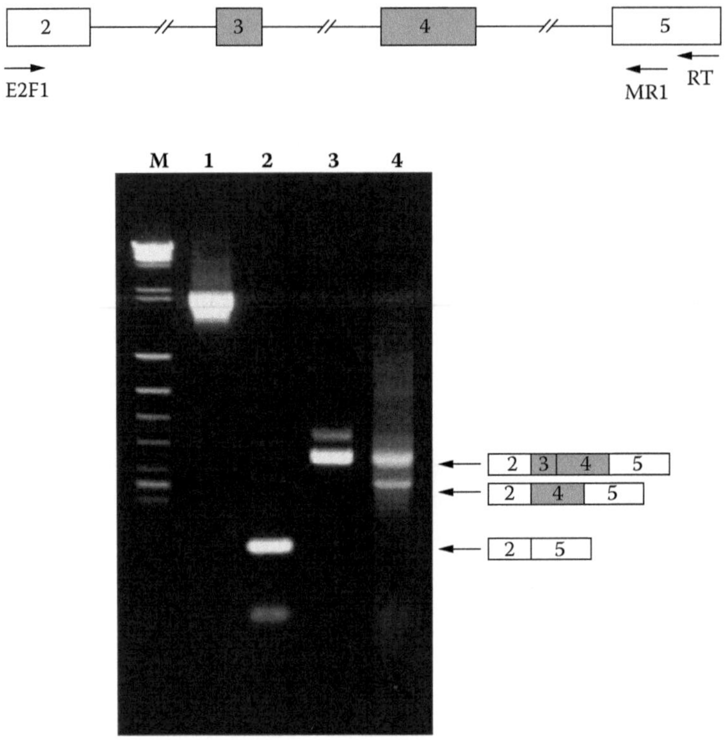

FIGURE 4.9 In vitro splicing products of *Bmdsx* pre-mRNAs. Pre-mRNA derived from a minigene construct containing exons 2–5 of *Bmdsx* was spliced in a reaction with (lane 4) or without (lane 1) HeLa nuclear extract. The resulting transcripts were reverse-transcribed with an RT primer, and the cDNAs were PCR-amplified with primers E2F1 and MR1. The same RT-PCR reactions were carried out using poly(A)+RNA extracted from testes (lane 2) and ovaries (lane 3) as templates. Arrows indicate the position and direction of primers using the RT-PCR reactions. M represents the DNA size marker. Cited from Suzuki et al. (2001).

Recently, Suzuki et al. (2008) discovered that a *cis* element (CE1) in exon 4 of the *Bmdsx* gene is essential for male-specific splicing. Furthermore, they found a protein interacting with CE1, BmPSI, the ortholog of the *Drosophila* P-element somatic inhibitor (PSI), which contains a KH domain. An RNAi experiment with the dsRNA of *Bmpsi* induced an increase of female-type *Bmdsx* mRNA in cultured cells derived from males (ZZ). They concluded that BmPSI is the male-specific splicing inhibitor of *Bmdsx*, that is, the direct upstream factor. The *Bmpsi* gene is not located on the sex chromosome but on an autosome, chromosome 20. Thus, the mechanism linking the W chromosome with *Bmpsi* in the sex determination cascade is still unknown (Figure 4.10).

OTHER AUTOSOMAL SEX-DETERMINING GENES

The sequence of *Sxl* is conserved in dipteran and lepidopteran insects, including *B. mori* (Niimi 2006; Traut et al. 2006). However, sex-dependent expression is not observed in *Sxl* homologues in *B. mori*. As described below, the *B. mori* genome does not carry a gene homologous to *tra*, the target gene of SXL. Further, the sex-specific splicing of *Bmdsx* is not regulated by TRA but by BmPSI, a very different protein as explained above. Considering this evidence, it is probable that *Sxl* is not a sex determinant in *B. mori*.

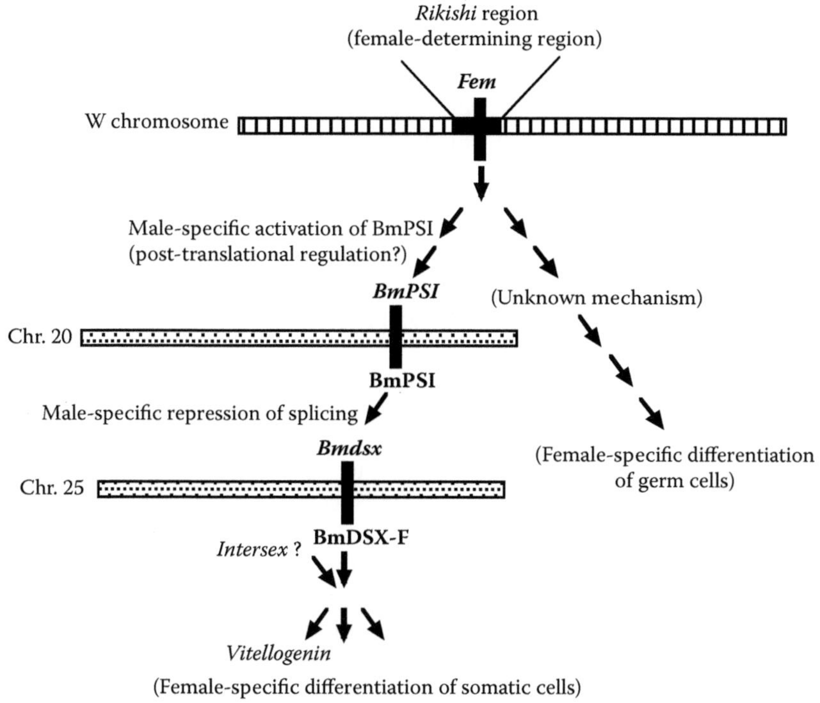

FIGURE 4.10 Hypothetical mechanism underlying the determination of femaleness by the W chromosome in somatic and germ line cells. *Fem* determines the sex in both somatic and germ line cells. The BmPSI-*Bmdsx* system regulates sexual differentiation in somatic cells depending on an unknown signal from the W chromosome.

Suzuki et al. (2005) showed that the ectopic expression of the female-type *Bmdsx* transcript induced only partial sex reversal in transgenic animals. In particular, the female-type of *Bmdsx* induced only a small amount of vitellogenin mRNA (Suzuki et al. 2005). Therefore, we can speculate that another molecule cooperates with BmDSX to bring about the female expression of genes such as *vitellogenin*. In *Drosophila*, the female-type DSX protein has been reported to interact with a partner protein encoded by the *intersex* (*ix*) gene, which activates transcription of the yolk protein gene (*Yp*; Garrett-Engele 2002). The WGS assembly of *Bombyx* contains a homologue of *ix* (Mita et al. 2004; Siegal and Baker 2005). It is possible that this gene interacts with BmDSX to bring about female-specific regulation of vitellogenin and other genes, though no such evidence has been found to date.

WGS sequences for *Bombyx* (Mita et al. 2004; Xia et al. 2004) do not contain a homologue of *tra*, which encodes an essential regulator of *Drosophila dsx*, though a *tra-2* (cofactor of *tra*) homologue was found in the *Bombyx* genome. This is consistent with the hypothesis that *Bmdsx* is not regulated by a splicing activator but by the splicing repressor, BmPSI (Suzuki et al. 2001; Suzuki et al. 2008).

A transgenic experiment (Suzuki et al. 2005) demonstrated that the development of female germ line cells in the ovary was basically unaffected by the female-type transgene of *Bmdsx*. This result suggests that the sex of the germ line cells is determined in a manner different from that of somatic cells in *Drosophila*, where control of germ line sex determination is also very different from that of somatic cells. Several genes including *Sxl*, *bam*, *otu*, *ovo*, and *snf* regulate sex determination of the germ line in this species (Casper and Van Doren 2006). Nevertheless, although orthologs of *Sxl*, *otu*, *ovo*, and *snf* can be found in *Bombyx* WGS contig sequences (Mita et al. 2004), a recognizable ortholog of *bam*, which is an important gene for determination of female germ cells in *Drosophila*, is absent in *Bombyx*. This suggests that the sexual differentiation pathway in *Bombyx* germ cells is different from that in *Drosophila*. Because various polyploids and deletions/translocations of the W chromosome always determine the sex of somatic and germ cells simultaneously, the unidentified genes for sex determination in germ cells are certainly also under control of the W chromosome, as are the sex-determining genes in somatic cells.

ANALYSIS OF THE FUNCTION OF SEX CHROMOSOMES AND *WOLBACHIA* INFECTION IN SEX DETERMINATION OF OTHER LEPIDOPTERAN INSECTS

As described above, it is essential to examine the translocations for the study of sex chromosomes to uncover the functions encoded in different regions/segments. In general, however, it is not easy to isolate chromosomal translocations in lepidopteran insects because their chromosomes are usually small and numerous and cannot be distinguished from each other. Moreover, few useful genetic markers are available for identifying translocations when they occur. However, in a pyralid moth, *Ephestia kuehniella*, a number of sex-chromosome aberrations including translocations were isolated and used in subsequent studies (Traut and Rathjens 1973; Marec and Mirchi 1990; see below). Similar to *B. mori*, the chromosome constitution of *E. kuehniella* is WZ for females and ZZ for males. The role of Z and W chromosomes in sex determination in *E. kuehniella* is not clear. Using a light microscope, W-Z bivalents can be recognized in the pachytene stage by the presence of a heterochromatic thread, which is a feature of the *E. kuehniella* W chromosome (Traut and Rathjens 1973). A sex chromatin body (SB), deduced to consist of condensed copies of the W chromosome, has been observed in the highly polyploid interphase nuclei of lepidopteran females (Ennis 1976; Traut and Marec 1996). In *E. kuehniella*, Traut and Rathjens (1973) found that fusion of the W chromosome with an autosome is accompanied by fragmentation of the otherwise cohesive SB, composed of heterochromatin. Therefore, SB morphology was successfully used as a cytogenetic marker to isolate T(W;A) chromosomes induced by γ-irradiation (Traut and Rathjens 1973; Rathjens 1974). The structure of each T(W;A) chromosome was analyzed in oocytes at the pachytene stage (Traut, Weith, and Traut 1986). Moreover, some T(W;A) chromosomes have been characterized using autosomal marker genes located on the translocated autosome (Traut, Weith, and Traut 1986).

With the aim to isolate a translocation of the Z chromosome onto the W chromosome, that is, the T(W;Z) chromosome, in *E. kuehniella*, Marec and Mirchi (1990) crossed γ-ray irradiated Z^{dz+}/W females to *dz*/*dz* (dark central area of the forewings) males and screened for Z^{dz+} females, which are exceptions to the normal pattern of sex-linked inheritance in this species. From Z^{dz+} females, they isolated twelve sex chromosome mutant strains with Z chromosomes bearing additional sequences which included a $+^{dz}$ gene (Z^{dz+} segment) and revealed that Z^{dz+} segments were translocated onto the W chromosome in four of the twelve lines. The four T(W;Z) translocations were experimentally confirmed in Marec and Traut (1994). In next four of the twelve lines, the Z^{dz+} segment fused with a fragment of the W chromosome as shown in Marec et al. (2001). Interestingly, males of two lines with the Z^{dz+} segment containing a piece of the W chromosome do not survive embryogenesis, whereas males in the other lines carrying the separate Z^{dz+} chromosomes with or without a piece of the W chromosome are viable. Based on these results, Marec et al. (2001) suggested that a part of the *E. kuehniella* W chromosome carries/contains a male-killing factor that interacts with the normal sex determination pathway.

It is well known that symbiotic (or parasitic) bacteria *Wolbachia* affect sex determination and reproduction in various insects and arthropods. In Lepidoptera, *Wolbachia* infection in crambid moths, *Ostrinia furnacalis* and *O. scapulalis*, causes male-specific lethality (Kageyama and Traut 2004; Sakamoto et al. 2007), whereas in the sulfur butterfly, *Eurema hecabe*, it changes the sex of genetically male (ZZ) individuals to females (Hiroki et al. 2002). This means that *Wolbachia* cross-talks with the sex-determining pathway through an unknown mechanism. The molecular analysis of *Wolbachia*-induced reproductive manipulation may be a useful method for studies on sex determination as well as the sex-determining role of sex chromosomes in Lepidoptera.

CONCLUSION AND PROSPECTS

The female/male sex chromosome constitutions in lepidopteran insects are basically WZ/ZZ or Z/ZZ depending on the species and strains (Traut, Sahara, and Marec 2007). All strains in the domesticated silkworm *B. mori* and its wild species *B. mandarina* possess WZ/WW sex chromosomes. Among W chromosomes of lepidopteran species, that of *Bombyx* is the most thoroughly analyzed. However, nobody has successfully identified the female determinant. The WGS analysis of *B. mori* is complete, and the build2 assembly of the genome can be accessed at http://sgp.dna.affrc.go.jp/KAIKObase/. However, this analysis was performed using male genomic DNA, and chromosome-wide sequencing of the W chromosome has not yet been carried out. Thus far, it is known that the W chromosome is composed of transposable elements and no protein-coding genes have been found (see Chapter 3 for a review on the W chromosome structure, composition, and occurrence in other Lepidoptera). Whether or not the female determinant is an unknown protein-coding gene will be resolved by an in-depth analysis of the W chromosome and sex determination mechanisms. This feature of the W chromosome remains the biggest frontier in the genomics of the silkworm and lepidopterans.

REFERENCES

Abe, H., T. Fujii, N. Tanaka, et al. 2008. Identification of the female-determining region of the W chromosome in *Bombyx mori*. *Genetica*. 133:269–82.

Abe, H., M. Kanehara, T. Terada, et al. 1998a. Identification of novel random amplified polymorphic DNAs (RAPDs) on the W chromosome of the domesticated silkworm, *Bombyx mori*, and the wild silkworm, *B. mandarina*, and their retrotransposable element-related nucleotide sequences. *Genes Genet. Syst.* 73:243–54.

Abe, H., M. Kobayashi, and H. Watanabe. 1990. Mosaic infection with a densonucleosis virus in the midgut epithelium of the silkworm, *Bombyx mori*. *J. Invertebr. Pathol.* 55:112–17.

Abe, H., K. Mita, Y. Yasukochi, T. Oshiki, and T. Shimada. 2005a. Retrotransposable elements on the W chromosome of the silkworm *Bombyx mori*. *Cytogenet. Genome Res.* 110:144–51.

Abe, H., F. Ohbayashi, T. Shimada, T. Sugasaki, S. Kawai, and T. Oshiki. 1998b. A complete full-length non-LTR retrotransposon, BMC1, on the W chromosome of the silkworm, *Bombyx mori*. *Genes Genet. Syst.* 73:353–58.

Abe, H., F. Ohbayashi, T. Shimada, et al. 2000. Molecular structure of a novel *gypsy*-Ty3-like retrotransposon (*Kabuki*) and nested retrotransposable elements on the W chromosome of the silkworm *Bombyx mori*. *Mol. Gen. Genet.* 263:916–24.

Abe, H., M. Seki, F. Ohbayashi, et al. 2005b. Partial deletions of the W chromosome due to reciprocal translocation in the silkworm *Bombyx mori*. *Insect Mol. Biol.* 14:339–52.

Abe, H., T. Shimada, T. Yokoyama, T. Oshiki, and M. Kobayashi. 1995. Identification of random amplified polymorphic DNA on the W chromosome of the Chinese 137 strain of the silkworm, *Bombyx mori* (in Japanese with English summary). *J. Seric. Sci. Jpn.* 64:19–22.

Abe, H., T. Sugasaki, T. Terada, et al. 2002. Nested retrotransposons on the W chromosome of the wild silkworm *Bombyx mandarina*. *Insect Mol. Biol.* 11:307–14.

Adams, D.S., T.H. Eickbush, R.J. Herrera, and P.M. Lizardi. 1986. A highly reiterated family of transcribed oligo(A)-terminated, interspersed DNA elements in the genome of *Bombyx mori*. *J. Mol. Biol.* 187:465–78.

Bhadra, M.P., U. Bhadra, J. Kundu, and J.A. Birchler. 2005. Gene expression analysis of the function of the male-specific lethal complex in *Drosophila*. *Genetics* 169:2061–74.

Casper, A., and M. Van Doren. 2006. The control of sexual identity in the *Drosophila* germline. *Development* 133:2783–91.

Cohen, B., M.E. McGuffin, C. Pfeifle, D. Segal, and S.M. Cohen. 1992. *apterous*, a gene required for imaginal disc development in *Drosophila* encodes a member of the LIM family of developmental regulatory proteins. *Genes Dev.* 6:715–29.

Dahlsveen, I.K., G.D. Gilfillan, V.I. Shelest, et al. 2006. Targeting determinants of dosage compensation in *Drosophila*. *PLoS Genet.* 2:e5.

Diaz-Benjumea, F.J., and S.M. Cohen. 1993. Interaction between dorsal and ventral cells in the imaginal disc directs wing development in *Drosophila*. *Cell* 75:741–52.

Doira, H. 1978. Genetic stocks of the silkworm. In *The silkworm, an important laboratory tool*, ed. Y. Tazima, 53–81. Tokyo: Kodansha Ltd.

Ennis, T.J. 1976. Sex chromatin and chromosome numbers in Lepidoptera. *Can. J. Cytol.* 18:119–30.

Fujii, T., H. Abe, S. Katsuma, et al. 2008. Mapping of sex-linked genes onto the genome sequence using various aberrations of the Z chromosome in *Bombyx mori*. *Insect Biochem. Mol. Biol.* 38:1072–79.

Fujii, T., and T. Shimada. 2007. Sex determination in the silkworm, *Bombyx mori*: A female determinant on the W chromosome and the sex-determining gene cascade. *Semin. Cell Dev. Biol.* 18:379–88.

Fujii, T., N. Tanaka, T. Yokoyama, et al. 2006. The female-killing chromosome of the silkworm, *Bombyx mori*, was generated by translocation between the Z and W chromosomes. *Genetica* 127:253–65.

Fujii, T., T. Yokoyama, O. Ninagi, et al. 2007. Isolation and characterization of sex chromosome rearrangements generating male muscle dystrophy and female abnormal oogenesis in the silkworm, *Bombyx mori*. *Genetica* 130:267–80.

Garrett-Engele, C.M., M.L. Siegal, D.S. Manoli, et al. 2002. *intersex*, a gene required for female sexual development in *Drosophila*, is expressed in both sexes and functions together with *doublesex* to regulate terminal differentiation. *Development* 129:4661–75.

Gilfillan, G.D., T. Straub, E. de Wit, F. Greil, et al. 2006. Chromosome-wide gene-specific targeting of the *Drosophila* dosage compensation complex. *Genes Dev.* 290:858–70.

Goldschmidt, R., and K. Katsuki. 1931. Vierte Mitteilung über erblichen Gynandromorphism von *Bombyx mori* L. *Biol. Zbl.* 48:685–99.

Goldsmith, M.R. 1995. Genetics of the silkworm: Revisiting an ancient model system. In *Molecular model systems in the lepidoptera*, ed. M.R. Goldsmith and A.S. Wilkins, 21–76. New York: Cambridge Univ. Press.

Goldsmith, M.R., T. Shimada, and H. Abe. 2005. The genetics and genomics of the silkworm, *Bombyx mori*. *Annu. Rev. Entomol.* 50:71–100.

Gotter, A.L., J.D. Levine, and S.M. Reppert. 1999. Sex-linked *period* genes in the silkmoth, *Antheraea pernyi*: Implications for circadian clock regulation and the evolution of sex chromosomes. *Neuron* 24:953–65.

Hashimoto, H. 1933. The role of the W-chromosome in the sex determination of *Bombyx mori* (in Japanese). *Jpn. J. Genet.* 8:245–47.

Hashimoto, H. 1948. Sex-limited zebra, an X-ray mutation in the silkworm (in Japanese with English summary). *J. Seric. Sci. Jpn.* 16:62–64.

Hashimoto, H. 1953. Genetical studies of *Bombyx mori* L. on the lethal gene which affects the male (in Japanese with English summary). *J. Seric. Sci. Jpn.* 22:200–204.

Hasselmann, M., T. Gempe, M. Schiøtt, C.G. Nunes-Silva, M. Otte, and M. Beye. 2008. Evidence for the evolutionary nascence of a novel sex determination pathway in honeybees. *Nature* 454(7203):519–22.

Hiroki. M., Y. Kato, T. Kamito, and K. Miura. 2002. Feminization of genetic males by a symbiotic bacterium in a butterfly, *Eurema hecabe* (Lepidoptera: Pieridae). *Naturwissenschaften* 89:167–170.

International Silkworm Genome Consortium. 2008. The genome of a lepidopteran model insect, the silkworm *Bombyx mori*. *Insect Biochem. Mol. Biol.* 38:1036–45.

Kageyama, D., and W. Traut. 2004. Opposite sex-specific effects of *Wolbachia* and interference with the sex determination of its host *Ostrinia scapulalis*. *Proc. Biol. Sci.* 271:251–58.

Kimura, K., C. Harada, and H. Aoki. 1971. Studies on the W-translocation of yellow blood gene in the silkworm (*Bombyx mori*) (in Japanese with English summary). *Jpn. J. Breed.* 21:199–203.

Koike, Y., K. Mita, M.G. Suzuki, et al. 2003. Genomic sequence of a 320-kb segment of the Z chromosome of *Bombyx mori* containing a *kettin* ortholog. *Mol. Genet. Genomics* 269:137–49.

Maeda, T. 1939. Chiasma studies in the silkworm *Bombyx mori*. *Jap. J. Genet.* 15:118–27.

Marec, F., and R. Mirchi. 1990. Genetic control of the pest Lepidoptera: Gamma-ray induction of translocations between sex chromosomes of *Ephestia kuehniella* Zeller (Lepidoptera: Pyralidae). *J. Stored Prod. Res.* 26:109–16.

Marec, F., A. Tothová, K. Sahara, and W. Traut. 2001. Meiotic pairing of sex chromosome fragments and its relation to atypical transmission of a sex-linked marker in *Ephestia kuehniella* (Insecta: Lepidoptera). *Heredity* 87:659–71.

Marec, F., and W. Traut. 1994. Sex chromosome pairing and sex chromatin bodies in W-Z translocation strains of *Ephestia kuehniella* (Lepidoptera). *Genome* 37:426–35.

Mine, E., S. Izumi, M. Katsuki, and S. Tomino. 1983. Developmental and sex-dependent regulation of storage protein synthesis in the silkworm, *Bombyx mori*. *Dev. Biol.* 97:329–37.

Mita, K., M. Kasahara, S. Sasaki, et al. 2004. The genome sequence of silkworm, *Bombyx mori*. *DNA Res.* 11:27–35.

Morohoshi, S. 1980. The control of growth and development in *Bombyx mori*. XLI: Control of hormonal antagonistic balance regarding insect development by brain hormone. *Proc. Jpn. Acad. Ser. B*: 56:200–05.

Niimi, T., K. Sahara, H. Oshima, et al. 2006. Molecular cloning and chromosomal localization of the *Bombyx Sex-lethal* gene. *Genome* 49:263–68.

Niino, T., R. Eguchi, A. Shimazaki, and A. Shibukawa. 1988. Breakage by γ-rays of the *+i-lem* locus on the translocated 2nd chromosome in the sex-limited yellow cocoon silkworm. *J. Seric. Sci. Jpn.* 57:75–76.

Niino, T., T. Kanda, R. Eguchi, et al. 1987. Defects and structure of translocated chromosome in the sex-limited yellow cocoon strain of the silkworm, *Bombyx mori* (in Japanese with English summary). *J. Seric. Sci. Jpn.* 56:240–46.

Ohbayashi, F., T. Shimada, T. Sugasaki, et al. 1998. Molecular structure of the *copia*-like retrotransposable element *Yokozuna* on the W chromosome of the silkworm, *Bombyx mori*. *Genes Genet. Syst.* 73:345–52.

Ohbayashi, F.M., G. Suzuki, K. Mita, et al. 2001. A homologue of the *Drosophila doublesex* gene is transcribed into sex-specific mRNA isoforms in the silkworm, *Bombyx mori*. *Comp Biochem. Physiol. B Biochem. Mol. Biol.* 128:145–58.

Osanai, M., H. Takahashi, K.K. Kojima, et al. 2004. Essential motifs in the 3′ untranslated region required for retrotransposition and the precise start of reverse transcription in non-long-terminal-repeat retrotransposon SART1. *Mol. Cell. Biol.* 24:7902–13.

Promboon, A., T. Shimada, H. Fujiwara, and M. Kobayashi. 1995. Linkage map of random amplified polymorphic DNAs (RAPDs) in the silkworm, *Bombyx mori*. *Genet. Res.* 66:1–7.

Rathjens, B. 1974. Zur Funktion des W-Chromatins bei *Ephestia kuehniella* (Lepidoptera): Isolierung und Charaktesisierung von W-Chromatin-Mutanten. *Chromosoma* 47:21–44.

Rincón-Limas, D.E., C.-H. Lu, I. Canal, and J. Botas. 2000. The level of DLDB/CHIP controls the activity of the LIM-homeodomain protein Apterous: Evidence for a functional tetramer complex *in vivo*. *EMBO J.* 19:2602–14.

Saccone, G., A. Pane, and L.C. Polito. 2002. Sex determination in flies, fruitflies and butterflies. *Genetica* 116:15–23.

Sakamoto, H., D. Kageyama, S. Hoshizaki, and Y. Ishikawa. 2007. Sex-specific death in the Asian corn borer moth (*Ostrinia furnacalis*) infected with *Wolbachia* occours across larval development. *Genome* 50:645–52.

Sakurai, T., T. Nakagawa, H. Mitsuno, et al. 2004. Identification and functional characterization of a sex pheromone receptor in the silkmoth *Bombyx mori*. *Proc. Natl. Acad. Sci. U.S.A.* 101:16653–58.

Saito, H. 1994. Diapause response in *Samia cynthia* subspecies and their hybrids (Lepidoptera: Saturniidae). *Appl. Entomol. Zool.* 29:296–98.

Schütt, C., and R. Nöthiger. 2000. Structure, function and evolution of sex-determining systems in Dipteran insects. *Development* 127:667–77.

Shimada, T., H. Ebinuma, and M. Kobayashi. 1986. Expression of homeotic genes in *Bombyx mori* estimated from asymmetry of dorsal closure in mutant/normal mosaics. *J. Exp. Zool.* 240:335–42.

Shimada, T., H. Yamauchi, and M. Kobayashi. 1988. Diapause of the inter-specific F1 hybrids between *Antheraea yamamai* (Guerin-Meneville) and *A. pernyi* (G.-M.) (Lepidoptera: Saturniidae). *Jpn. J. Appl. Entomol. Zool.* 32:120–25 (In Japanese with English summary).

Siegal, M.L., and B.S. Baker. 2005. Functional conservation and divergence of *intersex*, a gene required for female differentiation in *Drosophila melanogaster. Dev. Genes Evol.* 215:1–12.

Straub, T., and P.B. Becker. 2007. Dosage compensation: the beginning and end of generalization. *Nat. Rev. Genet.* 8:47–57.

Suzuki, M.G., S. Funaguma, T. Kanda, et al. 2003. Analysis of the biological functions of a *doublesex* homologue in *Bombyx mori. Dev. Genes Evol.* 213:345–54.

Suzuki, M.G., S. Funaguma, T. Kanda, et al. 2005. Role of the male BmDSX protein in the sexual differentiation of *Bombyx mori. Evol. Dev.* 7:58–68.

Suzuki, M.G., S. Imanishi, N. Dohmae, et al. 2008. Establishment of a novel in vivo sex-specific splicing assay system to identify a *trans*-acting factor that negatively regulates splicing of *Bombyx* female exons. *Mol. Cell. Biol.* 28:333–43.

Suzuki, M.G., F. Ohbayashi, K. Mita, and T. Shimada. 2001. The mechanism of sex-specific splicing at the *doublesex* gene is different between *Drosophila melanogaster* and *Bombyx mori. Insect Biochem. Mol. Biol.* 31:1201–11.

Suzuki, M.G., T. Shimada, and M. Kobayashi. 1998. Absence of dosage compensation at the transcription level of a sex-linked gene in a female heterogametic insect, *Bombyx mori. Heredity* 81:275–83.

Suzuki, M.G., T. Shimada, and M. Kobayashi. 1999. *Bm kettin*, homologue of the *Drosophila kettin* gene, is located on the Z chromosome in *Bombyx mori* and is not dosage compensated. *Heredity* 82:170–79.

Tanaka, Y. 1916. Genetic studies in the silkworm. *J. Coll. Agric. Sapporo* 6:1–33.

Tazima, Y. 1941. A simple method of sex discrimination by means of larval markings in *Bombyx mori* (in Japanese). *J. Seric. Sci. Jpn.* 12:184–88.

Tazima, Y. 1944. Studies on chromosome aberrations in the silkworm. II. Translocation involving second and W-chromosomes (in Japanese with English summary). *Bull. Seric. Exp. Stn.* 12:109–81.

Tazima, Y. 1948. Translocation of the Z chromosome to the W chromosome of the silkworm, *Bombyx mori.* In *Oguma commemoration volume on cytology and genetics* (in Japanese), 88–99.

Tazima, Y. 1952. Inheritance of sex (in Japanese). In *Silkworm genetics*, ed. Y. Tanaka, 351–72. Tokyo: Shokabo.

Tazima, Y. 1964. *The genetics of the silkworm.* London: Academic Press.

Tazima, Y. 2001. *Improvement of biological functions in the silkworm.* Enfield, NH: Science Publishers, Inc. (Translated from Japanese).

Tazima, Y., C. Harada, and N. Ohta. 1951. On the sex discriminating method by colouring genes of silkworm eggs. I. Induction of translocation between the W and the tenth chromosomes (in Japanese with English summary). *Jpn. J. Breed.* 1:47–50.

Traut, W., and F. Marec. 1996. Sex chromatin in Lepidoptera. *Q. Rev. Biol.* 71:239–56.

Traut, W., T. Niimi, K. Ikeo, and K. Sahara. 2006. Phylogeny of the sex-determining gene *Sex-lethal* in insects. *Genome* 49:254–62.

Traut, W., and B. Rathjens. 1973. Das W-Chromosom von *Ephestia kuehniella* (Lepidoptera) und die Arbeitung des Geschlechtschromatins. *Chromosoma* 41:437–46.

Traut, W., K. Sahara, and F. Marec. 2007. Sex chromosomes and sex determination in Lepidoptera. *Sex. Dev.* 1:332–46.

Traut, W., A. Weith, and G. Traut. 1986. Structural mutants of the W chromosome in *Ephestia* (Insecta, Lepidoptera). *Genetica* 70:69–79.

Wu, C., S. Asakawa, N. Shimizu, et al. 1999. Construction and characterization of bacterial artificial chromosome libraries from the silkworm, *Bombyx mori. Mol. Gen. Genet.* 261:698–706.

Xia, Q., Z. Zhou, C. Lu, et al. 2004. A draft sequence for the genome of the domesticated silkworm (*Bombyx mori*). *Science* 306:1937–40.

Yamamoto, D., K. Fujitani, K. Usui, et al. 1998. From behavior to development: Genes for sexual behavior define the neuronal sexual switch in *Drosophila*. *Mech. Dev.* 73:135–46.

Yasukochi, Y. 1998. A dense genetic map of the silkworm, *Bombyx mori*, covering all chromosomes based on 1018 molecular markers. *Genetics* 150:1513–25.

Yasukochi, Y., L.A. Ashakumary, K. Baba, et al. 2006. A second-generation integrated map of the silkworm reveals synteny and conserved gene order between lepidopteran insects. *Genetics* 173:1319–28.

5 Evolutionary and Developmental Genetics of Butterfly Wing Patterns
Focus on Bicyclus anynana *Eyespots*

Patrícia Beldade and Suzanne V. Saenko

CONTENTS

INTRODUCTION

The beautiful color patterns decorating butterfly wings have been and continue to be inspirational for studies of a variety of biological issues, ranging from systematics and evolution to developmental genetics and biochemistry of pigmentation. The amazing diversity of these patterns, together with knowledge on their adaptive value and underlying developmental basis, make them a favorite system in the relatively new field of evolutionary developmental biology, or evo-devo (Nijhout 1991; Beldade and Brakefield 2002; McMillan, Monteiro, and Kapan 2002; Joron et al. 2006; Parchem, Perry, and Patel 2007; Wittkopp and Beldade 2009).

Wing pattern diversity is astonishing, and many of the more than seventeen thousand species of butterflies can be recognized based on their wing patterns. Dramatic variation has also been documented within species, between geographical populations and seasonal forms, between males and females, and between the dorsal and ventral surfaces of one same wing (examples in Nijhout 1991). Despite the dazzling panoply of colorful spots, bands, and stripes, most butterfly wing patterns can be recognized as derivations of a basic "nymphalid ground plan," which has been very useful to identify pattern element homologies within and across species (Nijhout 1991). In this ground-plan representation, different types of pattern elements such as eyespots, chevrons, and bands are organized in parallel series, with individual elements repeated along the anterior-posterior axis within wing compartments bordered by veins. Independent development and evolution of individual pattern elements have presumably facilitated the diversification of butterfly wing patterns.

The adaptive significance of wing patterns and their underlying genetic basis have long been the focus of studies in ecology and evolution and have provided important insights into how variation in natural populations is shaped by natural and sexual selection. In the last couple of decades, this focus has extended to the detailed analysis of the development processes and pathways involved in pattern formation and diversification. These will be the focus of this chapter.

Ecology and Evo-Devo of Butterfly Wing Patterns

Lepidopterans include many textbook examples of adaptive coloration polymorphisms ranging from industrial melanism in the peppered moth *Biston betularia* (Majerus 1998), to mimicry in swallowtails (*Papilio*) and passion-vine (*Heliconius*) butterflies (Joron and Mallet 1998), to seasonal color pattern changes or polyphenisms in a number of species (Brakefield and French 1999). Wing color patterns also provide a favorite example of the evolution of particular adaptive traits, namely, butterfly eyespots, whose concentric rings of color are thought to resemble vertebrate eyes and function in deflection or intimidation of predators (Stevens 2005). The ecological significance of natural variation in butterfly wing patterns has been associated not only with avoiding predation, but also with factors such as intraspecific recognition, sexual selection, and thermal regulation.

Contrasting with studies of the ecology and basic genetics of wing color patterns, analysis of the developmental and molecular genetics of pattern formation have become topics of intensive research interest relatively more recently. Evo-devo studies in butterflies have focused on a few target species, and used a diversity of approaches to provide what are largely complementary insights about the developmental and genetic mechanisms behind the formation of different aspects of wing patterns. For example, studies in *Papilio* butterflies provided a detailed analysis of the biochemical pathways of pigment production and their relationship to polymorphism in overall coloration (Koch, Behnecke, and ffrench-Constant 2000). Linkage analysis in species of *Heliconius*, involving both classical low-resolution crosses and, more recently, high-resolution genetic mapping, has characterized the genetics underlying variation in wing patterns composed of large patches of different colors (Joron et al. 2006). See Chapter 6 for a review on genetics of wing patterns in *Heliconius*. Analysis of developing wings of *Junonia* (*Precis*) *coenia* and *Bicyclus anynana* have unraveled much of what is known about the developmental processes and genetic pathways implicated in the formation of eyespots (Nijhout 1991; Beldade and Brakefield 2002; McMillan, Monteiro, and Kapan 2002). These pathways and processes will be examined in this chapter, with a focus on research in laboratory populations of *B. anynana*.

Evo-Devo of Eyespot Patterns in *Bicyclus anynana*

Efforts by Paul Brakefield two decades ago established the tropical nymphalid *B. anynana* as a laboratory system (Brakefield, Beldade, and Zwaan 2009). These butterflies are small enough that large populations can be reared and analyzed, and large enough that manipulating single individuals and specific tissues is no technical challenge. This enables the full integration of population-, organismal- and molecular-level analysis. Importantly, experimental tractability, including growing genomic resources and transgenic tools for this species (Ramos and Monteiro 2007; Beldade, McMillan, and Papanicolaou 2008; Beldade et al. 2009), can be combined with knowledge of ecology and natural variation.

The eyespots on *B. anynana* wings, in particular, have proven to be a valuable system for an integrated study of the evolutionary and developmental processes that shape morphological variation. They function in predator avoidance (Lyytinen et al. 2004; Brakefield and Frankino 2006) and mate choice (Robertson and Monteiro 2005; Costanzo and Monteiro 2007), and their morphology varies greatly across and within species (Beldade, Brakefield, and Long 2005; Brakefield and Roskam 2006). *B. anynana* populations maintained in the laboratory harbor genetic polymorphisms for different aspects of eyespot phenotypes and provide ideal material to analyze how variation in genotypes is translated into phenotypic variation, via development. The dramatic differences seen between seasonal forms in natural populations can also be induced in a laboratory setting to study how environmental cues establish alternative developmental trajectories (Figure 5.1). Furthermore, comparisons of the diversity of eyespot patterns in the circa 80 *Bicyclus* species can be done within the framework of morphological (Condamin 1973) and molecular (Monteiro and Pierce 2001) phylogenies. In terms of understanding underlying developmental mechanisms, eyespots are amenable to detailed analysis ranging from the genetic pathways and cellular signaling underlying pattern specification, to biochemistry of pigment production (Figure 5.2).

In this chapter, we focus on studies of laboratory populations of *B. anynana* and what they have taught us about the genetic and developmental underpinnings of the formation and diversification of eyespot patterns. We first address the mechanisms of eyespot formation, including models of cellular interactions in pupal wings, and the genetic pathways involved in different stages of eyespot development. The next section focuses on the mechanisms behind intraspecific variation in eyespot patterns, including an overview about the contribution of different components of those developmental processes implicated in eyespot formation, and the type of more genome-wide search made possible with new genomic resources. The third section focuses on aspects of the evolutionary diversification of eyespot patterns with emphasis on the origin and modification of these novel traits and the diversification of serially repeated structures. We end by discussing some future avenues of research.

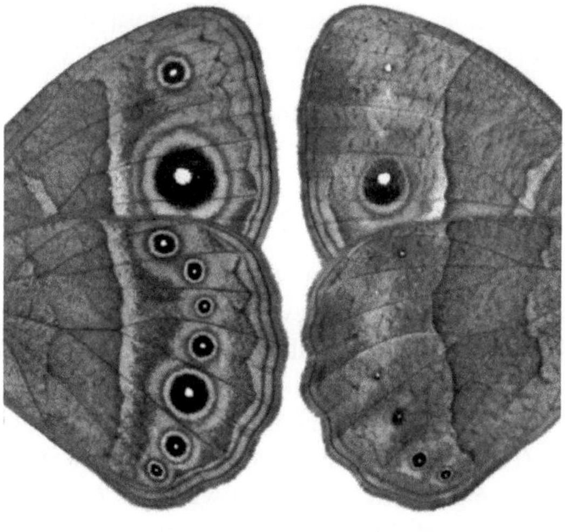

FIGURE 5.1 ***A color version of this figure follows page 176.*** Phenotypic plasticity in *Bicyclus anynana* wing pattern development. Two distinct seasonal forms occur in the wild and can be mimicked in the laboratory by rearing larvae at 27°C or 20°C. The emerging adults will then resemble the natural wet-season (left) and dry-season (right) forms, respectively. The distal section of the ventral surface of both forewings and hindwings is shown. The dorsal surface (not visible) does not show plasticity in relation to rearing temperature. The adaptive significance and underlying physiological basis of the alternative phenotypes is discussed in the text.

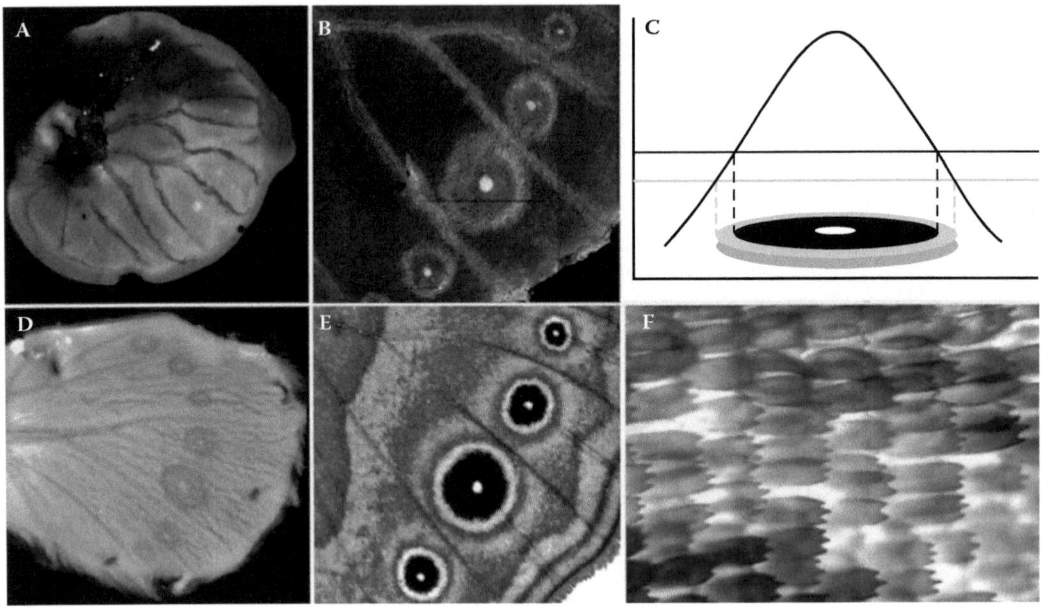

FIGURE 5.2 ***A color version of this figure follows page 176.*** Eyespot development in *Bicyclus anynana*. (a) Eyespot formation starts during the last larval instar, when future eyespot centers are established, presumably by the action of genes such as *Distal-less* (red) and *engrailed* (green), the protein products of which are detected here on a hindwing using immunofluorescence technique. Note that *Distal-less* is also detected in the margin, and *engrailed* throughout the posterior compartment, as is characteristic of insect wing development. (b) The same genes continue to be coexpressed in eyespot foci in the early pupal wings, and their protein products are also detected in the cells that form the inner (*Distal-less*) and the outer (*engrailed*) rings. (c) The formation of the color rings presumably relies on a signal-response mechanism whereby focal cells produce a signaling molecule that diffuses away and forms a concentration gradient (curve) to which the neighboring cells respond in a threshold-like fashion (horizontal lines). (d) The epidermal cells become fated to produce a particular color pigment shortly before adult eclosion. This late pupal hindwing shows mature scales in the white center, and an outer golden ring. The black inner ring and brown background scales will mature later. (e) The adult wing pattern is composed of serially repeated eyespots; shown here is a section of an adult hindwing with four eyespots. (f) Color patterns are formed by the arrangement of scales, each bearing one particular (black, yellow, or brown) pigment.

DEVELOPMENTAL GENETICS OF EYESPOT FORMATION

Of all types of different color pattern elements that can be recognized in butterfly wings, eyespots are undeniably those whose underlying development is best understood. In the early 1980s, pioneering work by Fred Nijhout used surgical manipulations of developing pupal wings of *J. coenia* to establish groundbreaking developmental models of eyespot formation. These and other experiments analyzing the evolution and development of butterfly wing patterns were summarized in his inspiring book (Nijhout 1991). In the mid-1990s, work in the lab of Sean Carroll on expression patterns of candidate genes in developing wings of *J. coenia* and *B. anynana* started to explore the genetic pathways involved in eyespot formation (Carroll et al. 1994; Brakefield et al. 1996; Figure 5.2a,b).

Butterfly wings are formed by two epidermal membranes nourished and supported by veins. They start to develop during the first larval instar as slight enlargements of the epidermal cells in the lateral meso- and metathoracic regions. Venation pattern develops and the wing imaginal discs greatly increase in size in the last larval stage. Scale maturation and pigment deposition occur in late pupal wings shortly before adult eclosion. Wing color patterns are formed by the arrangement of partially overlapping, monochromatic scales on a single cell layer. This two-dimensional nature simplifies modeling of the underlying developmental interactions that occur in late larval and early pupal wings and are thought to determine scale maturation and pigmentation long before any coloration is visible. Here, we review data on the molecular and genetic bases of different stages of eyespot development, from prepatterning in larval wing discs, to determination of color rings during the early pupal stage, to actual pigment production in late pupal wings.

Models of Eyespot Formation: Signal and Response Mechanism

Surgical manipulations of presumptive eyespots in pupal wings are facilitated by cuticular landmarks that enable the localization of the centers of the eyespots on the dorsal surface of the forewings. Such manipulations have established that these centers, called foci, have "eyespot-organizing" properties. Transplanting an eyespot focus into an eyespot-less position on the wing typically results in the formation of an ectopic eyespot at the host site, whereas damaging foci during the sensitive early pupal period typically reduce or completely eliminate the corresponding eyespots in adult wings (Nijhout 1980; French and Brakefield 1995). Results such as these led to the proposal of developmental models whereby the eyespot focus acts as an organizer by signaling to the neighboring cells via the production of a diffusible morphogen (Figure 5.2c). Diffusion of this signal away from the focus presumably forms a concentration gradient in the wing epidermis; the surrounding cells respond to the signal concentration in a threshold-like fashion and become fated to produce a particular color pigment. Alternative models of eyespot formation have also been proposed whereby foci degrade, rather than produce, the organizing signal (French and Brakefield 1992), or additional morphogen sources appear in concentric eyespot rings (Dilão and Sainhas 2004).

Despite strong experimental support for a signal-response mechanism of eyespot formation, little is known about the identity and mode of action of the actual genes involved in signaling and response. However, information on the genetic pathways involved in eyespot development has been increasing, and a number of genes have been implicated in different stages of eyespot development, which are discussed below.

Genetic Pathways Implicated in Establishment of Eyespot Foci in Late Larval Wings

Studies of gene expression patterns in developing larval and pupal wings of *B. anynana* and *J. coenia* implicated a number of pathways in eyespot development. These studies targeted candidate genes from pathways involved in the development of insect wings and extensively studied in *Drosophila melanogaster.* Genes involved in the compartmentalization of developing insect wings (such as *apterous* in the dorsal compartment, *engrailed* and *cubitus interruptus* in the posterior and anterior compartments, respectively, and *Distal-less* and *wingless* along the wing margin) perform similar functions in flies and butterflies, and some have also been redeployed to regulate different stages of eyespot formation (Carroll et al. 1994; Figure 5.3 and Table 5.1).

Establishment of the location of the eyespot focal cells takes place during the final larval instar in each eyespot-bearing, vein-delimited wing compartment. To date, the earliest known event associated with this process is the upregulation of the gene *Notch*, first in the intervein midline and subsequently in the future focal cells (see Table 5.1 for references). Expression of *Notch* is followed by the activation of *Distal-less*, and slightly later by the upregulation of *engrailed* and *spalt* (Figures 5.2a and 5.4a,b). Simultaneously, expression of genes from the Hedgehog-signaling pathway is activated in (*patched* and *cubitus interruptus*) or around (*hedgehog*) eyespot foci. Later, the ecdysone receptor protein, presumably upregulated by Distal-less, appears in the focal cells. The idea that these genes act in the determination of eyespot foci is strengthened by data on their expression patterns in *B. anynana* laboratory populations with altered eyespot morphology. These will be described in more detail in the next section. Exactly how these genes regulate each other and the downstream pathways leading to focal signaling and eyespot formation is not known. However, modeling approaches

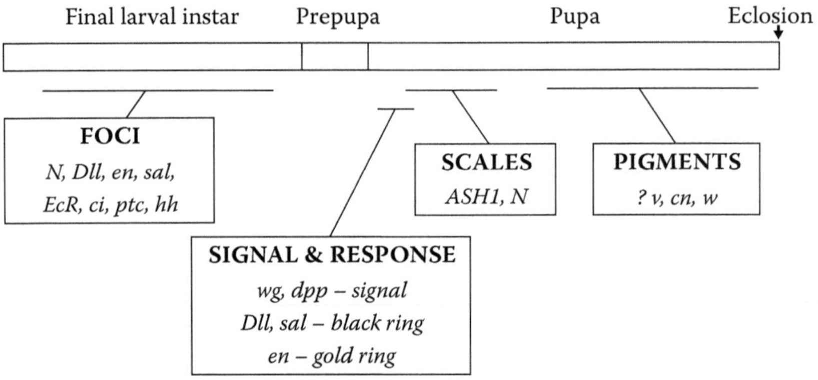

FIGURE 5.3 Genes associated with different stages of eyespot formation. The bar represents the consecutive stages in *B. anynana* development and the lines underneath it refer to those stages of eyespot formation that have been examined so far. Genes whose expression has been detected in association with wing pattern development are listed in the boxes (details in Table 5.1). Data for the stages of focal determination, signal-response interactions, and scale maturation were mostly gathered from studies in *B. anynana* and *J. coenia*. Data on pigmentation genes, on the contrary, have focused on species that do not have eyespots. The role of these genes in eyespot formation still needs to be investigated.

TABLE 5.1
Genes Implicated in Butterfly Wing Pattern Formation

Gene Name	Abbreviation	Molecular Function*	References
Achaete-scute homologue	*ASH1*	Transcription factor	Galant et al. 1998
Cinnabar	*cn*	Kynurenine 3-monooxygenase	Reed and Nagy 2005
Cubitus interruptus	*ci*	Transcription factor	Keys et al. 1999
Decapentaplegic	*dpp*	Morphogen	Monteiro et al. 2006
Distal-less	*Dll*	Transcription factor	Carroll et al. 1994; Brakefield et al. 1996
Ecdysone Receptor	*EcR*	Nuclear receptor with transcription factor activity	Koch et al. 2003
Engrailed	*en*	Transcription factor	Keys et al. 1999; Brunetti et al. 2001
Hedgehog	*hh*	Morphogen	Keys et al. 1999
Notch	*N*	Transmembrane receptor with transcription factor activity	Reed and Serfas 2004
Patched	*ptc*	Transmembrane receptor for *Hedgehog*	Keys et al. 1999
Spalt	*sal*	Transcription factor	Brunetti et al. 2001; Reed, Chen, and Nijhout 2007
Ultrabithorax	*Ubx*	Transcription factor	Lewis et al. 1999
Vermilion	*v*	Tryptophan 2,3-dioxygenase	Reed and Nagy 2005; Beldade, Brakefield, and Long 2005
White	*w*	Pigment precursor transporter	Reed and Nagy 2005
Wingless	*wg*	Morphogen	Monteiro et al. 2006

* Relevant molecular function as per Gene Ontology classification, available also in FlyBase

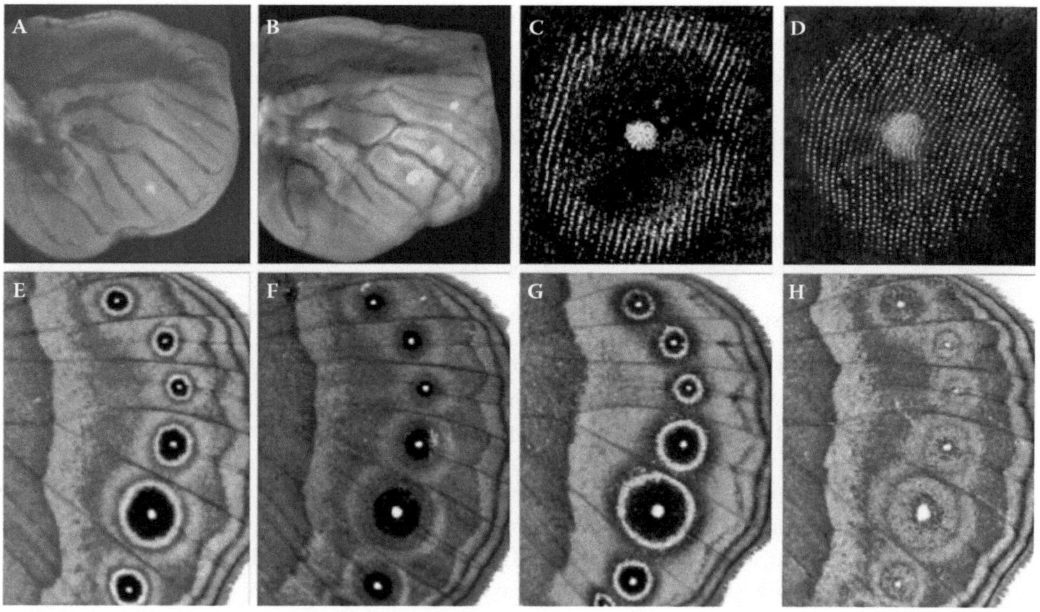

FIGURE 5.4 ***A color version of this figure follows page 176.*** Mutations of large effect on *Bicyclus anynana* wing patterns. (a–d) Engrailed (green) and Distal-less (red) proteins are detected in larval (a,b) and pupal (c,d) forewings. (e–h) Detail of the ventral surface of the adult hindwings from different laboratory populations. (a) In wild-type *B. anynana* there are two eyespots on the adult forewing, and the centers of these are already visible in late larval wings. (b) In *Spotty* mutants two additional foci appear in larval wings whose location matches the two extra eyespots on adult forewings. (c) Eyespot rings are established during the early pupal stage, and the engrailed protein is typically detected in an area corresponding to the outer golden ring. (d) In *Goldeneye* mutants, engrailed is expressed throughout the eyespot field in a pattern that corresponds to the adult eyespot color composition (see h). (e) Pigmentation pattern of wild-type *B. anynana*. (f) The *melanine* mutant has overall darkened coloration. (g) The *Band* mutant has a lighter distal half of the wing. (h) The *Goldeneye* mutant has altered eyespot color composition.

have been proposed to describe the genetic regulatory hierarchy of this process based on knowledge about these pathways in *D. melanogaster* (Evans and Marcus 2006; Marcus and Evans 2008).

Candidate Genes for Signal and Response Components of Eyespot Formation

Even though models for eyespot formation involving focal signaling and epidermal response were proposed several decades ago, still little is known about the molecular nature of those components. Recently, however, some progress has been made using knowledge from morphogens involved in *Drosophila* wing patterning. A study by Antónia Monteiro and colleagues showed that the proteins Wingless and the phosphorylated form of Smad, the signal transducer in the Decapentaplegic pathway, are detected in the cells of the presumptive eyespot field in early pupal wings of *B. anynana* (see Table 5.1 for references). This is consistent with the suggestion that both Wingless and Decapentaplegic are candidate eyespot signals, but functional analysis will be necessary to confirm their role in eyespot development.

Despite the lack of data on the mode of action of the focal signal or the identity of its immediate targets, some of the genes of the epidermal response cascade have been identified. Analysis of expression patterns in *B. anynana* showed that a number of genes expressed in eyespot foci in larval wings continue to be active during early pupal development, and some become upregulated in the cells of the future eyespot rings. Notably, *engrailed* is expressed in the area that corresponds to the outer (golden) ring of the adult eyespot, and *Distal-less* and *spalt* in the inner (black) ring (Figures 5.2b and 5.4c). The transcription factors encoded by these genes might be directly involved in the interpretation of focal signal(s) and activation of pigment biosynthesis pathways as their expression domains perfectly correlate with adult color rings, both in the wild-type *B. anynana* as well as in mutants with disturbed color composition as, for example, *Goldeneye* (Brunetti et al. 2001; Figure 5.4d). The same transcription factors are expressed in concentric rings in the eyespot fields of other butterfly species, but in different relative spatial domains that correlate with divergent adult eyespot color schemes (Brunetti et al. 2001).

Pigment Production and Scale Maturation in Late Pupal Wings

Most studies have focused on the cellular interactions and genetic pathways during the early stages of eyespot development. Much less is known about the downstream processes, whereby scale-forming cells interpret prepattern information to produce specific color pigments late in pupal development.

Four major classes of pigments have been characterized in butterfly wing scales: melanins (red-brown to black), ommochromes and pterins (yellow to red), and flavonoids (white to red and blue). The timing of scale maturation correlates with pigment production, with dark, melanized scales typically maturing last in most butterfly species (Nijhout 1991; Reed and Nagy 2005). In *B. anynana* eyespots, for example, the cells at the white center of the eyespot are the first to mature, followed by the yellow scales of the outer ring, and finally the black scales of the middle ring (Koch et al. 2000; Figure 5.2d–f).

Recent studies analyzed levels of expression of several pigmentation candidate genes in association with wing sections of different colors in *Vanessa cardui* and *Heliconius erato* (Reed and Nagy 2005; Reed, McMillan, and Nagy 2008). Some of these genes (e.g., *vermillion*) are also expressed in *B. anynana* developing wings (Beldade, Brakefield, and Long 2005) and are potentially involved in making the different color rings of the eyespot. Laboratory pigmentation mutants in this species will be invaluable to study the expression of these and other candidate genes and explore their relationship to different aspects of coloration, from the whole wing to large sections thereof, to eyespot rings (Figure 5.4e–h).

MECHANISTIC BASIS OF VARIATION IN EYESPOT PATTERNS

The experiments summarized in the previous section identified a number of pathways and processes involved in the formation of eyespots. The extent to which specific components of those pathways (i.e., individual genes) and of those processes (e.g., focal signal strength and epidermal response thresholds) can, and do, contribute to variation in eyespot morphology is a key issue in evo-devo research. Laboratory populations of *B. anynana* with distinct wing pattern phenotypes have been used to analyze the genetic, physiological, and developmental basis of variation in eyespot patterns. Crucial insights gained from such studies are summarized below.

Genetic and Physiological Underpinnings of Phenotypic Plasticity

B. anynana have clearly distinct seasonal forms that differ in life history and morphology, including wing color patterns. The alternative phenotypes confer adaptive advantage to the individuals living in the alternating wet and dry seasons in this butterfly's natural environment. Whereas inconspicuous brown butterflies rely on crypsis against the background of dry leaves to avoid predation in the dry season, bright eyespots in wet-season butterflies are thought to attract the attention of predators to the wing margin and away from the fragile body when the butterflies rest in a leafy green background (Brakefield and Frankino 2009).

In the laboratory, butterflies reared at warm or cool temperatures have phenotypes resembling the natural wet and dry forms, respectively (Figure 5.1). These have been used to start probing how differences in the environment influence developmental processes and lead to divergent adult phenotypes (Brakefield, Kesbeke, and Koch 1998). Moreover, artificial selection on variation segregating in a laboratory population produced butterflies with constitutive expression of the wet (large, conspicuous eyespots) or the dry (small eyespots) wing patterns across developmental temperatures (Brakefield et al. 1996; Koch, Brakefield, and Kesbeke 1996). Endocrine signals, namely ecdysteroids, have long been known to play an important role in insect development and the control of polyphenisms (Riddiford 1995; Nijhout 2003a). Measurements of ecdysone titers in pupae from the selection lines mentioned above showed clear differences in the timing of hormone release, with higher titers of ecdysteroids shortly after pupation associated with larger ventral eyespots (Koch, Brakefield, and Kesbeke 1996; Brakefield, Kesbeke, and Koch 1998). Furthermore, injections and infusions of this hormone in young pupae from the small eyespot line resulted in adults with larger ventral eyespots (Brakefield, Kesbeke, and Koch 1998).

Despite all the available knowledge about the role of ecdysteroids in lepidopteran development, the identity and mode of action of the eyespot-related genetic pathways up- and downstream of the hormones remain largely unknown. We do know that wet- and dry-season phenotypes show differences in the levels of expression of the *Distal-less* transcription factor (Brakefield et al. 1996), which has been shown to contribute to variation in dorsal eyespot size (Beldade, Brakefield, and Long 2002). However, the precise relationship between the timing of ecdysone release in the pupal haemolymph and the levels of *Distal-less* expression in the eyespot field is not fully understood (but see Koch et al. 2003). The new genomic tools available for a number of butterfly species, including *B. anynana* (Beldade, McMillan, and Papanicolaou 2008), will be crucial in advancing our understanding of the relationship between environment, genes, hormones, and development in producing alternative wing pattern phenotypes.

Genetic Variation in Laboratory Populations of *B. anynana*

Laboratory populations of *B. anynana* not only mimic the natural seasonal forms but also harbor different types of genetic variation affecting eyespot patterns (Brakefield, Beldade, and Zwaan 2009). Artificial selection experiments have explored heritable genetic variation segregating in a large laboratory population, and produced gradual and progressive changes in such traits as eyespot size, color composition, shape, and position (McMillan, Monteiro, and Kapan 2002; Beldade, Brakefield, and Long 2005). A number of spontaneous mutations with large effect on a variety of aspects of wing patterns have also been isolated and maintained in stable stocks. These, too, can affect not only eyespot size (e.g., *Bigeye* has overall enlarged eyespots; Brakefield et al. 1996), color composition (e.g., *Goldeneye* has no black ring; Figure 5.4h; Brunetti et al. 2001), and shape (e.g., *Cyclops* has elongated, fused eyespots; Brakefield et al. 1996), but also other aspects of wing coloration (e.g., *melanine* and *Band*; Figure 5.4f–g).

The extent to which the loci carrying alleles of extreme effect are the same as those harboring alleles of subtler effect that contribute to segregating, quantitative variation in laboratory and natural populations is not known. Mutant stocks and selection lines, however, offer the opportunity to characterize the genetic architecture of wing pattern variation and to identify additional developmental pathways and processes involved in color pattern formation and variation.

Variation in Developmental Processes

Work on laboratory populations of *B. anynana* has enabled characterization of variation across consecutive stages of eyespot formation, from expression of prepatterning genes in late larval wings, to the signal-response components in early pupal wings, to scale maturation and pigment deposition in late pupae.

A number of mutant stocks and selection lines show altered patterns of eyespot-associated gene expression in late larval wings. For example, in mutants with additional (e.g. *Spotty* mutant;

Figure 5.4b) or lost (e.g. *3+4* and *Missing*) eyespots, there is a perfect match between the gain/loss of eyespots in the adults, and the gain/loss of expression of focal marker genes *Notch*, *Distal-less* and *engrailed* in larval wings (Brakefield et al. 1996; Monteiro et al. 2003; Reed and Serfas 2004; Monteiro et al. 2007). Moreover, the appearance of the areas of gene expression also resembles eyespot shape (e.g. in the *Cyclops* mutant elongated areas of *Distal-less* expression match elongated eyespot foci; Brakefield et al. 1996) and eyespot size (e.g. individuals from artificial selection lines with reduced or enlarged dorsal eyespots show quantitative differences in the expression of *Distal-less* and *engrailed*; Beldade, Brakefield, and Long 2005). Other eyespot morphology variants do not associate with changes in marker gene expression in larval wings, but do correlate with patterns of gene expression at later stages of eyespot development. For example, altered color composition in the *Goldeneye* mutant correlates with expression pattern of *engrailed* in presumptive eyespots in pupal wings (Brunetti et al. 2001; Saenko et al. 2008; Figures 5.4d,h). These results suggest a number of candidate pathways for harboring allelic variation contributing to phenotypic variation in eyespot morphology.

Transplants of the eyespot-inducing centers between pupae from selection lines with divergent eyespot morphologies identified variation both in the strength of the focal signal, and in the threshold levels of epidermal response. For example, quantitative variation in eyespot size seems to be due largely to variation in focal signal (Monteiro, Brakefield, and French 1994), whereas the properties of the epidermis explain differences in eyespot color composition (Monteiro, Brakefield, and French 1997a) and shape (Monteiro, Brakefield, and French 1997b). Little is known about the way such differences in signal-response components influence scale maturation and pigment production in *B. anynana* eyespots. Because rates of scale maturation correlate with scale color, heterochronic changes in scale development have been proposed as a mechanism for generating variation in butterfly wing patterns and have been described in *B. anynana* and *Papilio glaucus* mutants (Koch et al. 2000).

Differences in patterns of gene expression per se do not identify the loci that are responsible for those differences or for the associated differences in adult phenotype. Indeed, variation in expression of any specific gene can be due to allelic variation at that locus (*cis*) or to variation in another gene involved in regulating the expression of the first gene (*trans*). The extent to which *cis* and *trans* factors contribute to variation in gene expression is an area of active research in relation to morphological diversification. Here, we will discuss experiments that attempted to identify the genes responsible for variation in butterfly wing patterns.

Identification of Genes Responsible for Wing Pattern Variation

Candidate developmental genes play an important role in the quest for identifying which loci contribute to variation in all sorts of phenotypes, including color patterns in butterfly wings. Whereas genome-wide mapping techniques have become increasingly powerful, especially as decreasing costs of genotyping allow monitoring more markers in larger mapping panels, it remains a challenge to move from mapped genetic regions to identifying single loci responsible for phenotypic variation. This is especially true for nonclassical model organisms like butterflies, which do not have fully sequenced genomes or very rich information on patterns of DNA sequence polymorphisms.

Work on *B. anynana* laboratory populations has shown that candidate genes from pathways implicated in eyespot development can contribute to inter-individual variation in eyespot morphology. DNA sequence polymorphisms in *Distal-less*, a key gene in eyespot development, cosegregate with quantitative variation in eyespot size in recombinant progeny from crosses between selection lines with divergent phenotypes (Beldade, Brakefield, and Long 2002). However, the same gene, whose pattern of expression also changes in association with variation in eyespot number, did not show close linkage with the locus *Missing*, which affects that aspect of eyespot patterns (Monteiro et al. 2007). This suggests that some gene upstream of *Distal-less* (possibly *Notch*) is responsible for the *Missing* phenotype. Tests of the contribution of a number of other candidate genes to variation in different aspects of eyespot morphology are necessary. Also for other types of pattern elements in other species, close linkage between candidate developmental genes and wing color pattern variation has been reported (Kronforst et al. 2006; Clark et al. 2008).

Genes within the developmental pathways shown to be involved in eyespot formation are clearly good candidates for contributing to variation in eyespot patterns. Other candidates are suggested by the analysis of pleiotropic mutations that affect both eyespot morphology and other more conserved developmental processes whose genetic basis is well understood in model organisms (Saenko et al. 2008). However, direct tests of the contribution of candidate genes to variation in wing patterns require sequence and polymorphism information. This type of information was very scarce until recently, when EST projects boosted gene discovery in a few target butterfly species (Beldade, McMillan, and Papanicolaou 2008; Papanicolaou et al. 2008), including *B. anynana* (Beldade et al. 2006; Long, Beldade, and Macdonald 2007; Beldade et al. 2009), and laid down the grounds for the development of new genomic tools. These tools will allow the fine-scale genetic dissection of wing pattern variation through a detailed analysis of candidate genes and pathways already identified, or via a less-biased genome-wide approach, or a combination of both. Exciting new comparative analysis showing conserved synteny for a number of orthologous loci betweeen the model *Bombyx mori* and the butterflies *H. melpomene* (Pringle et al. 2007) and *B. anynana* (Beldade et al. 2009), and between the tobacco hornworm *Manduca sexta*, *B. mori*, and *H. melpomene* (Sahara et al. 2007) suggests that future efforts in mapping of butterfly wing pattern variation will be able to rely on linkage information available for other lepidopterans to facilitate moving from mapped genomic regions to a testable number of candidate genes (Beldade et al. 2009).

DIVERSIFICATION OF EYESPOT PATTERNS

It is not know to what extent the underlying mechanisms discussed in relation to variation in lab populations of *B. anynana* can explain variation in eyespots across more and less distantly related species, nor to what extent it will be possible to find parallels in the molecular mechanisms underlying diversification in eyespots and in other types of pattern elements. Here we will focus on how butterfly wing patterns, and eyespots in particular, have made important contributions to our understanding of key mechanisms behind morphological diversification.

Much of the diversity in animal morphologies can be accounted for by two processes: the origin and modification of evolutionary novelties (e.g., the turtle shell and the bird feather), and the diversification of serially repeated structures (e.g., vertebrate teeth and insect body segments). Butterfly wings capture both processes beautifully: their color patterns are an evolutionary novelty, and are composed of different types of serially homologous pattern elements that have diversified between repeats and across species. Below we will discuss what is known about the genetic underpinnings of these two aspects of diversification in relation to butterfly eyespot patterns.

EVOLUTION OF MORPHOLOGICAL INNOVATIONS

One of the touchstones of evo-devo is that conserved genetic pathways are used in association with different developmental stages and tissues both within and across species. The formation of novel traits seems to rely largely on the redeployment of genes and pathways shared among lineages. Butterfly wings offer many and appealing examples of such co-option. The parallels between the genetic cascades well studied in *D. melanogaster* and involved in butterfly wing development have been documented in relation to scale formation (*achaete-scute* and *Notch* in fruitfly bristles and in butterfly scales; Galant et al. 1998; Reed 2004) and coloration (e.g., the ommochrom pathway in fruit fly eyes and butterfly wings; Beldade, Brakefield, and Long 2005; Reed and Nagy 2005), and in relation to specific pattern elements (e.g., the *hedgehog* pathway in fruit fly wings and butterfly eyespots; Keys et al. 1999).

Commonalities between the mechanisms for eyespot formation and those involved in embryonic development (by analysis of pleiotropic mutants affecting eyespot morphology and embryogenesis; Saenko et al. 2008) or response to wounding (by analysis of damage-induced eyespot formation; Brakefield and French 1995; Monteiro et al. 2006) offer unique opportunities to exploit information available for model systems to increase our knowledge about eyespot evo-devo. Embryonic development and wound healing are extensively studied in model organisms and are thought to be relatively conserved. Analysis of such well-described pathways in the context of eyespot development offers not only the possibility for translating knowledge from model systems into a deeper understanding of lineage-specific traits, but also the opportunity to study the evolution of key developmental pathways in the context of their involvement in novel functions (Beldade and Saenko 2009).

Mechanisms of Individualization of Serial Repeats

The nymphalid ground plan is probably the best illustration of the modular nature of butterfly wing patterns. Independence between pattern elements of different series (e.g., eyespots and chevrons) and correlations between serially repeated elements (e.g., two eyespots on the same wing surface) have been documented for a number of different species (Paulsen and Nijhout 1993; Nijhout 2003b; Allen 2008). Moreover, the genes expressed in association with the eyespot field show a nearly identical pattern of expression in relation to all eyespots, and typically are not expressed in association with other types of pattern elements (Figure 5.2b,c). Whether the serially repeated pattern elements appeared at once and then diversified, or whether individual repeats were added at different times, is still an open question (Monteiro 2008). So far, however, progress has been made in understanding the genetic and developmental basis of the diversification of serially homologous eyespots.

Despite strong correlations between serially repeated traits, artificial selection experiments in *B. anynana* have uncovered great flexibility for individual changes in eyespot morphology (Beldade, Koops, and Brakefield 2002). Furthermore, candidate genes implicated in such quantitative changes (Beldade, Brakefield, and Long 2002), as well as a number of spontaneous mutations of large phenotypic effect (Beldade, French, and Brakefield 2008; Monteiro et al. 2003) have been shown to affect only subsets of the serially repeated eyespots. The extent to which the development and evolution of individual eyespots is compartmentalized in *B. anynana* mutants led to the proposal of a genetic model for this compartmentalization in which eyespot-specific regulatory regions, each controlling the expression of the key "eyespot gene," is associated with each of the future eyespots (McMillan, Monteiro, and Kapan 2002; Monteiro et al. 2003). A very fine analysis of the regulatory sequences and/or functions of candidate eyespot genes will be needed to verify this model.

The extent to which the signal and response components of eyespot formation are compartmentalized across the wing surface is another crucial aspect for investigation. Transplantation of presumptive eyespot foci between pupae from artificial selection lines showed that whereas localized changes in the size of individual eyespots rely mostly on localized changes in focal signal strength, simultaneous changes in the size of multiple eyespots depend mostly on epidermal response sensitivities (Beldade, French, and Brakefield 2008). These results reflect potential differences in the degree of compartmentalization of the signal-response components of eyespot formation, and illustrate how the properties underlying the development of different aspects of eyespot morphology can impact evolutionary diversification of serial repeats (Allen et al. 2008).

PERSPECTIVES

The work on *B. anynana* laboratory populations discussed here provides important insights into the mechanisms of eyespot development and has implicated several genes and signaling pathways in this process. There are, however, still many gaps in our understanding of eyespot formation. Some stages of eyespot development have received little to no attention (see Figure 5.3), and we are still far from understanding how all the genes and processes we do know about interact with each other and are regulated by environmental factors to produce variation in eyespot patterns.

Detailed functional analysis of eyespot genes, from candidate morphogens to pigment enzymes, is now possible as transgenic techniques are being developed for *B. anynana* and other butterflies (Ramos and Monteiro 2007). Nonheritable transformation of wing epidermal cells using in vivo DNA electroporation (Golden et al. 2007) or viral vectors (Lewis et al. 1999; Lewis and Brunetti 2006) has been reported for different species, and successful germ line transformation with transposable element vectors has been developed for *B. anynana* (Marcus, Ramos, and Monteiro 2004). The latter is particularly exciting given that it can be combined with recent technical developments (namely, laser-mediated heat shock) allowing for precise spatio-temporal control of transgene expression (Ramos et al. 2006) and opening up the possibility of testing gene function via inducible knockout or ectopic activation (Ramos and Monteiro 2007).

These techniques, together with the growing amount of gene sequence and polymorphism information available for *B. anynana* and other lepidopteran species, hold much promise for closing the gaps in our understanding of wing pattern development and for establishing parallels, or documenting differences, between pattern formation and variation across species. Once the genetic and developmental mechanisms have been described for a variety of species, it will be exciting to go back to natural populations and integrate knowledge of ecological pressures to understand how those mechanisms impact the evolutionary diversification of wing patterns.

REFERENCES

Allen, C.E. 2008. The "eyespot module" and eyespots as modules: Development, evolution, and integration of a complex phenotype. *J. Exp. Zoolog. B Mol. Dev. Evol.* 310:179–90.

Allen, C.E., P. Beldade, B.J. Zwaan, P.M. Brakefield. 2008. Differences in the selection response of serially repeated color pattern characters: standing variation, development, and evolution. *BMC Evol. Biol.* 8:94.

Beldade, P., and P.M. Brakefield. 2002. The genetics and evo-devo of butterfly wing patterns. *Nat. Rev. Genet.* 3:442–52.

Beldade, P., P.M. Brakefield, and A.D. Long. 2002. Contribution of *Distal-less* to quantitative variation in butterfly eyespots. *Nature* 415:315–18.

Beldade, P., P.M. Brakefield, and A.D. Long. 2005. Generating phenotypic variation: prospects from "evo-devo" research on *Bicyclus anynana* wing patterns. *Evol. Dev.* 7:101–07.

Beldade, P., V. French, and P.M. Brakefield. 2008. Developmental and genetic mechanisms for evolutionary diversification of serial repeats: Eyespot size in *Bicyclus anynana* butterflies. *J. Exp. Zoolog. B Mol. Dev. Evol.* 310:191–201.

Beldade, P., K. Koops, and P.M. Brakefield. 2002. Developmental constraints versus flexibility in morphological evolution. *Nature* 416:844–47.

Beldade, P., W.O. McMillan, and A. Papanicolaou. 2008. Butterfly genomics eclosing. *Heredity* 100:150–57.

Beldade, P., S. Rudd, J.D. Gruber, and A.D. Long. 2006. A wing expressed sequence tag resource for *Bicyclus anynana* butterflies, an evo-devo model. *BMC Genomics* 7:130.

Beldade, P., S.V. Saenko, N. Pul, and A.D. Long. 2009. A gene-based linkage map for *Bicyclus anynana* butterflies allows for a comprehensive analysis of synteny with the lepidopteran reference genome. *PLoS Genet.* 5: e1000366.

Beldade, P., and S.V. Saenko. 2009. Conserved developmental processes and the evolution of novel traits: wounds, embryos, veins, and butterfly eyespots. In *Animal Evolution*, M Telford & T Littlewood (eds). Novartis Foundation. (in press)

Brakefield, P.M., P. Beldade, and B.J. Zwaan. 2009. The African butterfly *Bicyclus anynana*: Evolutionary genetics and evo-devo. In *Emerging model organisms: A laboratory manual,* Vol. 1, ed. R.R. Behringer, A.D. Johnson, and R.E. Krumlauf. Cold Spring Harbor: Cold Spring Harbor Laboratory. pp. 291–330.

Brakefield, P.M., and W.A. Frankino. 2009. Polyphenisms in Lepidoptera: Multidisciplinary approaches to studies of evolution. In *Phenotypic plasticity in insects. Mechanisms and consequences*, ed. D.W. Whitman and T.N. Ananthakrishnan. Oxford: Oxford Univ. Press. pp. 121–152.

Brakefield, P.M., and V. French. 1995. Eyespot development on butterfly wings: The epidermal response to damage. *Dev. Biol.* 168:98–111.

Brakefield, P.M., and V. French. 1999. Butterfly wings: The evolution of development of colour patterns. *Bioessays* 21:391–401.

Brakefield, P.M., J. Gates, D. Keys, et al. 1996. Development, plasticity and evolution of butterfly wing patterns. *Nature* 384:236–42.

Brakefield, P.M., F. Kesbeke, and P.B. Koch. 1998. The regulation of phenotypic plasticity of eyespots in the butterfly *Bicyclus anynana*. *Am. Nat.* 152:853–60.

Brakefield, P.M., and J.C. Roskam. 2006. Exploring evolutionary constraints is a task for an integrative evolutionary biology. *Am. Nat.* 168, Suppl. 6:S4–13.

Brunetti, C.R., J E. Selegue, A. Monteiro, V. French, P.M. Brakefield, and S.B. Carroll. 2001. The generation and diversification of butterfly eyespot color patterns. *Curr. Biol.* 11:1578–85.

Carroll, S.B., J. Gates, D.N. Keys, et al. 1994. Pattern formation and eyespot determination in butterfly wings. *Science* 265:109–14.

Clark, R., S.M. Brown, S.C. Collins, C.D. Jiggins, D.G. Heckel, and A.P. Vogler. 2008. Colour pattern specification in the Mocker swallowtail *Papilio dardanus*: The transcription factor *invected* is a candidate for the mimicry locus *H*. *Proc. Biol. Sci.* 275:1181–88.

Condamin, M. 1973. *Monographie du Genre Bicyclus (Lepidoptera, Satyridae).* Dakar: Inst. Fond. Afr. Noire.

Costanzo, K., and A. Monteiro. 2007. The use of chemical and visual cues in female choice in the butterfly *Bicyclus anynana*. *Proc. Biol. Sci.* 274:845–51.

Dilão, R., and J. Sainhas. 2004. Modelling butterfly wing eyespot patterns. *Proc. Biol. Sci.* 271:1565–69.

Evans, T.M., and J.M. Marcus. 2006. A simulation study of the genetic regulatory hierarchy for butterfly eyespot focus determination. *Evol. Dev.* 8:273–83.

French, V., and P.M. Brakefield. 1992. The development of eyespot patterns on butterfly wings: Morphogen sources or sinks? *Development* 116:103–09.

French, V., and P.M. Brakefield. 1995. Eyespot development on butterfly wings: The focal signal. *Dev. Biol.* 168:112–23.

Galant, R., J.B. Skeath, S. Paddock, D.L. Lewis, and S.B. Carroll. 1998. Expression pattern of a butterfly *achaete-scute* homolog reveals the homology of butterfly wing scales and insect sensory bristles. *Curr. Biol.* 8:807–13.

Golden, K., V. Saji, N. Markwarth, B. Chen, and A Monteiro. 2007. In vivo electroporation of DNA into the wing epidermis of a butterfly. *J. Insect Sci.* 7:53.

Joron, M., C.D. Jiggins, A. Papanicolaou, and W.O. McMillan. 2006. *Heliconius* wing patterns: An evo-devo model for understanding phenotypic diversity. *Heredity* 97:157–67.

Joron, M., and J.L.B. Mallet. 1998. Diversity in mimicry: Paradox or paradigm? *Trends Ecol. Evol.* 13:461–66.

Keys, D.N., D.L. Lewis, J.E. Selegue, et al. 1999. Recruitment of a *hedgehog* regulatory circuit in butterfly eyespot evolution. *Science* 283:532–34.

Koch, P.B., B. Behnecke, and R.H. ffrench-Constant. 2000. The molecular basis of melanism and mimicry in a swallowtail butterfly. *Curr. Biol.* 10:591–94.

Koch, P.B., P.M. Brakefield, and F. Kesbeke. 1996. Ecdysteroids control eyespot size and wing color pattern in the polyphenic butterfly *Bicyclus anynana* (Lepidoptera: Satyridae). *J. Insect Physiol.* 42:223–30.

Koch, P.B., U. Lorenz, P.M. Brakefield, and R.H. ffrench-Constant. 2000. Butterfly wing pattern mutants: Developmental heterochrony and co-ordinately regulated phenotypes. *Dev. Genes Evol.* 210:536–44.

Koch, P.B., R. Merk, R. Reinhardt, and P. Weber. 2003. Localization of ecdysone receptor protein during colour pattern formation in wings of the butterfly *Precis coenia* (Lepidoptera: Nymphalidae) and co-expression with Distal-less protein. *Dev. Genes Evol.* 212:571–84.

Kronforst, M.R., L.G. Young, D.D. Kapan, C. McNeely, R.J. O'Neill, and L.E. Gilbert. 2006. Linkage of butterfly mate preference and wing color preference cue at the genomic location of *wingless*. *Proc. Natl. Acad. Sci. U.S.A.* 103:6575–80.

Lewis, D.L., and C.R. Brunetti. 2006. Ectopic transgene expression in butterfly imaginal wing discs using vaccinia virus. *Biotechniques* 40:48, 50, 52.

Lewis, D.L., M.A. DeCamillis, C.R. Brunetti, et al. 1999. Ectopic gene expression and homeotic transformations in arthropods using recombinant Sindbis viruses. *Curr. Biol.* 9:1279–87.

Long, A.D., P. Beldade, and S.J. Macdonald. 2007. Estimation of population heterozygosity and library construction-induced mutation rate from expressed sequence tag collections. *Genetics* 176:711–14.

Lyytinen, A., P.M. Brakefield, L. Lindstrom, and J. Mappes. 2004. Does predation maintain eyespot plasticity in *Bicyclus anynana*? *Proc. Biol. Sci.* 271:279–83.

Marcus, J.M., and T.M. Evans. 2008. A simulation study of mutations in the genetic regulatory hierarchy for butterfly eyespot focus determination. *Biosystems*. 93:250–255.

Marcus, J.M., D.M. Ramos, and A. Monteiro. 2004. Germline transformation of the butterfly *Bicyclus anynana*. *Proc. Biol. Sci.* 271, Suppl. 5:S263–65.

Majerus, M.E.N. 1998. *Melanism: Evolution in action*. Oxford: Oxford Univ. Press.

McMillan, W.O., A. Monteiro, and D.D. Kapan. 2002. Development and evolution on the wing. *Trends Ecol. Evol.* 17:125–33.

Monteiro A. 2008. Alternative models for the evolution of eyespots and of serial homology on lepidopteran wings. *Bioessays* 30:358–66.

Monteiro, A.F., P.M. Brakefield, and V. French. 1994. The evolutionary genetics and developmental basis of wing pattern variation in the butterfly *Bicyclus anynana*. *Evolution* 48:1147–57.

Monteiro, A., P.M. Brakefield, and V. French. 1997a. Butterfly eyespots: The genetics and development of the color rings. *Evolution* 51:1207–16.

Monteiro, A., P.M. Brakefield, and V. French. 1997b. The genetics and development of an eyespot pattern in the butterfly *Bicyclus anynana*: Response to selection for eyespot shape. *Genetics* 146:287–94.

Monteiro, A., B. Chen, L.C. Scott, et al. 2007. The combined effect of two mutations that alter serially homologous color pattern elements on the fore and hindwings of a butterfly. *BMC Genet.* 8:22.

Monteiro, A., G. Glaser, S. Stockslager, N. Glansdorp, and D. Ramos. 2006. Comparative insights into questions of lepidopteran wing pattern homology. *BMC Dev. Biol.* 6:52.

Monteiro, A., and N.E. Pierce. 2001. Phylogeny of *Bicyclus* (Lepidoptera: Nymphalidae) inferred from COI, COII, and *EF-1* alpha gene sequences. *Mol. Phylogenet. Evol.* 18:264–81.

Monteiro, A., J. Prijs, M. Bax, T. Hakkaart, and P.M. Brakefield. 2003. Mutants highlight the modular control of butterfly eyespot patterns. *Evol. Dev.* 5:180–87.

Nijhout, H.F. 1980. Pattern formation on lepidopteran wings: Determination of an eyespot. *Dev. Biol.* 80:267–74.

Nijhout, H.F. 1991. *The development and evolution of butterfly wing patterns*. Washington, DC: Smithsonian Inst. Press.

Nijhout, H.F. 2003a. The development and evolution of adaptive polyphenisms. *Evol. Dev.* 5:9–18.

Nijhout, H.F. 2003b. Polymorphic mimicry in *Papilio dardanus*: Mosaic dominance, big effects, and origins. *Evol. Dev.* 5:579–92.

Papanicolaou, A., S. Gebauer-Jung, M.L. Blaxter, W.O. McMillan, and C.D. Jiggins. 2008. ButterflyBase: A platform for lepidopteran genomics. *Nucleic Acids Res.* 36:D582–87.

Parchem, R.J., M.W. Perry, and N.H. Patel. 2007. Patterns on the insect wing. *Curr. Opin. Genet. Dev.* 17:300–08.

Paulsen, S.M., and H.F. Nijhout. 1993. Phenotypic correlation structure among elements of the color pattern in *Precis coenia* (Lepidoptera, Nymphalidae). *Evolution* 47:593–618.

Pringle, E.G., S.W. Baxter, C.L. Webster, A. Papanicolaou, S.F. Lee, and C.D. Jiggins. 2007. Synteny and chromosome evolution in the lepidoptera: Evidence from mapping in *Heliconius melpomene*. *Genetics* 177:417–26.

Ramos, D.M., F. Kamal, E.A. Wimmer, A.N. Cartwright, and A. Monteiro. 2006. Temporal and spatial control of transgene expression using laser induction of the hsp70 promoter. *BMC Dev. Biol.* 6:55.

Ramos, D.M., and A. Monteiro. 2007. Transgenic approaches to study wing color pattern development in Lepidoptera. *Mol. Biosyst.* 3:530–35.

Reed, R.D. 2004. Evidence for Notch-mediated lateral inhibition in organizing butterfly wing scales. *Dev. Genes Evol.* 214:43–46.

Reed, R.D., P.H. Chen, and H.F. Nijhout. 2007. Cryptic variation in butterfly eyespot development: The importance of sample size in gene expression studies. *Evol Dev* 9:2–9.

Reed, R.D., W.O. McMillan, and L.M. Nagy. 2008. Gene expression underlying adaptive variation in Heliconius wing patterns: Non-modular regulation of overlapping cinnabar and vermilion prepatterns. *Proc. Biol. Sci.* 275:37–45.

Reed, R.D., and L.M. Nagy. 2005. Evolutionary redeployment of a biosynthetic module: expression of eye pigment genes vermilion, cinnabar, and white in butterfly wing development. *Evol. Dev.* 7:301–11.

Reed, R.D., and M.S. Serfas. 2004. Butterfly wing pattern evolution is associated with changes in a Notch/Distal-less temporal pattern formation process. *Curr. Biol.* 14:1159–66.

Riddiford, L.M. 1995. Hormonal regulation of gene expression during lepidopteran development. In *Molecular model systems in the Lepidoptera*, ed. M.R. Goldsmith and A.S. Wilkins, 1st ed., 293–322. Cambridge: Cambridge Univ. Press.

Robertson, K.A., and A. Monteiro. 2005. Female *Bicyclus anynana* butterflies choose males on the basis of their dorsal UV-reflective eyespot pupils. *Proc. Biol. Sci.* 272:1541–46.

Saenko, S.V., V. French, P.M. Brakefield, and P. Beldade. 2008. Conserved developmental processes and the formation of evolutionary novelties: Examples from butterfly wings. *Philos. Trans. R. Soc. Lond. B Biol. Sci.* 363:1549–55.

Sahara, K., A. Yoshido, F. Marec, et al. 2007. Conserved synteny of genes between chromosome 15 of *Bombyx mori* and a chromosome of *Manduca sexta* shown by five-color BAC-FISH. *Genome* 50: 1061–65.

Stevens, M. 2005. The role of eyespots as anti-predator mechanisms, principally demonstrated in the Lepidoptera. *Biol. Rev. Camb. Philos. Soc.* 80:573–88.

Wittkopp, P.J., and P. Beldade. 2009. Development and evolution of insect pigmentation: Genetic mechanisms and the potential consequences of pleiotropy. *Sem. Cell. Dev. Biol.* 20: 65–71.

6 Prospects for Locating Adaptive Genes in Lepidopteran Genomes

A Case Study of Butterfly Color Patterns

Simon W. Baxter, Owen McMillan, Nicola Chamberlain, Richard H. ffrench-Constant, and Chris D. Jiggins

CONTENTS

BACKGROUND

The Lepidoptera is a diverse clade that has long attracted the attention of biologists interested in ecological and evolutionary processes. This dates back to classic evolutionary genetic studies of natural populations, which have contributed significantly to our current understanding of the natural world (Clarke and Sheppard 1960; Kettlewell 1973; Ford 1975). Nonetheless, such studies were limited by the fact that the link between genotype and phenotype was essentially an intractable "black box." In contrast, we are now in an era in which it is becoming increasingly feasible to clone genes with major phenotypic effects in even the most poorly studied genomes, offering an unprecedented opportunity to understand phenotypic evolution at a molecular level (Feder and Mitchell-Olds 2003). Consequently, there is now considerable recent research interest in identifying genes controlling major phenotypic traits in the Lepidoptera, and several recent studies have made significant progress in documenting the genetic basis of phenotypic traits ranging from insecticide resistance, through dispersal ability, to morphological traits such as color pattern (Gahan, Gould, and Heckel 2001; Daborn et al. 2002; Hanski and Saccheri 2006; Joron et al. 2007).

Heliconius wing patterns are of considerable interest to evolutionary biologists as an example of recent adaptive radiation in a trait that is well understood at an ecological level (Joron et al. 2007). *Heliconius melpomene* and *H. erato* are comimetic neotropical butterflies that share wing patterns within local populations, yet show pattern diversity between geographic regions. Although *H. melpomene* and *H. erato* cannot interbreed, it is possible to cross races displaying variable wing patterns from within each species. Historical crosses have uncovered what appears to be a remarkably simple genetic system in which only a few loci control the qualitative presence or absence of *Heliconius* wing pattern elements (Sheppard et al. 1985). Thus, an apparently simple genetic system underlies a rapid adaptive radiation, and multiple cases of evolutionary convergence. A third species, *Heliconius numata*, has an unusual "supergene" system in which a single Mendelian locus controls whole-wing pattern polymorphism (Joron et al. 2007). Knowledge of the molecular basis of wing pattern formation therefore offers an exciting opportunity to break open the black box that links naturally variable genotypes with the phenotypes that affect fitness. In particular we would like to know how allelic changes at single loci control pattern development and the kinds of molecular changes that underlie such variation. In addition, this will offer an unusual opportunity to study the sequence variability around loci that have been subject to repeated bouts of strong selection in nature. Importantly, all of these evolutionary changes have occurred multiple times involving both divergence and convergence, offering a chance to test the repeatability of any patterns that are uncovered.

This chapter describes the molecular resources now available for *Heliconius* and how they are being deployed to identify the genes controlling wing pattern variability. These methods could be used to pinpoint Mendelian or quantitative trait loci of interest segregating in other lepidopteran species. More broadly, we also discuss the implications of chromosomal synteny within the Lepidoptera and the opportunities it presents for identifying—or excluding—candidate genes controlling major phenotypes. The major goal of this chapter is therefore to summarize what we have learned so far in our work on *Heliconius* that will be of general interest and utility for gene finding in the Lepidoptera.

RESOURCES AVAILABLE FOR *HELICONIUS*

Over the last five years, new molecular resources have been generated for *Heliconius* butterflies, including bacterial artificial chromosome (BAC) libraries, genome survey sequences (GSSs), expressed sequence tag (EST) libraries, microsatellite markers and ButterflyBase, an online database. These resources are being developed in parallel in the three focal species, *H. melpomene*, *H. erato*, and *H. numata*, with the aim of facilitating comparative analysis of the evolution of these genomes.

MICROSATELLITE MARKERS

In common with other lepidopterans, microsatellite markers have proven difficult to develop in large numbers for *Heliconius* (Mavárez and González 2006). Nonetheless, around twenty markers have been mapped in each of the two focal species, *H. erato* and *H. melpomene*, providing a set of loci that can be used to identify chromosomal linkage groups (LGs) in new mapping experiments (Jiggins et al. 2005; Kapan et al. 2006). In most cases markers developed for *H. melpomene* work well in the closely related *H. numata*. The number of loci that are in common between the two species is small, so these markers have not proven particularly useful for comparative linkage analysis. Nonetheless, a panel of thirteen loci have proven to be useful for population genetic experiments in the *H. melpomene* and *H. cydno* group, and have been used for hybrid zone studies, identification of cryptic species and studies of hybrid speciation (Mavárez et al. 2006)

Genomic Libraries

Large insert genomic BAC libraries are now available for three *Heliconius* species: *H. melpomene*, *H. numata*, and *H. erato*. Average insert sizes and genome coverage are approximately 120 kb and 8× for *H. melpomene* and 115 kb and 7.5× for *H. numata* (available from Amplicon Express). For *H. erato* two libraries are available, constructed using different restriction enzymes, with average insert sizes of 153 kb and 175 kb (C. Wu, pers. comm.), giving an overall genome coverage of 16× (Papa et al. 2008). These libraries are all available gridded onto high-density nylon membranes for screening. In the case of *H. melpomene* the whole library has also been fingerprinted and assembled into 2,091 contigs (plus singletons; Baxter et al. 2008).

Genomic Sequence Surveys

The 18,000 clones of the *H. melpomene* BAC library have been sequenced at both ends, giving a total of 32,528 high-quality single-pass genomic sequences (GSSs). These sequences offer a random snapshot representing 7 percent of the entire genome and include a large number of repeated elements and putative gene sequences. A clustering of the reads using the PartiGene pipeline (Parkinson et al. 2004) resulted in 23,420 sequence clusters and singletons. The larger clusters represent repeated elements containing up to 161 sequence reads. For a preliminary survey of repeated elements in the *Heliconius* genome, see Papa et al. (2008). In total 2,133 clusters have a translated BLAST similarity at a bit score greater than 80 to at least one member of the UniProt protein database, and 4,121 clusters show a tBLASTx hit to the *Bombyx* genome at the same cutoff bit score (Ramen contigs from KAIKObase http://sgp.dna.affrc.go.jp/pubdata/genomicsequences.html) and therefore must contain fragments of genes or other conserved elements. These sequences therefore represent a useful resource for gene and marker identification, for identification of repeat elements from the *Heliconius* genome, and for anchoring BAC contigs with sequence-based markers.

cDNA Libraries

Building good complementary DNA (cDNA) libraries is an important initial step in the development of genomic resources and a starting point for any study designed to identify functionally important genes. Libraries should be (1) composed of full-length cDNA, (2) directional (i.e., oriented relative to the five-prime and three-prime UTRs), and (3) normalized. Normalization is the process whereby the abundance of transcripts is equalized across the transcriptome. It is recommended because transcript abundance varies over several orders of magnitude, and a handful of highly expressed genes often account for the bulk of mRNA within a cell (Bonaldo, Lennon, and Soares 1996; Hillier et al. 1996). Thus, without normalization, most recombinant clones in the cDNA library contain an insert from one of these highly expressed genes, many of which are general "housekeeping" genes such as actins, tubulins, and ribosomal proteins. Normalization therefore facilitates the discovery of rare transcripts, some of which may play an important role in ecological and developmental processes (Bouck and Vision 2007).

Library construction has become far easier over the last decade, and there are a number of robust commercially available kits that couple cDNA isolation, normalization, and library construction. We have had good success recently with the Trimmer-Direct cDNA normalization kit (Evrogen, Moscow, Russia). Nonetheless, construction of a good cDNA library requires expertise, and the process can also be "farmed out" to a third party. Although the upfront costs are higher when using a third party, this can be the most cost-effective strategy (and the shortest path to success) in situations where technical expertise and experience are limiting.

Transcriptomic Sequence Surveys

Transcriptomic sequences are an invaluable tool for gene discovery, for annotation of genomic sequence, and for marker development. Sequencing costs are now relatively low (and falling), and with a modest investment, cDNA libraries can be mined to provide the first glimpse of the transcriptome from specific tissues. In addition, ESTs generated from randomly sequencing clones of a cDNA library comprise a large and rapidly growing source of taxonomically diverse genomic information (Bouck and Vision 2007). As such, ESTs are an important bridge to the genomic resources of the major model organisms and a foundation for studies of adaptive change, comparative mapping (Pringle et al. 2007), and expression studies (Reed, McMillan, and Nagy 2007; Bouck and Vision 2007). EST sequencing in *H. erato* and *H. melpomene* has been undertaken primarily using different developmental stages of wing disks (Papanicolaou et al. 2005). For *H. erato*, we have now sequenced about 18,000 ESTs from a cDNA library constructed using wing tissue collected at various stages during development. The library was not normalized and, as a consequence, for many genes we have multiple sequence reads including about 20 genes that have been sequenced 75 times or more. This redundancy increases cost and slows gene discovery but has two important benefits. First, the redundancy means that we typically assay a much longer region of the transcribed molecule, which helps in downstream gene identification and characterization. Second, because our library was made from many individuals, the overlapping sequences reveal a wealth of SNPs that can be used in future population genomic studies. The 18,000 *H. erato* ESTs fell into 6,564 clusters representing between 3,000 and 5,000 distinct gene objects, including genes known to be involved in insect wing development, signaling, and pigment production (Papanicolaou et al. 2007). In addition, there was a large subset of coding regions, some with fairly large predicted peptides, that did not show clear homology to genes in *Drosophila melanogaster* or *Bombyx mori* (the only Lepidoptera with a fully sequenced genome) and may be novel or highly divergent genes in butterflies.

The next generation of sequence technologies provides a more rapid and complementary strategy for characterizing transcriptional diversity and promises a rapid development of functional genomic tools for Lepidoptera (Toth et al. 2007; Hudson 2008; Vera et al. 2008). Importantly, for expression characterization these technologies do not require the production of a library, and the template for pyrosequencing reactions is approximately 1–4 μg of double-stranded cDNA. The cDNA can be normalized if required, and the rapidity with which EST data can be generated is impressive. For example, a single 8-hour run on a 454 pyrosequencing FLX machine (Roche Diagonostics, Burgess Hill, West Sussex, UK) can generate nearly 100 megabases of sequence comprising roughly 400,000 sequence reads averaging between 200 and 250 base pairs (bp; Hudson 2008). Single-read accuracies are reasonably high, except for homopolymer runs (Margulies et al. 2005; Huse et al. 2007). However, because there are multiple overlapping reads, consensus accuracies are much higher (Wicker et al. 2006). The first such experiment in *H. melpomene* has generated nearly 6,500 consensus sequences of average length 1,050 bp from a single run of normalized cDNA. This represents a similar number of genes as were generated from the large EST sequencing project in *H. erato*, although the short sequence reads can make assembly of overlapping sequences problematic where there is abundant intraspecific polymorphism or when alternative splicing is common (Trombetti et al. 2007). A combination of traditional Sanger sequencing and data from the next-generation 454 sequencing platforms should allow for high-quality alignments to be generated for a large proportion of the *Heliconius* transcriptome in a relatively short time, and similar sequencing projects are also now under way for *H. numata* and *H. erato*.

BUTTERFLYBASE

The large amount of data generated during even moderate-scale EST projects poses serious organizational challenges. Vector and poor-quality sequence must be removed, sequences need to be compared to each other and "clustered" into groups that likely come from the same locus, and sequences within these groups need to be assembled into tiles of overlapping sequences (i.e., contigs). Moreover, these data need to be assembled into a relational database that houses additional annotation information. At the very least, such a database should include information on (1) sequence similarity to genes in other organisms, (2) robust protein translation, and (3) possible biological role and presence of conserved protein domains. In addition, information on SNPs, expression patterns, and possible gene interactions should also be considered. There are now a variety of automated pipelines for the analysis of raw sequence data (see below) and a number of species-specific Lepidoptera databases are available. As an example, SpodoBase (http://bioweb.ensam.inra.fr/spodobase/) was designed as an integrated database to house EST data generated on the fall armyworm, *Spodoptera frugiperda*.

Species-specific databases such as SpodoBase, KAIKObase, and SilkDB are valuable resources but differ widely in methods of analysis and accessibility. As a result, the community is not taking full advantage of the diversity that has long made Lepidoptera exceptional models to study the origins and maintenance of functional variation. ButterflyBase has been established as a first attempt to coordinate and organize genomic information being generated in Lepidoptera (Papanicolaou et al. 2007). This relational database makes available all lepidopteran EST data in a clustered and annotated format. As of August 2007, the database houses 273,077 mRNA sequences clustered into 70,867 gene objects, most of which also have predicted protein sequences. This contrasts with only 6,907 protein sequences held in GenBank for all of the Lepidoptera, many of which are partial and fragmented. All of the sequences have been processed using a common analysis pipeline (Parkinson et al. 2004), making the data directly comparable between species. These data are searchable both using text searches against databased similarity annotation, and using BLAST. Several BLAST tools are available including the original blastall options (BLASTn, tBLASTx, etc.) and the newer algorithms psi-BLAST and msBLAST. The latter is designed for searching using short protein sequences obtained from proteomics studies.

Of course, there are limitations to the current level of data analysis, and, in particular, the use of an automated pipeline for analysis of the raw sequence data does have disadvantages. Primarily, there are errors in clustering of EST sequences: Distinct sequences from the same gene fail to cluster together because of allelic variation or sequencing errors, or conversely, closely related gene paralogs may be clustered together falsely. In the *H. erato* data set, for example, over 6,500 clusters can be grouped into 2,000 "superclusters," each composed of a number of individual clusters showing some sequence similarity. Intraspecific variation is high in *H. erato* (there is on average a polymorphism every 25 base pairs) and closer inspection suggests that at least some of these superclusters are composed of allelic variants of the same locus. Any automated pipeline will make such errors, which need to be taken into account when using the data from ButterflyBase. In the future the aim is to overcome such problems by developing a database with an element of manual curation similar to FlyBase, in which such errors are systematically corrected by users. There is already a feature on ButterflyBase for users to add annotations to gene objects by means of an annotation wiki, and we encourage the use of this feature by visitors to the site.

GENETIC ANALYSIS USING CROSSES

Genomic sequences provide a valuable set of tools, but the real value of the *Heliconius* system is the ability to carry out genetic experiments using crosses to identify naturally occurring allelic variants controlling wing patterns. Crosses between different *H. melpomene* and *H. erato* races are, therefore, a vital resource that enables genetic analysis of trait segregation. By crossing individuals of the same species displaying different wing patterns, dominant and recessive traits can be observed segregating in broods.

The haploid genome of lepidopterans typically contains a large but variable number (generally 28 to 32) of holokinetic chromosomes (Suomalainen 1969). The chromosomal system of Lepidoptera has a very useful feature that greatly facilitates genetic analysis of specific traits (characteristic features of lepidopteran chromosomes are reviewed in Chapter 3). Crossing over between homologous chromosomes occurs during meiosis and gamete formation in males, but not in females (Maeda 1939; Suomalainen, Cook, and Turner 1973; Turner and Sheppard 1975; Traut 1977). Each female-derived chromosome remains intact in her offspring, with all genes and markers on the same chromosome completely linked. Homologous chromosomes segregate during egg formation, and progeny are equally likely to inherit either one, allowing a direct assessment of linkage without the complication of exchange between chromosomes contributed by the two parents; whereas in male Lepidoptera, chromosomal crossing over occurs during spermatogenesis (Maeda 1939; Turner and Sheppard 1975; Traut 1977). Using molecular markers, these factors can be utilized in a genetic cross to assign a phenotypic locus to a chromosome using female informative markers, and then to a specific position using male informative markers.

Performing crosses between different phenotypic strains enables segregation of the trait of interest. This is essential for later positioning the locus on a linkage map. The most important consideration is to use single-pair mating experiments, as it can be extremely difficult to determine segregation patterns where mass crosses are involved. For example, after analysis of mass crosses, Eberle and Jehle (2006) concluded baculovirus resistance in the codling moth *Cydia pomonella* was autosomal and incompletely dominant. However, using a single-pair cross system, resistance was later demonstrated to be due to sex-linked inheritance (Asser-Kaiser et al. 2007).

Several crossing strategies can be implemented, depending on the segregation state of the trait of interest, be it dominant or recessive. Heckel et al. (1999) used a backcrossing strategy to follow segregation of a single locus encoding recessive resistance to *Bacillus thuringiensis* (*Bt*) toxin in diamondback moth, *Plutella xylostella*. Here, a homozygous recessive resistant individual was crossed to a homozygous susceptible, and the F_1 females backcrossed to the resistant strain. This created a "female informative" mapping family, which was used to assign markers to LGs. Concurrently, F_1 males were backcrossed to the resistant strain to create "male informative" mapping families in order to generate a recombination linkage map for particular LGs (Heckel at al. 1999; Baxter et al. 2005; Figure 6.1A).

In *Heliconius* we have used F_2 families obtained by crossing color pattern races collected from different parts of South America (Jiggins et al. 2005; Tobler et al. 2005; Kapan et al. 2006). The main reason for an F_2 intercross design rather than a backcross design is that both dominant and recessive traits from both parental strains can be mapped in the same family (Figure 6.1B). In the alternative backcross design, up to four distinct mapping families would have been needed to study the same loci (male- and female-informative in both directions). Furthermore, provided there is sufficient polymorphism both within and between lines, it is possible to construct a linkage map of markers segregating from both parental lines in a single F_2 cross design, taking advantage of the lack of crossing over in females to assign markers to LGs (Yasukochi 1998; Jiggins et al. 2005).

The number of F_2 or backcross progeny obtained from each single-pair cross experiment can be vitally important. In our study, a single large brood (140–180 progeny) has been used to generate preliminary data to determine the chromosome and approximate region in which the wing pattern locus lies. A minimum of around 70 individuals is needed to generate an integrated linkage map from a single F_2 family (Jiggins et al. 2005). Subsequently, tightly linked markers can be mapped in additional broods segregating for the same phenotypes in order to carry out finer-scale linkage mapping.

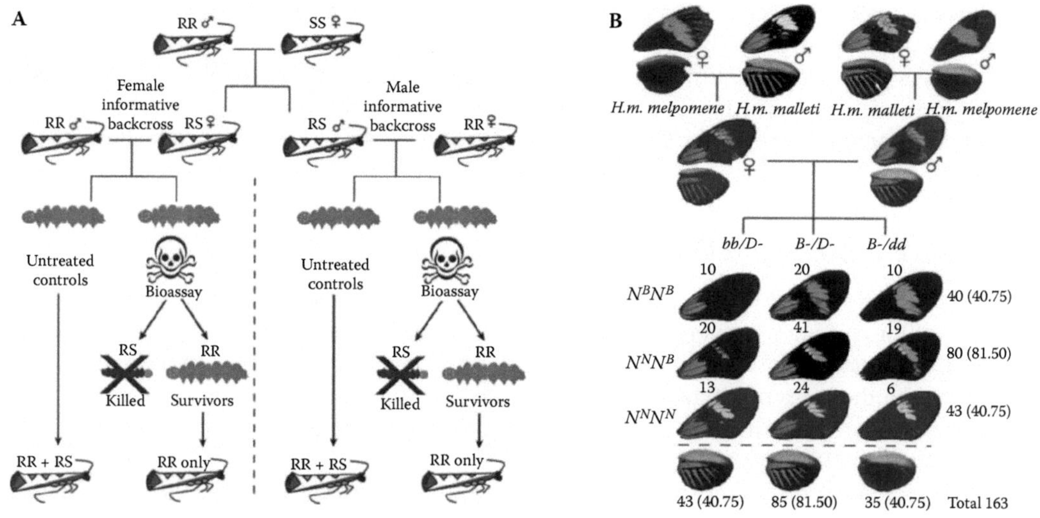

FIGURE 6.1 ***A color version of this figure follows page 176.*** Cross designs for genetic mapping. (A) A backcross enabling segregation of recessive resistance to Bt Cry1A toxins in diamondback moth, *Plutella xylostella*. Resistant (RR) and susceptible (SS) individuals are crossed in single-pair matings to produce Bt susceptible F_1 heterozygotes (RS). Two-phase crosses, using an F_1 female and F_1 male, are generated by independent backcrosses to the resistant strain. The F_2 larvae are either (i) reared to adult moths as untreated controls or (ii) bioassayed using a discriminating dose of Bt toxin, which kills heterozygous susceptible individuals. Due to the absence of crossing over during oogenesis, AFLP markers can be used to identify the LG that contains the resistance factor in the female-informative cross. Here, the resistance LG inherited from the F_1 female should segregate 1R:1S in the untreated control population and be fixed in the bioassay survivors. The male-informative cross can then be used to genetically map the resistance locus, by creating a linkage map with AFLP markers, as crossing over occurs during spermatogenesis. Figure adapted from Heckel et al. (1999) with author permission. (B) Crosses between different races of *Heliconius melpomene* displaying different wing patterns (brood 44). Dorsal butterfly wings are *H. m. melpomene* from French Guiana, displaying the *HmB* phenotype (*B*, dominant red forewing band) and *H. m. malleti* from Ecuador, displaying phenotypes *HmD* (*D*, dominant red hind wing rays plus forewing proximal region) and *HmN* (*N*, incompletely dominant yellow forewing band). F_1 progeny are heterozygous ($BbDdN^NN^B$) for all three phenotypes. These patterns segregate among 163 F_2 progeny and display 9 phenotypes. Observed frequencies do not significantly differ from expected ratios of 1:2:1, shown in brackets. N^NN^N, N^NN^B, N^BN^B are homozygous present, heterozygous and homozygous absent for *HmN*. *bb/D-* lacks the *HmB* phenotype, *B-/D-* are heterozygous for both red phenotypes, and *B-/dd* lacks the *HmD* phenotype. A "-" symbol represents a genotype masked by dominance. Also see Sheppard et al. (1985), page 449.

CREATING LINKAGE MAPS USING AFLPs

Amplified fragment length polymorphism (AFLP) was first described as a method for creating DNA fingerprints, regardless of the origin or structural complexity of the genome in question (Vos et al. 1995). More than a decade later, this method has become established as an important molecular tool for linkage mapping studies, quantitative trait locus (QTL) identification, phylogenetic reconstruction, analysis of population genetic structure and for transcriptome analysis using cDNA (for reviews, see Bensch and Akesson 2005; Meudt and Clarke 2007). Through the analysis of genetic markers such as AFLPs segregating in a backcross family, different chromosomes can be identified with a female-informative backcross. Subsequently, male-informative backcrosses can be used to determine distances between markers known to be on the same LG via recombination frequency.

By scoring large numbers of AFLP markers in a subset of the progeny available, whole-genome linkage maps for both *H. melpomene* and *H. erato* have been assembled (Jiggins et al. 2005; Kapan et al. 2006). Over 200 markers were scored to mark all 21 chromosomes of the 292 Mb *H.*

melpomene genome, plus about 20 microsatellite markers. The total genetic map is 1,600 cM, with 1 cM representing approximately 180 kb. *H. erato* has a larger genome size of 400 Mb, and 300+ AFLP markers were used to estimate the 2,000 cM genome size (1 cM ≈ 200 kb). Although AFLP markers alone are not useful for comparative mapping studies because their sequence identity is unknown, they provide a backbone onto which gene-based markers including candidate genes can be mapped.

CANDIDATE GENE APPROACH

Molecular and cellular pathways studied in model organisms can provide a list of genes considered as candidates for causing a phenotype of interest in the Lepidoptera. Once linkage maps have been assembled, candidate genes can be systematically mapped to chromosomes to test the hypothesis of association with the locus of interest. Conservation of gene function is such that this approach has proven remarkably successful in identifying genes of interest, although it necessarily has limitations (Haag and True 2000).

In the case of insecticide resistance, genes encoding proteins that interact with the insecticide, for example receptors or detoxification enzymes, may be considered as candidate genes for resistance. Candidate genes that do not genetically map to a chromosome or locus conferring resistance can be excluded from playing direct roles in resistance (Baxter et al. 2008). In this, way a candidate resistance gene (a cadherin-like protein) was successfully mapped to a locus (*BtR-4*) conferring 40–80 percent of resistance to insecticidal *Bt* toxin Cry1Ac in *Heliothis virescens* (Gahan, Gould, and Heckel 2001).

Nonetheless, there are clear limitations to this approach, most notably, that studies are confined to genes that have already been studied and that can be readily cloned in the species of interest. In addition, techniques are needed that make no *a priori* assumptions about the nature of the phenotype being studied, in order to identify the genes directly responsible for important field-based phenotypes rather than simply cataloging genes putatively involved in downstream pathways. Our work on *Heliconius* butterflies aims to identify genes controlling pattern polymorphism in poorly characterized genomes and where candidate gene approaches have so far failed. Numerous genes that are either in pigmentation pathways or from signaling pathways known to be expressed in insect wing development have been genetically mapped, including *wingless*, *Dopa-decarboxylase*, *Distal-less*, *invected*, *patched*, *vermilion*, *scalloped*, *cubitus interruptus*, *scarlet*, *white*, *apterous*, and so forth (Joron et al. 2007). None of these candidates was closely linked to any of the color pattern switch genes. By necessity we have therefore been applying positional cloning techniques that make no assumptions about the nature of pattern regulation.

IDENTIFYING AFLP MARKERS LINKED TO TRAITS OF INTEREST

There are two methods to map an AFLP marker close to a locus of interest; first, by chance in the construction of a linkage map; second, by site-directed AFLP bulked segregant analysis. Jiggins et al. (2005) identified a single AFLP marker "*CA_CTAa41*" (henceforth a41) closely linked to the incompletely recessive color pattern locus, *HmYb*, encoding a yellow hind wing bar in *H. melpomene cythera*. Sequencing this AFLP marker enabled genotyping assays to be performed in many similar broods. Ultimately, a41 was found to be one cM from the *HmYb* locus (Joron et al. 2007), highlighting the fact that randomly generated markers can, by chance, be linked to traits of interest.

However, to target specific regions of the genome, it is inefficient to score large numbers of individuals for every marker. Instead, AFLP bulk segregant analysis permits just a few samples to be screened with many different AFLP primer combinations to identify markers located in a particular region of interest (Baxter et al. 2008). The technique involves pooling AFLP preamplification reactions into two bulks that differ by one phenotypic trait. Since the individuals in a mapping family share the same alleles, the two bulks should only differ in markers linked to the genomic region of interest. Both parents of the mapping family and the two pools are screened, requiring only four samples to be scored for each AFLP primer combination.

In this way we have targeted the loci controlling red wing pattern elements in *H. melpomene*. Crosses were performed between two races of *H. melpomene* displaying very different red and black wing phenotypes: *HmB*, a red forewing patch, and *HmD*, red hind wing rays and proximal forewing base. Both loci are dominant and are known to map to the same LG (18). F_2 crosses showed segregation of these phenotypes in a ratio of 1 *HmB-/dd* : 2 *HmB-/D-* : 1 *Hmbb/D-*. As these markers are dominant, dashes are shown where the phenotypes may be homozygous or heterozygous, but are masked by the dominant allele. AFLP templates were prepared individually for F_2 progeny lacking either the forewing spot (*Hmbb*) or the rayed phenotype (*Hmdd*). Once successful AFLP products were identified from individual samples, the recessive phenotypes were pooled into two phenotypic bulks and screened with 256 AFLP primer combinations. In most cases, it was expected both pools would contain identical AFLP banding patterns, as they were siblings. However, bands present in only one bulk were likely to be tightly linked with wing color loci. Nineteen such bands were identified, and were analyzed in all individuals to assess the segregation pattern. From this, ten bands were confirmed positive: Five were associated with the absence of *HmB* phenotype, and five were associated with the absence of *HmD* phenotype. This high rate of false positives (9 of 19) is probably not unexpected given the vagaries of the AFLP PCR process from pooled DNA. The ten positive bands were excised from acrylamide gels and sequenced, and genotyping assays were performed on several similar broods to create a recombinational linkage map. All markers were located within a 7 cM region, and demonstrated *Hmb* and *Hmd* both mapped to the same chromosomal region (Figure 6.2A–C).

From AFLPs to BAC Tile Paths

AFLP markers linked to traits of interest can be sequenced and used to screen BAC libraries by screening clones arrayed onto membrane filters. Ultimately, a tiled path of BAC clones encompassing the locus is required to ensure all candidate genes are sequenced. Through fine-scale linkage mapping of large broods, it is important to identify recombination events in the BAC tile path on either side of the walk, to ensure the entire region surrounding the locus has been detected. Estimates of centimorgan distances, based on an AFLP linkage maps, were 180 kb/cM for *H. melpomene* and 280 kb/cM for *H. erato* (Jiggins et al. 2005; Kapan et al. 2006). Hence, an AFLP 1 cM from the locus of interest may still be several BAC clone lengths away.

To identify the gene or factor controlling the *HmYb* phenotype, the *a41* AFLP marker was used as a probe to screen the *H. melpomene* BAC genomic library. Clone sequencing and subsequent rounds of screening yielded a tile path of greater than 500 kb, containing 6 BAC clones. Similarly, for the *HmB* and *HmD* loci, a tile path of around 600 kb has now been identified that includes recombination events either side of the *HmB* locus (Baxter et al. 2008; Figure 6.2D–E).

Nonetheless, there are a number of difficulties in going from linked AFLP markers to a complete inclusive tile path for the gene of interest, and this has proven to be perhaps the most difficult step so far. A major difficulty is genetic heterogeneity in the BAC library, which has meant that fingerprint contigs are not as extensive as they could have been (Baxter et al. 2008). Indeed in some areas of the genome we have clearly identified two parallel contigs mapping to the same location, which most likely represent divergent alleles sampled in the library. Inspection of the fingerprint data shows little or no band sharing between these clones. In short, the *H. melpomene* fingerprint contigs are generally not sufficiently large to cover more than two to three BAC lengths, meaning that repeated rounds of library screening have been required to assemble a tile path covering a wider region. Thus, it would have been preferable to have made the BAC library from more homozygous source material. Nonetheless, the availability of end sequences for the entire library has greatly accelerated the chromosomal walk, as the end sequences for a particular contig are immediately available to be used as probes for library screening.

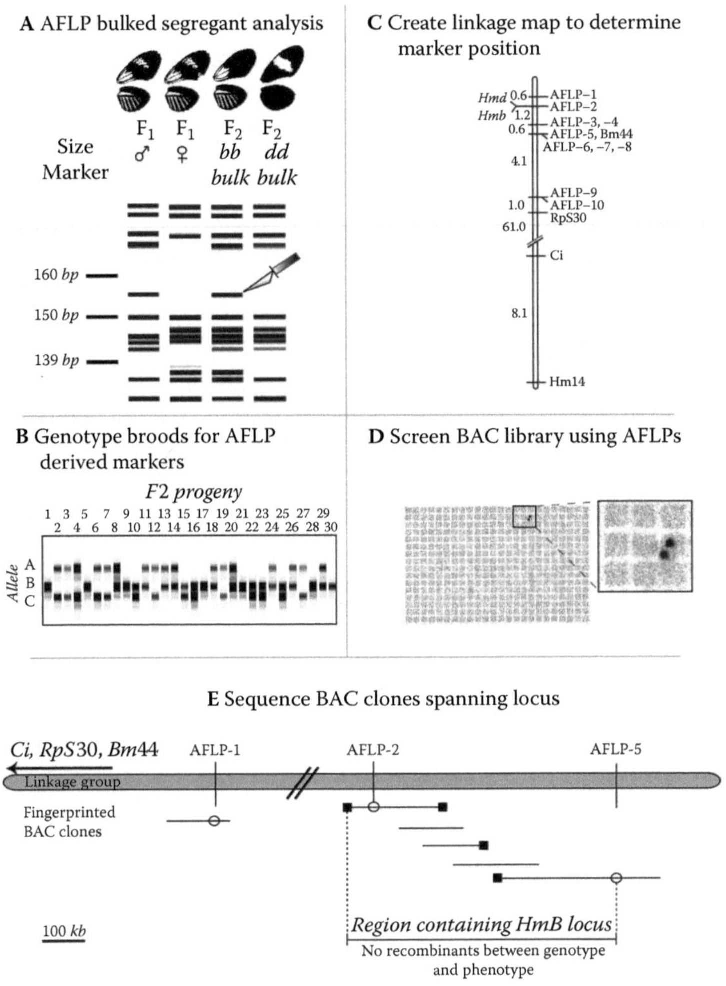

FIGURE 6.2 Identifying genes of major effect in Lepidoptera, using *Heliconius melpomene* brood 44 (Figure 6.1B) as an example. (A) Perform AFLP analysis on F_1 parents and F_2 bulked segregant pools, which are homozygous for *Hmbb/D-* or *HmB-/dd* phenotypes. Locate bands present in the father and one bulk only, and confirm they are genetically linked to the phenotypic locus, by performing AFLP analysis on individuals. AFLP bands segregating with the phenotype are excised from acrylamide gels and sequenced. (B) Develop genotyping assays from sequenced AFLP bands and genotype many broods that are segregating for the same phenotypes. (C) Create a linkage map using AFLP genotypes to determine marker order. Ten AFLP markers are shown, along with three genes, *cubitus interruptus*, *Ribosomal Protein S30*, *Bm44*, and one microsatellite, *Hm14*. Marker AFLP-2 is 0 cM from the B and D loci. (D) Screen a genomic BAC library to isolate clones containing AFLP markers. Enlargement shows a region of the BAC filter containing a positive clone, spotted in duplicate. (E) Sequence the minimum BAC tile path across the genomic region containing the gene(s) of interest. Here, markers AFLP-1, -2 and -5 identified nonoverlapping BAC contigs (open circles). Gaps in the BAC walk were joined between AFLP-2 and -5 by rescreening the BAC library using new probes developed from BAC-end sequences (black squares). Recombinational linkage mapping using BAC end sequence markers and AFLP markers identified the region containing the *HmB* locus.

Comparative Mapping of the Radiation

Before actually identifying the genes' controlling pattern, molecular markers developed in these mapping studies can be used for fine-scale comparisons of the position of patterning genes in different lineages. Where species are sufficiently distantly related that they cannot be crossed together, such as *H. melpomene* and *H. erato*, there has been considerable speculation regarding whether their convergent patterns are controlled by homologous genes (Mallet 1989; Nijhout 1991). Only now that linked molecular markers are available, for which homology can be firmly established, is it possible to compare directly the genetic architecture between species. In fact it turns out that there is a much greater sharing of genetic architecture than had been previously imagined. Not only are major loci with similar phenotypic effects in *H. melpomene* and *H. erato* apparently homologous, but in the third species, *H. numata*, very divergent patterns are also controlled by this same region (Joron et al. 2007; Baxter et al. 2008). Of course a major caveat is that the resolution of our mapping studies currently does not resolve the comparison down to the level of single genes. Nonetheless, these data clearly imply some form of shared patterning regulation across at least three *Heliconius* species. This is exciting as it also means that positional cloning efforts in different species are complementary and will facilitate efforts in others.

COMPARATIVE GENOMIC SEQUENCE DATA

We have now obtained several hundred kilobases of sequence at the *yb* locus of *H. melpomene* and homologous *Cr* locus of *H. erato*. Direct-sequence comparisons show gene-for-gene preservation of gene order between *H. melpomene* and *H. erato* (Figure 6.3). Clearly, this is only a snapshot of a section of the genome of two *Heliconius* species, but nevertheless it is interesting to note that even in two fairly diverged members of different *Heliconius* clades, gene order so far appears conserved. There is therefore no evidence as yet that the strong adaptive selection pressures acting on this region of the genome have led to reorganization of the chromosomal structure. Further sequencing in *Heliconius* species should answer the question of whether

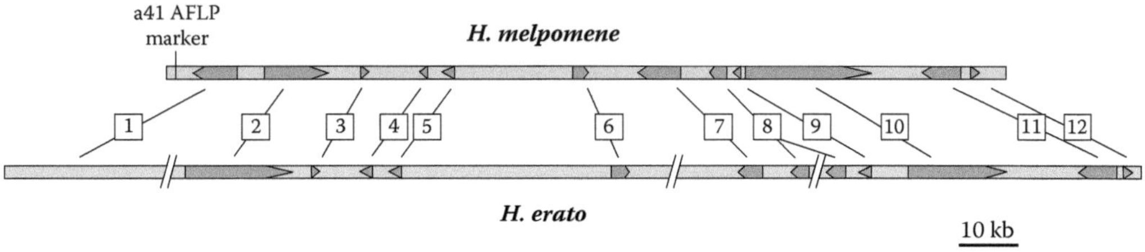

FIGURE 6.3 Comparison of genomic sequence near the *H. melpomene yb* and *H. erato Cr* loci. The *H. melpomene* BAC library was screened using AFLP marker a41, located 1 cM from the *yb* locus controlling a yellow hind wing bar. A gene isolated from this region (*Rab geranylgeranyl transferase beta subunit*) was used as a probe to identify the homologous region in *H. erato*. The gene order and orientation appear conserved; however, *H. erato* encodes multiple genes for endonuclease and reverse transcriptase absent from *H. melpomene* (not shown). Confirmation of gene order will require sequencing of gaps in the *H. erato* BAC sequence. 1. Acylamino acid–releasing enzyme (*Tribolium castaneum*). 2. Acylamino acid–releasing enzyme (*Tribolium castaneum*). 3. Hypothetical protein similar to CG10949, CG3838, CG3919. 4. Trehalase 1 (*Omphisa fuscidentalis*). 5. Trehalase 1 (*O. fuscidentalis*). 6. B9 Protein (*Xenopus laevis*). 7. Similar to CG5098-PA. 8. Similar to CG3184-PA (*T. castaneum*); note that this gene spans a gap in the *H. erato* BAC sequence. 9. Similar to CG18292-PA (*D. melanogaster*) 10. Similar to CG2519-PB (*D. melanogaster*). 11. Unkempt (*Apis mellifera*) 12. Histone H3 (*Mus musculus*). Duplications of the trehalase (4 and 5) and acylamino acid–releasing enzyme (1 and 2) genes are present in both species, and are likely to have occurred in a common ancestor. Although gene order appears conserved, the intronic and intergene sequences in *H. erato* tend to be larger than in *H. melpomene*, so that in *H. erato* gene density appears lower overall.

this genome snapshot is representative of the level of preservation of gene order and synteny in *Heliconius* genomes as a whole.

The data gathered indicate that intron and intergenic region size account for the fact that the *H. melpomene* genome is predicted to be 74 percent of the genome size of its comimic *H. erato* (Tobler et al. 2005). The fact that the sequence of the *Cr* region in *H. erato* is incomplete currently limits our ability to test this difference statistically.

IDENTIFYING WING-PATTERNING GENES—MULTIPLE LINES OF EVIDENCE

Convincing evidence for the role of a particular gene in controlling the *HmYb* wing-patterning phenotype will ideally require transgenic work to knock out the genetic factor controlling the wing phenotype, followed by genetic rescue. Until these research techniques become routine in Lepidoptera, three lines of evidence are being pursued to support identification of the genes or factors controlling wing patterning: (1) population genetic data, (2) expression data, and (3) antibody staining. We are sequencing genes in the *HmYb* and *HmB/HmD* BAC walk in *H. melpomene* and *H. erato* populations caught on either side of hybrid zones. We expect to observe fixed differences for the switch that ultimately controls wing color and regions in close linkage. Expression of patterning genes is likely to occur in developing larval or pupal wing discs (development of wing patterns is also discussed in this chapter). Performing quantitative analysis on gene expression between different races on genes identified from the BAC walks may also provide information on expression timing. There is, of course, the possibility that the switch gene itself may be due to expression of an RNA molecule, and a BAC tiling expression array experiment is under way to investigate which regions of the genomic sequence are expressed in wing tissue. Finally, antibody preparations specific to the strongest candidates generated from population genetic and expression data will be prepared to investigate protein localization across the wing and between developmental stages.

PATTERNS OF SYNTENY IN THE MACROLEPIDOPTERA

Lepidoptera commonly have between 28 and 32 chromosomes, although numbers can vary between 5 and 223 (Suomalainen 1969; Robinson 1971; De Prins and Saitoh 2003; chromosome numbers are also discussed in Chapter 3). An important recent finding has been that broad-scale patterns of chromosomal synteny are strongly conserved between *H. melpomene* ($n = 21$) and *B. mori* ($n = 28$), despite differences in chromosomal number (Pringle et al. 2007). Given the phylogenetic distance between these species, this pattern is likely to hold true across the macrolepidoptera. Such comparisons have only recently become possible since the publication of a detailed linkage map of cDNA derived markers for *Bombyx* (Yasukochi et al. 2006; see Chapter 2 for *Bombyx* linkage maps).

Previous maps based on AFLPs or microsatellite markers did not provide the conserved loci that can be readily identified in distantly related species. The large EST data sets for *Heliconius* allowed us to design PCR primers for large numbers of genes identified as putative homologues of markers already mapped in *Bombyx*. We concentrated on ribosomal proteins, as these are commonly single copy and widely distributed around the genome (Heckel 1993). They are also strongly represented in EST libraries and are therefore relatively easy to develop as markers (Papanicolaou et al. 2005; Beldade, McMillan, and Papanicolaou 2008). This set of markers now provides a toolbox of anchor loci for marking chromosomes and comparative linkage mapping across the butterflies.

As has been noted previously, the lack of crossing over in female Lepidoptera facilitates linkage mapping, and, in particular, allows studies of synteny through rapid assignment of markers to LGs using the method of forbidden recombinants (Heckel 1993). We used restriction fragment length polymorphisms (RFLPs) or length variation scored on agarose gels to assign markers to linkage groups in a single reference brood. In total, 73 orthologous markers have now been mapped in both *H. melpomene* and *B. mori*, and patterns of synteny are completely conserved (Pringle et al. 2007). All markers found to be linked in *Bombyx* were similarly syntenic in *Heliconius* (Figure 6.4). In

H. melpomene linkage group	*B. mori* linkage group	Number of genes in common
1	4	4
2	16	2
3	6	2
4	21	3
5	3	2
6	9	6
7	2, 11	7
8	25	1
9	7	3
10	5, P	5
11	15	6
12	8, 20	4
13	14, 22	5
14	19	1
15	17	3
16	18	1
17	13, 24	7
18	23, U	5
19	12	2
20	10	2
21	1	2
N/A	26	0
21 Chromosomes	28 Chromosomes	73

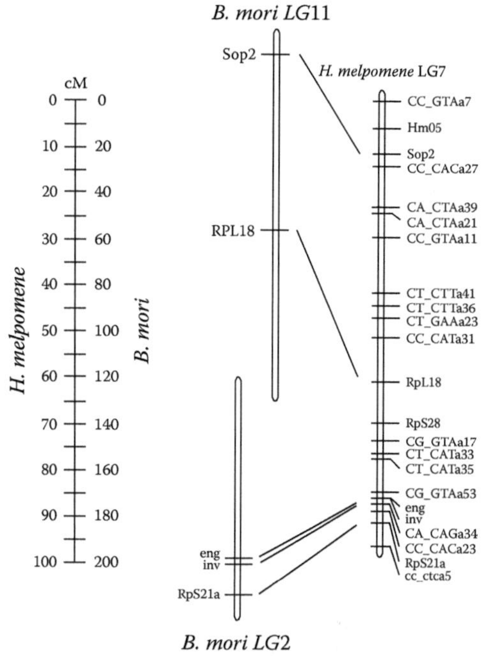

FIGURE 6.4 Conservation of synteny between *Heliconius melpomene* and *Bombyx mori.* A total of 73 markers shared between the species show complete conservation of synteny. There are six possible chromosome fusions, as expected given the overall chromosome number of 21 in *H. melpomene* as compared with 28 in *B. mori.* A linkage map is shown for one of those chromosomes, *H. melpomene* LG7. Note the different scale for the two species. Figure adapted from Pringle et al. (2007) with *Bombyx* data derived from Yasukochi et al. (2006).

addition, there was evidence of six chromosomal fusions in which markers on two distinct LGs in *B. mori* were found on the same LG in *H. melpomene*. This is expected given that derived heliconiines have 21 chromosome pairs, whereas *B. mori* has 28 (Brown et al. 1992). Both are most likely derived from an ancestral number of 31 (De Prins and Saitoh 2003), so additional chromosomal fusions must be present in both lineages but have not yet been detected. This is not unexpected given the small number of markers mapped so far.

This striking conservation of synteny is interesting in its own right as an evolutionary phenomenon. Further, given the wide variation in chromosome number among extant lepidopterans, it is surprising that there has not been greater interchromosomal rearrangement of genes between *Bombyx* and *Heliconius*. For the purposes of identifying genes under selection, however, this result is exciting because it implies that a fully assembled *Bombyx* genome, now in progress (see Chapter 2 for the status of the *Bombyx* genome sequence), will be an invaluable resource for predictive gene finding in taxa even as distant as the butterflies. Of course, the value of this should not be overstated; microsynteny is likely to be far less well conserved, and we would predict a high rate of inversions and intrachromosomal rearrangements between divergent species. This has certainly proven to be the case for fully sequenced comparative genomes such as *Caenorhabditis elegans* and *Caenorhabditis briggsae*, which initially showed high levels of conserved macrosynteny, but at finer scales differ by a high rate of genomic rearrangement (Blaxter 2003).

Thus, in any lepidopteran species where crosses can be obtained, it should be possible to map representative markers from each of the *Bombyx* chromosomes, and thereby assign a phenotypic marker to an LG with putative homology to a *Bombyx* chromosome. Comparison with the *Bombyx* genome would then give additional candidate genes from the same chromosome for mapping and to localize further the gene of interest. As the *Bombyx* genome assembly improves, this is likely to become a valuable tool across the Lepidoptera. Currently genetic data are only available for macrolepidopterans, and therefore in the future it will be important to obtain comparisons of synteny across more basal taxa.

CONCLUSIONS

Heliconius butterflies offer a remarkable opportunity to study multiple recent adaptive radiations. Over the last few years we have developed genomic resources for these butterflies including a backbone of mapped genes conserved across the macrolepidoptera that should prove useful for comparative genetic studies. In addition, we have demonstrated the utility of a method for pinpointing genetic markers associated with a particular genetic trait using bulk-segregant AFLP analysis. In combination with large insert genomic BAC libraries, this method can facilitate cloning and sequencing of a target region in any genome. These methods offer exciting prospects for locating genes of interest in any lepidopteran genome.

ACKNOWLEDGMENTS

We thank other members of the *Heliconius* consortium who have contributed to this work, including Mathieu Joron, Sean Humphray, Alexie Papanicolaou, Riccardo Papa, and Jim Mallet. We also thank the Biotechnology and Biological Sciences Research Council, National Science Foundation, National Environment Research Council (NERC), Leverhulme Trust, Smithsonian Tropical Research Institute, and the Royal Society for funding; and Autoridad Nacional del Medio Ambiente (ANAM) and Ministerio del Ambiente for permits to collect in Panama and Ecuador, respectively.

REFERENCES

Asser-Kaiser, S., E. Fritsch, K. Undorf-Spahn, et al. 2007. Rapid emergence of baculovirus resistance in codling moth due to dominant, sex-linked inheritance. *Science* 317:1916–18.

Baxter, S., R. Papa, N. Chamberlain, et al. 2008. Convergent evolution in the genetic basis of Müllerian mimicry in *Heliconius* butterflies. *Genetics* 180:1567–77.

Baxter, S.W., J.-Z. Zhao, L.J. Gahan, A.M. Shelton, B.E. Tabashnik, and D.G. Heckel. 2005. Novel genetic basis of field-evolved resistance to Bt toxins in *Plutella xylostella*. *Insect Mol. Biol.* 14:327–34.

Beldade, P., W.O. McMillan, and A. Papanicolaou. 2008. Butterfly genomics eclosing. *Heredity* 100:150–57.

Bensch, S., and M. Akesson. 2005. Ten years of AFLP in ecology and evolution: Why so few animals? *Mol. Ecol.* 14:2899–2914.

Blaxter, M. 2003. Comparative genomics: Two worms are better than one. *Nature* 426:395–96.

Bonaldo, M.F., G. Lennon, and M.B. Soares. 1996. Normalization and subtraction: Two approaches to facilitate gene discovery. *Genome Res.* 6:791–806.

Bouck, A., and T. Vision. 2007. The molecular ecologist's guide to expressed sequence tags. *Mol. Ecol.* 16:907–24.

Brown, K.S., T.C. Emmel, P.J. Eliazar, and E. Suomalainen. 1992. Evolutionary patterns in chromosome numbers in neotropical Lepidoptera. I. Chromosomes of the Heliconiini (Family Nymphalidae: Subfamily Nymphalinae). *Hereditas* 117:109–25.

Clarke, C.A., P.M. Sheppard. 1960. The evolution of mimicry in the butterfly *Papilio dardanus*. *Heredity* 14:163–73.

Daborn, P.J., J.L. Yen, M. Bogwitz, et al. 2002. A single P450 allele associated with insecticide resistance in global populations of *Drosophila*. *Science* 297:2253–56.

De Prins, J., and K. Saitoh. 2003. Karyology and sex determination. In *Lepidoptera, moths and butterflies: Morphology, physiology, and development*, ed. N. P. Kristensen, 449–68. Berlin: Walter de Gruyter.

Eberle, K.E., and J.A. Jehle. 2006. Field resistance of codling moth against *Cydia pomonella* granulovirus (CpGV) is autosomal and incompletely dominant inherited. *J. Invertebr. Pathol.* 93:201–06.

Feder, M.E., and T. Mitchell-Olds. 2003. Evolutionary and ecological functional genomics. *Nat. Rev. Genet.* 4:651–57.

Ford, E.B. 1975. *Ecological genetics*. London: Chapman and Hall.

Gahan, L.J., F. Gould, and D.G. Heckel. 2001. Identification of a gene associated with Bt resistance in *Heliothis virescens*. *Science* 293:857–60.

Haag, E.S., and J.R. True. 2000. From mutants to mechanisms? Assessing the candidate gene paradigm in evolutionary biology. *Evolution* 55:1077–84.

Hanski, I., and I. Saccheri. 2006. Molecular-level variation affects population growth in a butterfly metapopulation. *PLoS Biol.* 4:e129.

Heckel, D.G. 1993. Comparative linkage mapping in insects. *Annu. Rev. Entomol.* 38:381–408.

Heckel, D.A., L.J. Gahan, Y. Liu, and B.E. Tabashnik. 1999. Genetic mapping of resistance to *Bacillus thuringiensis* toxins in diamondback moth using biphasic linkage analysis. *Proc. Natl. Acad. Sci. U.S.A.* 96:8373–77.

Hillier, L.D., G. Lennon, M. Becker, et al. 1996. Generation and analysis of 280,000 human expressed sequence tags. *Genome Res.* 6:807–28.

Hudson, M.E. 2008. Sequencing breakthroughs for genomic ecology and evolutionary biology. *Mol. Ecol. Res.* 8:3–17.

Huse, S.M., J.A. Huber, H.G. Morrison, M.L. Sogin, and D.M.Welch. 2007. Accuracy and quality of massively-parallel DNA pyrosequencing. *Genome Biol.* 8:R143.

Jiggins, C.D., J. Mavarez, M. Beltrán, J.S. Johnston, and E. Bermingham. 2005. A genetic map of the mimetic butterfly, *Heliconius melpomene*. *Genetics* 171:557–70.

Joron, M., R. Papa, M. Beltrán, et al. 2007. A conserved supergene locus controls colour pattern diversity in *Heliconius* butterflies. *PLoS Biol.* 4:e303.

Kapan, D.D., N. Flanagan, A. Tobler, et al. 2006. Localization of Müllerian mimicry genes on a dense linkage map of *Heliconius erato*. *Genetics* 173:735–57.

Kettlewell, H.B.D. 1973. *The evolution of melanism. The study of a recurring necessity*. Oxford: Blackwell.

Maeda, T. 1939. Chiasma studies in the silkworm *Bombyx mori*. *Jpn. J. Genet.* 15:118–27.

Mallet, J. 1989. The genetics of warning colour in Peruvian hybrid zones of *Heliconius erato* and *H. melpomene*. *Proc. R. Soc. Lond. B Biol. Sci.* 236:163–85.

Margulies, M., M. Egholm, W.E. Altman, et al. 2005. Genome sequencing in microfabricated high-density picolitre reactors. *Nature* 437:376–80.

Mavárez, J., and M. González. 2006. A set of microsatellite markers for *Heliconius melpomene* and closely related species. *Molecular Ecology Notes* 6:20–23.

Mavárez, J., C. Salazar, E. Bermingham, C. Salcedo, C.D. Jiggins, and M. Linares. 2006. Speciation by hybridization in *Heliconius* butterflies. *Nature* 411:868–71.

Meudt, H.M., and A.C. Clarke. 2007. Almost forgotten or latest practice? AFLP applications, analyses and advances. *Trends Plant Sci.* 12:106–17.

Nijhout, H.F. 1991 *The development and evolution of butterfly wing patterns*. Washington, DC: Smithsonian Institution Press.

Papa, R., C.M. Morrison, J. R. Walters, et al. 2008. Highly conserved gene order and numerous novel repetitive elements in genomic regions linked to wing pattern variation in *Heliconius* butterflies. *BMC Genomics* 9:345.

Papanicolaou, A., S. Gebauer-Jung, M.L. Blaxter, W.O. McMillan, and C.D. Jiggins. 2007. ButterflyBase: A platform for lepidopteran genomics. *Nucleic Acids Res.* 36:D582–87.

Papanicolaou, A., M. Joron, M.L. Blaxter, W.O. McMillan, and C.D. Jiggins. 2005. Genomic tools and cDNA derived markers for butterflies. *Mol. Ecol.* 14:2883–97.

Parkinson, J., A. Anthony, J. Wasmuth, R. Schmid, A. Hedley, and M. Blaxter. 2004. PartiGene—constructing partial genomes. *Bioinformatics* 20:1398–1404.

Pringle, E.G., S. Baxter, C.L. Webster, A. Papanicolaou, S.F. Lee, and C.D. Jiggins. 2007. Synteny and chromosome evolution in the Lepidoptera: Evidence from mapping in *Heliconius melpomene*. *Genetics* 177:417–26.

Reed, R.D., W.O. McMillan, and L.M. Nagy. 2007. Gene expression underlying adaptive variation in *Heliconius* wing patterns: Non-modular regulation of overlapping cinnabar and vermilion patterns. *Proc. Biol. Sci.* 275:37–45.

Robinson, R. 1971. *Lepidoptera genetics*. Oxford: Pergamon Press.

Sheppard, P.M., J.R.G. Turner, K.S. Brown, W.W. Benson, and M.C. Singer. 1985. Genetics and the evolution of Müllerian mimicry in *Heliconius* butterflies. *Philos. Trans. R. Soc. Lond. B Biol. Sci.* 308:433–613.

Suomalainen, E. 1969. Chromosome evolution in the Lepidoptera. *Chromosomes Today* 2:132–38.

Suomalainen, E., L.M. Cook, and J.R.G. Turner. 1973. Achiasmatic oogenesis in the heliconiine butterflies. *Hereditas* 74:302–04.

Tobler, A., D.D. Kapan, N.S. Flanagan, et al. 2005. First generation linkage map of the warningly-colored butterfly *Heliconius erato*. *Heredity* 94:408–17.

Toth, A.L., K. Varala, T.C. Newman, et al. 2007. Wasp gene expression supports an evolutionary link between maternal behavior and eusociality. *Science* 318:441–44.

Traut, W. 1977. A study of recombination, formation of chiasmata and synaptonemal complexes in female and male meiosis of *Ephestia kuehniella* (Lepidoptera). *Genetica* 47:135–42.

Trombetti, G.A., R.J.P. Bonnal, E. Rizzi, G. De Bellis, and L. Milanesi. 2007. Data handling strategies for high throughput pyrosequencers. *BMC Bioinformatics* 8, Suppl.1:S22.

Turner, J.R.G., and P.M. Shepard. 1975. Absence of crossing-over in female butterflies (*Heliconius*). *Heredity* 34:265–69.

Vera, J.C., C.W. Wheat, H.W. Fescemyer, et al. 2008. Rapid transcriptome characterization for a non-model organism using 454 pyrosequencing. *Mol. Ecol.* 17:1636–47.

Vos, P., R. Hogers, M. Bleeker, et al. 1995. AFLP: A new technique for DNA fingerprinting. *Nucleic Acids Res.* 23:4407–14.

Wicker, T., E. Schlagenhauf, A. Graner, T.J. Close, B. Keller, and N. Stein. 2006. 454 sequencing put to the test using the complex genome of barley. *BMC Genomics* 7:275.

Yasukochi, Y. 1998. A dense genetic map of the silkworm, *Bombyx mori*, covering all chromosomes based on 1018 molecular markers. *Genetics* 150:1513–25.

Yasukochi, Y., L.A. Ashakumary, K. Baba, A. Yoshido, and K. Sahara. 2006. A second-generation integrated map of the silkworm reveals synteny and conserved gene order between lepidopteran insects. *Genetics* 173:1319–28.

7 Molecular and Physiological Innovations of Butterfly Eyes

Marilou P. Sison-Mangus and Adriana D. Briscoe

CONTENTS

INTRODUCTION

Color vision is a complex trait that can impact the survivorship of short-lived insects like the Lepidoptera. Within this order, the color vision systems are diverse and are best known among butterflies, which are classified into five families. Several recent reviews have focused on the eyes of the basal papilionid (i.e., *Papilio xuthus*) and pierid butterfly (i.e., *Pieris rapae*) lineages (Stavenga and Arikawa 2006; Wakakuwa, Stavenga, and Arikawa 2007). Both of these groups have eyes that differ from each other and from the other butterfly families in terms of the copy number of the opsin genes that encoded the visual pigments, their spatial expression pattern, and the distribution of lateral filtering pigments. Only one study to date has examined the visual pigments in a riodinid butterfly, *Apodemia mormo* (Frentiu et al. 2007). This chapter focuses on recent advances in our understanding of the unique visual system of lycaenid butterflies, with a special emphasis on the sexually dimorphic retina of *Lycaena rubidus* (Lycaeninae) and the color vision behavior of *Polyommatus icarus* (Polyommatinae). It is clear from character mapping of opsin genes and their expression patterns on a phylogeny of butterfly families that all butterfly eyes are derived from a much simpler eye that resembles the nymphalid eye (Briscoe 2008). Hence, to put the innovations of the lycaenid butterfly visual system into an evolutionary framework, we begin by describing the much simpler visual system of nymphalid butterflies. We then trace the molecular changes in the opsin genes and their expression patterns, and the physiological changes in the visual receptors they encode. Lastly, we discuss the potential behavioral outcomes of the unique eye design of lycaenids. In the course of the review, we mentioned some fertile areas of interest for future study.

ANATOMY OF THE BUTTERFLY OMMATIDIUM

The butterfly compound eye consists of thousands of ommatidia (Yagi and Koyama 1963). Each ommatidium contains a cornea, a crystalline cone, and nine photoreceptor cells (R1–9; Figure 7.1A), along with primary and secondary pigment cells (not shown). The microvilli of each photoreceptor cell form rhabdomeres, which fuse together to form the cylindrical optical structure, a rhabdom. The rhabdom acts as an optical waveguide (Nilsson, Land, and Howard 1988), which extends from the crystalline cone to the basal lamina. The microvillar arrangement that makes up the rhabdom can vary depending on the species. In the simplest case, rhabdomeres of R1–8 have approximately the same length and contribute more or less equally to the rhabdom, and the R9 cell contributes a few microvilli at the base of the rhabdom, producing a tiered structure (Briscoe et al. 2003). At the

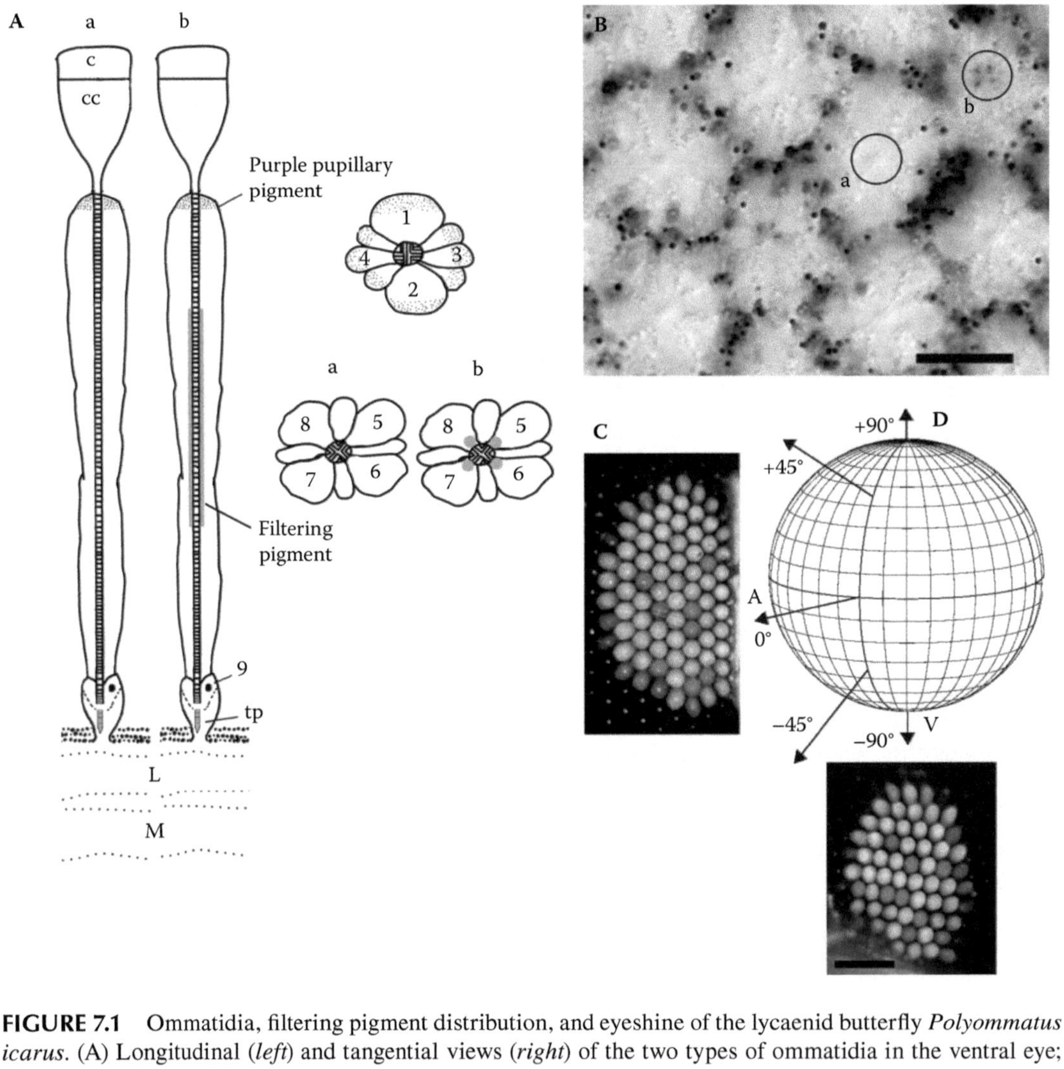

FIGURE 7.1 Ommatidia, filtering pigment distribution, and eyeshine of the lycaenid butterfly *Polyommatus icarus*. (A) Longitudinal (*left*) and tangential views (*right*) of the two types of ommatidia in the ventral eye; nonpigmented (a) and red-pigmented (b). Purple pupillary pigments are also present distally in all R1–R8 photoreceptor cells regulating the amount of light entering each ommatidium. (B) Red-filtering pigment in the lateral eye is absent in some ommatidia (a) and present in others (b). Scale bar, 10 μm. (C) Eyeshine of a female. Ommatidia looking into the anterior (A) and ventral (V) direction reflect yellow (light gray) and red (dark gray); dorsal (D) direction. Scale bar, 50 μm. c, cornea; cc, crystalline cone; 9, the ninth photoreceptor; tp, tapetum; L, lamina; M, medulla. Modified from Sison-Mangus et al. (2008).

proximal end of the rhabdom are stacks of tracheolar cells filled with air that compose the tapetum, a structure that functions as an interference mirror and can bounce unabsorbed light back through the rhabdom, allowing visual pigments to reabsorb the returning light.

The tapetum, the visual pigments, and if present, the filtering pigments (Figure 7.1B) together are responsible for the colored glow (eyeshine; Figure 7.1C) seen in most butterfly eyes (Stavenga et al. 2001). The eyeshine reflectance spectrum is of interest physiologically because it can be used as a noninvasive *in vivo* probe of the absorbance spectrum of the visual pigments found in the ommatidia, particularly in the long-wavelength part of the spectrum, in a completely intact butterfly (Bernard and Miller 1970; Vanhoutte and Stavenga 2005). It can also be used to infer the existence of heterogeneously expressed yellow, orange, and red filter pigments, which modify the wavelengths of light available to excite the visual pigments (Arikawa and Stavenga 1997; Arikawa et al. 1999; Figure 7.1B,C). The butterfly tapetum structure, however, is not universal; it is absent in papilionids (Miller 1979), and variable in its distribution within pierids. Butterflies in the genus *Pieris* have it (e.g., Figure 2 in Briscoe and Bernard 2005), whereas those in the genus *Anthocharis* lack it (Takemura, Stavenga, and Arikawa 2007). For those species that lack the tapetum, it is still possible of course to measure spectral sensitivities of photoreceptor cells using intracellular recordings, and for those species with and without tapeta, histological sections provide the most direct evidence of filtering pigment distributions.

Visual Pigments of the Butterfly Eye

Visual pigments are the light-absorbing molecules located in the rhabdomeric microvilli of each photoreceptor cell. Butterfly visual pigments are composed of a rhabdomeric opsin protein (Briscoe 1998; Kitamoto et al. 1998), a member of the G protein–coupled receptor (GPCR) subfamily similar to vertebrate melanopsin (Provencio et al. 1998), covalently linked to a light-sensitive chromophore, 11-*cis*-3-hydroxyretinal (Smith and Goldsmith 1990). Because butterflies use only one type of chromophore, the absorbance spectrum maximum (λ_{max}) of the visual pigment depends on the amino acid residues of its opsin. Thus, it is the opsin protein that allows animals to see different wavelengths of light and is responsible for photosensory responses. Opsins are ubiquitous in all animals and have mediated the phototransduction cascades prior to the evolution of Metazoa (Terakita 2005; Plachetzki, Degnan, and Oakley 2007). Animal opsins have been classified according to the type of photoreceptor cell in which they are found (e.g., rhabdomeric or ciliary-type; Arendt 2003) and are also based on the specific G protein subtype that links proper GPCRs (Santillo et al. 2006).

Butterfly photoreceptor cells may also be roughly classified according to their sensitivity to ultraviolet (UV; 300–400 nm), blue (B; 400–500 nm), and long-wavelength (LW; 500–600 nm) light. Surveys of spectral sensitivity measurements using electroretinogram, intracellular, or epi-microspectrophotometric recordings suggest that most moth and nymphalid butterfly eyes contain at least one class of UV-, B-, and LW-sensitive photoreceptor cell (Briscoe and Chittka 2001). The nymphalid, *Vanessa cardui* (Nymphalinae), has photoreceptor cells with peak sensitivities at 360, 470, and 530 nm; the monarch, *Danaus plexippus* (Danaiinae), has photoreceptors cells with peak sensitivities at 340 nm, 435 nm, and 545 nm (Figure 7.2A; Stalleicken, Labhart, and Mouritsen 2006; Frentiu et al. 2007); and the sphingid moth, *Manduca sexta*, has photoreceptor cells with peak sensitivities at 357 nm, 450 nm, and 520 nm (Bennett and Brown 1985). Physiological data for the photoreceptor cells of the lycaenid eye are rare but are consistent with this general pattern. The adult eye of the thecline, *Narathura japonica*, for instance, has at least three classes of photoreceptor with peak sensitivities to 380 nm, 460 nm, and 560 nm light (Imafuku et al. 2007); the polyommatine, *Celastrina argiolus*, has at least three photoreceptors with peak sensitivities at 380 nm, 440 nm, and 560 nm (Eguchi et al. 1982); and the polyommatine, *Pseudozizeeria maho*, has at least three photoreceptors with peak sensitivities at 400 nm, 520 nm, and 560 nm (Eguchi et al. 1982). Using epi-microspectrophotometry, however, four photoreceptor spectral types were identified in the retina of butterflies in the genus *Lycaena* (Lycaeninae): *L. rubidus*, *L. heteronea*, *L. dorcas*, and

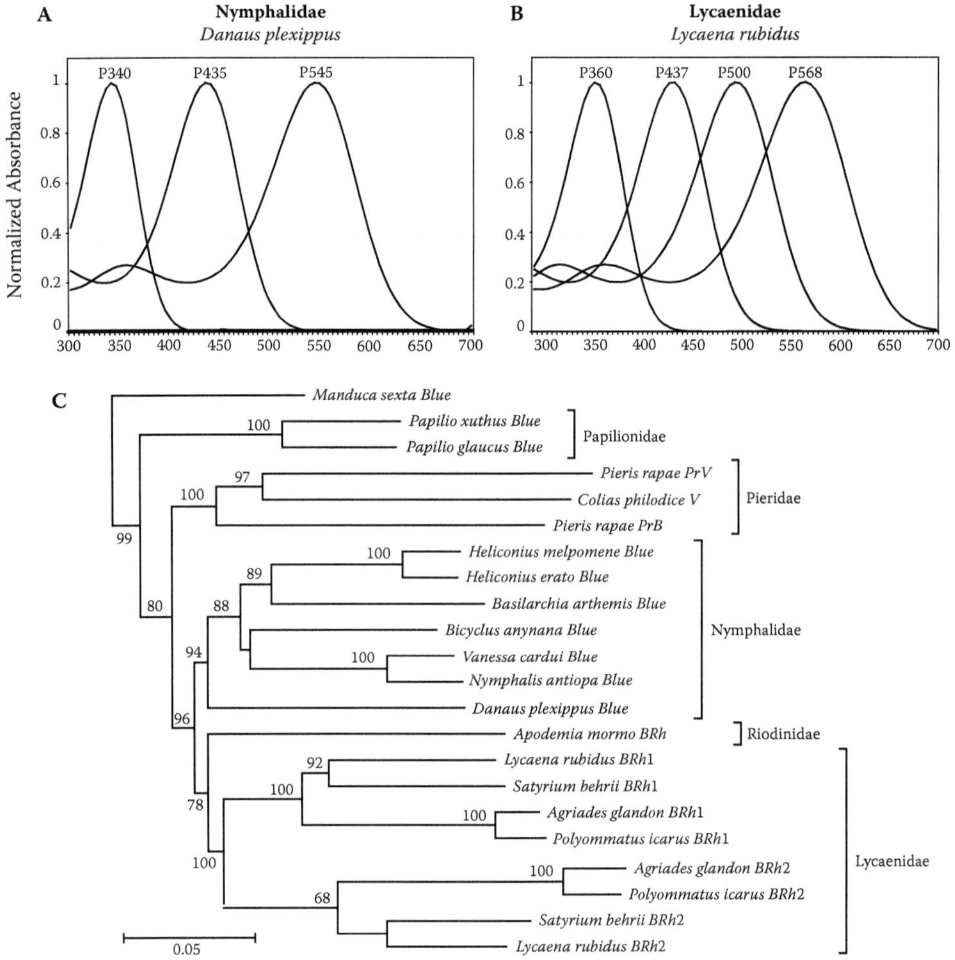

FIGURE 7.2 Representative nymphalid and lycaenid visual pigment absorption spectra and lepidopteran blue opsin phylogeny. (A) The monarch *Danaus plexippus* eye contains three visual pigments. P denotes maximum peak absorption (λ_{max}) of the visual pigment. (B) The ruddy-winged copper *Lycaena rubidus* eye contains four visual pigments. (C) Phylogeny of blue lepidopteran opsin genes based upon neighbor-joining analysis of 1,077 nucleotide sites, using Tamura-Nei distance and heterogeneous pattern of nucleotide substitution among lineages. Bootstrap values shown are based upon 500 maximum likelihood (ML) bootstrap replicates determined using the GTR+Γ+I model with estimated gamma shape parameter = 0.574 and proportion of invariant sites = 0.1474. GenBank accession numbers are as follows: *Manduca sexta* (Sphingidae; *Manop3*, AD001674); *Papilio xuthus* (*PxRh4*, AB028217); *Papilio glaucus* (*PglRh6*, AF077192); *Pieris rapae* (*PrB*, AB208675; *PrV*, AB208674); *Colias philodice* (*V*, AY918899); *Danaus plexippus* (*Blue*, AY605544); *Bicyclus anynana* (*BlueRh*, AY918894); *Heliconius erato* (*BlueRh*, AY918906); *Heliconius melpomene* (*BlueRh*, AY918897); *Basilarchia arthemis astyanax* (*BlueRh*, AY918902); *Nymphalis antiopa* (*BlueRh*, AY918893); *Vanessa cardui* (*BRh*, AY613987); *Apodemia mormo* (*BRh*, AY587906); *Polyommatus icarus* (*BRh1*, DQ402500; *BRh2*, DQ402501); *Agriades glandon* (*BRh1*, DQ402502; *BRh2*, DQ402503); *Satyrium behrii* (*BRh1*, DQ402498; *BRh2*, DQ402499); *Lycaena rubidus* (*BRh1*, AY587902; *BRh2*, AY587903). Phylogeny modified from Sison-Mangus et al. (2006).

L. nivalis have visual pigments with peak absorbances at 360 nm, 437 nm, 500 nm, and 568 nm (or 575 nm in *L. nivalis*), respectively (Figure 7.2B; Bernard and Remington 1991). Whereas these early physiological studies seem to suggest a similar number (3) of opsins in some nymphalids and lycaenids, it is now clear from molecular studies that the number of opsin genes and spectral receptors inferred from their spatial expression patterns in the eye differ strikingly between these groups of butterflies (see below).

THE BLUE OPSIN GENE HAS DUPLICATED IN LYCAENID BUTTERFLIES

The nymphalid butterflies *V. cardui* and *D. plexippus* have the least number of opsins, with eyes containing only a single copy of the UV, B, and LW opsin genes (Briscoe et al. 2003; Sauman et al. 2005). Although these cDNAs were originally cloned from eye-specific cDNA pools, BLAST searches of an EST library consisting of 9,484 unique cDNA sequences derived from monarch brain yielded the same result (Zhu, Casselman, and Reppert 2008). Deviating from nymphalids, *L. rubidus* (Lycaenidae) and *Apodemia mormo* (Riodinidae), considered to be sister taxa to the nymphalids (Campbell, Brower, and Pierce 2000), have four opsins each. However, *L. rubidus* has two copies of the B opsin gene (*BRh1*, encoding a P437 nm pigment, and *BRh2*, encoding a P500 nm pigment), a single copy of UV (*UVRh*, encoding a P360 pigment) and LW (*LWRh*, encoding a P568 pigment) opsin genes; whereas *A. mormo* has two LW copies (*LWRh1*, P505 and *LWRh2*, P600) and only a single copy of B (*BRh*, P450) and UV (*UVRh*, P340) genes (Sison-Mangus et al. 2006; Frentiu et al. 2007; Briscoe 2008). Strikingly, both butterflies have acquired visual pigments of similar λ_{max} in the blue-green range (500–505 nm), which are encoded by two different opsin gene family members, a *BRh2* (P500) opsin in *L. rubidus* and an *LWRh1* (P505) opsin in *A. mormo*. This suggests that these closely related butterflies, which diverged more than 70 million years ago (Wahlberg 2007), have co-opted different ancestral genes to achieve the same visual pigment physiology (i.e., wavelength of peak absorbance).

The blue-absorbing visual pigments of *P. rapae* are also noteworthy in this regard because they have λ_{max} of 425 nm and 453 nm, and are encoded by duplicate B opsin genes, *PrV* and *PrB* (Arikawa et al. 2005). Because the handful of papilionid and nymphalid eyes that had been investigated contained only one B opsin-encoding cDNA, we decided to investigate the evolutionary origins of the lycaenid and pierid gene duplications to see if they were independent of each other. To do this, we screened eye-specific cDNA libraries from ten additional butterfly taxa including lycaenids from the three largest (out of seven) lycaenid subfamilies (Lycaeninae, Polyommatinae, and Theclinae) and from three other butterfly families (Pieridae, Nymphalidae, and Riodinidae). We cloned a total of fourteen full-length blue opsin-encoding cDNAs from the ten taxa, including homologues of both *BRh1* and *BRh2* in all surveyed lycaenid subfamilies (Figure 7.2C). We detected only one blue opsin cDNA in each of the seven species of nymphalid surveyed. Phylogenetic analyses unambiguously indicated that the blue opsins of *L. rubidus* evolved independently from that of *P. rapae crucivora*, which is consistent with the very different λ_{max} of the pierid blue visual pigments compared with those of the lycaenid. Our results also indicated that the *L. rubidus* blue opsin gene duplication event occurred before the radiation of the coppers, hairstreaks, and blues (Lycaeninae+ Theclinae+Polyommatinae; Sison-Mangus et al. 2006). We subsequently investigated the remaining opsins of a lycaenid in the subfamily Polyommatinae, *P. icarus,* and found that like *L. rubidus*, besides the duplicate blue opsins, its eye contains one UV opsin mRNA and one LW opsin mRNA (Sison-Mangus et al. 2008). Although the electroretinogram studies of individual species from the Theclinae and Polyommatinae mentioned previously detected only three major spectral peaks in the eye, it is important to note that our molecular studies made it clear that the eyes of butterflies in all three lycaenid subfamilies contain four visual pigments and not three.

Why have duplicate blue opsins evolved in butterflies? The molecular evolution of a blue-green–absorbing visual pigment (P500) in *L. rubidus*, which is encoded by a blue opsin gene (*BRh2*) and red-shifted by 63 nm compared with the visual pigment (P437) encoded by its paralogue (*BRh1*) along with the green-absorbing visual pigment (P568 nm) encoded by the LW opsin gene *LWRh* (Sison-Mangus et al. 2006), might enhance color vision in the blue-green part of the light spectrum. We attempted to examine this hypothesis in the lycaenid *P. icarus* (see below). The duplication of blue opsins in *P. rapae crucivora*, on the other hand, may allow the animal to gain a violet receptor of blue opsin origin to discriminate better in the short wavelength light spectra, a hypothesis that remains to be tested behaviorally.

OPSIN SPATIAL EXPRESSION PATTERN IN THE NYMPHALID EYE

To appreciate how different the lycaenid eye is as a result of these blue opsin gene duplications, it is necessary first to describe the simpler nymphalid eye. The butterfly compound eye is subdivided into three domains, the dorsal rim area (DRA), and the dorsal and the ventral domains of the main retina. The DRA of butterflies, like that of many insects, is specialized for detecting polarized light (Labhart and Meyer 1999; Stalleicken, Labhart, and Mouritsen 2006). This area is typically composed of a few rows of ommatidia on the dorsal edge of the eye that have microvilli that differ in structure and orientation compared with those in the rest of the eye. The microvillar membranes or rhabdomeres of the ommatidia in the DRA are arranged at right angles to each other, in contrast to the more random orientation of the microvilli in the rhabdomeres of the main dorsal and ventral retina (Figure 7.3A,B; Briscoe et al. 2003; Reppert, Zhu, and White 2004). Like the main retina, the ommatidia of the DRA each contain nine photoreceptor cells (R1–9). Photoreceptor

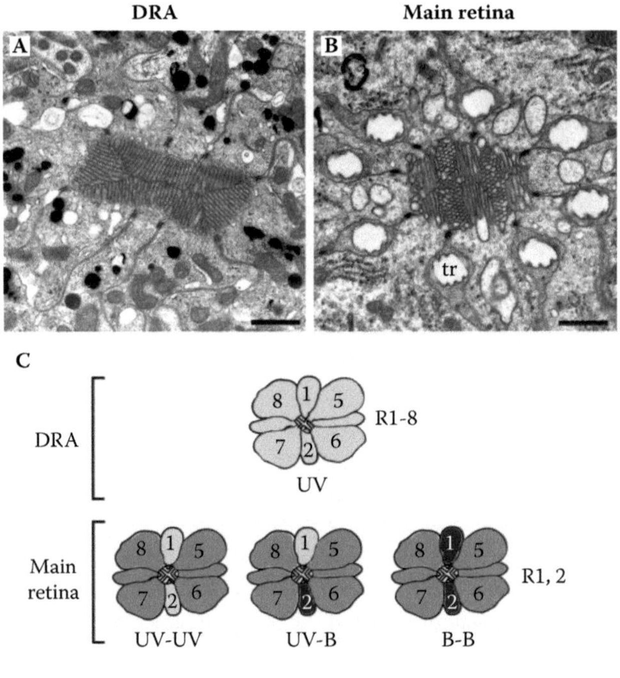

FIGURE 7.3 Rhabdomeric microvilli in the DRA and main retina of the butterfly and opsin expression patterns in the typical nymphalid eye. (A) DRA ommatidia of the lycaenid *Lycaena rubidus* are square shaped and have microvilli organized for polarized light detection. (B) Main retina ommatidia of *L. rubidus* are round shaped and have microvilli that are not organized for polarized light detection. Modified from Sison-Mangus et al. (2006). (C) Opsin expression in the DRA ommatidia of the nymphalid butterfly *Danaus plexippus* consists entirely of UV opsin in R1–R8 photoreceptor cells. By contrast, opsin expression in the main retina of *D. plexippus* consists of LW opsin in all R3–R8 cells, and either UV-UV, UV-B or B-B in the R1 and R2 photoreceptor cells.

subtype-specific patterns of opsin expression vary dramatically, however, between the main retina and DRA of the nymphalid eye (see below).

The opsin expression pattern among the ommatidia of the main retina of nymphalids is similar to that of the bee worker wherein the short-wavelength–sensitive opsins, UV and B, are expressed in the R1 and/or R2 cells, while the LW opsin is expressed in the six receptor cells R3–R8 (Spaethe and Briscoe 2005; Wakakuwa, Stavenga, and Arikawa 2007). Because UV and B opsin expression is restricted to R1 and R2 cells, three ommatidial subtypes based on the expression of these opsins in R1 and R2 cells have been identified (Figure 7.3C): UV-UV (type I ommatidia), UV-B (type II), B-B (type III). These patterns of opsin expression are exemplified by the sphingid moth *M. sexta* (White et al. 2003) and nymphalid butterflies *D. plexippus*, *V. cardui*, and *Heliconius erato* (Briscoe et al. 2003; Sauman et al. 2005; Zaccardi et al. 2006). The pattern of opsin expressed in the DRA is different from that of the main retina. In *D. plexippus*, using an antibody generated against a short peptide in the C-terminal region of the *Papilio glaucus* UV opsin, we found that only UV opsin is found in R1–R8 photoreceptor cells in the dorsal rim, whereas in the main retina, UV opsin is only expressed in R1 and/or R2 cells (Sauman et al. 2005). Input from the ultraviolet polarized light–sensitive DRA photoreceptor cells is important for directional orientation by migratory monarchs (Reppert et al. 2004; Sauman et al. 2005).

SEXUALLY DIMORPHIC RETINA OF *LYCAENA RUBIDUS*

In all nymphalids studied so far, the pattern of opsin mRNA expression in the eye does not vary between the sexes. By contrast, we found that the pattern of opsin mRNA expression in the main retina of the lycaenid *L. rubidus* eye is sexually dimorphic (the DRA has not yet been investigated), a pattern that confirmed an earlier epi-microspectrophotometric report of sexually dimorphic distribution of visual pigments in the eye of the same species (Bernard and Remington 1991). After cloning the opsin cDNAs from *L. rubidus* (*UVRh*, *BRh1*, *BRh2*, and *LWRh*), we found that all four opsin mRNAs are not only expressed in the eyes in a sex-specific manner, but are also distributed differentially in a dorso-ventral manner. The male dorsal retina expresses only *UVRh* and *BRh1*. New ommatidial subtypes dominate the male dorsal retina, because *BRh1* is expressed in the R3–R8 cells instead of the *LWRh* opsin that is commonly seen in the nymphalid eye (and all butterfly eyes examined to date). Moreover, the *UVRh-UVRh* (R1 and R2 cell) ommatidial type is dominant dorsally with a small number of *UVRh-BRh1* and *BRh1-BRh1* ommatidia in this part of the eye. The dorsal eye of the male is therefore likely color-blind in the red range and can only see short-wavelength light. The female dorsal retina is a different story, with *UVRh*, *BRh1*, and *LWRh* mRNAs expressed in this region (Figure 7.4A–H). Most strikingly, the two opsin genes *BRh1* and *LWRh* are coexpressed in the R3–R8 photoreceptor cells, making *L. rubidus* sexually dimorphic in this eye region (Figure 7.5E). To our knowledge, *L. rubidus* is the first insect species that has two visual pigments, one short-wavelength absorbing and one long-wavelength absorbing, coexpressed in the same photoreceptor cells (Sison-Mangus et al. 2006). Assuming that both visual pigments are involved in phototransduction, their coexpression in a single photoreceptor cell suggests that the receptors will have a broad sensitivity from blue to yellow spectral range. Intracellular recordings of these coexpressing opsins are needed to confirm this idea. Together with the UV receptors in R1 and R2 cells, this ommatidial type dominates the dorsal eye and implies that the *L. rubidus* female will be able to detect light spanning from UV to the long-wavelength region, in sharp contrast to the male dorsal eye, which is specialized for detection of UV to blue light. Interestingly, it is likely that the female dorsal eye is not specifically adaptive in itself, but rather reflects an intermediate step along the evolutionary pathway to developing the unique male dorsal eye (see discussion in Sison-Mangus et al. 2006 and below).

The ventral eye of both sexes, on the other hand, has a more typical opsin expression pattern; *LWRh* is expressed in R3–R8 cells, while the short-wavelength sensitive opsin mRNAs (*UVRh*, *BRh1*, *BRh2*) are expressed in the R1–R2 cells in a nonoverlapping fashion. However, because of

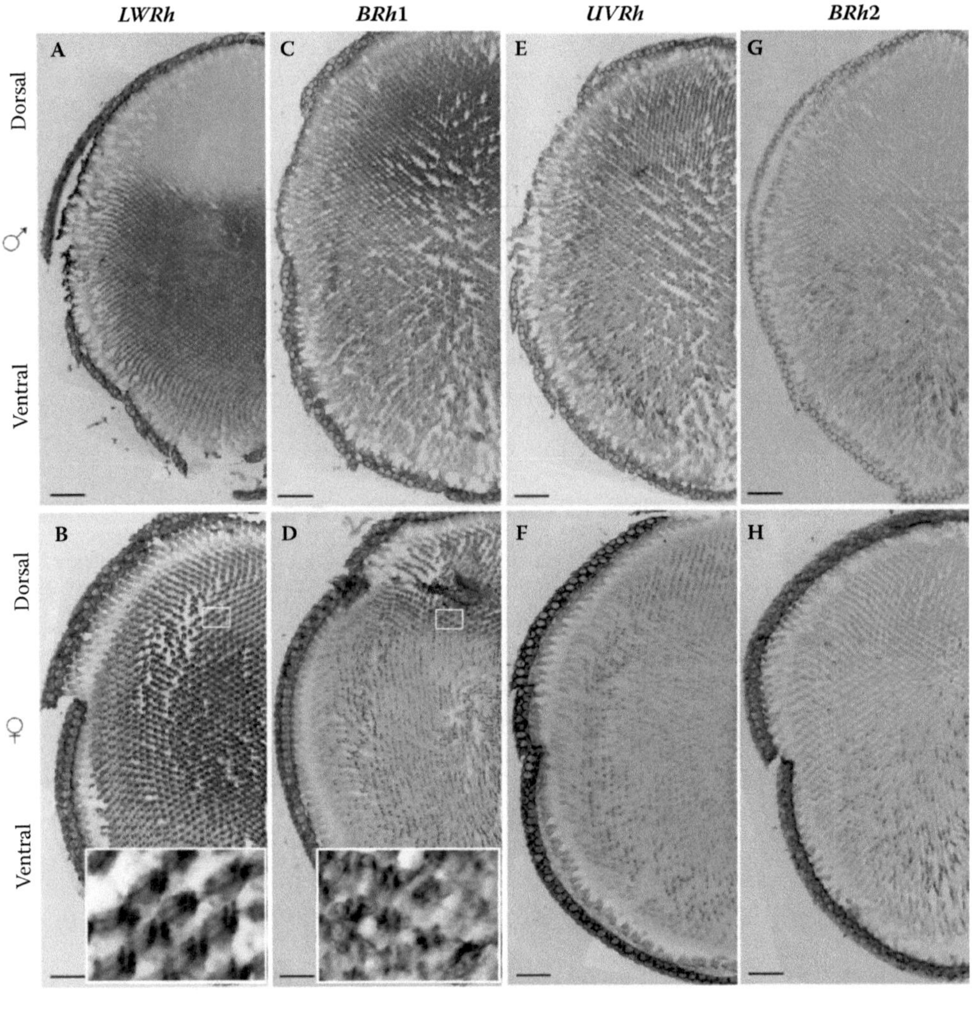

FIGURE 7.4 Sexually dimorphic *LWRh* opsin mRNA expression, coexpression of *LWRh* and *BRh1* opsin mRNAs in female dorsal eye, and dorso-ventral differences in *UVRh* and *BRh2* opsin mRNA expression in the adult retina of *Lycaena rubidus*. (A) *LWRh* is only expressed ventrally in males. (B) By contrast, *LWRh* is expressed uniformly across the retina in females. Inset: magnified view of *LWRh* mRNA expression in R3–R8 photoreceptor cells of female dorsal eye. (C) *BRh1* is expressed abundantly in the dorsal eye and less abundantly in the ventral eye in males. (D) Similarly, *BRh1* expression is more abundant in the dorsal area than in the ventral area in females. However, *BRh1* is coexpressed with *LWRh* in the dorsal eye in females. Inset: magnified view of *BRh1* mRNA expression in R3–R8 photoreceptor cells. *UVRh* expression is more abundant in the dorsal area than in the ventral area in both males (E) and females (F). *BRh2* mRNA expression is absent in the dorsal area and only seen in the ventral part of the retina in both males (G) and females (H). Scale bar = 100 μm.

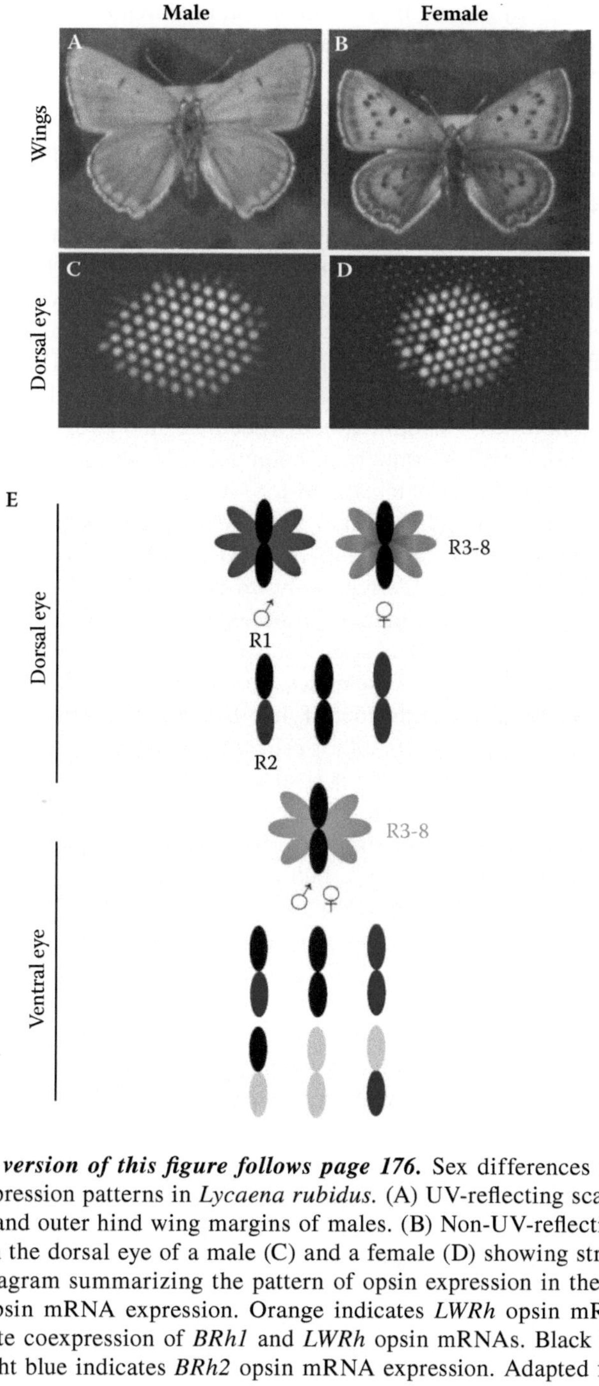

FIGURE 7.5 ***A color version of this figure follows page 176.*** Sex differences in wing color pattern, eyeshine, and opsin expression patterns in *Lycaena rubidus*. (A) UV-reflecting scales (iridescent purple) on the lower forewing and outer hind wing margins of males. (B) Non-UV-reflecting scales on wings of females. Eyeshine from the dorsal eye of a male (C) and a female (D) showing strongly sexually dimorphic coloration. (E) Diagram summarizing the pattern of opsin expression in the *L. rubidus* eye. Dark blue indicates *BRh1* opsin mRNA expression. Orange indicates *LWRh* opsin mRNA expression. Dark blue and orange indicate coexpression of *BRh1* and *LWRh* opsin mRNAs. Black indicates *UVRh* opsin mRNA expression. Light blue indicates *BRh2* opsin mRNA expression. Adapted from Sison-Mangus et al. (2006).

the presence of the duplicate blue opsin, *BRh2*, in the ventral retina, the ventral eye is equipped with six ommatidial types (UV-UV, UV-B1, UV-B2, B1-B1, B1-B2, B2-B2) in contrast to the typical three ommatidial types found in nymphalid butterfly eyes (UV-UV, UV-B, and B-B). Moreover, a pink filtering pigment is always found in the ommatidia in which *BRh2* is expressed. Presumably, this could result in an additional receptor, but behavioral experiments are needed to determine if it actually participates in color vision.

Are Sexually Dimorphic Retinas Correlated with Sexually Dimorphic Wings?

Many butterfly wing color patterns serve a function in the context of crypsis or mimicry. But in butterflies like the lycaenids, whose coloration varies between sexes, the difference in wing color suggests sexual selection. For example, the male *L. rubidus* displays bright, intense red-copper coloration, whereas the female appears dull brown (Figure 7.5A,B). Measurement of the dorsal wing reflectance spectrum of males indicates that they reflect light in the UV and red spectra, whereas the female reflects only in the red (Bernard and Remington 1991). Similar wing reflectance was also observed in the New Zealand species *L. salustris* (Meyer-Rochow 1991). Male lycaenids surveyed from three subfamilies, Polyommatinae, Theclinae, and Lycaeninae, reflect in the UV as well (Vertesy et al. 2006). UV is an important component of sexual signaling among butterflies (Silberglied 1979), and it has been demonstrated for blue lycaenids in mate choice (Burghardt et al. 2000; Knüttel and Fiedler 2001). In addition, male lycaenids are well known for displaying territoriality (Clark and Dickson 1971; McCubbin 1971; Atsatt 1981), a behavior that suggests strong male-male competition. The pattern of exclusively *BRh1* opsin mRNA expression in the R3–R8 photoreceptor cells, along with predominantly *UVRh* expression in the R1 and R2 cells (Figure 7.4 and Figure 7.5E), suggests that the male *Lycaena* dorsal eye is specialized for the detection of flickering moving objects, such as other airborne males. So far, there is no direct behavioral proof of the utility of the male dorsal eye of lycaenids. Identifying the circumstances under which the male dorsal eye outperforms the female eye is crucial for providing a connection to male-male competition.

The eye of the pierid *P. rapae crucivora* is also sexually dimorphic (Arikawa et al. 2005). However, its sexually dimorphic eye is not achieved through modification of the opsin expression but is achieved through a filter pigment strategy. Male and female *P. rapae crucivora* are indistinguishable by color in full-spectrum light, but when photographed with a UV filter, the wing reflectance varies between sexes. The female wing strongly reflects UV, whereas the male barely reflects UV; moreover, UV reflectance (more evident in female wings) is the main cue that would elicit sexual behavior in the male (Obara 1970). In light of these differences in wing patterns, it is interesting to note that *P. rapae crucivora* has three short-wavelength sensitive opsins, one UV and a pair of blue opsins (PrV and PrB); but one of the copies, PrV, is sensitive to violet. These short-wavelength opsin mRNAs are expressed independently in R1 and R2 cells, but only the usual three types of ommatidia are found, with PrV restricted to type II ommatidia. It is in these ommatidia that the fluorescing pigment is found in males but not in females. Because the pigment acts as a violet-absorbing spectral filter, the violet receptor sensitivity has been modified into a double-peak blue, with a small peak in the violet range and a high peak in the blue range. Given the animal's spectral set, the male may be able to acutely discriminate a conspecific female on the basis of her UV wing reflectance, corroborating the observations made on the mating behavior in this *Pieris* subspecies (Obara and Hidaka 1968; Obara 1970).

These pigment modifications strongly suggest that sexually dimorphic eyes evolved to accommodate communication via UV signaling by the opposite sex in *P. rapae crucivora* and by the same sex in *L. rubidus*. It would be interesting to evaluate the visual systems of other sexually dimorphic butterflies to determine whether similar patterns are observed. Butterflies from the genus *Colias*, for instance, have been studied extensively for UV signaling in the context of sexual selection (Silberglied and Taylor 1978; Silberglied 1979; Rutowski 1985). A violet sensitive opsin has been cloned from *Colias philodice*, which is homologous to that of *PrV* in *Pieris*

(Sison-Mangus et al. 2006), a good indication that this animal likely has three short-wavelength opsins. The presence of a violet sensitive opsin in *C. philodice* corroborates the finding of a violet receptor (λ_{max} = 400 nm) reported in another species, *C. erate* (Eguchi et al. 1982), suggesting that this receptor is also encoded by a violet-sensitive opsin. Another good candidate for examination is the sexually dimorphic nymphalid butterfly, *Hypolimnas bolina*, in which males posses bright, iridescent UV-reflecting markings in the dorsal wing (Kemp and Macedonia 2006) and females show preference for males with bright, iridescent markings that highly reflect in the UV (Kemp 2007).

COLOR VISION STUDIES IN BUTTERFLIES

The spectral diversity of visual pigments and their arrangement in the butterfly eye have the potential to be utilized for color vision. However, the mere occurrence of multiple receptors in the compound eye does not demonstrate color vision. To have color vision, a butterfly needs to discriminate objects of different colors irrespective of the intensity (Goldsmith 1990). Color vision requires the following: two visual pigments with distinct spectral sensitivity located in different photoreceptor cells, the presence of interneurons in the optic lobe with antagonistic input from these receptors, proper wiring in the brain to compare the signals from different stimuli, and the behavioral response of the animal being tested in color choice experiments (Goldsmith 1990; Kelber, Vorobyev, and Osorio 2003; Kelber 2006). Receptors produced by a visual pigment and a filtering pigment can participate in color discrimination because the presence of a filtering pigment can modify the absorption spectrum of a receptor.

Our knowledge of how the brains of nymphalid or lycaenid butterflies process color information is scant (Swihart 1968), and this is a fertile area for further study. The clearest demonstration of color vision in nymphalid and lycaenid butterflies has been shown through behavioral tests. The empirical testing and solid demonstration of true color vision in Lepidoptera involves training a naive animal to associate color with a food reward. Training or the ability of the animal to learn the rewarding color is an essential aspect of the behavioral test. This gives the experimenter a handle on the sensory competence of the animal, because learning indicates perception of color information. Once the animal has "learned" the rewarding color, it is then given a series of choices between the rewarding and unrewarding colors, while the light intensity of the color stimuli is manipulated (Kelber 1999; Kelber and Pfaff 1999; Zaccardi et al. 2006; Sison-Mangus et al. 2008) or the colors are compared with varying shades of gray (Kinoshita, Shimada, and Arikawa 1999). The animal has true color vision if it chooses the rewarding color independent of intensity.

ALTERNATIVE STRATEGIES FOR RED-GREEN COLOR VISION

Behavioral experiments that utilize this rigorous approach have shown that nymphalids and lycaenids, like other butterflies and moths, have true color vision; they can discriminate between pairs of colors of different wavelengths over intensity ratios ranging from 0.01 to 100 or when compared with gray (Kelber, Balkenius, and Warrant 2002; Table 7.1). The total range of colors that nymphalid and lycaenid butterflies can discriminate has not been exhaustively studied. From the experiments that have been performed, however, it is clear that different butterfly species apply different strategies to achieve color vision in the green-red range. For example, the nymphalid *V. atalanta*, whose eye contains the typical three visual pigments, UV-, B-, and LW-absorbing, can discriminate blue from orange light (440 vs. 620 nm) but is unable to discriminate yellow from orange light (590 vs. 620 nm; Zaccardi et al. 2006). In contrast, the nymphalid *H. erato*, whose eye also contains UV-, B-, and LW-absorbing visual pigments, can see in the red range (620 vs. 640 nm) despite having only a single LW opsin. It is with the aid of a heterogeneously distributed red filtering pigment that the animal produces a fourth, red-sensitive receptor (Zaccardi et al. 2006). The lycaenid *P. icarus*, on the other hand, can

TABLE 7.1
Species and Wavelengths Tested in Lepidopteran Color Vision Experiments

Family	Species	Wavelength (nm)/ Color Tested	Color Vision	Sources
Sphingidae	*Macroglossum stellatarum*	380 vs. 360	yes	Kelber and Henique 1999
		380 vs. 420	yes	
		380 vs. 470	yes	
		470 vs. 500	yes	
		500 vs. 420	yes	
		620 vs. 470	yes	
		470 vs. 620	yes	
		595 vs. 620	no	
Sphingidae	*Deilephila elpenor*	blue vs. shades of gray	yes	Kelber et al. 2002
		blue vs. shades of blue	no	
		yellow vs. shades of gray	yes	
		yellow vs. shades of yellow	yes	
Papilionidae	*Papilio aegeus*	430 vs. 590 and 640	yes	Kelber and Pfaff 1999
		640 vs. 430 and 590	yes	
Papilionidae	*Papilio xuthus*	red vs. yellow, green, blue	yes	Kinoshita, Shimada, and Arikawa 1999
		yellow vs. red, green, blue	yes	
		green vs. red, yellow, blue	yes	
		blue vs. red, yellow, green	no*	
Nymphalidae	*Heliconius erato*	590 vs. 440	yes	Zaccardi et al. 2006
		620 vs. 590	yes	
		620 vs. 640	yes	
Nymphalidae	*Vanessa atalanta*	620 vs. 440	yes	Zaccardi et al. 2006
		620 vs. 590	no	
Lycaenidae	*Polyommatus icarus*	450 vs. 590	yes	Sison-Mangus et al. 2008
		560 vs. 590	yes	
		570 vs. 590	no	

The color or wavelength first listed was used as the training and rewarding color.
* No preference for blue.

discriminate colors in the green range up to 560 nm when feeding (Sison-Mangus et al. 2008). It cannot discriminate colors in the red range (590 vs. 640 nm), however, despite having a red-reflecting ommatidium produced by the LW receptor and a red-filtering pigment (Figure 7.1; Sison-Mangus et al. 2008). The photoreceptors being used for this task are most likely the B2- and LW-absorbing visual pigments, which by homology are most similar to the P500 and P568 visual pigments of *L. rubidus*. Nevertheless, physiological data for *P. icarus* are needed to demonstrate more fully that it is, indeed, the duplicate blue opsin, in conjunction with the LW opsin, that is mediating this behavior.

The contrasting results between *P. icarus* and *H. erato* suggest that the impact of filtering pigments on butterfly color discrimination should be evaluated on a species-specific basis. Further behavioral studies, in conjunction with physiological and molecular studies on these and other butterfly species known to have lateral filtering pigments such as *P. rapae crucivora* (Wakakuwa et al. 2004), *D. plexippus* (Sauman et al. 2005), *Bicyclus anynana*, *Zizeeria maha* (Stavenga 2002), *Sasakia charonda*, and *Polygonium c-aureum* (Kinoshita, Sato, and Arikawa 1997), are needed to determine whether filtering pigments play a role in their color vision. Such studies should identify the number of opsins in the eye and establish whether lateral filtering pigments have an impact on the spectral sensitivity of individual photoreceptor cells. Direct measurement of the λ_{max} values of the reconstituted visual pigments, via transgenic expression of the opsins in *Drosophila* or cultured cells, together with *in vivo* measurements, would also provide crucial experimental evidence demonstrating the effects of the lateral filtering pigment on color vision.

CONCLUSION

The butterfly eye has been elegantly molded by evolution. The fact that lycaenid and nymphalid butterflies bear modifications of their visual systems through different mechanisms that lead to similar traits (e.g., red-green color vision) is highly suggestive of adaptive evolution. So far, gene duplication of the B and LW opsins and the addition of filter pigments are the most common strategies utilized by butterflies to achieve similar physiological and behavioral ends. The ease with which some butterflies have changed opsin expression patterns in a particular domain of their eye, as occurs among some sexually dimorphic butterflies in which such eye modifications are most pronounced, also suggests a lack of developmental constraints. It will be interesting to determine in the future whether or not some of the unique characteristics of the nymphalid and lycaenid eye described here turn out to be defining features of each of these families as has been the case for classical morphological characters.

ACKNOWLEDGMENTS

We wish to thank Steven Reppert for helpful comments on the manuscript. This work was supported in part by National Science Foundation Grant IOS-0819936.

REFERENCES

Arendt, D. 2003. Evolution of eyes and photoreceptor cell types. *Int. J. Dev. Biol.* 47:563–71.

Arikawa, K., D.G.W. Scholten, M. Kinoshita, and D.G. Stavenga. 1999. Tuning of photoreceptor spectral sensitivities by red and yellow pigments in butterfly *Papilio xuthus*. *Zool. Sci.* 16:17–24.

Arikawa, K., and D.G. Stavenga. 1997. Random array of colour filters in the eyes of butterflies. *J. Exp. Biol.* 200:2501–06.

Arikawa, K., M. Wakakuwa, X.D. Qiu, M. Kurasawa, and D.G. Stavenga. 2005. Sexual dimorphism of short-wavelength photoreceptors in the small white butterfly, *Pieris rapae crucivora*. *J. Neurosci.* 25:5935–42.

Atsatt, P.R. 1981. Lycaenid butterflies and ants—selection for enemy-free space. *Am. Nat.* 118:638–54.

Bennett, R.R., and P.K. Brown. 1985. Properties of the visual pigments of the moth *Manduca sexta* and the effects of two detergents, digitonin and chaps. *Vision Res.* 25:1771–81.

Bernard, G.D., and W.H. Miller. 1970. What does antenna engineering have to do with insect eyes? *IEEE Student J.* 8:2–8.

Bernard, G.D., and C.L. Remington. 1991. Color-vision in *Lycaena* butterflies: Spectral tuning of receptor arrays in relation to behavioral ecology. *Proc. Natl. Acad. Sci. U.S.A.* 88:2783–87.

Briscoe, A.D. 1998. Molecular diversity of visual pigments in the butterfly *Papilio glaucus*. *Naturwissenschaften* 85:33–35.

Briscoe, A.D. 2008. Reconstructing the ancestral butterfly eye: Focus on the opsins. *J. Exp. Biol.* 211:1805–13.

Briscoe, A.D., and G.D. Bernard. 2005. Eyeshine and spectral tuning of long wavelength-sensitive rhodopsins: No evidence for red-sensitive photoreceptors among five Nymphalini butterfly species. *J. Exp. Biol.* 208:687–96.

Briscoe, A.D., G.D. Bernard, A.S. Szeto, L.M. Nagy, and R.H. White. 2003. Not all butterfly eyes are created equal: Rhodopsin absorption spectra, molecular identification, and localization of ultraviolet-, blue-, and green-sensitive rhodopsin-encoding mRNAs in the retina of *Vanessa cardui*. *J. Comp. Neurol.* 458:334–49.

Briscoe, A.D., and L. Chittka. 2001. The evolution of color vision in insects. *Annu. Rev. Entomol.* 46:471–510.

Burghardt, F., H. Knuttel, M. Becker, and K. Fiedler. 2000. Flavonoid wing pigments increase attractiveness of female common blue (*Polyommatus icarus*) butterflies to mate-searching males. *Naturwissenschaften* 87:304–07.

Campbell, D.L., A.V.Z. Brower, and N.E. Pierce. 2000. Molecular evolution of the *wingless* gene and its implications for the phylogenetic placement of the butterfly family Riodinidae (Lepidoptera: Papilionoidea). *Mol. Biol. Evol.* 17:684–96.

Clark G.C., and C.G.C. Dickson. 1971. *Life histories of South African lycaenid butterflies.* Capetown: Purnell.

Eguchi, E., K. Watanabe, T. Hariyama, and K. Yamamoto. 1982. A comparison of electrophysiologically determined spectral responses in 35 species of Lepidoptera. *J. Insect Physiol.* 28:675–82.

Frentiu, F.D., G.D. Bernard, M.P. Sison-Mangus, A.V.Z. Brower, and A.D. Briscoe. 2007. Gene duplication is an evolutionary mechanism for expanding spectral diversity in the long-wavelength photopigments of butterflies. *Mol. Biol. Evol.* 24:2016–28.

Goldsmith, T.H. 1990. Optimization constraint and history in the evolution of eyes. *Q. Rev. Biol.* 65:281–322.

Imafuku, M., I. Shimizu, H. Imai, and Y. Shichida. 2007. Sexual difference in color sense in a lycaenid butterfly *Narathura japonica. Zool. Sci.* 24:611–13.

Kelber, A. 1999. Ovipositing butterflies use a red receptor to see green. *J. Exp. Biol.* 202:2619–30.

Kelber, A. 2006. Invertebrate colour vision. In *Invertebrate vision,* ed. E.J. Warrant and D.E. Nilsson, 250–90. Cambridge: Cambridge Univ. Press.

Kelber, A., A. Balkenius, and E.J. Warrant. 2002. Scotopic colour vision in nocturnal hawkmoths. *Nature* 419:922–25.

Kelber, A., and U. Henique. 1999. Trichromatic colour vision in the hummingbird hawkmoth *Macroglossum stellatarum* L. *J. Comp. Physiol. A* 184:535–41.

Kelber, A., and M. Pfaff. 1999. True colour vision in the orchard butterfly *Papilio aegeus. Naturwissenschaften* 86:221–24.

Kelber, A., M. Vorobyev, and D. Osorio. 2003. Animal colour vision—behavioural tests and physiological concepts. *Biol. Rev.* 78:81–118.

Kemp, D.J. 2007. Female butterflies prefer males bearing bright iridescent ornamentation. *Proc. Biol. Sci.* 274:1043–47.

Kemp, D.J., and J.M. Macedonia. 2006. Structural ultraviolet ornamentation in the butterfly *Hypolimnas bolina* L. (Nymphalidae): Visual, morphological and ecological properties. *Aust. J. Zool.* 54:235–44.

Kinoshita, M., M. Sato, and K. Arikawa. 1997. Spectral receptors of nymphalid butterflies. *Naturwissenschaften* 84:199–201.

Kinoshita, M., N. Shimada, and K. Arikawa. 1999. Colour vision of the foraging swallowtail butterfly *Papilio xuthus. J. Exp. Biol.* 202:95–102.

Kitamoto, J., K. Sakamoto, K. Ozaki, Y. Mishina, and K. Arikawa. 1998. Two visual pigments in a single photoreceptor cell: Identification and histological localization of three mRNAs encoding visual pigment opsins in the retina of the butterfly *Papilio xuthus. J. Exp. Biol.* 201:1255–61.

Knüttel, H., and K. Fiedler. 2001. Host-plant-derived variation in ultraviolet wing patterns influences mate selection by male butterflies. *J. Exp. Biol.* 204:2447–59.

Labhart, T., and E.P. Meyer. 1999. Detectors for polarized skylight in insects: A survey of ommatidial specializations in the dorsal rim area of the compound eye. *Microsc. Res. Tech.* 47:368–79.

McCubbin, C. 1971. *Australian butterflies.* Melbourne: Thomas Nelson.

Meyer-Rochow, V.B. 1991. Differences in ultraviolet wing patterns in the New Zealand lycaenid butterflies *Lycaena salustius*, *L. rauparaha*, and *L. feredayi* as a likely isolating mechanism. *J. R. Soc. New Zealand* 21:169–77.

Miller, W.H. 1979. Ocular optical filtering. In *Handbook of sensory physiology*, vol. VII/6A, ed. H. Autrum, 69–143. Berlin, Heidelberg, New York: Springer.

Nilsson, D.E., M.F. Land, and J. Howard. 1988. Optics of the butterfly eye. *J. Comp. Physiol. A* 162:341–66.

Obara, Y. 1970. Studies on the mating behavior of the white cabbage butterfly, *Pieris rapae cruvivora* Boisduval. III. Near ultraviolet reflection as the signal of intraspecific communication. *Z. Vergl. Physiol.* 69:99–116.

Obara, Y., and T. Hidaka. 1968. Recognition of the female by the male on the basis of ultra-violet reflection in the white cabbage butterfly *Pieris rapae crucivora* Boisduval. *Proc. Jpn. Acad.* 44:829–32.

Plachetzki, D.C., D.M. Degnan, and T.H. Oakley. 2007. The origins of novel protein interactions during animal opsin evolution. *PLoS ONE* 2:e1054.

Provencio, I., G.S. Jiang, W.J. De Grip, W.P. Hayes, and M.D. Rollag. 1998. Melanopsin: An opsin in melanophores, brain, and eye. *Proc. Natl. Acad. Sci. U.S.A.* 95:340–45.

Reppert, S.M., H.S. Zhu, and R.H. White. 2004. Polarized light helps monarch butterflies navigate. *Curr. Biol.* 14:155–58.

Rutowski, R.L. 1985. Evidence for mate choice in sulfur butterfly *Colias eurytheme*. *Z. Tierpsychol.* 70:103–14.

Santillo, S., P. Orlando, L. De Petrocellis, L. Cristino, V. Guglielmotti, and C. Musio. 2006. Evolving visual pigments: Hints from the opsin-based proteins in a phylogenetically old eyeless invertebrate. *Biosystems* 86:3–17.

Sauman, I., A.D. Briscoe, H.S. Zhu, et al. 2005. Connecting the navigational clock to sun compass input in monarch butterfly brain. *Neuron* 46:457–67.

Silberglied, R.E. 1979. Communication in the ultraviolet. *Annu. Rev. Ecol. Syst.* 10:373–98.

Silberglied, R.E., and O.R. Taylor. 1978. Ultraviolet reflection and its behavioral role in courtship of sulfur butterflies *Colias eurytheme* and *Colias philodice* (Lepidoptera, Pieridae). *Behav. Ecol. Sociobiol.* 3:203–43.

Sison-Mangus, M.P., G.D. Bernard, J. Lampel, and A.D. Briscoe. 2006. Beauty in the eye of the beholder: The two blue opsins of lycaenid butterflies and the opsin gene-driven evolution of sexually dimorphic eyes. *J. Exp. Biol.* 209:3079–90.

Sison-Mangus, M.P., A.D. Briscoe, G. Zaccardi, H. Knuttel, and A. Kelber. 2008. The lycaenid butterfly *Polyommatus icarus* uses a duplicated blue opsin to see green. *J. Exp. Biol.* 211:361–69.

Smith, C.W., and T.H. Goldsmith. 1990. Phyletic aspects of the distribution of 3-hydroxyretinal in the Class Insecta. *J. Mol. Evol.* 30:72–84.

Spaethe, J., and A.D. Briscoe. 2005. Molecular characterization and expression of the UV opsin in bumblebees: Three ommatidial subtypes in the retina and a new photoreceptor organ in the lamina. *J. Exp. Biol.* 208:2347–61.

Stalleicken, J., T. Labhart, and H. Mouritsen. 2006. Physiological characterization of the compound eye in monarch butterflies with focus on the dorsal rim area. *J. Comp. Physiol. A* 192:321–31.

Stavenga, D.G. 2002. Colour in the eyes of insects. *J. Comp. Physiol. A Neuroethol. Sens. Neural Behav. Physiol.* 188:337–48.

Stavenga, D.G., and K. Arikawa. 2006. Evolution of color and vision of butterflies. *Arthropod Struct. Dev.* 35:307–18.

Stavenga, D.G., M. Kinoshita, E.C. Yang, and K. Arikawa. 2001. Retinal regionalization and heterogeneity of butterfly eyes. *Naturwissenschaften* 88:477–81.

Swihart, S.L. 1968. Single unit activity in the visual pathway of the butterfly *Heliconius erato*. *Journal of Insect Physiology* 14:1589–1601.

Takemura, S.Y., D.G. Stavenga, and K. Arikawa. 2007. Absence of eye shine and tapetum in the heterogeneous eye of *Anthocharis* butterflies (Pieridae). *J. Exp. Biol.* 210:3075–81.

Terakita, A. 2005. The opsins. *Genome Biol.* 6:213.

Vanhoutte, K.J.A., D.G. Stavenga. 2005. Visual pigment spectra of the comma butterfly, *Polygonia c-album*, derived from in vivo epi-illumination microspectrophotometry. *J. Comp. Physiol. A* 191:461–73.

Vertesy, Z., Z. Balint, K. Kertesz, J.P. Vigneron, V. Lousse, and L.P. Biro. 2006. Wing scale microstructures and nanostructures in butterflies—natural photonic crystals. *J. Microsc. Oxford* 224:108–10.

Wahlberg, N. 2007. That awkward age for butterflies: Insights from the age of the butterfly subfamily Nymphalinae (Lepidoptera: Nymphalidae). *Syst. Biol.* 55:703–14.

Wakakuwa, M., D.G. Stavenga, and K. Arikawa. 2007. Spectral organization of ommatidia in flower-visiting insects. *Photochem. Photobiol.* 83:27–34.

Wakakuwa, M., D.G. Stavenga, M. Kurasawa, and K. Arikawa. 2004. A unique visual pigment expressed in green red and deep-red receptors in the eye of the small white butterfly *Pieris rapae crucivora*. *J. Exp. Biol.* 207:2803–10.

White, R.H., H.H. Xu, T.A. Munch, R.R. Bennett, and E.A. Grable. 2003. The retina of *Manduca sexta:* Rhodopsin expression the mosaic of green- blue- and UV-sensitive photoreceptors and regional specialization. *J. Exp. Biol.* 206:3337–48.

Yagi, N., and N. Koyama. 1963. *The compound eye of Lepidoptera: Approach from organic evolution.* Tokyo: Shinkyo Press.

Zaccardi, G., A. Kelber, M.P. Sison-Mangus, and A.D. Briscoe. 2006. Color discrimination in the red range with only one long-wavelength sensitive opsin. *J. Exp. Biol.* 209:1944–55.

Zhu, H., A. Casselman, and S.M. Reppert. 2008. Chasing migration genes: A brain expressed sequence tag resource for summer and migratory monarch butterflies (*Danaus plexippus*). *PLoS ONE* 3:e1345.

8 Lepidopteran Circadian Clocks
From Molecules to Behavior

Christine Merlin and Steven M. Reppert

CONTENTS

INTRODUCTION

Circadian clocks are endogenous timing mechanisms that control molecular, cellular, physiological, and behavioral rhythms in organisms ranging from cyanobacteria to humans. These genetically determined clocks are entrained (synchronized) to the 24-hour day primarily by the daily light-dark cycle. Many aspects of insect biology are affected by circadian clock–driven rhythmic behaviors, including the timing of egg hatching, adult eclosion, reproductive behavior, social interactions, and more remarkable circadian behaviors, such as time-compensated sun-compass orientation in migratory species (Saunders 2002).

At the molecular level, insect circadian clocks are composed of self-sustaining transcriptional feedback loops of a defined set of clock genes. The clockwork mechanism has been most extensively studied in *Drosophila melanogaster*, which has been considered the model for insect clockworks for many decades (Stanewsky 2003; Rosato, Tauber, and Kyriacou 2006). However, the recent discovery of a novel and ancestral circadian clock mechanism in the monarch butterfly, *Danaus plexippus*, has not only challenged the *Drosophila* model, but has provided a new twist on the evolution of clockwork mechanisms in insects in general (Zhu et al. 2008).

The study of lepidopteran circadian clocks has a rich history, which dates back to the elegant silk moth studies of Carroll Williams in the 1960s on brain regulation of the photoperiodic control of pupal diapause, and the seminal silk moth studies of James Truman and Lynn Riddiford in the 1970s on brain control of the timing of adult eclosion. Using brain transplantation experiments, Williams was the first to show that the photoreceptive mechanism for the circadian clock resides in the brain (Williams 1963; Williams and Adkisson 1964). Using a similar transplantation approach, Truman and Riddiford (1970) determined that the species-specific time of adult eclosion was inherent to the donor brain; that is, brain transplantation not only restored the daily rhythmicity of adult eclosion but also determined the phase of the rhythm in the host moths. This was the first direct demonstration in any animal that a circadian clock actually resides in the brain. Thus, these classic silk moth studies contributed substantially to our early understanding of the location of circadian clocks driving behaviors in insects.

Here we present an overview of the diverse types of physiological and behavioral rhythms occurring in the Lepidoptera, detail the molecular mechanism of lepidopteran circadian clocks, and define cellular clock location, with a special emphasis on the circadian clock of the monarch butterfly. To illustrate the critical nature of circadian timing in the Lepidoptera, we elaborate on two systems in which the circadian clock is used to drive the ecology of a species: its use in the pheromonal communication of moths (reviewed in Chapter 10), which is essential for appropriate reproductive responses, and its use in time-compensated sun-compass orientation, which is critical for the proper navigation of migratory monarch butterflies.

OVERVIEW OF LEPIDOPTERAN CIRCADIAN RHYTHMS

Events at different stages of the life cycle of Lepidoptera exhibit daily rhythms. Some of these events are part of developmental processes, including the daily timing of egg-hatching, larval-to-larval ecdysis, larval-to-pupal ecdysis, larval gut purging, and adult eclosion (Truman 1992). Adult rhythms include those in locomotion, flight, pheromonal communication, mating, and oviposition. These adult rhythms are usually restricted to a species-specific window of time (reviewed in Saunders 2002). Among the daily rhythms reported, many possess the ability to be maintained in constant conditions with a period close to 24 hours; that is, they are truly circadian. The involvement of circadian clocks in lepidopteran photoperiodism, mentioned briefly above, has been reviewed elsewhere (Vas Nunes and Saunders 1999; Saunders 2002) and is not discussed further in this chapter. Instead, we focus on clock control of the overt daily rhythms in diverse physiological and behavioral events in Lepidoptera (Table 8.1).

RHYTHMS DURING DEVELOPMENT

Egg hatching is a profoundly important event in the life history of Lepidoptera. Proper initiation of this program is critical because it frees the larva from the constraints of life in the egg to fulfilling its biological destiny in the outside world (Sauman and Reppert 1998). The behavioral program leading to the time of day of egg hatching, at least in several moth species, is under circadian control.

Circadian regulation of the time of day of egg hatching in Lepidoptera was first shown in the pink bollworm, *Pectinophora gossypiella*, by Minis and Pittendrigh (1968); eggs raised in a 12-hour light/12-hour dark (LD) cycle resulted in a hatching rhythm characterized by a peak in hatching just after dawn, whereas eggs raised in constant light or darkness led to an arrhythmic hatching pattern. This population rhythm manifests the main properties of a circadian rhythm, that is, it was entrained by light by midembryogenesis and, once entrained, the hatching rhythm persisted in constant conditions (Minis and Pittendrigh 1968). Similarly, egg-hatching circadian rhythms have been described in the Chinese oak silk moth, *Antheraea pernyi* (Riddiford and Johnson 1971; Sauman and Reppert 1996, 1998; Sauman et al. 1996), the domesticated silkworm, *Bombyx mori* (Shimizu and Matsui 1983), and the southwestern corn borer, *Diatraea grandiosella* (Takeda 1983), showing

TABLE 8.1
Representative Physiological and Behavioral Circadian Rhythms in Lepidoptera

Circadian Rhythms	Species	References
	Development	
Egg-hatching	*Antheraea pernyi*	Sauman et al. 1996; Sauman and Reppert 1998
	Bombyx mori	Shimizu and Matsui 1983
	Diatraea grandiosella	Takeda 1983
	Pectinophora gossypiella	Minis and Pittendrigh 1968
Gut purge	*Samia cynthia ricini*	Mizoguchi and Ishizaki 1982
Adult eclosion	*Antheraea pernyi*	Truman and Riddiford 1970
	Bombyx mori	Truman 1972
	Diatraea grandiosella	Takeda 1983
	Danaus plexippus	Froy et al. 2003
	Hyalophora cecropia	Truman and Riddiford 1970
	Hyphantria cunea	Morris and Takeda 1994
	Reproduction	
Pheromone production/emission	*Agrotis segetum*	Rosén 2002
Pheromone reception/ attraction behavior	*Agrotis segetum*	Rosén, Han, and Löfstedt 2003
	Spodoptera littoralis	Merlin et al. 2007; Silvegren, Löfstedt, and Rosén 2005
Mating behavior	*Spodoptera littoralis*	Silvegren, Löfstedt, and Rosén 2005
Sperm retention	*Spodoptera littoralis*	Bebas, Cymborowski, and Giebultowicz 2002; Bebas et al. 2002
Sperm release	*Cydia pomonella*	Giebultowicz and Brooks 1998
	Lymantria dispar	Wergin et al. 1997
	Spodoptera littoralis	Syrova, Sauman, and Giebultowicz 2003
Oviposition	*Agrotis segetum*	Byers 1987
	Cydia pomonella	Riedl and Loher 1980
	Ostrinia nubilalis	Skopik and Takeda 1980
	Pectinophora gossypiella	Minis 1965
	Locomotion	
Flight	*Antheraea pernyi*	Truman 1974
	Hyalophora cecropia	
	Samia cynthia ricini	
	Migration	
Time-compensated sun-compass orientation	*Danaus plexippus*	Froy et al. 2003; Mouritsen and Frost 2002; Perez, Taylor, and Jander 1997
	Aphrissa starira	Oliveira, Srygley, and Dudley 1998
	Phoebis argante	

that a functioning circadian clock is ticking during early development. To our knowledge, an egg-hatching circadian rhythm has not yet been described in butterflies.

Transplantation experiments in *A. pernyi* have shown that the circadian clock controlling egg-hatching behavior resides in the brain, and that a humoral factor, perhaps a novel hormone termed "hatchin," mediates this circadian regulation (Sauman and Reppert 1998). Elucidation of such a "chronoactive" diffusible substance in an experimentally tractable insect may aid identification of a similar substance in other animals.

In contrast to egg hatching, larval-to-larval and larval-to-pupal ecdyses are not directly controlled by a circadian clock in Lepidoptera but instead are associated with apparent circadian rhythms of endocrine events (hormone release), as exemplified in the silk moths *Samia cynthia ricini* (Fujishita and Ishizaki 1981, 1982) and *B. mori* (Sakurai 1983). However, larval gut purging, which occurs at the end of the feeding period in the fifth instar larva to evacuate the gut in preparation for pupation, has been shown to be timed by a circadian clock in *S. c. ricini* (Mizoguchi and Ishizaki 1982).

As mentioned previously, another important life event in Lepidoptera, whose timing is controlled by a circadian clock, is adult eclosion, which is the time of emergence of the fully matured adult from its pupal case. In fact, the population rhythm of adult eclosion has been used extensively as a reliable marker of circadian function in many insect species. Among the Lepidoptera, this includes the silk moths *A. pernyi* and *B. mori*, the cecropia moth, *Hyalophora cecropia*, the fall webworm, *Hyphantria cunea*, the corn borer, *D. grandiosella*, and the monarch butterfly, *D. plexippus* (Truman 1972; Truman and Riddiford 1970; Takeda 1983; Morris and Takeda 1994; Froy et al. 2003). The adults of these Lepidoptera emerge at species-specific times of day. For example, *B. mori*, monarch butterflies, and *H. cecropia* emerge in the early portion of the light period; *A. pernyi* emerges later in the day; and *D. grandiosella* and *H. cunea* emerge close to dusk. For all of these species, the eclosion rhythms free-run with a period length close to 24 hours.

Rhythms in Reproductive Behaviors and Physiology

Reproductive behaviors have been extensively studied in moths, which display pheromone-mediated stereotypical behaviors at species-specific times of day. Female moths produce and emit sex pheromones in a time-of-day–specific fashion, which is synchronized with the male rhythm of sensitivity to pheromones, a phenomenon usually occurring during the night (Cardé and Minks 1997). The circadian nature of these behaviors has been studied in the turnip moth, *Agrotis segetum* (Rosén 2002; Rosén, Han, and Löfstedt 2003), and the cotton leafworm, *Spodoptera littoralis* (Silvegren, Löfstedt, and Rosén 2005), in which the production/emission of sex pheromones by females and the pheromonal reception/attraction behavior by males are both controlled by a circadian clock, leading to coordinated rhythms in mating behaviors (further detailed below).

Less apparent circadian rhythms in reproduction in moths include physiological rhythms of sperm retention in the testis and sperm acidification, and sperm release from the testis as described in the codling moth, *Cydia pomonella*, in the gypsy moth, *Lymantria dispar*, and in *S. littoralis* (Wergin et al. 1997; Giebultowicz and Brooks 1998; Bebas, Cymborowski, and Giebultowicz 2002; Bebas et al. 2002; Syrova, Sauman, and Giebultowicz 2003). The critical aspect of sperm release rhythms in moths is exemplified by the male sterility caused by rhythm disruption under constant light (Riemann and Ruud 1974; Giebultowicz, Ridway, and Imberski 1990). Oviposition rhythms have also been shown in *A. segetum*, *C. pomonella*, the European corn borer, *Ostrinia nubilalis*, and *P. gossypiella* (Minis 1965; Riedl and Loher 1980; Skopik and Takeda 1980; Byers 1987).

Rhythms in Locomotion and Migration

The most common mode of adult locomotion in the Lepidoptera is flight. Flight activity in individual animals has been shown to be under circadian control in some moth species, including *A. pernyi*, *H. cecropia*, and *S. c. ricini*, with a brief burst of flight at the beginning of the night and a broader bout of activity later in the night (Truman 1974). Flight behavior rhythms have only been examined in a few species, as strong masking influences of constant light or darkness on locomotion increase the difficulty of conducting these experiments. However, it is likely that many other Lepidoptera display rhythmic flight that is under circadian control.

Being able to monitor locomotor rhythms in individual moths or butterflies would be the ideal way of tracking the output activity of the circadian clock. Utilizing locomotion assays would diminish the problem of desynchronization among individual rhythms that is inherent in monitoring population rhythms, such as in egg hatching and adult eclosion, which are currently the only reliable means of monitoring rhythmic output in Lepidoptera.

A fascinating example of the involvement of a circadian clock in a vital behavior is its use in time-compensated sun-compass orientation exploited by monarch butterflies for proper navigation during their fall migration (Perez, Taylor, and Jander 1997; Mouritsen and Frost 2002; Froy et al. 2003). The circadian clock provides time compensation that allows the butterflies to continually correct their flight direction relative to skylight parameters (sensed by the sun compass) to maintain a fixed flight bearing in the southerly direction as the sun moves across the sky during the day (further detailed below). The neotropical pierids *Aphrissa statira* and *Phoebis argante* also use a time-compensated sun compass for orientation during migration (Oliveira, Srygley, and Dudley 1998).

MOLECULAR MECHANISMS OF LEPIDOPTERAN CLOCKS

In *Drosophila* and mammals, the intracellular clock mechanism involves transcriptional feedback loops that drive persistent rhythms in mRNA and protein levels of key clock components (Reppert and Weaver 2002; Stanewsky 2003). The negative transcriptional feedback loop is essential for clockwork function and in *Drosophila* involves the transcription factors clock (CLK) and cycle (CYC), which drive the expression of the *period* (*per*) and *timeless* (*tim*) genes. The resultant PER and TIM proteins heterodimerize and translocate back into the nucleus, where PER inhibits CLK:CYC–mediated transcription. TIM appears to regulate PER protein stability and nuclear transport and is also necessary for photic responses that entrain the circadian clock. *Drosophila* cryptochrome (CRY) is a blue-light photoreceptor involved in this photic entrainment (Emery et al. 1998, 2000; Stanewsky et al. 1998). In contrast to *Drosophila*, less attention has been directed at the clockwork mechanism in other, nondrosophilid insect species. Although clock genes and proteins have been used as clock markers in many Lepidoptera, rigorous studies of their clockwork mechanisms have been restricted to two species, the silk moth *A. pernyi* and the monarch butterfly, *D. plexippus*.

The Molecular Clock Mechanism of the Silk Moth *Antheraea pernyi*

The first *per* homologue cloned outside of the fruitfly was in the lepidopteran *A. pernyi* (Reppert et al. 1994). This study was a catalyst for increasing research on the lepidopteran clockwork mechanism. A clock feedback loop has been constructed in vitro from *A. pernyi* components in *Drosophila* Schneider 2 (S2) cells using overexpressed clock proteins (CLK, CYC, PER, and TIM) and the relevant *per* gene promoter element (Chang et al. 2003). Utilizing luciferase reporter gene assays, CLK:CYC heterodimers were shown to activate transcription through an E-box enhancer element in the *per* promoter. PER repressed its own transcription by inhibiting CLK:CYC–mediated transcription, and TIM modestly increased the inhibitory activity of PER. Thus, the in vitro transcriptional feedback loop constructed from *A. pernyi* clock components shows the main features of the *Drosophila* clock. This suggested that the clockwork mechanism originally described in *Drosophila* was likely to be widespread among insects. However, recent studies of the molecular mechanism of the monarch butterfly circadian clock have challenged this hypothesis.

The Monarch Butterfly CRY-Centric Clock: A Prototype of the Lepidopteran Circadian Clock

The molecular clock mechanism in the monarch butterfly, *D. plexippus* (Figure 8.1), has been investigated as part of an effort to understand the vital role that the circadian clock plays in its spectacular migration. Interestingly, in addition to a *Drosophila*-like CRY, designated CRY1, which functions

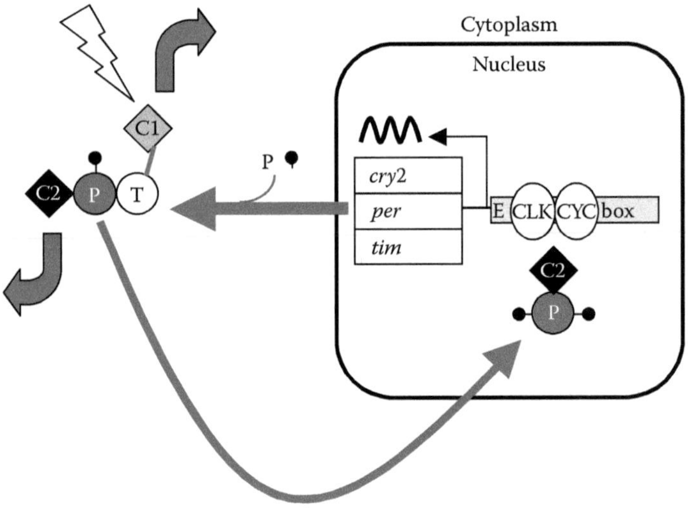

FIGURE 8.1 Proposed monarch butterfly circadian clock mechanism. The main gear of the clock mechanism is an autoregulatory transcription feedback loop in which CLK and CYC heterodimers drive the transcription of the *per, tim*, and *cry2* genes through E-box enhancer elements; CACGTG E-box elements are found within the 1.5 kb 5' flanking regions of the butterfly *per, tim*, and *cry2* genes. TIM (T), PER (P), and CRY2 (C2) form complexes in the cytoplasm and CRY2 is shuttled into the nucleus where it inhibits CLK:CYC–mediated transcription. PER is progressively phosphorylated and likely helps translocate CRY2 into the nucleus. CRY1 (C1) is a circadian photoreceptor that, upon light exposure (lightning bolt), causes TIM degradation, allowing light to gain access to the central clock mechanism. The thick gray arrows represent functions for CRY1 and for CRY2 that may connect the circadian clock to various outputs, including the sun compass. Modified from Zhu et al. (2008).

primarily as a circadian photoreceptor (Zhu et al. 2005, 2008; Song et al. 2007), monarch butterflies also express a second *cry* gene encoding a vertebrate-like, light-insensitive protein designated CRY2 (Zhu et al. 2005, 2008). CRY2 appears to function as a major transcriptional repressor of the core clock feedback loop in monarchs (Figure 8.1). In addition to being a potent repressor of CLK:CYC–mediated transcription in cell culture (Zhu et al. 2005), monarch CRY2 translocates to the nucleus in putative clock cells in monarch brain (discussed below) at the appropriate time for transcriptional repression (Zhu et al. 2008). Monarch PER does not inhibit CLK:CYC–mediated transcription, but it does stabilize CRY2 and may help translocate CRY2 into the nucleus (Zhu et al. 2008; Figure 8.1).

The finding of two functionally distinct *cry* genes in the butterfly, along with database searches, led to the recognition of the presence of insect *cry2* genes in other Lepidoptera, including *A. pernyi, B. mori*, and *S. littoralis* (Zhu et al. 2005; Merlin et al. 2007; Yuan et al. 2007), as well as in every other nondrosophilid insect so far examined (Zhu et al. 2005; Rubin et al. 2006; Yuan et al. 2007). All insect CRY2 proteins evaluated (including those of the bee and the beetle, see below) are potent repressors of CLK:CYC–mediated transcription in cell culture (Yuan et al. 2007).

The discovery of insect CRY2 provides new insight into the evolution of circadian clocks in this class of arthropods. *Drosophila* expresses CRY1 only, whereas other insects (mosquitoes, butterflies, and moths) express both CRY1 and CRY2. Surprisingly, the honeybee *Apis mellifera* and the beetle *Tribolium castaneum* express only CRY2 (Zhu et al. 2005; Rubin et al. 2006). This suggests that the core circadian oscillator has evolved in the insect lineage producing three categories of insect clocks (Yuan et al. 2007): the ancestral clock (apparent in mosquitoes/Lepidoptera) in

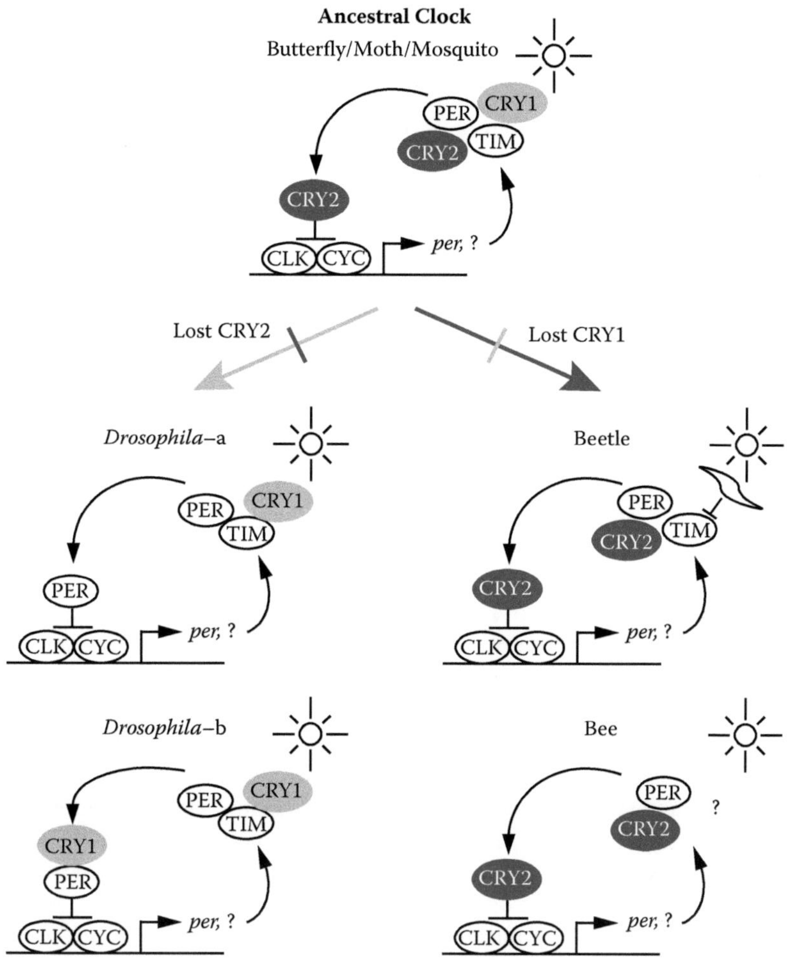

FIGURE 8.2 Insect clockwork models. With the existence of two functionally distinct CRYs in insects, three major types of clockwork models occur: the ancestral clock (apparent in mosquitoes, the monarch butterfly, and silk moths) in which both CRY1 and CRY2 exist and function differently within the clockwork; a derived clock (the *Drosophila* form) in which CRY2 has been lost and CRY1 functions in the central brain clock only as a circadian photoreceptor (panel a; Emery et al. 1998) or in peripheral clocks as both a photoreceptor and central clock component (panel b; Ivanchenko, Stanewsky, and Giebultowicz 2001; Krishnan et al. 2001; Levine et al. 2002); and a derived clock in which CRY1 has been lost and only CRY2 exists and functions within the clockwork, as in beetles and bees. In beetles, CRY2 acts as a transcriptional repressor of the clockwork and light input may be mediated through the degradation of TIM. In bees, which lack TIM (Rubin et al. 2006), CRY2 acts as a transcriptional repressor and novel light input pathways (?) are used to entrain the clock. Modified from Yuan et al. (2007).

which both CRY1 and CRY2 exist and function differentially within the clockwork; a derived clock (the *Drosophila* form) in which CRY2 has been lost and only CRY1 functions; and a derived clock (found in beetles and bees) in which CRY1 has been lost and only CRY2 exists and functions within the clockwork (Figure 8.2). Interestingly, CRY2 proteins in beetles and bees are not light sensitive in culture, as assessed either by degradation of CRY2 or by derepression of inhibitory transcriptional activities, suggesting that these species have novel light input pathways to their circadian clocks as both lack CRY1 (Yuan et al. 2007). In a broader context, the ancestral circadian clockwork of monarch butterflies may be the prototype of a novel clock mechanism shared by nondrosophilid insects that express both *cry1* and *cry2*.

CELLULAR CLOCKS IN THE CENTRAL NERVOUS SYSTEM

To understand how a circadian clock controls behavioral outputs, it is essential to understand where the cellular clock resides in the brain, in addition to the clock molecular machinery itself. In the Lepidoptera, the location of cellular clocks has been most extensively investigated in *A. pernyi*, the monarch butterfly, and, to a lesser extent in *B. mori* and the tobacco hornworm, *Manduca sexta*.

In *A. pernyi*, *per* RNA and protein are colocalized and their levels oscillate in two pairs of cells in the pars lateralis (PL) in the dorsolateral region of each brain hemisphere; these cells also coexpress TIM (Sauman and Reppert 1996). Based on the results of earlier transplantation and ablation studies (Truman 1972, 1974), these PL cells are in the correct location to house circadian clocks. An enigmatic aspect of PER and TIM staining in these cells is that it remains cytoplasmic over the course of the day, and there is no evidence of nuclear translocation (see below). The initial identification of an oscillating antisense *per* transcript in the PL cells (Sauman and Reppert 1996) was later shown to originate from a locus on the female-specific W chromosome (Gotter, Levine, and Reppert 1999). Thus, the antisense *per* transcript in *A. pernyi* is not essential for clock function, as male moths (ZZ) exhibit normal molecular and behavioral rhythmicity (Gotter, Levine, and Reppert 1999).

In monarchs, using a strategy that relied on the coexpression of the main clock proteins probed with monarch-specific antibodies (PER, TIM, CRY1, and CRY2), four cells in the PL (two in each hemisphere) were identified as the putative location of a circadian clock (Figure 8.3A; Sauman et al. 2005; Zhu et al. 2008). Four PL cells in each brain hemisphere express TIM and CRY2, and two of them coexpress PER and CRY1; these two cells in each PL thus contain the proteins necessary for the core clockwork mechanism (Figure 8.3A).

The brain localization studies in *A. pernyi* and the monarch butterfly indicate that the location of clock cells in the PL is conserved in lepidopteran brains. Moreover, in *B. mori*, four large type-1a_1 neurosecretory cells in each PL region express several clock proteins, including PER, the *Drosophila*-like CRY, and CYC (Závodská, Sauman, and Sehnal 2003; Sehadova et al. 2004). In addition, PER-positive cells in the PL of *M. sexta* that are homologous to the PER-positive cells in the silk moth and butterfly PL cells have also been mapped as type-1a_1 cells (Wise et al. 2002).

Interestingly, monarch CRY2 is the only clock protein so far examined in any lepidopteran brain that rhythmically accumulates in the nuclei of PL cells at the appropriate times for transcriptional repression, thus closing the transcriptional feedback loop in vivo (Figure 8.3B; Zhu et al. 2008). The relevance of its role in transcriptional repression is reinforced by the presence in vivo of nuclear CRY2 at the time of maximal repression of the *per* RNA oscillation (Figure 8.3C).

PER, TIM, CRY1, and CRY2 are also colocalized in the monarch brain in large neurosecretory cells in the pars intercerebralis (PI), but the circadian control of PER levels is less obvious there (Sauman et al. 2005; Zhu et al. 2008). The role of these PI cells is unclear, but they could be part of a circadian network leading to migratory behaviors or physiological processes not yet identified.

In addition to clocks in cells located in the central brain, it has been shown in *A. pernyi* that photoreceptor cells of the eye possess a circadian clock, in which not only *per* mRNA and protein levels oscillate, but PER also translocates in and out of the nucleus (Sauman and Reppert 1996). In *A. pernyi* photoreceptor cells, *per* mRNA levels peak in the early night, followed by PER nuclear translocation several hours later, as does the *Drosophila* homologue (Siwicki et al. 1988; Zerr et al. 1990). This could explain the repressive activity of *A. pernyi* PER on CLK:CYC–mediated transcription observed in S2 cells (Chang et al. 2003; see above). In contrast, none of the clock proteins examined in the monarch butterfly are expressed in the eye (Sauman et al. 2005; Zhu et al. 2008).

Considering the similarities in the molecular mechanism of the circadian clock between *A. pernyi* and the monarch butterfly and its cellular localization among *A. pernyi*, the monarch butterfly, and *B. mori*, it is reasonable to propose that the monarch circadian clock constitutes the prototype of the circadian clock in the Lepidoptera. But CRY2 brain localization and nuclear entry in the PL need to be assessed in other Lepidoptera, such as *A. pernyi* and *B. mori*, to confirm this hypothesis.

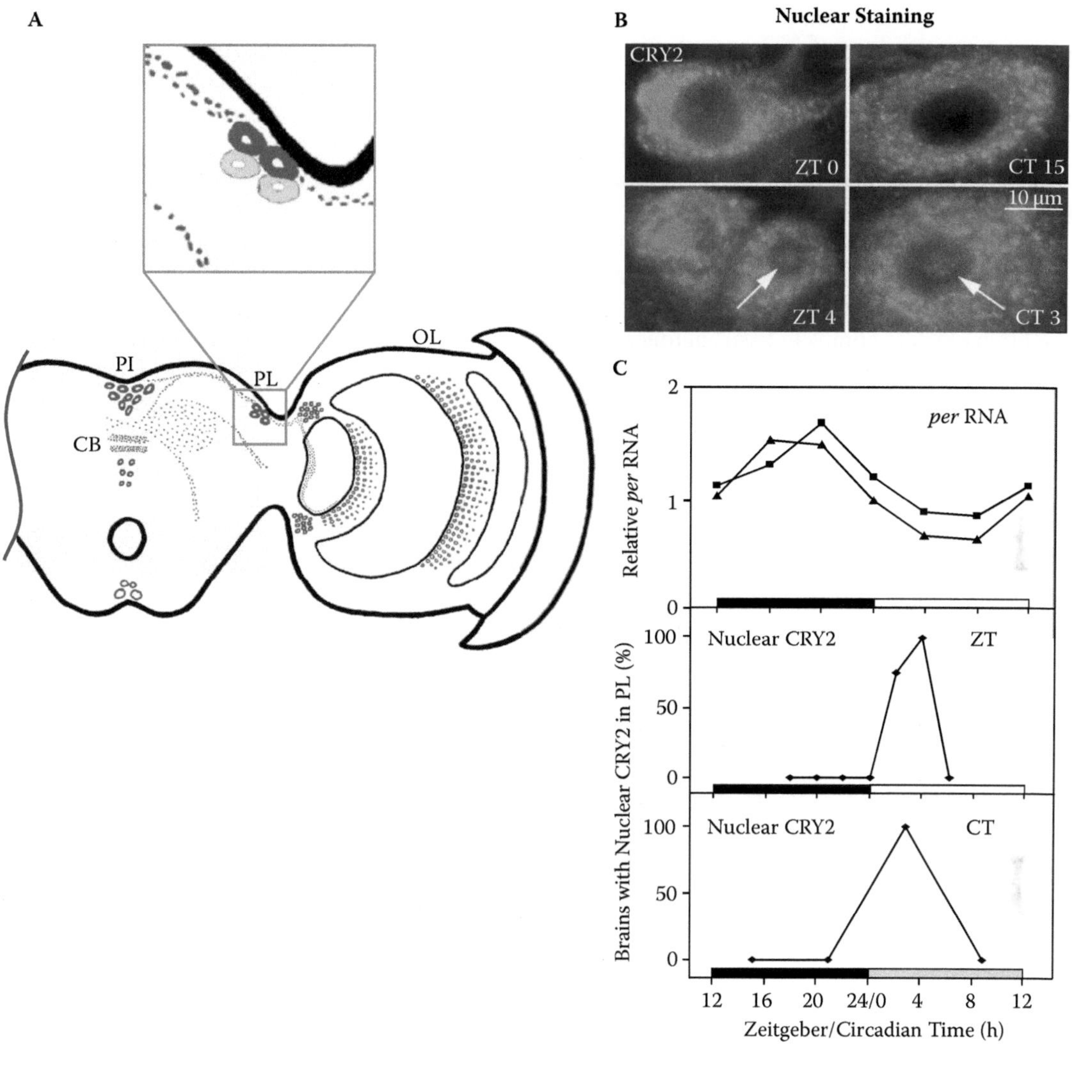

FIGURE 8.3 ***A color version of this figure follows page 176.*** Cellular location of clock cells in monarch butterfly brain. (A) Schematic diagram of a partial frontal brain section illustrating the topography of CRY2-positive cells and neuronal projections, revealed by using monarch-specific antibodies. (OL) Optic lobe; (PL) pars lateralis; (PI) pars intercerebralis; (CB) central body. Top, enlarged PL regions showing the four clock protein-positive cells; the two red cells coexpress PER, TIM, CRY1, and CRY2, while the two pink cells coexpress TIM and CRY2. Modified from Zhu et al. (2008). (B) CRY2 nuclear staining in PL cells. Top left, zeitgeber time [ZT] 0; bottom left, ZT4; top right, circadian time [CT] 15; bottom right, CT3. CRY2 staining was not found in the nucleus at ZT0 or CT15, but it was found in the nucleus in PL at ZT4 and CT3 (arrows). From Zhu et al. (2008). (C) Comparison of *per* RNA levels in the brain with temporal patterns of CRY2 nuclear staining in PL. *Per* RNA levels for two sets of dissected brains without photoreceptors (black and blue lines) collected at 4-hour intervals over 24 hours in LD (upper). Semiquantitative assessment of nuclear CRY2 immunostaining in PL at seven ZT (middle) and CT (lower) times plotted as a percentage of brains examined (n = 4–5 brains for each time point). From Zhu et al. (2008).

CELLULAR CLOCKS IN PERIPHERAL TISSUES

Although it is clear that circadian clocks located in the brain control a variety of behavioral rhythms in insects, it is important to note that clocks located in peripheral organs can regulate many physiological processes in a tissue-dependent manner (for review, see Giebultowicz 1999, 2000, 2001), an idea that originally emerged from studies in *Drosophila*. Using *per*-driven luciferase activity, Plautz et al. (1997) elegantly demonstrated that *per* oscillates in vivo in fruitfly peripheral organs such as the antenna, the proboscis, the wing, and the leg. When cultured in vitro, the bioluminescence in these organs remained rhythmic for several days, revealing the circadian autonomy of these peripheral clocks.

The occurrence of peripheral clocks in the Lepidoptera has been demonstrated by identifying rhythmic expression of clock genes and proteins in organs and cells as well as physiological output rhythms in peripheral organs. Yet, their degree of autonomy ranges from a complete dependency on the brain clock to complete tissue autonomy. In *A. pernyi*, *per* mRNA and protein circadian expression as well as PER nuclear translocation have been observed in the midgut epithelium of embryos and first instar larvae (Sauman et al. 1996; Sauman and Reppert 1998). However, PER movement into the nuclei of midgut epithelial cells was disrupted by head ligations, thus demonstrating that the PER oscillation in midgut was dependent on a substance in the brain. In contrast, the reproductive physiology of male moths is regulated by autonomous peripheral circadian clocks, which control sperm release from the testis to the upper vas deferens (UVD) (Wergin et al. 1997; Giebultowicz, Bell, and Imberski 1988; Giebultowicz et al. 1989; Giebultowicz and Brooks 1998; Syrova, Sauman, and Giebultowicz 2003) and control sperm retention in the UVD (Bebas, Cymborowski, and Giebultowicz 2002; Bebas et al. 2002). Finally, in moths, evidence has been published recently for physiological rhythms in pheromonal reception and the occurrence of a peripheral antennal clock, which could be responsible for the rhythms in this organ (Flecke et al. 2006; Iwai et al. 2006; Merlin et al. 2006, 2007; Schuckel, Siwicki, and Stengl 2007).

TIMING IS EVERYTHING: BIOLOGICAL SIGNIFICANCE OF CIRCADIAN REGULATION

To illustrate the impact that circadian clocks can have on moth and butterfly behaviors in an ecological context, we consider in further detail two systems: the pheromone communication system in moths and time-compensated sun-compass orientation in migratory monarch butterflies.

MATING BEHAVIORS: CIRCADIAN CONTROL OF THE MOTH PHEROMONE COMMUNICATION SYSTEM

In most insect species, especially nocturnal species such as moths, mating behavior is mediated by sex pheromones (sexual communication in Lepidoptera is discussed in Chapter 10). It occurs at a specific time of the day, corresponding to a specific timing of pheromone communication. Female calling behavior and pheromone production/release, used to attract conspecific males, as well as male attraction to pheromones, are restricted to species-specific synchronous times of day (Raina and Menn 1987). In nature, this temporal species-specificity is particularly relevant for the isolation of sympatric species that use similar pheromones.

The temporal control of insect mating behavior is well documented in nocturnal Lepidoptera; mating occurs during a few defined hours of the night. Daily rhythms of female pheromone production/emission and male pheromone responsiveness have been documented in many species, but their control by a circadian clock has been less extensively investigated. However, recent data in the moths *A. segetum* and *S. littoralis* (Rosén 2002; Rosén, Han, and Löfstedt 2003; Silvegren, Löfstedt, and Rosén 2005; Merlin et al. 2007) provide us with an understanding of how the circadian control of mating behavior in moths occurs. In *A. segetum*, female pheromone production and male pheromone behavioral responsiveness are endogenously regulated, and both peak during the same temporal window at night (Rosén 2002; Rosén, Han, and Löfstedt 2003). The impact of such a synchrony between sexes on the success of mating has been demonstrated in *S. littoralis* (Silvegren, Löfstedt, and Rosén 2005).

Males detect sex pheromones by specialized receptors located in the antennae, which are the site of a peripheral reception necessary for the male behavioral response to pheromones; removing the antennae leads to a lack of attraction (Rosén, Han, and Löfstedt 2003). In contrast to *A. segetum*, in which no circadian rhythm has been found in the physiological response of the antennae to pheromones (Rosén, Han, and Löfstedt 2003), *S. littoralis* exhibits a response rhythm in the antennae (Merlin et al. 2007), in accordance with previous evidence of daily changes in the sensitivity of pheromone-sensitive units (sensilla) in the antennae of *M. sexta* (Flecke et al. 2006). Moreover, the study of the temporal and spatial expression patterns of clock genes and clock proteins in moth antennae, including *S. littoralis*, revealed that circadian clocks are present in the olfactory organ at the base of the odorant-sensitive sensilla (Merlin et al. 2006, 2007; Schuckel, Siwicki, and Stengl 2007).

Is the circadian rhythm of pheromone reception controlled by these circadian antennal clocks and directly involved in the rhythmic male behavioral response to pheromones? Based on data available in *Drosophila* showing that antennal clocks are necessary and sufficient for olfaction rhythms (Tanoue et al. 2004), this hypothesis is realistic, but only studies conducted on isolated olfactory organs will establish the involvement of these clocks in the physiological rhythms generated in moth antennae. More challenging, but nevertheless essential, will be to disrupt the moth antennal clocks genetically in vivo, which would determine whether they are indeed responsible for the behavioral rhythms of pheromone attraction.

Navigational Capabilities: Time-Compensated Sun-Compass Orientation in the Migratory Monarch Butterfly

Eastern North American monarch butterflies accomplish an extraordinary migration each fall to overwintering sites in central Mexico (Urquhart 1987; Brower 1995); the biological basis of their navigation is gradually being elucidated.

The circadian clock appears to play an essential role in navigation during migration by providing the timing component of time-compensated sun-compass orientation (Perez, Taylor, and Jander 1997; Mouritsen and Frost 2002; Froy et al. 2003). The circadian clock allows monarchs to compensate for the movement of the sun across the sky during the day, so that the butterflies can constantly correct their flight direction to south/southwest (Figure 8.4A, upper panel). Evidence that monarch butterflies use a time-compensated sun compass has been most convincingly shown using a flight simulator, which allows the study of flight trajectories from tethered butterflies during sustained flight (Mouritsen and Frost 2002). The importance of the circadian clock in regulating the time-compensated component of flight orientation has been shown in clock-shift experiments, in which the timing of the daily light-dark cycle is either advanced or delayed, causing predictable alterations in the direction of flight (Perez, Taylor, and Jander 1997; Mouritsen and Frost 2002; Froy et al. 2003). Moreover, when the clock is disrupted experimentally by constant light, monarch butterflies lose their ability to correct their flight direction relative to the sun position, instead flying toward the sun (Figure 8.4A, lower panel; Froy et al. 2003).

Understanding how the clock communicates with the sun compass and how directional information is integrated with the motor system will be key elements to provide a comprehensive view of the navigational capabilities of monarch butterflies. Information about the neural pathways connecting the circadian clock to the sun compass is emerging. Indeed, a CRY1-staining neural pathway appears to connect the circadian clock in the PL to polarized light input entering the brain from the dorsal rim area photoreceptor cells of the compound eye (Figure 8.4B), which is important for sun-compass navigation (Sauman et al. 2005; Reppert, Zhu, and White 2004).

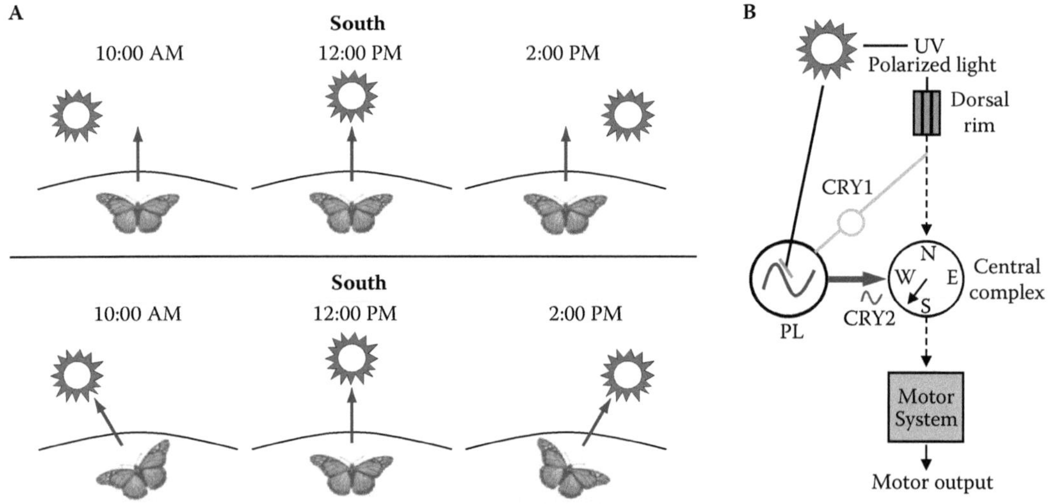

FIGURE 8.4 Time-compensated sun-compass orientation. (A) Lost without a clock. (*Upper panel*) Eastern North American monarch butterflies use a time-compensated sun compass to orient south during their fall migration. The circadian clock allows the butterflies to compensate for the movement of the sun. The migrants are thereby able to maintain a constant bearing in the southerly direction over the course of the day even though the sun has moved relative to their path of migration. (*Lower panel*) Monarch butterflies follow the sun without a functioning circadian clock. A broken circadian clock would disrupt migration south, and the butterflies would not be able to travel successfully to their overwintering grounds. (B) Model of the proposed clock-compass pathways in monarch butterfly brain. The circadian clock in the pars lateralis (PL) is entrained by light through CRY1, which also connects the circadian clock to axons arising from UV/polarized light–sensitive photoreceptors of the dorsal dim area of the eye (Sauman et al. 2005). CRY2-positive fiber pathways may directly link the circadian clock to the sun compass (in central complex), which ultimately controls motor output. Modified from Zhu et al. (2008).

The sun compass likely resides in the central complex for integrating skylight information (including polarized light; Vitzthum, Muller, and Homberg 2002; Heinze and Homberg 2007), a midline structure of the brain composed of the central body, which is subdivided into upper and lower parts, and the protocerebral bridge. Interestingly, in monarch butterflies a dense arborization of CRY2 staining has been found in the central body (Figure 8.3A), which is under circadian control, with intense staining in the middle of the night and little to no staining in the middle of the subjective day (Zhu et al. 2008). Importantly, other clock proteins (PER, TIM, and CRY1) are not expressed there. In addition, the CRY2-positive arborizations in the central complex are likely to arise from cells in the PL, the PI, or both regions, as suggested by the CRY2-staining neural pathway observed (Figure 8.3A). CRY2 could either be marking the clock-compass neural pathway and/or regulating rhythmic neural activity in the central complex.

CONCLUSIONS

As emphasized in this chapter, Lepidoptera have made major contributions to circadian biology and have expanded our knowledge of insect clockwork mechanisms. The centerpiece is the recent discovery of the critical clockwork component, CRY2, in all insects examined so far, with *Drosophila* being the exception. In contrast to *Drosophila* circadian clocks, in which PER is the main transcriptional repressor, the transcriptional repressive activity of most other insects may rely on CRY2. Moreover, to date CRY2 is the only protein shown to accumulate in vivo in the nuclei of lepidopteran clock cells. Determining if CRY2 is essential in vivo for generating appropriate behavioral outputs remains to be tested, however. This goal can be realistically pursued using *B. mori*, because this

9 Lepidopteran Chemoreceptors

Kevin W. Wanner and Hugh M. Robertson

CONTENTS

INTRODUCTION

The ability to sense and respond to chemical stimuli in the environment was a fundamental evolutionary development. Many complex insect behaviors, including foraging, host seeking and selection, feeding, oviposition, nesting, hygiene, mating, and kin selection behaviors, are mediated by the chemical senses. Research during the last decade is revealing an equally complex molecular genetic and neural system for detecting and discriminating chemical stimuli. Much of the recent progress toward understanding the molecular mechanisms of the chemical senses has been facilitated by the whole genome sequence of *Drosophila melanogaster* completed in 2000 (Adams et al. 2000). For the first time, insect chemoreceptor (Cr) genes were identified using bioinformatics tools to search the fruit fly genome (Clyne et al. 1999; Gao and Chess 1999; Vosshall et al. 1999). The rapid advance of whole genome sequencing is helping to extend this progress broadly to nonmodel and emerging model species such as the silkworm *Bombyx mori* (Goldsmith, Shimada, and Abe 2005; see Chapter 2 for silkworm genomics). The first drafts of the silkworm genome sequence were published in 2004 (Mita et al. 2004; Xia et al. 2004), providing an insight into the chemoreceptor sequences of the Lepidoptera (Wanner et al. 2007; Wanner and Robertson 2008). Although this field is progressing rapidly, it remains in its infancy. When an insect genome is sequenced, one of the first steps toward chemosensory research is the annotation of its chemoreceptors. Almost all species of Lepidoptera are herbivorous and therefore represent a large proportion of the insect pests of food and fiber crops worldwide. An analysis of Lepidoptera Cr genes will not only provide interesting insights into the evolution of plant host interactions (see Chapter 11 for evolution of host range), but it will also provide new gene targets for the development of insect management tools. In this chapter we will review the phylogenetics of the lepidopteran Cr superfamily within the broad context of knowledge gained from *D. melanogaster* research.

WHAT ARE CHEMORECEPTORS?

Chemical stimuli are first detected at the periphery by specialized sensory neurons that transduce the information into nerve impulses that are processed by the central nervous system (CNS). In a landmark study, Buck and Axel (1991) discovered a large family of G-protein coupled receptors (GPCRs) specifically expressed in rat olfactory epithelium. GPCRs are members of a receptor superfamily that commonly initiate secondary messenger signaling systems in response to activation by a specific ligand. Genomic research indicates that the GPCR superfamily has provided the genetic starting material for chemoreceptor evolution several times within independent lineages representing mammals, nematodes, and echinoderms (Buck and Axel 1991; Robertson and Thomas 2006; Sea Urchin Genome Sequencing Consortium 2006; Thomas and Robertson 2008). With expectations that insects would be no different, Clyne et al. (1999) employed an algorithm on the fruit fly genome to predict genes encoding proteins with seven transmembrane domains, a canonical feature of GPCRs. A family of putative odorant receptors was identified from the first 15 percent of the *D. melanogaster* genome sequence available before the WGS sequencing of this genome by Celera Genomics (Rockville, Maryland). Vosshall et al. (1999) used a similar approach to find the same family, and Clyne, Warr, and Carlson (2000) identified an additional, distantly related family of candidate gustatory receptors. The term *chemoreceptor* is used to describe a superfamily, collectively referring to both the odorant receptor (Or) and gustatory receptor (Gr) families. A total of approximately sixty genes in each family have now been identified from *D. melanogaster* (Robertson, Warr, and Carlson 2003), and the function of many have been characterized based on their ligand specificity and expression in specific olfactory and gustatory neurons (summarized in several recent reviews, including Rutzler and Zwiebel 2005; Thorne, Bray, and Amrein 2005; Hallem, Dahanukar, and Carlson 2006). However, recent studies have demonstrated that the topology of insect Ors is opposite that of GPCRs; the N-terminus of Ors is intracellular and C-terminus is extracellular, calling into question their designation as GPCRs (Benton et al. 2006; Wistrand, Käll, and Sonnhammer 2006; Lundin et al. 2007). Furthermore, and somewhat surprisingly, little is known about the signaling pathway downstream of the Ors and Grs, and fruit fly mutations in the signaling pathway have yet to be reported. Therefore, the exact mechanism of insect chemosensory signal transduction remains to be determined. Two recent studies were published that provide an interesting insight into the functional nature of insect odorant receptors. These two studies reported evidence that insect Ors function by a novel mechanism, themselves acting as heteromeric ligand-gated ion channels, giving an explanation for the observed rapid activation kinetics (Sato et al. 2008; Wicher et al. 2008). Remarkably, Wicher et al. (2008) propose that in addition to direct activation of the heteromeric Or cation channel by an odor, the Or channel can also be activated by cAMP via interaction with a G protein, resulting in slower and prolonged activation kinetics.

Although the peripheral sensory system includes several different gene families, molecular genetic experiments with the model *D. melanogaster* indicate the Crs play a preeminent role in the detection and discrimination of chemical stimuli. The sensillum lymph that surrounds sensory neurons includes transport proteins (odorant binding proteins, OBPs) and biotransformation enzymes (cytochrome P450s, glutathione-S-transferases, and esterases) (e.g., Leal 2005; Rutzler and Zwiebel 2005). Evidence from fruit flies and moths indicate that some OBPs may interact with the receptors rather than simply having a passive role in transporting ligands (Xu et al. 2005; Grosse-Wilde, Svatos, and Krieger 2006; Smith 2007). Insects have fewer odorant receptors but many more OBPs compared with mammals; therefore, insect Ors and OBPs may act in a combinatorial mechanism to increase ligand coding capacity. Sensory neuron membrane protein 1 (SNMP1) was identified from moth antennae prior to the discovery of the *Drosophila* Ors (Rogers et al. 1997), but its function remained uncertain. Recently the *Drosophila* homologue SNMP1 has been demonstrated to play a role in Or-mediated signal transduction (Benton, Vannice, and Vosshall 2007). Clearly, signal transduction at the periphery will be influenced by several different gene families and their potential interactions. However, it is also clear from *Drosophila* research that the Crs play a preeminent role (Hallem, Ho, and Carlson 2004; Hallem, Dahanukar, and Carlson 2006), a view now reinforced by the fact that insect Ors may themselves act as ion channels (Sato et al. 2008; Wicher et al. 2008).

In many cases the relationship between sensory input and resulting behavior is complex and context dependent. In other cases, however, the sensory detection of particular stimuli is more directly linked to behavior, including sex pheromones that mediate mating behavior and sweet and bitter tastes that mediate food acceptance and aversion. In many (but not all) cases *Drosophila* Ors can be expressed in an "empty" olfactory neuron that subsequently responds with the same electrophysiological profile as the wild-type neuron natively expressing the same Or (Hallem, Ho, and Carlson 2004; Hallem, Dahanukar, and Carlson 2006). A Lepidoptera pheromone receptor has also been functionally expressed in this *Drosophila* system (Syed et al. 2006). Furthermore, the pheromone receptors of distantly related insect species BmOr1 (Nakagawa et al. 2005) and DmOr67d (Ha and Smith 2006) can be juxtaposed and the resulting transgenic male fruit flies become attracted to the silkworm sex pheromone bombykol rather than their own aggregation and sex pheromone vacenyl acetate (Kurtovic, Widmer, and Dickson 2007). A relatively direct link between the activation of gustatory sensory neurons responsible for detecting sweet or bitter compounds and feeding stimulation or avoidance behavior has also been demonstrated using transgenic fruit flies (reviewed in Scott 2005; Ebbs and Amrein 2007). Therefore, evolutionary changes to chemoreceptor sequences can significantly impact insect behavior and their ecology, making the phylogenetic and evolutionary analysis of this gene family a fascinating endeavor.

PHYLOGENETICS OF LEPIDOPTERAN CHEMORECEPTORS

Olfaction versus Gustation, Ors versus Grs

The distinction between the Or and Gr families can be made based on their phylogenetic relatedness and their gene expression patterns, and functionally by the class of ligand that they detect. Insect Or genes are expressed in specific subsets of olfactory neurons whose axons project to the antennal lobe; Gr genes are usually expressed in gustatory neurons whose axons project to taste processing centers of the CNS such as the subesophageal ganglion. Ors tend to detect volatile hydrophobic odors; Grs (from the limited data available) tend to detect soluble chemicals acquired from contact with a substrate. However, these distinctions between olfaction and gustation are not always perfectly clear. Insect antennae primarily function as olfactory organs, but they retain some taste sensilla with gustatory neurons that project to the subesophageal ganglion (Jørgensen et al. 2006). Some Gr genes, including the fruit fly carbon dioxide receptors and their homologues in mosquitoes (Jones et al. 2007; Kwon et al. 2007; Lu et al. 2007; Robertson and Kent 2008), are expressed in olfactory neurons located on fly antennae or mosquito palps. To date there are no examples of an Or gene expressed in gustatory neurons.

Insect Ors are proposed to have evolved from a Gr lineage during the transition to a terrestrial environment by an arthropod ancestor (Robertson, Warr, and Carlson 2003). Gene duplication and gene loss combined with sequence divergence have resulted in weak phylogenetic relationships between distantly related insects. For example, if several Cr sequences are selected at random from different insect orders and their protein sequences aligned, few if any common amino acid motifs can be identified. In fact, with only a few exceptions, there are almost no amino acid residues in common between Crs from different insect orders. To illustrate this point we choose AgOr1/AgGr1, AmOr11/AmGr4, BmOr1/BmGr18, DmOr35a/DmGr8a, and TcOr6/TcGr140 representing the orders Diptera, Coleoptera, Hymenoptera, and Lepidoptera (*Anopheles gambiae*, Ag; *Apis melifera*, Am; *Bombyx mori*, Bm; *Drosophila melanogaster*, Dm; and *Tribolium castaneum*, Tc, respectively) (Figures 9.1 and 9.2). The ligand specificity of many more Ors has been characterized compared with Grs; therefore, we were able to choose as examples four Ors that have been functionally characterized. AgOr1 responds to a component of human sweat (Hallem et al. 2004); AmOr11 responds to the main component of the queen honeybee pheromone (Wanner et al. 2007); BmOr1 responds to the silkworm sex pheromone (Sakurai et al. 2004; Nakagawa et al. 2005; Syed et al. 2006); and DmOr35a responds to six carbon alcohols (Hallem, Ho, and Carlson 2004; Wanner et al. 2007). The aligned Ors have only seven amino acids in common, while the aligned Grs have only six (Figures 9.1B and 9.2B). In pairwise alignments the amino acid identity of these Ors ranges from 13 to 23 percent, while the Grs ranges from 7 to 15 percent (Figure 9.3B).

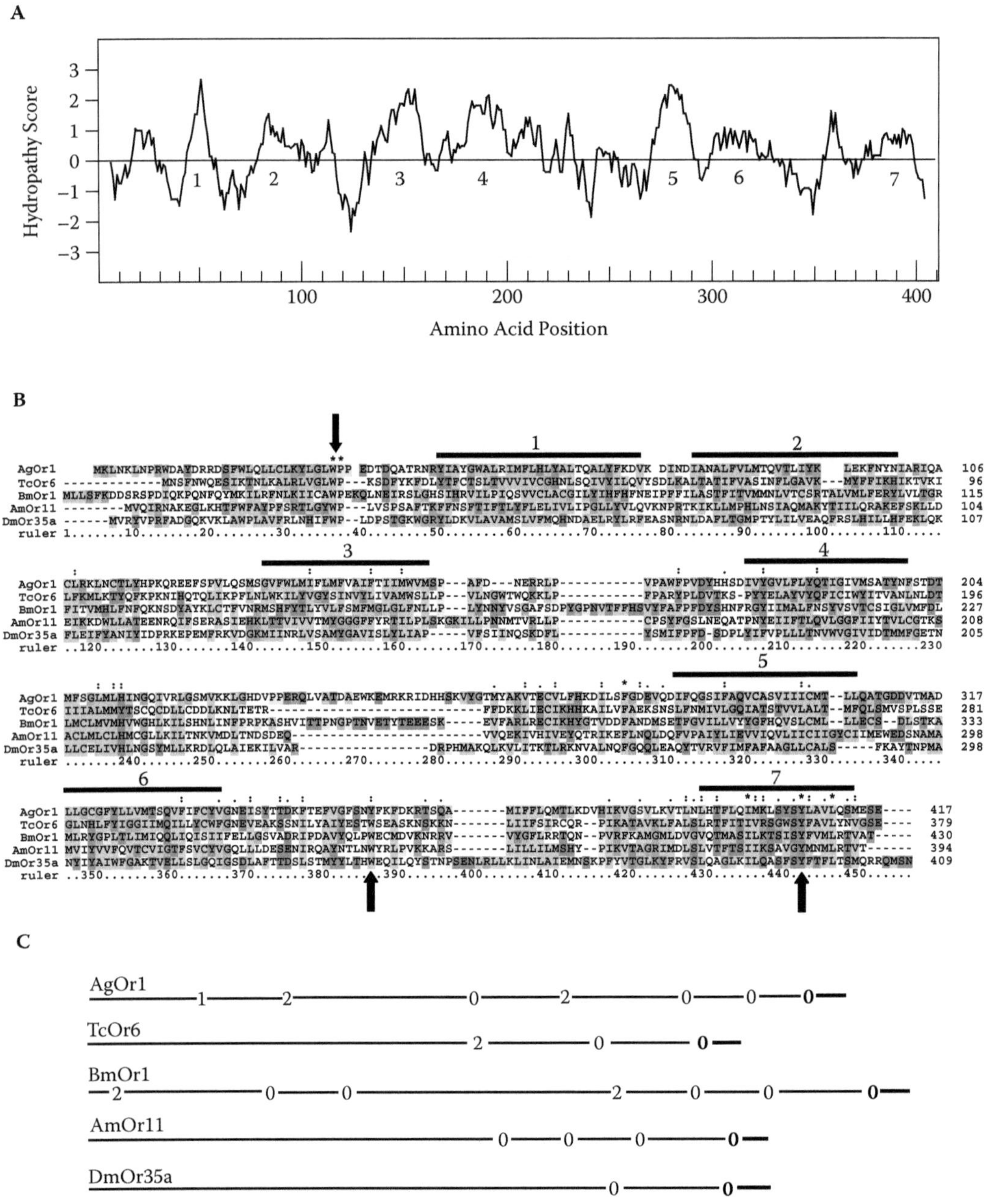

FIGURE 9.1 Molecular characteristics of insect Ors. (A) Hydropathy plot (Kyte-Doolittle method) of DmOr35a illustrating seven potential transmembrane domains. (B) ClustalX multiple alignment of five insect Or sequences representing four different insect orders (AgOr1, *Anopheles gambiae*, Diptera; TcOr6, *Tribolium castaneum*, Coleoptera; BmOr1, *Bombyx mori*, Lepidoptera; AmOr11, *Apis mellifera*, Hymenoptera; and DmOr35a, *Drosophila melanogaster*, Diptera). Bars with numbers illustrate the seven putative transmembrane domains. Arrows indicate conserved amino acid residues. (C) Gene structure of the five Ors illustrating their diversity. Solid lines represent exons (scaled relative to each other); numbers represent the location of intron/exon boundaries and their splice phase (phase 0, 1, or 2; see text for explanation of splicing phase). A darker solid line is used to highlight the short, last exon typically spliced in the 0 phase.

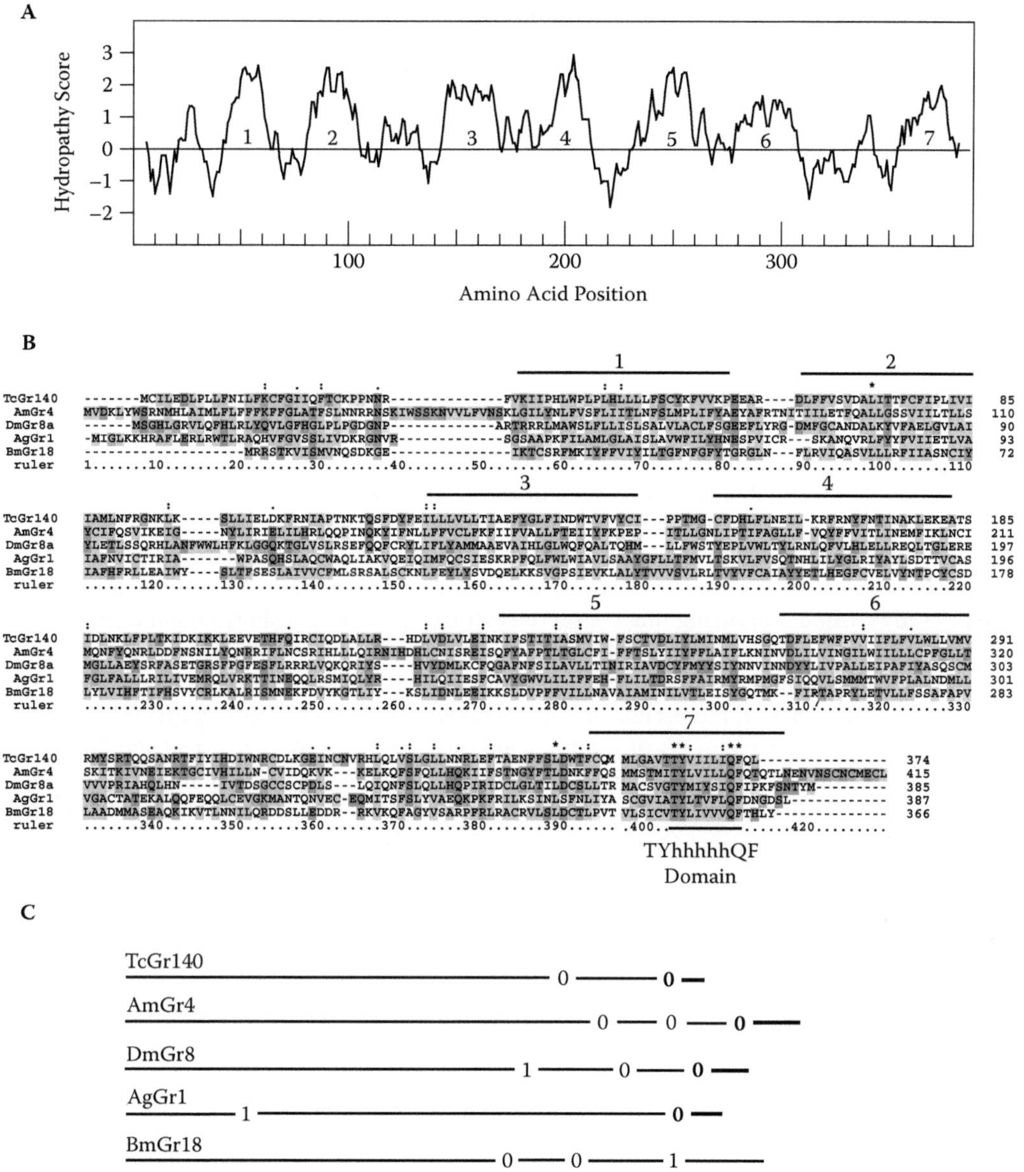

FIGURE 9.2 Molecular characteristics of insect Grs. (A) Hydropathy plot of AgGr1 illustrating seven potential transmembrane domains. (B) ClustalX alignment of five insect Gr sequences representing four different insect orders (AgGr1, *Anopheles gambiae*, Diptera; TcGr140, *Tribolium castaneum*, Coleoptera; BmGr18, *Bombyx mori*, Lepidoptera; AmGr4, *Apis mellifera*, Hymenoptera; and DmGr35a, *Drosophila melanogaster*, Diptera). Bars with numbers illustrate the seven putative transmembrane domains. The conserved TYhhhhhQF domain at the C-terminus is also illustrated. (C) Gene structure of the five Grs illustrating their diversity. Solid lines represent exons (scaled relative to each other); numbers represent the location of intron/exon boundaries and their splice phase (phase 0, 1, or 2). A darker solid line is used to highlight the short, last exon typically spliced in the 0 phase.

Despite this extreme sequence divergence, Or and Gr sequences do retain some similar characteristics. Hydropathy plots can be used to map hydrophobic protein domains and thus predict putative transmembrane spanning regions. Insect Ors and Grs both have seven potential transmembrane-spanning domains (DmOr35a and AgGr1 in Figures 9.1A and 9.2A, for example) discernable in protein alignments as regions with high frequencies of hydrophobic amino acids (Figures 9.1B and 9.2B). Consistent with their divergent amino acid sequences, Or and Gr gene structures are also highly variable in terms of the number, location, and splicing phase of the introns (Figures 9.1C and 9.2C). Insect Or and Gr genes typically have anywhere from one to as many as nine or ten introns, with just a few having lost all introns (in contrast to the mammalian odorant receptors, which have no introns in their coding regions). Despite this diversity, most insect Crs have a short, last exon spliced in the zero phase (Figures 9.1C and 9.2C), a feature in common between Ors and Grs (intron/exon boundaries can be spliced in one of three phases relative to a codon, between two codons [phase 0], between the first and second base pairs of a codon [phase 1], or between the second and third base pairs of a codon [phase 2]). The almost ubiquitous presence of this final intron in insect chemoreceptors implies an important role that remains obscure.

It is also this C-terminal region of insect Ors and Grs that exhibits some of the only discernable amino acid conservation within and between the two protein families (Figures 9.1B and 9.2B). Both Ors and Grs tend to have a tyrosine (Y) amino acid conserved in the same location in the last exon, the eighth residue after the phase 0 splice site. A serine (S) or threonine (T) residue often precedes the conserved tyrosine. Following this conserved tyrosine residue is a short region of hydrophobic amino acids, particularly conserved in the Grs. The Grs (but not the Ors) also have conserved glutamine (Q) and phenylalanine (F) residues following the hydrophobic region, creating a conserved TYhhhhhQF motif, where h represents any hydrophobic amino acid (Figure 9.2B). The Or family in particular has two other regions of amino acid conservation, a WP pair near the N-terminus, and a W in the third intracellular loop between transmembrane domains 6 and 7 (Figure 9.1B). Remarkably, the function of these conserved residues remains unknown. Although the diversity is incredibly high (the identity of Ors and Grs is typically only 6 to 12 percent, Figure 9.3B), there remain sufficient conserved characters to separate Ors and Grs using phylogenetic techniques, as evidenced in Figure 9.3A, where the ten Cr sequences representing different insect orders group into their Or and Gr families.

Gustatory Receptor Family

In *Drosophila* Gr genes are predominantly expressed in gustatory neurons on taste organs (Clyne, Warr, and Carlson 2000; Wang et al. 2004; Thorne, Bray, and Amrein 2005; Slone, Daniels, and Amrein 2007). However, recent research indicates that compared with Or genes, some Grs can be expressed more broadly and may function in physiological processes (Thorne and Amrein 2008). Some Grs may have been co-opted for other functions, or their function outside of the chemosensory system could reflect an ancestral role in the internal physiological detection and monitoring of nutrient chemicals. At this time we can only speculate, but it is interesting to note parallels with the vertebrate gustatory system, where Gr genes have been found to be expressed in the gastrointestinal tract (Wu et al. 2002). Another interesting parallel is the trend for both insect and vertebrate genomes that encode a small number of relatively well-conserved Grs dedicated to the detection of essential nutrients such as sugars, and large numbers of expanded and highly divergent gustatory receptors believed to be dedicated to the detection of bitter compounds (Go 2006; Shi and Zhang 2006). Our efforts to annotate the silkworm Gr family indicate that the Lepidoptera Gr family will follow a similar trend.

Recently, we completed the annotation of sixty-five Gr genes from the silkworm genome (Wanner and Robertson 2008). Only three other lepidopteran Grs have been published to date, HvCr1, 4, and 5 (Krieger et al. 2002), identified from *Heliothis virescens* genomic DNA sequence privately held by Bayer Corporation. Originally termed Crs, these three sequences belong to the Gr family and

A

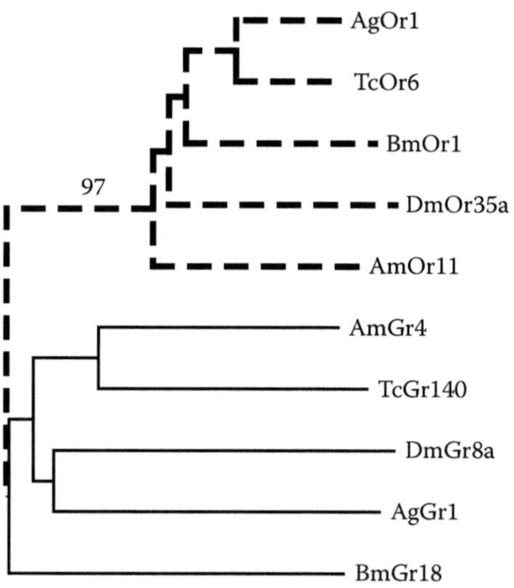

B

	AgOr1	TcOr6	BmOr1	DmOr35a	AmOr11	AmGr4	TcGr140	DmGr8a	AgGr1	BmGr18
AgOr1	**0**	23	14	14	17	10	11	8	8	7
TcOr6	23	**0**	15	13	14	10	11	11	8	9
BmOr1	14	15	**0**	14	15	12	8	7	6	9
DmOr35a	14	13	14	**0**	14	9	10	9	9	9
AmOr11	17	14	15	14	**0**	9	11	11	9	8
AmGr4	10	10	12	9	9	**0**	14	12	13	11
TcGr140	11	11	8	10	11	14	**0**	14	10	11
DmGr8a	8	11	7	9	11	12	14	**0**	15	7
AgGr1	8	8	6	9	9	13	10	15	**0**	11
BmGr18	7	9	9	9	8	11	11	7	11	**0**

FIGURE 9.3 Insect Ors and Grs form distinct families within the Cr superfamily. (A) Phylogenetic tree (corrected distance) of the five Ors and five Grs illustrated in Figures 9.1 and 9.2, representing four insect orders (Ag, *Anopheles gambiae*, Diptera; Tc, *Tribolium castaneum*, Coleoptera; Bm, *Bombyx mori*, Lepidoptera; Am, *Apis mellifera*, Hymenoptera; and Dm, *Drosophila melanogaster*, Diptera). Despite their amino acid diversity, the Or family members group together separate from the Gr family (bootstrap value equals 97 of 100 replicates). (B) Amino acid identity matrix of the five Ors and five Grs illustrated in Figures 9.1B and 9.2B.

possess typical Gr traits illustrated in Figure 9.2. The phylogenetic relationships of the BmGrs with HvCr1, 4, and 5 along with a selection of Grs from other insect orders are illustrated in Figure 9.4. *B. mori* Grs1–10 form three conserved lineages with Grs from other insect orders; these are the only Gr lineages conserved between different insect orders (Robertson and Wanner 2006; Wanner and Robertson 2008). BmGrs1–3 are orthologs of the two to three carbon dioxide receptors functionally characterized in flies (Jones et al. 2007; Kwon et al. 2007; Lu et al. 2007; Robertson and Kent 2008). BmGrs4–8, along with HvCr1 and 5, are members of the insect sugar receptor subfamily based on functional characterization of the *Drosophila* sugar receptors (e.g., Slone et al. 2007; Dahanukar et al. 2007; Kent and Robertson, 2009). BmGrs9 and 10, along with HvCr4, are orthologs of the DmGr43a protein (Figure 9.4; Robertson, Warr, and Carlson 2003), a lineage that is conserved within endopterygote insects including honeybees (AmGr3; Robertson and Wanner, 2006), mosquitoes (AgGr25 and AaGr20; Hill et al. 2002; Kent, Walden, and Robertson 2008), and beetles (TcGr20–28 and 183, Tribolium Genome Sequencing Consortium 2008). The DmGr43a orthologs form a highly confident single lineage, but their function remains unknown.

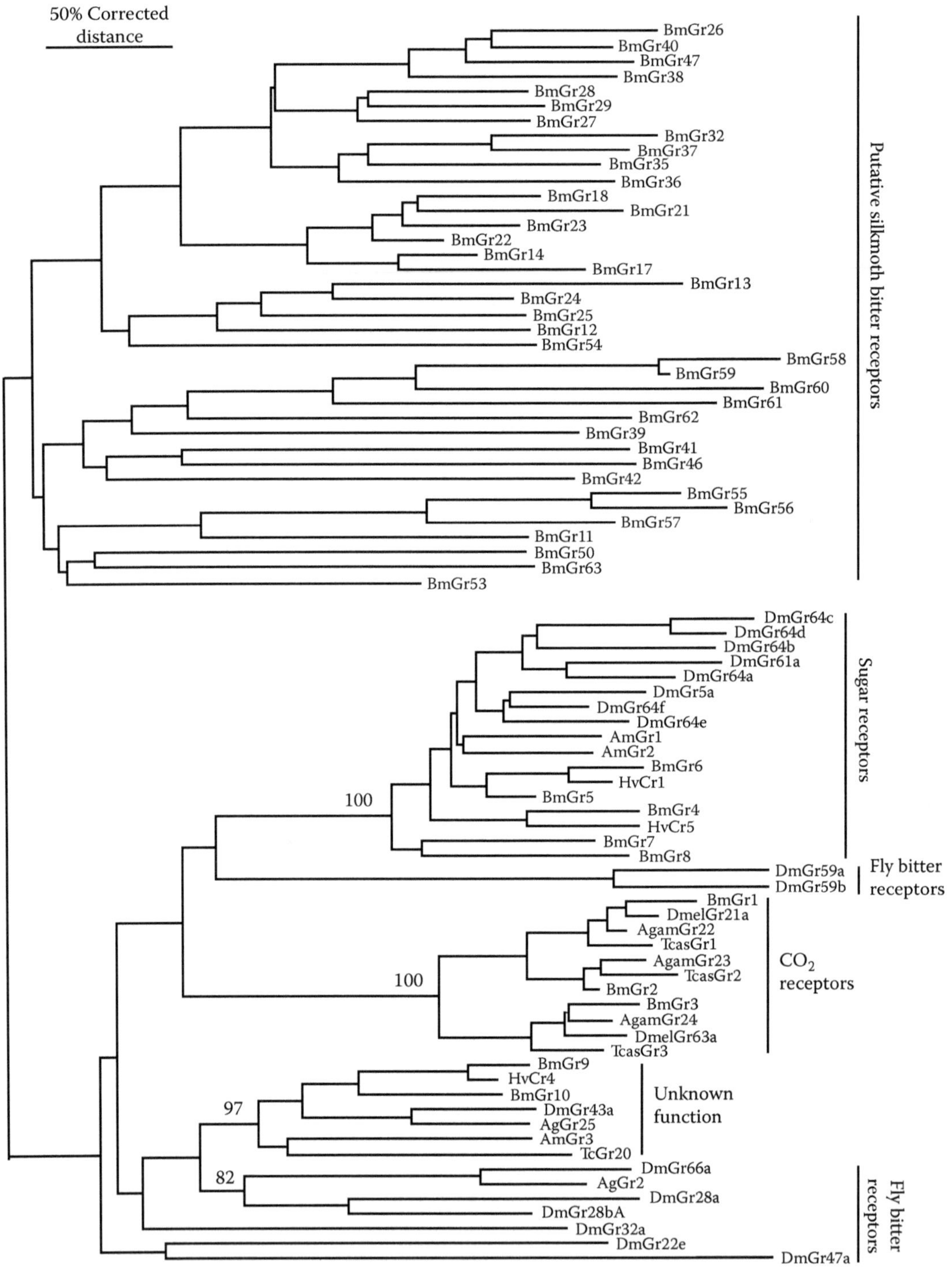

FIGURE 9.4 Phylogenetic tree relating the 65 BmGrs and 3 published *Heliothis virescens* Crs, HvCr1, 4, and 5 (Krieger et al. 2002). Representative fly (Dm), mosquito (Ag), beetle (Tc), and bee (Am) orthologs of the conserved sugar, CO_2, and DmGr43a lineages are included to illustrate their conservation between different insect orders. A selection of putative fly bitter receptors (Robertson, Warr, and Carlson 2003) are included to illustrate their lack of relationship to the putative moth bitter receptors. The clusters of carbon dioxide, sugar, Dm43a orthologs, and candidate bitter taste receptor lineages are indicated with bars on the right. Bootstrap support for major branches is shown as the percentage of 10,000 replications of uncorrected distance analysis.

One of the more fascinating aspects arising from the analysis of the 65 BmGrs is the apparent monophyletic expansion of the remaining 55 Grs to form a subfamily distinct from those of other insects (Figure 9.4). The amino acid sequences of these 55 Grs are very diverse, commonly sharing less than 10 percent amino acid identity. Like other insect Grs, the very end of the C-terminus has the most conserved region including the TYhhhhhQF motif (Figure 9.2B; Wanner and Robertson 2008). A consequence of their sequence diversity, their monophyletic origin cannot be supported statistically using distance-based phylogenetic analyses (Figure 9.4). However, their monophyletic origin can be supported using a Bayesian analysis (91 percent bootstrap value; Wanner and Robertson 2008). Additionally and quite remarkably, 54 of these 55 Grs have an identical and unique gene structure (illustrated by the gene structure of BmGr18 in Figure 9.2C), further supporting a common origin for this large silkworm-specific Gr subfamily (the 55th gene has lost all three introns). Members of this subfamily have a long N-terminal exon encoding more than 50 percent of the Gr, much like many Grs in other insects (e.g., Robertson, Warr, and Carlson 2003), followed by an unusually short exon encoding approximately 25 amino acids, a third exon encoding approximately 50 amino acids, and a final relatively long C-terminal exon (Figure 9.2C). Further supporting the uniqueness of this moth Gr lineage, the typical phase 0 spliced intron/exon boundary near the C-terminus is absent (Figure 9.2C). Contact chemoreception is used to recognize and select host plants for feeding and oviposition as well as kin and mate recognition. In some cases host selection by phytophagous insects is a balance between phagostimulating nutrients and deterring bitter compounds, but in other cases specific "sign" stimuli are used (Chapman 2003; see also Chapter 11). Therefore, some of these 55 Grs may detect specific cues from their host plants or mates, but the majority may be dedicated to detecting bitter, deterrent plant chemicals.

With the exception of the apparent monophyletic expansion of putative bitter receptors, the Gr family in the silkworm is generally consistent with that of other insects. Their total number of 65 fits well with that of other insects such as mosquitoes (60–79 Grs; Hill et al. 2002; Kent, Walden, and Robertson 2008), fruit flies (56–60 Grs; Robertson, Warr, and Carlson 2003; Gardiner et al. 2008), and parasitoid wasps (58 Grs; H.M. Robertson, unpublished results). One difference is the apparent absence of alternatively spliced silkworm Grs, something that has been found in the Gr families for some other insects examined, including *Drosophila*, *Anopheles*, *Aedes*, and *Tribolium*. In these cases alternative splicing can result in moderate increases in the number of different Gr proteins (4–7 additional proteins in *Drosophila* spp.) or quite large increases (29–35 additional proteins in *Anopheles* and *Aedes* mosquitoes and *Tribolium* beetles) (e.g., Kent, Walden, and Robertson 2008; Tribolium Genome Sequencing Consoritum 2008).

Odorant Receptor Family

Insect Or genes are expressed in olfactory neurons found on olfactory organs such as the antenna. To date, insect genomes have been found to encode anywhere from about 60 to 250 Or genes (e.g., Robertson, Warr, and Carlson 2003; Tribolium Genome Sequencing Consortium, 2008). Studies in *Drosophila* have determined that in most cases an olfactory neuron expresses only one of the about 60 *Drosophila* Or genes (along with the ubiquitous partner Or83b described further in the next paragraph), although there are a few exceptions. Furthermore, the axons of all olfactory neurons expressing the same Or gene project to the same glomerulus in the antennal lobe, providing a neurological mechanism to amplify and code odor signals detected in the periphery (Vosshall, Wong, and Axel 2000). This recent progress in the molecular and neurological mechanisms of *Drosophila* olfaction has been summarized in several recent reviews (e.g., Rutzler and Zwiebel 2005; Hallem, Dahanukar, and Carlson 2006; Vosshall and Stocker 2007).

With one exception, Ors from different insect orders have no discernable orthology. Early Cr research discovered a single Or lineage that is highly conserved between different insect orders. The ortholog of this lineage in *Drosophila* is termed Or83b, and *Drosophila* research has demonstrated that it acts as a partner and chaperone for most other Ors (e.g., Benton et al. 2006). This lineage is commonly referred to as DmOr83b orthologs, or as Or2 in the Lepidoptera, Coleoptera, and Hymenoptera (Krieger et al. 2003). Reflecting its critical functional role, insect genomes that have been sequenced encode only one copy of the Or83b subfamily, and the olfactory ability of transgenic fruit flies lacking Or83b is broadly impaired (Larsson et al. 2004). In vitro assays similarly demonstrate that the honeybee ortholog AmOr2 is required for the response of AmOr11 to the queen pheromone 9-ODA (Wanner et al. 2007), and beetle olfaction can be impaired by interfering with the expression of TcOr2 by RNAi (Engsontia et al. 2008). Most recently, evidence suggests that Or83b forms an ion channel regulated by its partner Or that imparts ligand specificity (Sato et al. 2008; Wicher et al. 2008).

Other than the Or83b lineage, no other orthologous Or lineage has been identified between species representing different insect orders. Flies and mosquitoes represent two different suborders of the Diptera, an early phylogenic split within this order. A phylogenic analysis of their Ors revealed only low levels of orthology, with some Or lineages expanding in flies (represented by *Drosophila*) and others expanding in mosquitoes (represented by *Anopheles gambiae*; Hill et al. 2002). This lack of amino acid similarity has hindered the experimental discovery of insect Ors with the exception of the well-conserved *Drosophila* Or83b orthologs. For example, 66 lepidopteran Ors have been discovered only after whole genome sequencing, 48 from the *B. mori* genome (Wanner et al. 2007) and 18 from *H. virescens* genome sequence (Krieger et al. 2002). However, a BLASTp search of GenBank reveals 6 additional Or83b orthologs discovered experimentally from six species of Lepidoptera (*Antheraea pernyi*, *Helicoverpa assulta*, *Helicoverpa zea*, *Mamestra brassicae*, *Spodoptera litura*, and *Spodoptera exigua*), all sharing greater than 85 percent amino acid identity. This higher level of identity allows relatively easy experimental discovery from insect species whose genomes have not been sequenced.

An analysis of the 68 published lepidopteran Ors (Figure 9.5 and Wanner et al. 2007) reveals higher levels of potential orthology between *B. mori* (family Bombycidae) and *H. virescens* (family Noctuidae) compared with similar analyses performed within the Diptera (Hill et al. 2002; Robertson, Warr, and Carlson 2003) and Hymenoptera (*A. mellifera* and *Nasonia vitripennis*; K.W. Wannner and H.M. Robertson, unpublished results). When comparing flies and mosquitoes or wasps and bees, large suborder-specific expansions of Or lineages predominate. Although numbering less than half, the 18 Hvir Ors are interspersed within several silkworm lineages, including several potential orthologous pairs (Figure 9.5). The number of orthologous relationships between these two species will likely increase as more *H. virescens* sequences are identified and published. However, as additional lepidopteran genomes are sequenced in the future, we will expect to see family-specific expansions in the Lepidoptera, but not to the extent of those seen between flies and mosquitoes, for example. BmOrs45–48, three of which are expressed at relatively higher levels in female antennae, may be one such family-specific expansion. In fact, family-specific Or expansions within the Lepidoptera may point toward interesting differences in odor detection that are biologically significant.

The potential for greater Or orthology within the Lepidoptera compared with other insects may simply reflect the fact that there has been less evolutionary time. The majority of extant Lepidoptera have evolved much more recently in the last 100 million years (phylogeny of Lepidoptera is discussed in Chapter 1), compared with flies plus mosquitoes and bees plus wasps, whose divergence is believed to have occurred closer to 200 million years ago (Grimaldi and Engel 2005). However, the fact that almost all species of Lepidoptera feed on plants may also reflect a more uniform chemosensory environment, at least compared with the Diptera, Coleoptera, and Hymenoptera, which exhibit a much broader range of life histories.

Two immediate practical applications result from increased Or orthology within the Lepidoptera. First, it makes their experimental discovery using homology-based approaches much easier, and second, it raises the possibility of conserved functions. With this in mind, Wanner et al. (2007) screened the expression patterns of all 48 known silkworm Ors for sex-biased expression in the antennae. Several Ors including BmOrs 1 and 3–6 had been previously identified as candidate sex

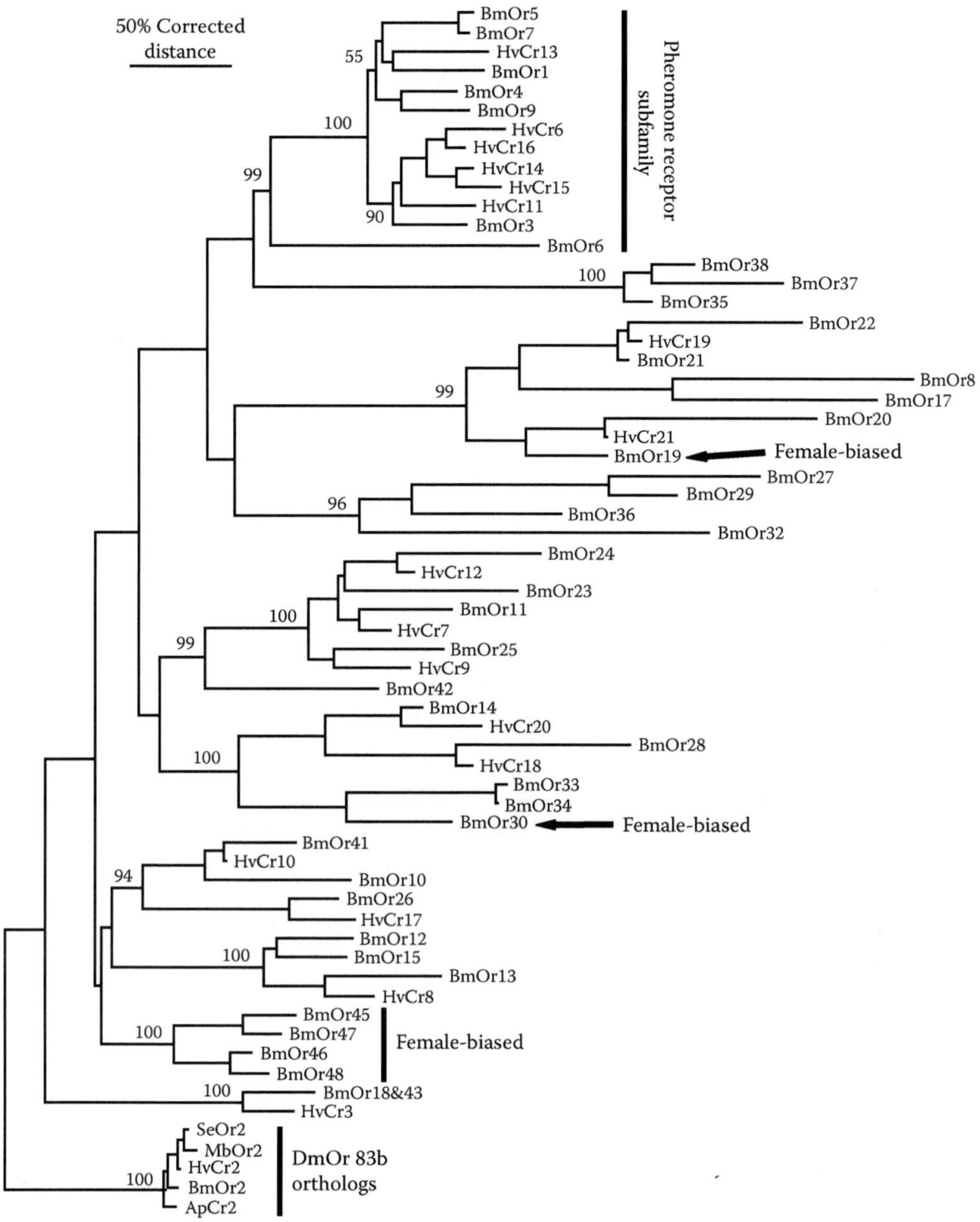

FIGURE 9.5 Phylogenetic tree relating the published *Bombyx mori* (Wanner et al. 2007) and *Heliothis virescens* Ors (Krieger et al. 2002). The tree is rooted using lepidopteran orthologs of the DmOr83b lineage (*Spodoptera exigua*, Se; *Mamestra brassicae*, Mb; *H. virescens*, Hv; *B. mori*, Bm; and *Antheraea pernyi*, Ap). The sex pheromone receptor subfamily is highlighted by a bar on the right-hand side, as is the DmOr83b lineage and a subfamily of Ors (BmOr45–48) with female-biased expression. BmOrs 19 and 30 (arrows) also exhibit significant female-biased expression.

pheromone receptors based on male-biased expression patterns (Sakurai et al. 2004; Nakagawa et al. 2005). Two receptors, BmOrs 19 and 30, exhibited large female-biased expression patterns, while BmOrs 45–47 were moderately female-biased (Wanner et al. 2007). Female-biased expression suggests that these receptors are detecting odors that mediate behaviors important to female moth biology, such as the detection of host plant odors for oviposition or the detection of male-produced aphrodisiac pheromones. As more Or sequences are identified from different moth species, the conservation of Ors that might function in female-specific behaviors can be tested.

Sex pheromones produced by female moths that attract male moths from long distances for mating have been a focus of research efforts during the last century (Schneider 1992). Hundreds of moth pheromones have been chemically identified and synthesized, an effort largely directed toward developing ecologically friendly pest management tools based on lures to attract and monitor or mass-trap male moths or to disrupt their ability to locate females. The discovery and characterization of the first moth sex-pheromone receptors has been received with great interest. The first candidate pheromone receptors were discovered by Krieger et al. (2004). Using an approach that combined antennal cDNA library screening with access to genomic sequence, they identified 18 *H. virescens* Ors, 6 of which formed a more highly conserved subfamily that exhibited biased expression in male moth antennae (HvCrs 6, 11, and 13–16). With access to the recently completed silkworm genome, Sakurai et al. (2004) similarly identified a group of 5 related silkworm Ors (BmOrs 1 and 3–6) with biased expression in male moth antennae. Significantly, BmOrs 1 and 3–6 clustered together with HvCr6, 11, and 13–16 in a phylogenetic analysis (Nakagawa et al. 2005; Krieger et al. 2005). A subsequent analysis of silkworm Ors revealed a total of 7 BmOrs within this pheromone receptor subfamily (Figure 9.5; Wanner et al. 2007). The fact that all pheromone receptors identified to date form a conserved subfamily indicates that the sex pheromone receptors may be broadly conserved within the Lepidoptera, a fact that would have significant applied and scientific impact (sexual communication of Lepidoptera is reviewed in Chapter 10).

The phylogenetic analysis in Figure 9.5 indicates the presence of at least two pheromone receptor lineages within the ancestors of the bombycids and noctuids, one expanding in the *B. mori* lineage (BmOrs 1, 4, 5, 7, and 9 group with HvOr13) and the other expanding in the *H. virescens* lineage (HvCrs 6, 11, and 14–16 group with BmOr3). BmOrl responds to the main pheromone component bombykol, while BmOr3 responds to bombykal (Nakagawa et al. 2005). HvCrs 13, 14, and 16 respond to components of the *H. virescens* sex pheromone components in vitro, whereas HvirCr13 responds specifically to the main component Z-11-hexadecenal (Grosse-Wilde et al. 2007). This differential expansion of pheromone receptors in two different moth families fits well with the overall pattern of Or evolution typified by differential Or gene expansions and contractions, and this trend may prove to be common with the sex pheromone receptor lineage.

SUMMARY AND FUTURE PERSPECTIVES

Adult moths use their chemical senses to find and select mates, and both adult and larval Lepidoptera use their chemical senses to search for and select their host plants. While the CNS processes sensory information, the first critical step is the actual detection of the chemical stimuli by the peripheral sensory system. As the first critical step in a chain leading to important and complex behaviors, the molecular basis of chemosensation at the periphery was a highly sought research goal. The discovery of the first mammalian Crs in the rat peripheral sensory system (Buck and Axel 1991) earned a Nobel prize. Global attempts to translate this discovery to insects were fruitless, and only the advent of whole genome sequencing revealed the first family of insect Crs nine years later. Continued elucidation of the molecular mechanisms of chemical detection at the periphery in the *Drosophila* model, coupled with the development of new functional genomics tools, is providing a platform to extend this research broadly to diverse insect groups including the Lepidoptera.

Identifying and cloning Cr genes and analyzing their function in nonmodel organisms presented two significant hurdles. The first hurdle, identifying the Cr genes, is being overcome by new massively parallel, high-throughput DNA sequencing technologies such as 454 pyrosequencing (Hudson 2008 reviews new DNA sequencing technology and its applications to ecology). Lower cost DNA sequencing is continuing to accelerate whole genome and whole transcriptome sequencing efforts, both of which will increase Cr sequence representation in public databases such as the National Center for Biotechnology Information (NCBI). The second hurdle, functional analysis in nonmodel species, is being overcome with the use of cellular-based receptor activation assays that allow receptor ligand specificity to be analyzed in vitro (Nakagawa et al. 2005; Grosse-Wilde, Svatos, and Krieger 2006; Grosse-Wilde et al. 2007; Kiely et al. 2007; Wanner et al. 2007). Combined, these two approaches have the potential to identify and characterize conserved Cr lineages that mediate important behaviors in the Lepidoptera.

Increased Cr sequence representation in the Lepidoptera along with gene expression profiling to identify sex- and tissue-specific expression patterns will provide insights into Cr lineages that mediate mate and host-plant selection. Grs expressed in contact chemosensilla located on female moth front tarsi and ovipositors may be one example. Female-produced sex pheromones are believed to contribute to speciation in the Lepidoptera (Cardé and Haynes 2004), and for the first time it is now feasible to trace the origins and evolution of the Or lineage responsible for detecting the sex pheromones. Many male moths also produce sex pheromones (also termed aphrodisiac pheromones) that are detected by female moths during mating. Male moth sex pheromone gland morphology is diverse and believed to have evolved several times independently within the Lepidoptera (Cardé and Haynes 2004). Experimental approaches are now in place to identify the Ors expressed in female moth antennae that detect male-produced sex pheromones, and their origins can be studied (for details, see Chapter 10).

Additionally, the higher levels of receptor protein identity observed between different taxonomic families within the Lepidoptera will facilitate experimental approaches designed to discover Ors from important pest species. For example, sex pheromones are used widely in integrated pest mangement (IPM) programs as lures to trap moths or to disrupt their mating. It is now feasible to attempt to clone putative sex pheromone receptors from many different moth families using degenerate PCR primers. In vitro assays can be used to screen chemical analogs that may act as pheromone agonists and/or antagonists that could improve pheromone-based lures. One interesting question is whether the observed levels of Or homology between *B. mori* (Bombycidae) and *H. virescens* (Noctuidae) will be indicative of Or homology more broadly in the Lepidoptera. If some or many Or lineages have retained specificity for a class of ligands (such as the sex pheromone receptor subfamily), a brute force effort to deorphanize all of the silkworm Ors might also predict the ligand-binding specificity of homologous Ors from many economically important pest species. From an ecological perspective it will be interesting to survey lineage-specific Or expansions and contractions in response to shifts in the type and breadth of host species utilized (generalist vs. specialists), patterns already seen in *Drosophila* species (McBride 2007; McBride, Arguello, and O'Meara 2007).

An interesting outcome of our analysis of the silkworm Gr family is the apparent monophyletic expansion of 55 Grs, many of which may function as bitter receptors. Based on the extremely divergent sequences, this lineage appears to be old. As more Gr sequences become available from the Lepidoptera, it will be interesting to trace the phylogenetic roots of this Gr lineage. Evolution of the insect Gr family appears to be quite responsive to habitat changes. Only 10 Grs have been found in the honeybee genome, apparently due to the loss and/or lack of expansion of most of the putative bitter receptors (Robertson and Wanner 2006). The flour beetle Gr family, on the other hand, provides a remarkable expansion of most Gr lineages, yielding an astounding total of 215 Gr genes encoding potentially 245 Gr proteins (Tribolium Genome Sequencing Consortium 2008). The expansion of a single Gr lineage in the Lepidoptera may point toward an ancestral habitat shift toward herbivory. Although entirely speculative at this time, future advances in lepidopteran genomics promises to provide clues to evolutionary patterns in the Cr superfamily.

REFERENCES

Adams, M.D., S.E. Celniker, R.A. Holt, et al. 2000. The genome sequence of *Drosophila melanogaster*. *Science* 287:2185–95.

Benton, R., S. Sachse, S.W. Michnick, and L.B. Vosshall. 2006. Atypical membrane topology and heteromeric function of *Drosophila* odorant receptors *in vivo*. *PLoS Biol.* 4:e20.

Benton, R., K.S. Vannice, and L.B. Vosshall. 2007. An essential role for a CD36-related receptor in pheromone detection in *Drosophila*. *Nature* 450:289–93.

Buck, L., and R. Axel. 1991. A novel multigene family may encode odorant receptors: A molecular basis for odor recognition. *Cell* 65:175–87.

Cardé, R.T., and K.F. Haynes. 2004. Structure of the pheromone communication channel in moths. In *Advances in insect chemical ecology*, ed. R.T. Cardé and J.G. Millar, 283–332. Cambridge: Cambridge Univ. Press.

Chapman, R.F. 2003. Contact chemoreception in feeding by phytophagous insects. *Annu. Rev. Entomol.* 48:455–84.

Clyne, P.J., C.G. Warr, and J.R. Carlson. 2000. Candidate taste receptors in *Drosophila*. *Science* 287:1830–34.

Clyne, P.J., C.G. Warr, M.R. Freeman, D. Lessing, J. Kim, and J.R. Carlson. 1999. A novel family of divergent seven-transmembrane proteins: Candidate odorant receptors in *Drosophila*. *Neuron* 22:327–38.

Dahanukar, A., Y.T. Lei, J.Y. Kwon, and J.R. Carlson. 2007. Two Gr genes underlie sugar reception in *Drosophila*. *Neuron* 56:503–16.

Ebbs, M.L., and H. Amrein. 2007. Taste and pheromone perception in the fruit fly *Drosophila melanogaster*. *Pflugers Arch.* 454:735–47.

Engsontia, P., A.P. Sanderson, M. Cobb, K.K. Walden, H.M. Robertson, and S. Brown. 2008. The red flour beetle's large nose: An expanded odorant receptor gene family in *Tribolium castaneum*. *Insect Biochem. Mol. Biol.* 38:387–97.

Gao, Q., and A. Chess. 1999. Identification of candidate *Drosophila* olfactory receptors from genomic DNA sequence. *Genomics* 60:31–39.

Gardiner, A., D. Barker, R.K. Butlin, W.C. Jordan, and M.G. Ritchie. 2008. *Drosophila* chemoreceptor gene evolution: Selection, specialization and genome size. *Mol. Ecol.* 17:1648–57.

Go, Y. 2006. Lineage-specific expansions and contractions of the bitter taste receptor gene repertoire in vertebrates. *Mol. Biol. Evol.* 23:964–72.

Goldsmith, M.R., T. Shimada, and H. Abe. 2005. The genetics and genomics of the silkworm, *Bombyx mori*. *Annu. Rev. Entomol.* 50:71–100.

Grimaldi, D., and M.S. Engel. 2005. *Evolution of the insects*. New York: Cambridge Univ. Press.

Grosse-Wilde, E., T. Gohl, E. Bouché, H. Breer, and J. Krieger. 2007. Candidate pheromone receptors provide the basis for the response of distinct antennal neurons to pheromonal compounds. *Eur. J. Neurosci.* 25:2364–73.

Grosse-Wilde, E., A. Svatos, and J. Krieger. 2006. A pheromone-binding protein mediates the bombykol-induced activation of a pheromone receptor *in vitro*. *Chem. Senses* 31:547–55.

Ha, T.S., and D.P. Smith. 2006. A pheromone receptor mediates 11-cis-vaccenyl acetate-induced responses in *Drosophila*. *J. Neurosci.* 26:8727–33.

Hallem, E.A., A. Dahanukar, and J.R. Carlson. 2006. Insect odor and taste receptors. *Annu. Rev. Entomol.* 51:113–35.

Hallem, E.A., A.N. Fox, L.J. Zwiebel, and J.R. Carlson. 2004. Olfaction: Mosquito receptor for human-sweat odorant. *Nature* 427:212–13.

Hallem, E.A., M.G. Ho, and J.R. Carlson. 2004. The molecular basis of odor coding in the *Drosophila* antenna. *Cell* 117:965–79.

Hill, C.A., A.N. Fox, R.J. Pitts, et al. 2002. G-protein-coupled receptors in *Anopheles gambiae*. *Science* 298:176–78.

Hudson, M. 2008. Sequencing breakthroughs for genomic ecology and evolutionary biology. *Mol. Ecol. Resources* 8:3–17.

Jones, W.D., P. Cayirlioglu, I.G. Kadow, and L.B. Vosshall. 2007. Two chemosensory receptors together mediate carbon dioxide detection in *Drosophila*. *Nature* 445:86–90.

Jørgensen, K., P. Kvello, T.J. Almaas, and H. Mustaparta. 2006. Two closely located areas in the suboesophageal ganglion and the tritocerebrum receive projections of gustatory receptor neurons located on the antennae and the proboscis in the moth *Heliothis virescens*. *J. Comp. Neurol.* 496:121–34.

Kent, L.B., and H.M. Robertson. 2009. Evolution of the sugar receptors in insects. *BMC Evol. Biol.* 9:41.

Kent, L.B., K.K. Walden, and H.M. Robertson. 2008. The Gr family of candidate gustatory and olfactory receptors in the yellow-fever mosquito *Aedes aegypti*. *Chem. Senses* 33:79–93.

Kiely, A., A. Authier, A.V. Kralicek, C.G. Warr, and R.D. Newcomb. 2007. Functional analysis of a *Drosophila melanogaster* olfactory receptor expressed in Sf9 cells. *J. Neurosci. Methods* 159:189–94.

Krieger, J., E. Grosse-Wilde, T. Gohl, et al. 2004. Genes encoding candidate pheromone receptors in a moth (*Heliothis virescens*). *Proc. Natl. Acad. Sci. U.S.A.* 101:11845–50.

Krieger, J., E. Grosse-Wilde, T. Gohl, and H. Breer. 2005. Candidate pheromone receptors of the silkmoth *Bombyx mori*. *Eur. J. Neurosci.* 21:2167–76.

Krieger, J., O. Klink, C. Mohl, K. Raming, and H. Breer. 2003. A candidate olfactory receptor subtype highly conserved across different insect orders. *J. Comp. Physiol. A Neuroethol. Sens. Neural Behav. Physiol.* 189:519–26.

Krieger, J., K. Raming, Y.M. Dewer, et al. 2002 A divergent gene family encoding candidate olfactory receptors of the moth *Heliothis virescens*. *Eur. J. Neurosci.* 16:619–28.

Kurtovic, A., A. Widmer, and B.J. Dickson. 2007. A single class of olfactory neurons mediates behavioural responses to a *Drosophila* sex pheromone. *Nature* 446:542–46.

Kwon, J.Y., A. Dahanukar, L.A. Weiss, and J.R. Carlson. 2007. The molecular basis of CO_2 reception in *Drosophila*. *Proc. Natl. Acad. Sci. U.S.A.* 104:3574–78.

Larsson, M.C., A.I. Domingos, W.D. Jones, et al. 2004. Or83b encodes a broadly expressed odorant receptor essential for *Drosophila* olfaction. *Neuron* 43:703–14.

Leal, W.S. 2005. Pheromone reception. *Top. Curr. Chem.* 240:1–36.

Lu, T., Y.T. Qiu, G. Wang, et al. 2007. Odor coding in the maxillary palp of the malaria vector mosquito *Anopheles gambiae*. *Curr. Biol.* 17:1533–44.

Lundin, C., L. Käll, S.A. Kreher, et al. 2007. Membrane topology of the *Drosophila* OR83b odorant receptor. *FEBS Lett.* 581:5601–04.

McBride, C.S. 2007. Rapid evolution of smell and taste receptor genes during host specialization in *Drosophila sechellia*. *Proc. Natl. Acad. Sci. U.S.A.* 104:4996–5001.

McBride, C.S., J.R. Arguello, and B.C. O'Meara. 2007. Five *Drosophila* genomes reveal nonneutral evolution and the signature of host specialization in the chemoreceptor superfamily. *Genetics* 177:1395–1416.

Mita, K., M. Kasahara, S. Sasaki, et al. 2004. The genome sequence of silkworm, *Bombyx mori*. *DNA Res.* 11:27–35.

Nakagawa, T., T. Sakurai, T. Nishioka, and K. Touhara. 2005. Insect sex-pheromone signals mediated by specific combinations of olfactory receptors. *Science* 307:1638–42.

Robertson, H.M., and L.B. Kent. 2009. Evolution of the gene lineage encoding the carbon dioxide receptor in insects and other arthropods. *J. Insect Science*. In press.

Robertson, H.M., and J.H. Thomas. 2006. The putative chemoreceptor families of *C. elegans* (January 06, 2006). In *WormBook*, ed. The *C. elegans* Research Community, doi/10.1895/wormbook.1.66.1, http://www.wormbook.org.

Robertson, H.M., and K.W. Wanner. 2006. The chemoreceptor superfamily in the honeybee *Apis mellifera*: Expansion of the odorant, but not gustatory, receptor family. *Genome Res.* 16:1395–1403.

Robertson, H.M., C.G. Warr, and J.R. Carlson. 2003. Molecular evolution of the insect chemoreceptor superfamily in *Drosophila melanogaster*. *Proc. Natl. Acad. Sci. U.S.A.* 100: 14537–42.

Rogers, M.E., M. Sun, M.R. Lerner, and R.G. Vogt. 1997. Snmp-1, a novel membrane protein of olfactory neurons of the silk moth *Antheraea polyphemus* with homology to the CD36 family of membrane proteins. *J. Biol. Chem.* 272:14792–99.

Rutzler, M., and L.J. Zwiebel. 2005. Molecular biology of insect olfaction: Recent progress and conceptual models. *J. Comp. Physiol. A. Neuroethol. Sens. Neural. Behav. Physiol.* 191:777–90.

Sakurai, T., T. Nakagawa, H. Mitsuno, et al. 2004. Identification and functional characterization of a sex pheromone receptor in the silkmoth *Bombyx mori*. *Proc. Natl. Acad. Sci. U.S.A.* 101:16653–58.

Sato, K., M. Pellegrino, T. Nakagawa, et al. 2008. Insect olfactory receptors are heteromeric ligand-gated ion channels. *Nature* 452:1002–06.

Schneider, D. 1992. 100 years of pheromone research, an essay on Lepidoptera. *Naturwissenschaften* 79:241–50.

Scott, K. 2005. Taste recognition: Food for thought. *Neuron* 48:455–64.

Sea Urchin Genome Sequencing Consortium. 2006. The genome of the sea urchin *Strongylocentrotus purpuratus*. *Science* 314:941–52.

Shi, P., and J. Zhang. 2006. Contrasting modes of evolution between vertebrate sweet/umami receptor genes and bitter receptor genes. *Mol. Biol. Evol.* 23:292–300.

Slone, J., J. Daniels, and H. Amrein. 2007. Sugar receptors in *Drosophila*. *Curr. Biol.* 17:1809–16.

Smith, D.P. 2007. Odor and pheromone detection in *Drosophila melanogaster*. *Pflugers Arch.* 454:749–58.

Syed, Z., Y. Ishida, K. Taylor, D.A. Kimbrell, and W.S. Leal. 2006. Pheromone reception in fruit flies expressing a moth's odorant receptor. *Proc. Natl. Acad. Sci. U.S.A.* 103:16538–43.

Thomas, J.H., and H.M. Robertson. 2008. The *Caenorhabditis* nematode chemoreceptor gene families. *BMC Biol.* 6:42.

Thorne, N., and H. Amrein. 2008. Atypical expression of *Drosophila* gustatory receptor genes in sensory and central neurons. *J. Comp. Neurol.* 506:548–68.

Thorne, N., S. Bray, and H. Amrein. 2005. Function and expression of the *Drosophila* Gr genes in the perception of sweet, bitter and pheromone compounds. *Chem. Senses* 30:i270–72.

Tribolium Genome Sequencing Consortium. 2008. The genome of the developmental model beetle and pest *Tribolium castaneum*. *Nature* 452:949–55.

Vosshall, L.B., H. Amrein, P.S. Morozov, A. Rzhetsky, and R. Axel. 1999. A spatial map of olfactory receptor expression in the *Drosophila* antenna. *Cell* 96:725–36.

Vosshall, L.B., and R.F. Stocker. 2007. Molecular architecture of smell and taste in *Drosophila*. *Annu. Rev. Neurosci.* 30:505–33.

Vosshall, L.B., A.M. Wong, and R. Axel. 2000. An olfactory sensory map in the fly brain. *Cell* 102:147–59.

Wang, Z., A. Singhvi, P. Kong, and K. Scott. 2004. Taste representations in the *Drosophila* brain. *Cell* 117:981–91.

Wanner, K.W., A.S. Nichols, K.K. Walden, et al. 2007. A honey bee odorant receptor for the queen substance 9-oxo-2-decenoic acid. *Proc. Natl. Acad. Sci. U.S.A.* 104:14383–88.

Wanner, K.W., and H.M. Robertson. 2008. The gustatory receptor family in the silkworm moth *Bombyx mori* is characterized by a large expansion of a single lineage of putative bitter receptors. *Insect Mol. Biol.* 17:621–29.

Wicher, D., R. Schäfer, R. Bauernfeind, et al. 2008. *Drosophila* odorant receptors are both ligand-gated and cyclic-nucleotide-activated cation channels. *Nature* 452:1007–11.

Wistrand, M., L. Käll, and E.L. Sonnhammer. 2006. A general model of G protein-coupled receptor sequences and its application to detect remote homologs. *Protein Sci.* 15:509–21.

Wu, S.V., N. Rozengurt, M. Yang, et al. 2002. Expression of bitter taste receptors of the T2R family in the gastrointestinal tract and enteroendocrine STC-1 cells. *Proc. Natl. Acad. Sci. U.S.A.* 99:2392–97.

Xia, Q., Z. Zhou, C. Lu, et al. 2004. A draft sequence for the genome of the domesticated silkworm (*Bombyx mori*). *Science* 306:1937–40.

Xu, P., R. Atkinson, D.N. Jones, and D.P. Smith. 2005. *Drosophila* OBP LUSH is required for activity of pheromone-sensitive neurons. *Neuron* 45:193–200.

10 Sexual Communication in Lepidoptera

A Need for Wedding Genetics, Biochemistry, and Molecular Biology

Fred Gould, Astrid T. Groot, Gissella M. Vásquez, and Coby Schal

CONTENTS

INTRODUCTION

Why study sexual communication in Lepidoptera?

In night-flying moths, highly specific, long distance, pheromonal communication is essential for mating success and reproductive isolation of species. Emission of two or more volatile compounds by females, in precise ratios, is typically required to attract conspecific males (e.g., Cardé and Haynes 2004). Although there are thousands of moth species with unique pheromone blends (e.g., Cork and Lobos 2003; Witzgall et al. 2004; El-Sayed 2008), the evolutionary processes that resulted in this diversity of sexual communication signals and species are not understood.

A rare female with a mutation leading to an alteration in pheromone blend is expected to have lower mating success than normal females unless males do not discriminate between the typical and altered blends (Butlin and Trickett 1997). And, in almost all published studies, normal males do discriminate against females with atypical pheromonal signals (e.g., Zhu et al. 1997). Similarly, a male with a mutation that results in response to an altered female pheromone blend is expected to be less efficient at finding typical females. Evidence of this lower efficiency comes from studies of moth genotypes that differ in pheromone responses (e.g., Linn et al. 1997). This selection against new male and female mating traits, when they are at low frequency, is expected to result in stabilizing selection that could constrain the evolutionary diversification of moth mating communication systems (Butlin and Trickett 1997; Phelan 1997).

Hypotheses to explain the evolution of new mating communication signals and responses have invoked the possibility that alteration of the signal and response are pleiotropically controlled by the same genes (Hoy, Hahn, and Paul 1977), or that males do not prefer normal females over those with altered blends. Published research to date does not support either of these assumptions (Butlin and Ritchie 1989; Butlin and Trickett 1997). Nevertheless, the impressive diversity of chemical mixtures used by moths for sexual communication stands as evidence that evolution of novel signal/response systems has not been stymied.

There is debate among evolutionary biologists who conclude that diversification is highly unlikely when there is stabilizing selection (e.g., Coyne, Barton, and Turelli 1997) and others who find evidence that stochastic events (i.e., genetic drift) could result in diversification, even in the face of stabilizing selection (Wade and Goodnight 1998). Both of these groups agree that the likelihood of such diversification would be influenced by (1) the number of genes involved in the initial divergence, (2) the magnitude of effect of each gene on fitness-related phenotypes, and (3) allelic interactions affecting fitness-related phenotypes (Coyne and Orr 1998; Wade and Goodnight 1998; Dieckmann and Doebeli 1999; Kondrashov and Kondrashov 1999; Whitlock and Phillips 2000).

The major premise of this chapter is that combining a detailed understanding of quantitative genetics, biochemistry, and molecular biology of the signals and responses used by moths will enable us to better understand how this system diversified, and could serve as a model for studying evolution of other traits that appear to be under stabilizing selection. Molecular and biochemical studies alone can tell us how many and which enzymes are in the biosynthesis pathways leading to production of a pheromone blend. They can also tell us which specific molecules are needed for males to perceive these blends. Quantitative genetic studies on their own could tell us a lot about how many genes affect variation in signals and responses within and between species. However, it is only by combining genetic and molecular studies that we will be able to understand just what kinds of changes (e.g., single nucleotide polymorphisms in open reading frames, cis- or trans-regulatory changes) in which genes led to diversification of moth mating systems. These types of data can also inform the ongoing debate about whether changes in open reading frame sequences or in regulatory sequences have been more critical to the evolution of ecological adaptation and diversification (Carroll 2005; Hoekstra and Coyne 2007).

In this chapter we present an overview of what is and is not known about the genetics, biochemistry, and molecular biology of female moth pheromone production and male response. In doing so, we show that even though there are major gaps in our knowledge, overall, a great deal is known about these aspects of sexual communication in moths, and that by combining the knowledge in these areas we could make major steps forward in our understanding of evolution. At the end of this chapter we point out a few potential avenues for future research.

QUANTITATIVE GENETIC STUDIES

Female Pheromones

The state of knowledge regarding quantitative genetics of female pheromone blends has been reviewed in detail (e.g., Löfstedt 1990, 1993; Linn and Roelofs 1995; Butlin 1995; Phelan 1997; Roelofs and Rooney 2003; Cardé and Haynes 2004), so we will only present a selective overview.

Early genetic studies of sexual communication systems focused on determining whether the same genes that controlled signal production also controlled signal perception in the opposite sex through pleiotropic effects (sometimes referred to as genetic coupling). Although an early empirical study found evidence supporting the possibility of genetic coupling of acoustic mate communication (Hoy, Hahn, and Paul 1977), later studies of both acoustic and chemical sexual communication indicate that such coupling is very rare (see Butlin and Ritchie 1989). One study that did find a genetic correlation between male and female signal/response traits in offspring from field-collected insects determined that these correlations broke down after randomized mating in the laboratory (Gray and Cade 1999). This indicated that gametic disequilibrium and not pleiotropy (or strong physical gene linkage) had caused the correlation. Given the lack of evidence for genetic coupling, recent efforts have focused on understanding the genetic architecture of variation in signal production and response that would allow coevolution of male and female aspects of sexual signaling in insects (see Butlin and Trickett 1997; Phelan 1997).

Studies of the genetic architecture of differences in sexual communication between two races of *Ostrinia nubilalis* stand out as the most complete for any insect species. The ratios of the acetate pheromone components *E*11–14:OAc and *Z*11–14:OAc, produced by females of two races of *O. nubilalis*, differ dramatically (the E strain with a 97:3 ratio of E to Z acetates, and the Z strain with a 1:99 ratio), and males of each race prefer females of that race. The differences in pheromone blend between the races appear to be mostly controlled by a single autosomal gene (Klun 1975), although multiple genes with tight physical linkage cannot be ruled out. Other modifier genes have smaller impacts on the blend ratio (Löfstedt et al. 1989). The major gene that controls pheromone blend is not linked to the genetic region that controls male behavioral response to the pheromone (e.g., Linn et al. 1999). Recent genetic analysis of *Ostrinia scapulalis*, which also has a Z and an E race in Japan, demonstrates that the difference in this species' races also segregates as if it is mostly controlled by a single gene (Takanashi et al. 2005). The *O. scapulalis* system is especially interesting because it is one of the few in which the males have not been found to discriminate between the pheromone blends of the two races (Takanashi et al. 2005).

Studies of F_2 and backcross progeny from hybridization of other lepidopteran species have also uncovered evidence suggestive of single-gene control of production of specific pheromone component ratios. In most of these cases, as with *O. nubilalis*, the difference between the species is simply in the isomeric forms of the pheromone components, and both isomers are derived from the same precursor. A noteworthy exception, where one gene controls a change in ratios of less-related pheromone components, involves a laboratory-derived mutant of *Trichoplusia ni* (Haynes and Hunt 1990; Jurenka et al. 1994; Zhu et al. 1997). In this case, a genetic change in a chain-shortening enzyme is hypothesized to cause the altered component ratios, and as in other cases, the normal males were less attracted to the mutant female pheromone blend (Zhu et al. 1997). In the other cases mentioned, increases in one pheromone component result in a decline in only one other component.

Although the literature emphasizes examples of single-gene control, there are also cases where multiple genes may be involved in sexual communication differences between races and species (Cardé and Haynes 2004). Most studies supporting multiple-gene control involve simple segregation analyses that find no evidence for single-gene control, or find that single-gene explanations are not sufficient to account for all of the genetic variation observed (e.g., Teal and Tumlinson 1997). Unfortunately, these qualitative findings have little explanatory power.

To better understand traits controlled by multiple genes, our laboratory has utilized a genetic approach called Quantitative Trait Locus analysis (QTL; Remington et al. 1999) for assessing the number of loci responsible for differences between *Heliothis virescens* and *H. subflexa* in their complex pheromone blends (Sheck et al. 2006). These two species differ in ratios or presence/absence of at least ten compounds and therefore present a rich system for analysis. In our first study, the two moth species were hybridized and then backcrossed to *H. subflexa*. Pheromone glands from female progeny of these backcrosses were analyzed, and DNA from each moth was subjected to amplified fragment length polymorphism (AFLP) marker analysis (Sheck et al. 2006) to construct a genetic map of all thirty-one chromosomes (Gahan, Gould, and Heckel 2001). Because the F_1 individual used in the cross was a female, there was no recombination (Heckel 1993), so each chromosome from *H. virescens* remained intact as a single linkage group (LG). This crossing design enabled us to correlate the presence of specific chromosomes from the nonrecurrent parent (*H. virescens*), with the ratios of compounds in the pheromone glands of individual backcross progeny females. The basic results from these crosses are shown in Figure 10.1.

This study demonstrated that at least five chromosomes were involved in determining the difference between the two species in the composition of the blend in the pheromone gland, implying that at least five loci affect the difference between blends of the two species (Sheck et al. 2006). In two cases, a specific chromosome significantly impacted the relative amount of only a single compound, but other chromosomes affected the relative amounts of two to four compounds; and in four cases a single compound was affected by more than one chromosome. Moreover, females that had both *H. virescens* chromosomes 4 and 22 had the least amount of the three acetate esters that are only found in *H. subflexa*. This coupling of genetic control of the amounts of all three acetates suggests that the same metabolic process affects the production of each.

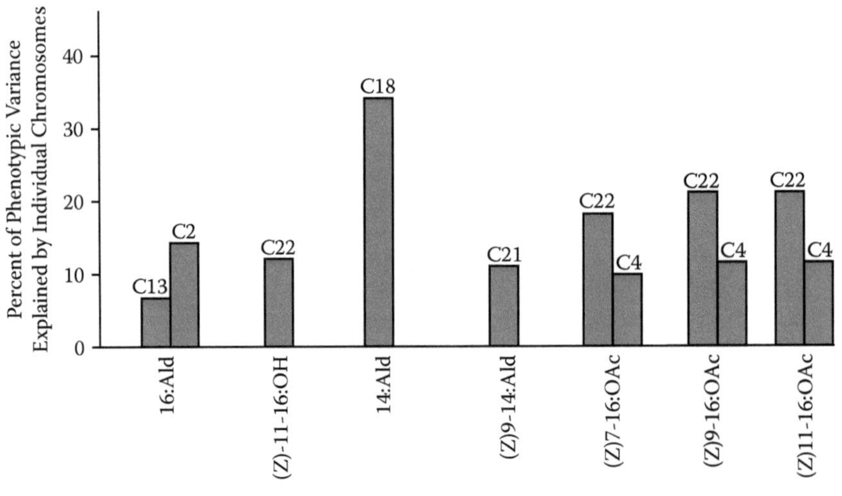

FIGURE 10.1 Results of QTL analysis of pheromone components in *Heliothis virescens* and *H. subflexa*. The percent of variation in the amount of specific *Heliothis* pheromone compounds in backcross females that can be explained by the presence/absence of specific chromosomes from *H. virescens* (out of thirty autosomes and one sex chromosome). Chromosomes are numbered with the prefix C. This backcross entailed mating of F_1 females (*H. virescens* X *H. subflexa*) to *H. subflexa* males. From data in Sheck et al. (2006).

We are conducting more QTL studies using backcrosses to both *H. virescens* and *H. subflexa* in an attempt to gain a better understanding of genes with smaller phenotypic effects that could impact evolutionary processes. Although the increased number of crosses and the higher sample size of backcross female offspring are expected to reveal more QTL that affect the pheromone blends, QTL studies are just one step toward the understanding of the evolutionary forces and genetic pathways that resulted in diversification of pheromone blends. Ultimately, it will be necessary to move from QTL analyses or other quantitative genetic methods to the molecular level to determine the types of genes and mutations that were involved in the diversification in moth mating systems.

Male Response

In comparison to what we know about the genetics of pheromone blends, very little is known about the genetics of male moth response to pheromones. The most detailed and fascinating published result in this area is a study by Cossé et al. (1995), who found that males from the two *O. nubilalis* pheromone races differed in signals transmitted by olfactory receptor neurons (ORNs) after stimulation with each of the two pheromone components, but that these differences in ORNs' responses to the pheromones mapped to a completely different genomic location than the males' actual behavioral response. This result emphasizes the point that studying the genetics or the receptors alone could lead to erroneous conclusions. Beyond the studies on genetics of male *Ostrinia* species and race response to pheromones, we could find published genetic experiments on only one other pair of moths. In crosses between *Ctenopseustis obliquana* and *C. herana*, male perception of pheromone blends was mostly sex linked (Hansson, Löfstedt, and Foster 1989). Studies of genetics of male response in more species groups are clearly needed to determine if this pattern is common.

MOLECULAR AND BIOCHEMICAL STUDIES OF PHEROMONE BLENDS

Biochemical Analyses of Pheromone Synthesis

The sex pheromones of many moths are even-numbered C_{10}–C_{18} straight-chain, unsaturated derivatives of fatty acids, with the carbonyl carbon modified to form an oxygen-containing functional group (alcohol, aldehyde, or acetate ester). Free saturated fatty acids are produced *de novo* and converted to their acyl-CoA thioesters before being incorporated into glycerolipids or converted to pheromone (Foster 2005). Pheromone precursor acids appear to be stored mostly in triacylglycerols, with lesser amounts associated with other glycerolipids and phospholipids (Foster 2005). During periods of high pheromone biosynthesis, triacylglycerols are hydrolyzed to release stored fatty acids, which can then be converted to pheromone (Foster 2005). The most common fatty acids produced in Lepidoptera pheromone glands are stearic acid (18:CoA), palmitic acid (16:CoA), and myristic acid (14:CoA; Jurenka 2003). These acids can subsequently be reduced to alcohols (OH) via fatty acid reductase (Morse and Meighen 1987). Alcohols are converted to aldehydes by alcohol oxidase, and to acetate esters (OAc) by acetyltransferase. Conversely, aldehydes can be converted to alcohols by aldehyde reductase, while acetates can be converted (back) to alcohols by acetate esterase (Tumlinson and Teal 1987; Roelofs and Wolf 1988). The most commonly observed pathways of Lepidoptera sex pheromone biosynthesis are schematically depicted in Figure 10.2.

What We Know and Do Not Know about Enzymes and Genes Involved in Pheromone Biosynthesis

In contrast to the extensive biochemical studies to elucidate pheromone biosynthetic pathways, only a few studies have been conducted to identify molecularly the genes encoding the enzymes involved in these pathways. The desaturases stand out as the major exception, and most research has focused

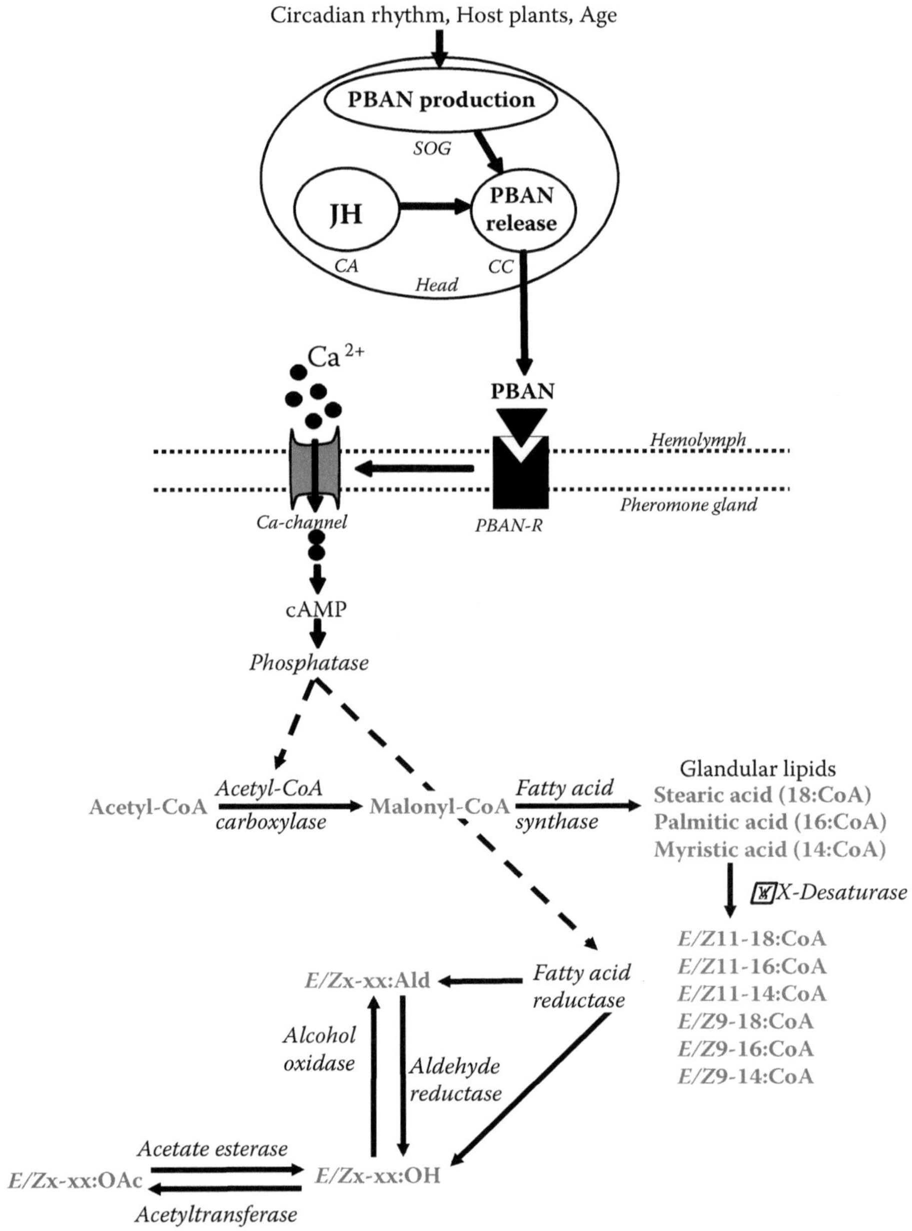

FIGURE 10.2 Schematic view of enzymes involved in the likely pathway of sex pheromone biosynthesis in most Lepidoptera. All possible enzymes are shown, although production of cAMP is not apparent in *B. mori* (Hull et al. 2007). Δx-desaturase may be Δ5, Δ9, Δ10, Δ11, Δ12, Δ13, and Δ14. PBAN, pheromone biosynthesis activating neuropeptide; SOG, suboesophagial ganglion; JH, juvenile hormone; CA, corpora allata; CC, corpora cardiaca; OH, alcohol; Ald, aldehyde; OAc, acetate ester. Adapted from Jurenka (2003), Rafaeli (2005), and Ohnishi, Hull, and Matsumoto (2006).

on identifying desaturase genes in a number of moth species (e.g., Knipple et al. 1998, 2002; Tsfadia et al. 2008). Only one additional enzyme has been molecularly characterized so far, a fatty acyl reductase in *Bombyx mori* (Moto et al. 2003). Below we give an overview of studies that have characterized enzymes and enzymatic reactions involved in pheromone biosynthesis in moths.

Acetyl-CoA Carboxylase

Acetyl-CoA carboxylase (ACCase) catalyzes the ATP-dependent carboxylation of acetyl-CoA to malonyl-CoA in the rate-limiting step of long-chain fatty acid biosynthesis (Pape, Lopez-Casillas, and Kim 1988). In *Helicoverpa armigera* and *Plodia interpunctella* females, when the activity of ACCase was inhibited, sex pheromone biosynthesis was inhibited as well, indicating that ACCase is a key regulatory enzyme in the pheromone biosynthetic pathway in these moth species (Eliyahu, Applebaum, and Rafaeli 2003; Tsfadia et al. 2008). Evidence from incorporation of labeled isotopes into sex pheromone components indicates that the activity of ACCase is influenced by pheromone biosynthesis activating neuropeptide (PBAN) in a number of moth species (e.g., Jurenka, Jacquin, and Roelofs 1991), but so far it has not been determined whether production of this enzyme is upregulated in response to PBAN treatment (Rafaeli 2005).

Fatty Acid Synthase

Animal fatty acid synthase (FAS) is the largest known multifunctional protein, having the most catalytic domains (Wakil, Stoops, and Joshi 1983). In insects, elongation reactions in integumental microsomal fractions have been studied in the housefly (Gu et al. 1997), German cockroach (Juárez 2004), and in triatomine bugs (Juárez and Fernández 2007), primarily in the context of production of long-chain methyl-branched fatty acids, alcohols, and hydrocarbons. The multifunctional enzyme FAS uses malonyl-CoA, acetyl-CoA, and NADPH to synthesize saturated fatty acids in two-carbon increments; methylmalonyl-CoA is used to insert a methyl branch in the aliphatic chain. The principal end-products of FAS in most lepidopteran systems, demonstrated with labeling studies with acetate, are palmitic acid (16:0) and stearic acid (18:0; e.g., Jurenka, Jacquin, and Roelofs 1991). FASs have been sequenced in a number of insect species (e.g., *Aedes aegypti*, accession XM_001658958 and XM_001654917; *Drosophila melanogaster*, accession number NM_134904). No FAS enzymes involved in sex pheromone biosynthesis of moths have been identified or characterized.

Chain-Shortening Enzymes

Changes in the substrate specificities of chain-shortening enzymes can lead to diversification of pheromone blends, as demonstrated in populations of *Zeiraphera diniana* (Baltensweiler and Priesner 1988), *Argyrotaenia velutinana* (Roelofs and Jurenka 1996), and *Agrotis segetum* (Wu et al. 1998). Of particular note is an in vitro enzyme assay study of a mutant line of *T. ni*, which produced elevated amounts of Z9–14:OAc, a minor component of the pheromone blend of normal *T. ni* females (Haynes and Hunt 1990). Jurenka et al. (1994) demonstrated that whereas pheromone glands of normal females mostly shorten Z11–16:CoA to Z7–12:CoA with two rounds of chain shortening, the pheromone glands of mutant females shorten Z11–16:CoA by only one round, to Z9–14:CoA. Chain-shortening enzymes have not been characterized or sequenced in insects, but they are presumably similar to vertebrate peroxisome enzymes (Bjostad and Roelofs 1983).

Desaturases

Integral membrane desaturases are ubiquitous in eukaryotic cells, where they play a primary role in the homeostatic regulation of physical properties of lipid membranes in response to cold (Tiku et al. 1996). In female moth pheromone biosynthesis, desaturases introduce a double bond into the saturated fatty acid chain or a second double bond into monounsaturated fatty acids. Moth pheromone desaturases, including Δ5, Δ9, Δ10, Δ11, Δ12, Δ13, and Δ14, have different regio- and stereo-specificities. Several desaturases have been sequenced and characterized by expressing them in yeast cells lacking an endogenous desaturase in order to elucidate their specific role in the sex pheromone biosynthetic pathway (Knipple et al. 1998; Matoušková, Pichová, and Svatoš 2007).

Δ9-Acyl-CoA desaturases occur commonly in animal and fungal tissues (Liu et al. 1999), which suggests that these desaturases are ancestral and serve general functions in organisms. This may explain why the Δ9-desaturase sequences are highly conserved in animals (Rodriguez et al. 1992). Two Δ9-desaturase groups have been identified and characterized in pheromone glands of moth species: one with a substrate preference of $C_{16} > C_{18}$, and the other with a substrate preference of $C_{18} > C_{16}$ (Rosenfield et al. 2001). Thus, it seems that the integral membrane desaturase gene family has evolved in Lepidoptera to function not only in normal cellular lipid metabolism, but also in pheromone biosynthesis (Knipple et al. 2002).

One phylogenetically related group of Δ11-desaturases that catalyzes the formation of Δ11 fatty acyl pheromone precursors is specifically expressed in lepidopteran sex pheromone glands (Knipple et al. 1998). Some amino acid positions in this desaturase group are hypervariable among species (Knipple et al. 2002). No function has been determined for the other three desaturase types that are also regularly found in sex pheromone glands (Knipple et al. 2002).

Some desaturase genes are transcribed in pheromone gland cells but are not translated to proteins (see Roelofs and Rooney 2003; Xue et al. 2007). For example, in *O. nubilalis* three Δ14 gene sequences and ten Δ11 desaturase genes have been found; but only one transcript, for a Δ11 desaturase, appears to be functional in this species, which uses *Z*11– and *E*11–14:OAc pheromone components. *Ostrinia furnacalis*, which uses *Z*12– and *E*12–14:OAc pheromone components, has two Δ14 desaturase genes and five Δ11 genes (Xue et al. 2007). However, in *O. furnacalis* only protein products of a Δ14 desaturase gene were found in the pheromone gland (Roelofs and Rooney 2003).

Fatty Acid Reductase

There are two routes for aldehyde pheromone biosynthesis in moths. The fatty acyl CoA pheromone precursor can be reduced to the corresponding alcohol by certain fatty acid reductases (FARs) and then oxidized to the corresponding aldehyde through an alcohol oxidase (e.g., Rafaeli 2005). Alternatively, aldehydes can be formed by direct action of a specific FAR on fatty acyl CoA. In the two races of the European corn borer (*O. nubilalis*), the distinct pheromonal blends appear to be determined by differences in the specificity of their respective fatty acyl reductase (Zhu et al. 1996): The FAR in the Z strain shows greater selectivity for Z11–14:Acyl, whereas in the E-strain there is greater selectivity for *E*11–14:Acyl. Unfortunately, the actual enzymes have not been identified or isolated.

Evidence for FAR activity was found in homogenates of *B. mori* pheromone glands by reduction of palmitoyl-CoA to the corresponding hexadecanol without the release of the aldehyde intermediate (Ozawa and Matsumoto 1996). Subsequently, Moto et al. (2003) identified an alcohol-generating FAR in *B. mori*. The sequence of this FAR showed homology with that of a plant FAR (jojoba), which converts seed wax fatty acids to their corresponding fatty alcohols (Metz et al. 2000). Ohnishi, Hull, and Matsumoto (2006) used dsRNA injections into pupae to silence the pheromone gland FAR in *B. mori*. Suppression of FAR expression reduced bombykol (alcohol pheromone) production to basal levels, confirming that FAR plays an important role in pheromone production in vivo (see Matsumoto et al. 2007). No other FARs have been identified from moth pheromone glands.

Aldehyde Reductase

Activity of aldehyde reductase has been detected in gland extracts of *Choristoneura fumiferana* (Morse and Meighen 1986). It is very difficult to prove that these enzymes first produce aldehydes that are then converted to alcohols because aldehyde reductases are also present that catalyze the reduction of the fatty aldehyde to the alcohol, so alcohols and not aldehydes are the major products (e.g., Fang, Teal, and Tumlinson 1995). The reverse reaction is catalyzed through alcohol oxidases. Both enzymes are more generally called alcohol dehydrogenases.

Alcohol Oxidase

Fatty alcohols are pheromone intermediates as well as pheromone components in the pheromone glands of many moth species, and alcohol oxidases catalyze the formation of aldehyde pheromones from these alcohols. Fang, Teal, and Tumlinson (1995) demonstrated that the oxidase in the cuticle of the pheromone gland of *Manduca sexta* converts alcohols of different chain length (C_{14}–C_{17}). Hoskovec et al. (2002) showed that the oxidase in *M. sexta* glands can also oxidize other primary alcohols, including aromatic, allylic, or heterocyclic compounds, although there is a strong preference for primary alcohols of benzylic, saturated, and allylic types (Luxová and Svatoš 2006). The overall substrate specificity closely resembled yeast alcohol dehydrogenase, but so far the enzyme has not been successfully isolated (Luxová and Svatoš 2006).

Acetyltransferase

This functional class of enzyme converts fatty alcohols to acetate esters in pheromone glands; it has been biochemically characterized in *C. fumiferana* (Morse and Meighen 1987) and *A. velutinana* (Jurenka and Roelofs 1989). In both species acetyltransferases were found only in the pheromone gland. Substrate preference assays conducted in vitro indicated specificity for the Z isomer in *A. velutinana* as well as in other tortricid moths, but not in *T. ni* (Noctuidae) or *O. nubilalis* (Pyralidae; Jurenka and Roelofs 1989). Remarkably, although acetate esters are common pheromone components in moths, no acetyltransferase genes have been cloned.

Acetate Esterase

Hydrolysis of esters occurs during pheromone synthesis as well as degradation (Ding and Prestwich 1986; Prestwich, Vogt, and Riddiford 1986). Acetate esterase activity in pheromone glands has been shown in *C. fumiferana* (Morse and Meighen 1987), *Hydraecia micacea*, *H. virescens*, and *H. subflexa* (Teal and Tumlinson 1987). In *H. subflexa* acetate esters are components of the pheromone blend (e.g., Groot et al. 2007), but in *H. virescens* acetates have never been found in the gland and they strongly antagonize attraction in an otherwise attractive blend (e.g., Groot et al. 2006). Teal and Tumlinson (1987) suggested that acetate esterase in *H. virescens* glands converts the acetates into alcohols as rapidly as the acetate esters are produced.

MOLECULAR AND BIOCHEMICAL ANALYSIS OF PHEROMONE RECEPTION

Male moth navigation toward receptive females is achieved through intermittently emitted trace quantities of female sex pheromones (Roelofs and Cardé 1977). Male moths intercept these chemical signals by means of trichoid sensilla (Kaissling and Priesner 1970). These specialized antennal cuticular hairs contain one to three specialized ORNs narrowly tuned to distinct pheromone compounds (e.g., Baker et al. 2004). The pheromone molecule enters the trichoid sensillum lumen through a cuticular pore tubule and is typically encapsulated by a pheromone-binding protein (PBP) that transports the hydrophobic molecule through the sensillum lymph and toward the ORN dendrite (reviewed in Leal 2005; Rützler and Zwiebel 2005; Vogt 2005). The PBP ejects the pheromone upon interaction with negatively charged sites at the dendritic membrane, allowing it to bind with pheromone receptor proteins (PRPs) located on the ORN dendritic surface (Leal 2005; Rützler and Zwiebel 2005). Coupling of the pheromone molecule with its receptor results in a local depolarization that spreads to an electrically sensitive region of the neuron where nerve impulses are elicited. The electrical signal travels through the ORN axon to the brain, where axons of pheromone-responsive ORNs converge into the macroglomerular complex in the antennal lobe for further processing (Mustaparta 1996). Resetting of the ORN is possible through degradation of the pheromone molecule upon release from PRPs by pheromone-degrading enzymes (PDEs). This signal inactivation is essential for pheromone plume resolution (Vickers 2006).

While little is known about the genetics of differences among species and populations in male responses to pheromones, recent breakthroughs in molecular biology have led to a much better understanding of the amino acid sequences and biochemical properties of PRPs and pheromone-processing proteins (i.e., PBPs, PDEs, and chemosensory proteins [CSPs]) involved in pheromone perception (Jurenka 2003; Knipple and Roelofs 2003; Leal 2005; Rützler and Zwiebel 2005; Vogt 2005; Gohl and Krieger 2006; Hallem, Dahanukar, and Carlson 2006; Sato et al. 2008; Wicher et al. 2008). A model for the role of these molecular components in pheromone signal processing and proposed transduction mechanisms is depicted in Figure 10.3.

Pheromone Receptor Proteins

PRPs are members of a divergent family of insect ORs that contain seven-transmembrane domains (Mombaerts 1999). These PRPs lack any sequence similarity to vertebrate G-protein coupled receptors (GPCRs), and exhibit an atypical membrane topology with the amino terminus located intracellularly (Benton et al. 2006). Insect ORs typically form a heteromeric complex composed of two subunits, a conventional and variable OR coupled with a highly conserved, ubiquitously expressed, Or83b coreceptor (for receptor phylogenetic relationships, see Chapter 9). Lepidopteran PRPs also appear to dimerize with an Or83b ortholog chaperone protein as suggested by *in situ* hybridization and heterologous expression of *B. mori* PRPs in *Xenopus* oocytes (Nakagawa et al. 2005). However, in another *in situ* hybridization study (Krieger et al. 2005), there was no clear coexpression of *B. mori* PRPs with the chaperone protein. Furthermore, in Flp-In T-REx293/Gα15 cells (Große-Wilde, Svatoš, and Krieger 2006) and *Drosophila* ab3A neurons (Syed et al. 2006), PRP was activated by the pheromone alone without expression of the chaperone protein (i.e., Or83b ortholog). Differences in labeling techniques and PRP processing by the different heterologous host cells used may explain these conflicting results. Recent electrophysiological and fluorescent optical experiments on heterologously expressed insect OR heteromeric complexes, including a *B. mori* PRP-Or83b ortholog complex, showed that they form a cation nonselective ion channel, directly gated by odor or pheromone binding to the OR (Sato et al. 2008). In addition to this ionotropic signal transduction pathway, a metabotropic pathway involving a cyclic-nucleotide-activated channel in the Or83b coreceptor has also been shown (Wicher et al. 2008). These recent findings indicate that functional PRPs require the presence of the "helper" protein for chemical signal transduction.

ORNs typically express only one conventional OR gene (Vosshall et al. 1999; Mombaerts 2004) that determines the ORN odorant response profile (Hallem, Ho, and Carlson 2004), and PRPs generally follow this one receptor–one ORN organization. However, unlike general insect ORs that typically bind to more than one ligand (e.g., Hallem, Ho, and Carlson 2004), PRPs are narrowly tuned to specific ligands (e.g., Große-Wilde et al. 2007).

Lepidopteran PRPs share little sequence similarity with general insect ORs, and possibly form a single lineage of proteins sharing a high degree of sequence similarity and exhibiting conserved functions as indicated by phylogenetic analyses of *B. mori* and *H. virescens* ORs (Krieger et al. 2005; Nakagawa et al. 2005; Wanner et al. 2007). These analyses also suggest that PRPs form two main lineages: one that expanded in the bombycids and the other in the noctuids (Figure 10.4). The higher degree of sequence identity within lineages suggests that these clusters may have arisen from ancestral pheromone receptor gene duplication events. As PRPs of more moth lineages are sequenced, this pattern may become more complex (for additional details of insect OR lineages see Chapter 9).

Candidate genes for *H. virescens* PRPs have been examined by J. Krieger and his colleagues. *H. virescens* olfactory receptors (HRs) were first identified by screening an antennal cDNA library with probes generated by an analysis of an *H. virescens* genomic database based on similarity to *Drosophila melanogaster* OR sequences (Krieger at al. 2002). Further screening of the antennal cDNA library with probes encoding HRs and other insect OR short regions allowed the identification

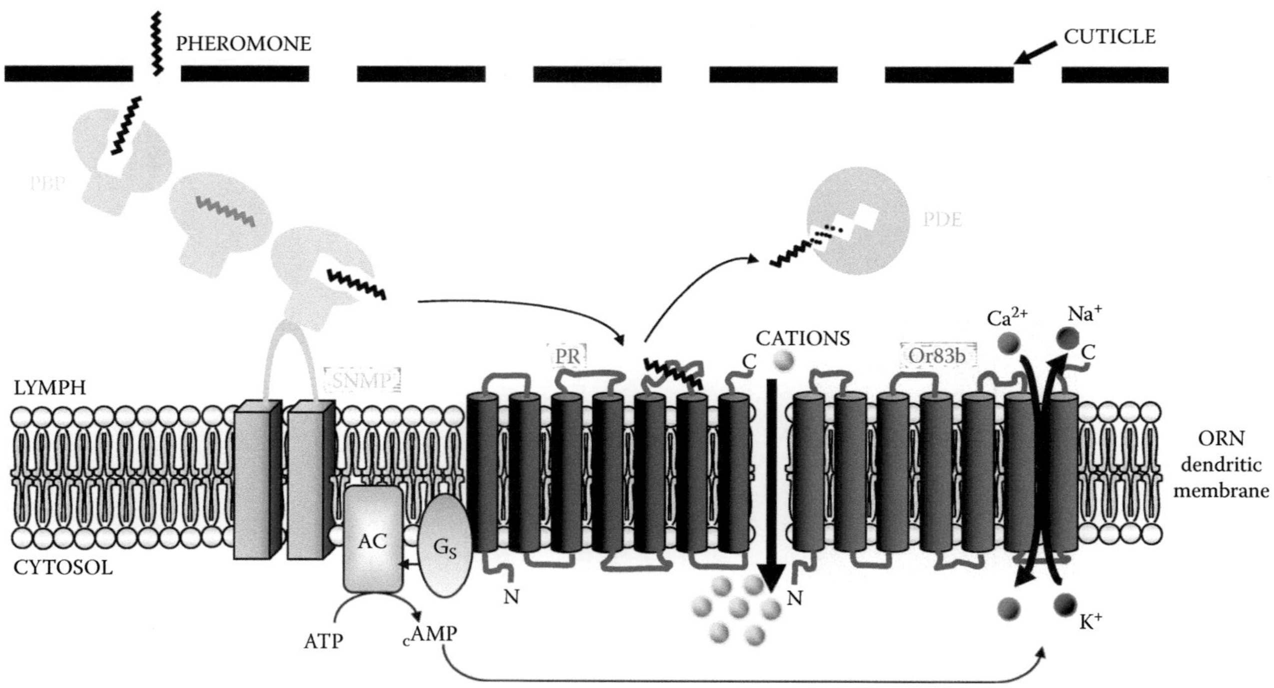

FIGURE 10.3 ***A color version of this figure follows page 176.*** Schematic model of pheromone reception in moths. A pheromone molecule entering the lumen of a trichoid sensillum through a cuticular pore is bound to a pheromone binding protein (PBP), which transports the pheromone to the dendritic membrane of the olfactory receptor neuron (ORN). A sensory neuron membrane protein (SNMP) binds to the PBP-pheromone complex, or the pheromone only, and directs the pheromone to the nearby pheromone receptor (PR). The pheromone binds to the PR-Or83b heteromeric complex resulting in either very rapid recognition by means of an ionotropic pathway, or slower yet prolonged detection via a metabotropic G protein–mediated signal amplification. Pheromone degrading enzymes (PDEs) inactivate unbound pheromones. (Modified from Rützler and Zwiebel, 2005.)

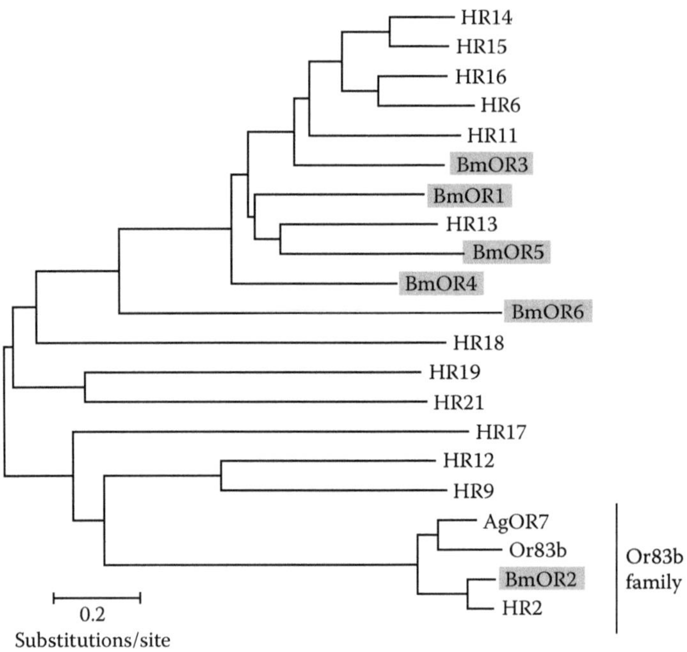

FIGURE 10.4 Phylogenetic tree of ORs for *Heliothis virescens* (HR), *Bombyx mori* (BmOR), and *Anopheles gambiae* (AgOR). From Nakagawa et al. (2005).

of four candidate PRPs (HR13, HR14, HR15, and HR16) that are exclusively expressed beneath pheromone-responsive sensilla trichoidea of male antennae, and share at least 40 percent amino acid identity (Krieger et al. 2004). HR13 has been shown to be expressed in neurons of sensilla trichoidea type A (Gohl and Krieger 2006). Moreover, immunohistochemical studies combined with functional analysis in a heterologous expression system clearly indicate that HR13 specifically interacts with Z11–16:Ald, the major pheromone blend component of *H. virescens* (Gohl and Krieger 2006; Große-Wilde et al. 2007).

Male-specific *B. mori* OR genes were isolated by differential screening of a male antenna cDNA library, and the first *B. mori* PRP, BmOR-1, was identified based on sequence similarity to other insect ORs (Sakurai et al. 2004). BmOR-1 is exclusively expressed in cells located beneath the long trichoid sensilla and has high homology to some *H. virescens* receptors. Ectopic expression of this receptor in female antennae and in *Xenopus* oocytes demonstrated its specificity for bombykol, the silk moth sex pheromone (Sakurai et al. 2004). Subsequent studies using different heterologous expression systems further corroborated this finding (Große-Wilde, Svatoš, and Krieger 2006; Syed et al. 2006). Similar *in situ* hybridization and heterologous expression studies identified BmOR-2 as the receptor for bombykal, an oxidized form of bombykol that does not elicit male-orientating behavior (Nakagawa et al. 2005).

Heterologous expression systems, including *Xenopus laevis* oocytes (Sakurai et al. 2004), modified HEK 293 cells (Große-Wilde, Svatoš, and Krieger 2006), and the *D. melanogaster* Δ*halo* mutant with an empty ab3A neuron (Dobritsa et al. 2003) and *Or67d-GAL4* mutant (Kurtovic, Widmer, and Dickson 2007), have been used successfully for the functional characterization of candidate moth PRPs in vivo. Additionally, these systems could be used in the future for comparative functional analyses between wild and mutated PRPs to determine if specific changes in pheromone receptor gene sequences affect ligand specificity.

Pheromone Binding Proteins

PBPs are members of the encapsulin family, proteins that solubilize hydrophobic compounds in aqueous environments (Vogt 2005). PBPs are α-helical proteins characterized by the presence of a major hydrophobic domain, a signal peptide, and six well-conserved cysteine residues forming three disulfide bridges (e.g., Sandler et al. 2000). Unlike other members of the OBP gene family, PBPs are expressed exclusively or predominantly in long sensilla trichoidea (e.g., Laue and Steinbrecht 1997), where they are produced by support cells and found at high concentration in the lumen (Steinbrecht, Ozaki, and Ziegelberger 1992). PBPs bind, encapsulate, and ferry pheromones to the external PRP loops on the ORN dendritic membrane, and protect them from PDEs as well (Krieger and Breer 1999; Leal 2005). Contact with dendritic membrane negatively charged sites leads to the formation of an additional C-terminal α-helix that fills the pheromone binding site and ejects the pheromone out of the PBP (Leal 2005, and references therein).

PBPs, first identified in *Antheraea polyphemus* (Vogt and Riddiford 1981), have been characterized from several moth species, allowing identification of multiple PBP subtypes displaying considerable diversity (32–92 percent amino acid identity; e.g., Abraham, Löfstedt, and Picimbon 2005). Phylogenetic analyses have shown that several duplication events appear to have given rise to specific subtypes (e.g., Robertson et al. 1999; Xiu, Zhou, and Dong 2008). Lepidoptera PBPs divide into three main groups, each comprising PBPs from various species, with noctuid PBPs forming three distinct groups (Figure 10.5) that may have arisen through two duplication events (Xiu and Dong 2007).

Cloning of PBP genes predates work on pheromone receptor genes; Krieger et al. (1993) cloned an *H. virescens* PBP over a decade ago. Functional assays using modified HEK 293 cells showed that an *H. virescens* PBP, HvirPBP2, increased HR13 sensitivity and specificity to Z11–16:Ald (Große-Wilde et al. 2007). However, the specificity of two other heterologously expressed PRPs, HR14 and HR16, was not increased in the presence of HvirPBP1 or HvirPBP2. Interestingly, heterologous expression of *BmOR1* in HEK 293 cells showed that BmorPBP increased specificity to bombykol (Große-Wilde, Svatoš, and Krieger 2006), whereas *BmOR1* expression in *Xenopus* oocytes and Δ*halo* mutants indicated that BmorPBP was not necessary for response to bombykol (e.g., Syed et al. 2006). The latter findings correspond with early studies in *M. sexta*–cultured ORN, suggesting that PBPs are not necessary for PRP response to pheromones (Stengl et al. 1992). Hence, the pheromone alone rather than a PBP-pheromone complex appears to activate the PRP. However, PBPs play a role in pheromone reception kinetics and sensitivity: (1) PBP pH-dependent conformational change is consistent with the millisecond time scale of pheromone peripheral perception events essential during male moth oriented navigation (Leal et al. 2005); (2) PBPs facilitate the diffusion of pheromones into the sensillar lymph and transport them selectively (Leal 2005; Syed et al. 2006); (3) PBPs screen out a subset of odorants and concentrate pheromones in the sensillum lymph (Pelosi 1996). By increasing the uptake of pheromone molecules, PBPs could lower the threshold for pheromone response (van den Berg and Zielgelberger 1991). Ligand binding ranges from very specific to very broad in PBPs (e.g., Rivière et al. 2003), thus only PBPs with high-binding specificity may be involved in pheromone component discrimination (Bette, Breer, and Krieger 2002; Maida, Ziegelberger, and Kaissling 2003). PBPs appear to be required for olfactory system sensitivity, and to some extent, specificity. However, the specific mechanisms involved remain to be determined.

Pheromone Degrading Enzymes

PDEs are thought to inactivate pheromones before they reach the PRPs if they are not bound to PBPs, and to degrade pheromone molecules that have already stimulated PRPs. PDEs can modify pheromone chemistry (Rybczynski, Reagan, and Lerner 1989), and degrade pheromone molecules on a millisecond timescale in vitro (Vogt, Riddiford, and Prestwich 1985), although degradation

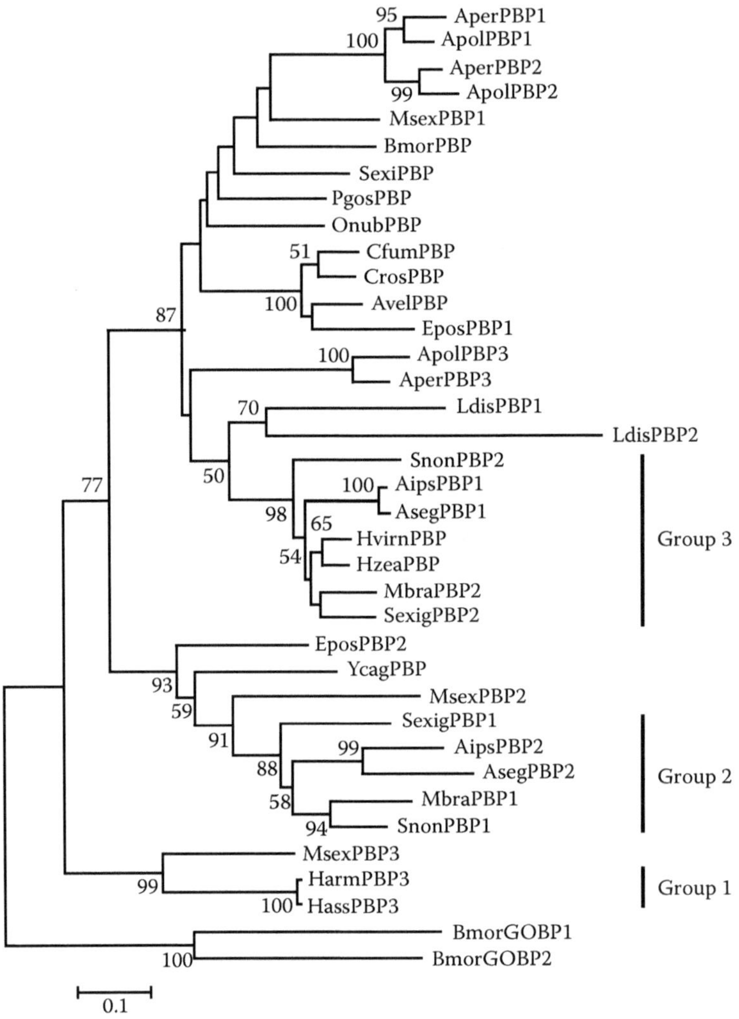

FIGURE 10.5 Phylogenetic analysis of amino acid PBP sequences. General odorant binding protein (GOBP) sequences are the outgroup. *Agrotis ipsilon* (Aips), *A. segetum* (Aseg), *Anthreraea pernyi* (Aper), *A. polyphemus* (Apol), *Argyrotaenia velutinana* (Avel), *Bombyx mori* (Bmor), *Choristoneura fumiferana* (Cfum), *C. rosaceana* (Cros), *Epiphyas postvittana* (Epos), *Helicoverpa armigera* (Harm), *H. assulta* (Hass), *Heliothis virescens* (Hvir), *H. zea* (Hzea), *Lymantria dispar* (Ldis), *Mamestra brassicae* (Mbra), *Manduca sexta* (Msex), *Ostrinia nubilalis* (Onub), *Pectinophora gossypiella* (Pgos), *Spodoptera exigua* (Sexig), *Sesamia nonagrioides* (Snon), *Yponomeuta cagnagellus* (Ycag). From Xiu and Dong (2007).

in vivo is slower, probably due to pheromone protection by PBPs (Kaissling 2001). Despite their important role in signal inactivation, knowledge on the molecular structures of PDEs is limited, and only a few genes have been identified and characterized (Vogt 2005).

A male antenna-specific esterase and a cytochrome P450 enzyme cDNA have been cloned in *A. polyphemus* (Ishida and Leal 2002) and *Mamestra brassicae* (Maïbèche-Coisne et al. 2002), respectively, but whether they function as PDEs is yet unclear. A previously characterized *A. polyphemus* sensillar esterase (Vogt, Riddiford, and Prestwich 1985), *ApolPDE*, has been isolated, cloned, and expressed in a baculovirus vector, allowing estimation of sensillar lymph concentration (approximately 20,000-fold lower than a PBP) and the study of pheromone inactivation kinetics (Ishida and Leal 2005). Hence, the *ApolPDE* sequence could be used to identify PDEs in other moth species.

In *B. mori*, an aldehyde oxidase preferentially expressed in male antennae catabolizes bombykal (Rybczynski, Vogt, and Lerner 1990). Recently, a partially sequenced *M. brassicae* aldehyde oxidase expressed exclusively in olfactory sensilla (Merlin et al. 2005) was used to identify putative aldehyde oxidase genes in *B mori* (Pelletier et al. 2007). A single gene selectively expressed in male *B. mori* antennae, *BmAOX2*, may code for the PDE involved in bombykal degradation; however, functional characterization is needed. Extended neural activity in ab3A sensilla heterologously expressing *BmorOR1* has been hypothesized to result from a lack of bombykol inactivation due to the absence of PDE in this system (Syed et al. 2006). This expression system could be used to test the role of *BmAOX2* in bombykal degradation.

Interactions among PRP, PBP, and PDE

It is likely that PBPs, PDEs, PRPs, and the central nervous system all play roles in specificity within species and the differences among species in male response to pheromones. The importance of each is likely to differ by compound and taxonomic lineage (Leal 2005; Rützler and Zwiebel 2005; Vogt 2005; Hallem, Dahanukar, and Carlson 2006). Two "layers of filters," PBPs and PRPs, are thought to be involved in male response specificity through a combinatorial process (Leal 2003, 2005). For example, heterologously expressed *BmOR1* responds to both bombykol and bombykal; however, *BmorPBP* increases specificity, possibly by selective delivery of bombykol to this receptor (Große-Wilde, Svatoš, and Krieger 2006). Moreover, higher *BmOR1* response has been observed in the presence of BmorPBP in ab3A empty cells, possibly through pheromone solubilization (Syed et al. 2006). Heterologous coexpression of heterospecific PBPs or OBPs with *BmOR1* would ultimately corroborate BmorPBP function.

As previously mentioned, *Or83b* homologues appear also to play a role in pheromone molecular recognition through interaction with PRPs. *Or83b* is involved in OR localization to ORN dendrites and heterodimer formation, which is essential for receptor responsiveness and signal transduction (Rützler and Zwiebel 2005; Wicher et al. 2008; Sato et al. 2008). Similarly, other molecules such as sensory neuron membrane proteins could interact with ligands (Vogt 2003; Benton, Vannice, and Vosshall 2007) or act as PBP receptors (Rogers et al. 1997; Rogers, Krieger, and Vogt 2001; Jacquin-Joly and Merlin 2004).

Processing of Signals

In male moths, pheromone response is governed by pheromonal excitation of peripheral olfactory pathways that activate behavioral circuits in the brain. Pheromone-induced electrical signals spread from the ORN dendrites to axons that project to enlarged glomeruli in the macroglomerular complex (MGC) of the antennal lobes, where signals are further processed and sent out through projection neurons to the protocerebrum (Vickers, Poole, and Linn 2005). It is known that each pheromone component of a blend is represented in a single MGC glomerulus and that the combinatorial pattern of activity across several glomeruli represents the pheromonal blend (Vickers and Christensen 2003). Moreover, ORNs expressing the same pheromone receptor (PR) gene are expected to converge onto one glomerulus (e.g., Datta et al. 2008). Unlike mammals, where ORN axons' convergence in the olfactory bulb is receptor dependent, insect ORs, and possibly PRPs, are not involved in guiding axon convergence into their cognate glomeruli (Dobritsa et al. 2003).

In *D. melanogaster* males, activation of the sex pheromone cis-vaccenyl acetate (cVA) receptor inhibits male-male courtship, whereas in females it promotes receptivity to males. Using an approach combining genetics and optical neural tracing, Datta et al. (2008) found that cVA activates a single glomerulus, which is innervated by post-synaptic projection neurons (PNs) exhibiting sexually dimorphic projections in the protocerebrum lateral horn. A male-specific transcription factor, Fruitless (*Fru*^{M}) in the glomerulus PNs and other *Fru*^{M}-expressing cells, controlled the formation of the male-specific axonal arbor in the lateral horn. Behavioral dimorphism could have resulted from third-order neurons receiving greater input from male PNs or restricting their synapses to the male-specific region of the glomerulus axon arbor. A similar sexually dimorphic neural circuit in the protocerebrum may occur in moth species. Moreover, between-species differences in male response to the same pheromone compound may be explained by a comparable interspecific anatomical dimorphism that may lead to either positive or antagonistic responses in closely related species perceiving the compound. Thus, it is feasible that central nervous system processes dictate these opposite responses.

FUTURE DIRECTIONS

In comparison with most other ecological traits of animals, we know quite a bit about the molecular and biochemical underpinnings of pheromone production and reception. Less is known about the genetic architecture of variation in sexual communication systems, but that knowledge base is increasing. Each of these areas of research is of interest on its own, but it is a synthesis of these three areas of knowledge that will provide more insight into the processes that gave rise to evolutionary diversification of pheromone-based sexual communication systems. A few researchers have begun this synthesis, but much more is needed. Below, we summarize a number of testable hypotheses and research approaches that could aid in this synthesis, and we describe some of the pioneering work in this area.

"Candidate Gene" versus "Genomic Network" Hypotheses

Based on what we know, it is possible that most of the genetic changes involved in diversification of pheromone-based systems were changes in the amino acid sequences of enzymes in pheromone production pathways, and in sequences of receptors and other proteins involved in pheromone detection. This could be considered the "candidate gene hypothesis." At the other extreme, it is possible that most of the evolutionarily important genetic changes that led to diversification were due to alterations in complex trans-acting genetic factors that regulate expression of these candidate genes or modify their effects on the phenotype. This could be considered the "genomic network hypothesis." Although one can find examples of other traits in which one of the two hypotheses hold, there has been considerable discussion about which of these two mechanisms is most important to evolution in general; however, the data base is limited (Wittkopp, Haerum, and Clark 2004, 2008; Carroll 2005; Sambandan et al. 2006; Hoekstra and Coyne 2007). Because candidate genes for male reception of pheromones are well described, and those for pheromone biosynthesis should soon be in hand, testing these hypotheses in the moth sexual communication system should be feasible and promises to help resolve this more general debate.

Molecular Analysis of Past Selection on Pheromone Production and Response

There are direct and indirect ways to test the two hypotheses defined above. One approach is to build databases of the genomic DNA sequences that code for the production of each of the candidate genes in many species from a genus or family of moths that have diversified in the components of their pheromone blends. This information can then be used to examine patterns of change in homologous genes of these species and to search for signatures of stabilizing/directional selection, drift, and gene duplication/loss. This was an onerous task a few years ago, but recent technological breakthroughs in sequencing make this more feasible today. As discussed above, Knipple et al. (2002) made a major breakthrough by using cDNA coding sequences for desaturases in a number of subfamilies of Lepidoptera as a means to assess such patterns. Their findings were complex, and patterns differed for the six grouping of desaturases examined. Overall, there was no signature of directional selection, but some amino acid substitutions in one group of desaturases are in positions that could alter the catalytic site of the enzyme. More detailed structure-function relationships are needed to examine this possibility. Comparisons of the desaturases of Lepidoptera and *D. melanogaster* indicate that gene duplication resulting in at least three of the desaturase groups occurred around 280 million years ago (Knipple et al. 2002).

The general pattern of coding sequence conservation found by Knipple et al. (2002) at least hints at the possibility that differences in expression of desaturase genes between species may have a more important role in species-to-species pheromone differences than do changes in coding sequence. The approach taken by D.C. Knipple and his colleagues could be applied to the reductases and to other enzymes in the pheromone biosynthetic pathway once they are identified.

In terms of male response, cDNA sequences coding for PRPs and OBPs in a variety of species have been analyzed for simple phylogenetic relatedness (Figures 10.4 and 10.5). Willett (2000) took this one step further and found evidence for directional selection on PBPs from *Choristoneura* species. Curiously, there was no relationship among species in the extent of directional change in the PBPs and changes in the pheromone blends. Willett (2000) postulates that a selection force unrelated to pheromone blend could have been selecting on the protein sequences.

We feel that these few studies point toward a useful direction for future research. At least at the peripheral sensory level, we have many good candidate genes involved in male moth pheromone reception, so the raw material is available to conduct more detailed phylogenetic analyses. A caveat is that these phylogenetic studies are clearly informative, but typically they are not definitive. Without careful analyses such as done by Willett (2000), it is easy to draw erroneous conclusions about selective factors.

W.L. Roelofs and his collaborators and colleagues have conducted pioneering work that used the *Ostrinia* genus to couple more fully phylogeny, gene transcription, and translation. The pheromone composition of eight *Ostrinia* species has been determined. The major components in all of the species are unsaturated acetates except for *O. latipennis*, which only uses Z11–14:OH (see Roelofs and Rooney 2003; Xue et al. 2007). As described in this chapter, the surprising result was that for each of the two species examined in detail, many more desaturase genes are transcribed than are translated. In the distantly related moth *B. mori*, it was also found that not all destaturase genes that were transcribed in pheromone gland cells produced active proteins (Moto et al. 2004). Although it is clear that desaturase production differences have been involved in diversification of *Ostrinia* pheromones, many of the differences appear to be due to posttranscriptional processes, so a simple candidate gene analysis could miss the important factors.

At the evolutionary and mechanistic levels, it becomes important to determine what type of changes in what genomic DNA sequences determine which desaturase mRNA transcripts are translated into functional proteins. At this point we cannot answer the question of how many genetic changes or genetic networks control differences among the species in enzymes involved in pheromone biosynthesis. There may be a single, small, critical gene sequence within the candidate gene, or a key cis regulatory element that controls most of the variation in the concentration and activity of a single enzyme; on the other hand, unrelated and unlinked genes could also regulate which mRNAs result in the production of active enzymes.

At the level of male response to pheromones, W.L. Roelofs' group again led the way in revealing the problem with assuming that changes in or around a candidate gene would be responsible for differences among males that respond to different pheromone blends. Their early work with *O. nubilalis* clearly showed that males of the E and Z races differed in the amplitude of the neuronal response spikes formed when exposed to the E and Z isomers of 11–14:OAc; the E strain had a higher amplitude spike when exposed to *E*11–14:OAc, and the Z strain had a higher amplitude spike to Z11–14:OAc (Roelofs et al. 1987). F_1 hybrids had spikes of intermediate amplitude, and F_2 male spikes indicated that the difference was inherited on a single autosome (Roelofs et al. 1987). The simplest hypothesis was that the differences in spike amplitude were responsible for, or at least would be correlated with, the differences in behavioral response. Wind tunnel analysis of F_2 offspring found that males who inherited the autosome that coded for the E race spike amplitude were no more likely to respond to the E strain pheromone blend than males who inherited the homologous chromosome coding for the Z race spike amplitudes. Further genetic analysis indicated that genes encoding the differential behavioral response were sex-linked (Roelofs et al. 1987). We still do not know what those sequences are, but it is clear that they are not in cis with genes that determine spike amplitude.

When W.L. Roelofs and his collaborators were doing this early genetic work, they were probably pleased that they were dealing with one autosome and one sex chromosome, because it made the genetic analysis feasible. Today, AFLP and microsatellite markers have made it much easier to localize loci that control pheromone responses in *O. nubilalis* (e.g., Dopman et al. 2005), and in the next few years it is reasonable to expect the entire nuclear genome of *O. nubilalis* to be sequenced, providing even more detailed information on the sequences differentiating the two *O. nubilalis* races. With greater information at the genomic level, we should certainly be able to identify alleles on the sex chromosome responsible for the differential male response. Similarly, we should be able to identify the enzyme-coding genes and regulatory sequences in the two races that code for changes in Z and E isomer ratios. Once the overall sequence differences are identified, it may be possible to determine which of the nucleotide differences are most important in altering the phenotypes. This kind of information will bring us much closer to understanding the evolutionary genetics of diversification in *Ostrinia* sexual communication.

Is the Genetic Architecture of Differences in Pheromone Blends and Responses the Same within and among Species?

A general question in evolutionary biology is whether the genes responsible for microevolution (changes within species) are the same as those that result in macroevolutionary changes (differences among species and higher taxa). Weber et al. (2008) reviewed a variety of studies in which a number of populations were selected equally for change in a trait. They conclude that for some traits, the genes responding to selection can differ completely between populations, but for other traits they always seem to be similar. In cases where the genetics of response is always similar, there is a higher likelihood that macroevolutionary changes will utilize the same genes as microevolutionary ones. If this is the case with moth sexual communication traits, then detailed studies on the origin of among population differences will be extremely helpful in understanding the macroevolution of diversity in moth sexual communication. On a cautionary note, however, in a study by Gleason and Ritchie (2004) the genetic regions found to affect differences in courtship song between two *Drosophila* species were not the same as the genetic regions associated with differences within *D. melanogaster*. If this is the case in other sexual communication systems, studies at the population level may not offer strong inference about macroevolutionary processes.

Coupling Molecular and Genetic Analyses with Lab and Field Studies of Behavior

Identifying which DNA sequence changes impact moth phenotypes is very important. However, if we are to understand how specific sequences have spread by natural selection, we must couple studies of sequence identification with tests of how they impact mating fitness of males and females. The ideal approach would be to examine effects of single genetic alterations on fitness of field-released individuals. While this may be possible in some species, it is not feasible with most moths. To get around this problem, Groot et al. (2006) estimated the fitness effect of genes that alter the amount of acetates in the pheromone blend of *H. subflexa* by using a combination of field and cage studies. The field studies estimated effects on long-distance attraction, while the cage studies measured probability of mating and sperm transfer. Similar kinds of approaches could be used to link phenotype to mating fitness in other moth systems. With these data in hand it would finally be possible to determine the intensity of selection for or against rare females and males with single mutations in genes coding for sexual communication traits.

CONCLUSION

Moth sexual communication systems are intriguing to molecular biologists, biochemists, and evolutionary biologists because of their precision, efficiency, and incredible diversity. Although much could be learned by researchers in each of these disciplines working independently, the time seems to have arrived when more interdisciplinary work could result in breakthroughs that would be important for the understanding of moth sexual communication, and more generally for our understanding of the evolutionary process as a whole.

REFERENCES

Abraham, D., C. Löfstedt, and J.F. Picimbon. 2005. Molecular characterization and evolution of pheromone binding protein genes in *Agrotis* moths. *Insect Biochem. Mol. Biol.* 35:1100–11.

Baker, T.C., S.A. Ochieng, A.A. Cossé, S.G. Lee, and J.L. Todd. 2004. A comparison of responses from olfactory receptor neurons of *Heliothis subflexa* and *Helitohis virescens* to components of their sex pheromone. *J. Comp. Physiol. A* 190:155–65.

Baltensweiler, W., and E. Priesner. 1988. Studien zum Pheromon-polymorphismus von *Zeiraphera diniana* Gn. (Lep., Tortricidae). *J. Appl. Entomol.* 106:217–31.

Benton, R., S. Sachse, S.W. Michnick, and L.B. Vosshall. 2006. Atypical membrane topology and heteromeric function of *Drosophila* odorant receptors *in vivo*. *PLoS Biol.* 4:e20.

Benton, R., K.S. Vannice, and L.B. Vosshall. 2007. An essential role for a CD36-related receptor in pheromone detection in *Drosophila*. *Nature* 450:289–93.

Bette, S., H. Breer, and J. Krieger. 2002. Probing a pheromone binding protein of the silkmoth *Antheraea polyphemus* by endogenous tryptophan fluorescence. *Insect Biochem. Mol. Biol.* 32:241–46.

Bjostad, L.B., and W.L. Roelofs. 1983. Sex pheromone biosynthesis in *Trichoplusia ni*: key steps involve delta-11 desaturation and chain shortening. *Science* 220:1387–89.

Butlin, R. 1995. Genetic variation in mating signals and responses. In *Speciation and the recognition concept: Theory and application*, ed. D.M. Lambert and H.G. Spencer, 327–66. Baltimore: Johns Hopkins Univ. Press.

Butlin, R., and M.G. Ritchie. 1989. Genetic coupling in mate recognition systems: What is the evidence? *Biol. J. Linn. Soc.* 37:237–46.

Butlin, R., and A.J. Trickett. 1997. Can population genetic simulations help to interpret pheromone evolution? In *Insect pheromone research: New directions*, ed. R.T. Cardé and A.K. Minks, 548–62. New York: Chapman and Hall.

Cardé, R.T., and K.F. Haynes. 2004. Structure of the pheromone communication channel in moths. In *Advances in insect chemical ecology*, ed. R.T. Cardé and J.G. Millar, 283–332. Cambridge: Cambridge Univ. Press.

Carroll, S.B. 2005. Evolution at two levels: On genes and form. *PLoS Biol.* 3:1159–66.

Cork, A., and E.A. Lobos. 2003. Female sex pheromone components of *Helicoverpa gelotopoeon*: First heliothine pheromone without (Z)-11-hexadecenal. *Entomol. Exp. Appl.* 107:201–06.

Cossé, A.A., M. Campbell, T.J. Glover, et al. 1995. Pheromone behavioral responses in unusual male European corn borer hybrid progeny not correlated to electrophysiological phenotypes of their pheromone-specific antennal neurons. *Experientia* 51:809–16.

Coyne, J.A., N.H. Barton, and M. Turelli. 1997. Perspective: A critique of Sewall Wright's shifting balance theory of evolution. *Evolution* 51:643–71.

Coyne, J.A., and H.A. Orr. 1998. The evolutionary genetics of speciation. *Philos. Trans. R. Soc. Lond. B Biol. Sci.* 353:287–305.

Datta, S.R., M.L. Vasconcelos, V. Ruta, et al. 2008. The *Drosophila* pheromone cVA activates a sexually dimorphic neural circuit. *Nature* 452:473–77.

Dieckmann, U., and M. Doebeli. 1999. On the origin of species by sympatric speciation. *Nature* 400:354–57.

Ding, Y., and G.D. Prestwich. 1986. Metabolic transformations of tritium labeled pheromone by tissues of *Heliothis virescens* moths. *J. Chem. Ecol.* 12:411–29.

Dobritsa, A.A., W. van der Goes van Naters, C.G. Warr, RA. Steinbrecht, and J.R. Carlson. 2003. Integrating the molecular and cellular basis of odor coding in the *Drosophila* antennae. *Neuron* 37:827–41.

Dopman, E.B., L. Perez, S.M. Bogdanowicz, and R.G. Harrison. 2005. Consequences of reproductive barriers for genealogical discordance in the European corn borer. *Proc. Natl. Acad. Sci. U.S.A.* 102:14706–11.

Eliyahu, D., S.W. Applebaum, and A. Rafaeli. 2003. Moth sex pheromone biosynthesis is inhibited by the herbicide diclofop. *Pestic. Biochem. Physiol.* 77:75–81.

El-Sayed, A.M. 2008. The Pherobase: Database of insect pheromones and semiochemicals. http://www.pherobase.com (accessed September 12, 2008).

Fang, N., P.E.A. Teal, and J.H. Tumlinson. 1995. PBAN regulation of pheromone biosynthesis in female tobacco hornworm moths, *Manduca sexta* (L.). *Arch. Insect Biochem. Physiol.* 29:35–44.

Foster, S.P. 2005. Lipid analysis of the sex pheromone gland of the moth *Heliothis virescens*. *Arch. Insect Biochem. Physiol.* 59:80–90.

Gahan, L.J., F. Gould, and D.G. Heckel. 2001. Identification of a gene associated with Bt resistance in *Heliothis virescens*. *Science* 293:857–60.

Gleason, J.M., and M.G. Ritchie. 2004. Do quantitative trait loci (QTL) for a courtship song difference between *Drosophila simulans* and *D. sechellia* coincide with candidate genes and intraspecific QTL? *Genetics* 166:1303–11.

Gohl, T., and J. Krieger. 2006. Immunolocalization of a candidate pheromone receptor in the antennae of the male moth, *Heliothis virescens*. *Invert. Neurosci.* 6:13–21.

Gray, D.A., and W.H. Cade. 1999. Quantitative genetics of sexual selection in the field cricket, *Gryllus integer*. *Evolution* 53:848–54.

Groot, A.T., J. Bennett, J. Hamilton, R.G. Santangelo, C. Schal, and F. Gould. 2006. Experimental evidence for interspecific directional selection on moth pheromone communication. *Proc. Nat. Acad. Sci. U.S.A.* 103:5858–63.

Groot, A.T., R.G. Santangelo, E. Ricci, C. Brownie, F. Gould, and C. Schal. 2007. Differential attraction of *Heliothis subflexa* males to synthetic pheromone lures in Eastern US and Western Mexico. *J. Chem. Ecol.* 33:353–68.

Große-Wilde, E., T. Gohl, E. Bouché, H. Breer, and J. Krieger. 2007. Candidate pheromone receptors provide the basis for the response of distinct antennal neurons to pheromonal compounds. *Eur. J. Neurosci.* 25:2364–73.

Große-Wilde, E., A. Svatoš, and J. Krieger. 2006. A pheromone-binding protein mediated the bombykol-induced activation of a pheromone *in vitro*. *Chem. Senses* 31:547–55.

Gu, P., W.H. Welch, L. Guo, K.M. Schegg, and G.J. Blomquist. 1997. Characterization of a novel microsomal fatty acid synthetase (FAS) compared to a cytosolic FAS in the housefly, *Musca domestica*. *Comp. Biochem. Physiol.* 118B:447–56.

Hallem, E.A., A. Dahanukar, and J.R. Carlson. 2006. Insect odor and taste receptors. *Annu. Rev. Entomol.* 51:113–35.

Hallem, E.A., M.G. Ho, and J.R. Carlson. 2004. The molecular basis of odor coding in the *Drosophila* antennae. *Cell* 117:965–79.

Hansson, B.S., C. Löfstedt, and S.P. Foster. 1989. Z-linked inheritance of male olfactory response to sex pheromone components in two species of tortricid moths, *Ctenopseusis obliquana* and *Ctenopseusis* sp. *Entomol. Exp. Appl.* 53:137–45.

Haynes, K.F., and R.E. Hunt. 1990. A mutation in the pheromonal communication system of the cabbage looper moth, *Trichoplusia ni*. *J. Chem. Ecol.* 16:1249–57.

Heckel, D.G. 1993. Comparative genetic linkage mapping in insects. *Annu. Rev. Entomol.* 38:381–408.

Hoekstra, H.E., and J.A. Coyne. 2007. The locus of evolution: Evo devo and the genetics of adaptation. *Evolution* 61:995–1016.

Hoskovec, M., A. Luxová, A. Svatoš, and W. Boland. 2002. Biosynthesis of sex pheromones in moths: Stereochemistry of fatty alcohol oxidation in *Manduca sexta*. *Tetrahedron* 58:9193–9201.

Hoy, R.R., J. Hahn, and R.C. Paul. 1977. Hybrid cricket auditory behavior: Evidence for genetic coupling in animal communication. *Science* 195:82–84.

Hull, J.J., R. Kajigaya, K. Imai, and S. Matsumoto. 2007. The *Bombyx mori* sex pheromone biosynthetic pathway is not mediated by cAMP. *J. Insect Physiol.* 53:782–93.

Ishida, Y., and W.S. Leal. 2002. Cloning of putative odorant-degrading enzyme and integumental esterase cDNAs from the wild silkmoth, *Antheraea polyphemus*. *Insect Biochem. Mol. Biol.* 32:1775–80.

Ishida, Y., and W.S. Leal. 2005. Rapid inactivation of a moth pheromone. *Proc. Natl. Acad. Sci. U.S.A.* 102:14075–79.

Jacquin-Joly, E., and C. Merlin. 2004. Insect olfactory receptors: Contributions of molecular biology to chemical ecology. *J. Chem. Ecol.* 30:2359–97.

Juárez, M.P. 2004. Fatty acyl-CoA elongation in *Blatella germanica* integumental microsomes. *Arch. Insect Biochem. Physiol.* 56:170–78.

Juárez, M.P., and G.C. Fernández. 2007. Cuticular hydrocarbons of triatomines. *Comp. Biochem. Physiol. A Mol. Integr. Physiol.* 147:711–30.

Jurenka, R. 2003. Biochemistry of female moth sex pheromones. In *Insect pheromone biochemistry and molecular biology*, ed G.J. Blomquist and R.C. Vogt, 54–80. Amsterdam: Elsevier.

Jurenka, R.A., K.F. Haynes, R.O. Adolf, M. Bengtsson, and W.L. Roelofs. 1994. Sex pheromone component ratio in the cabbage looper moth altered by a mutation affecting the fatty acid chain-shortening reactions in the pheromone biosynthetic pathway. *Insect Biochem. Mol. Biol.* 24:373–81.

Jurenka, R.A., E. Jacquin, and W.L. Roelofs. 1991. Control of the pheromone biosynthetic pathway in *Helicoverpa zea* by the pheromone biosynthesis activating neuropeptide. *Arch. Insect Biochem. Physiol.* 17:81–91.

Jurenka, R.A., and W.L. Roelofs. 1989. Characterization of the acetyltransferase used in pheromone biosynthesis in moths: Specificity for the Z isomer in Tortricidae. *Insect Biochem.* 19:639–44.

Kaissling, K.-E. 2001. Olfactory perireceptor and receptor events in moths: A kinetic model. *Chem. Senses* 26:125–50.

Kaissling, K.-E., and E. Priesner. 1970. Die Riechschwelle des Seidenspinners. *Naturwissenschaften* 57:23–28.

Klun, J.A. 1975. Insect sex pheromones: Intraspecific pheromonal variability of *Ostrinia nubilalis* in North America and Europe. *Environ. Entomol.* 4:891–94.

Knipple, D.C., and W.L. Roelofs. 2003. Molecular biological investigations of pheromone desaturases. In: *Insect pheromone biochemistry and molecular biology*, ed. G.J. Blomquist and R.C. Vogt, 81–106. London: Elsevier Academic Press.

Knipple, D.C., C.-L. Rosenfield, S. J. Miller, et al. 1998. Cloning and functional expression of a cDNA encoding a pheromone gland-specific acyl-CoA Δ^{11}-desaturase of the cabbage looper moth, *Trichoplusia ni. Proc. Natl. Acad. Sci. U.S.A.* 95:15287–92.

Knipple, D.C., C.-L. Rosenfield, R. Nielsen, K.M. You, and S.E. Jeong. 2002. Evolution of the integral membrane desaturase gene family in moths and flies. *Genetics* 162:1737–52.

Kondrashov, A.S., and F.A. Kondrashov. 1999. Interactions among quantitative traits in the course of sympatric speciation. *Nature* 400:351–54.

Krieger, J., and H. Breer. 1999. Olfactory reception in invertebrates. *Science* 286:720–23.

Krieger, J., H. Gaenssle, K. Raming, and H. Breer. 1993. Odorant binding proteins of *Heliothis virescens. Insect Biochem. Mol. Biol.* 23:449–56.

Krieger, J., E. Große-Wilde, T. Gohl, and H. Breer. 2005. Candidate pheromone receptors of the silkmoth *Bombyx mori. Eur. J. Neurosci.* 21:2167–76.

Krieger, J., E. Große-Wilde, T. Gohl, Y.M.E. Dewer, K. Raming, and H. Breer. 2004. Genes encoding candidate pheromone receptors in a moth (*Heliothis virescens*). *Proc. Natl. Acad. Sci. U.S.A.* 101:11845–50.

Krieger, J., K. Raming, Y.M.E. Dewer, S. Bette, S. Conzelmann, and H. Breer. 2002. A divergent gene family encoding candidate olfactory receptors of the moth *Heliothis virescens. Eur. J. Neurosci.* 16:619–28.

Kurtovic, A., A. Widmer, and B.J. Dickson. 2007. A single class of olfactory neurons mediates behavioural responses to a *Drosophila* sex pheromone. *Nature* 446:542–46.

Laue, M., and R.A. Steinbrecht. 1997. Topochemistry of moth olfactory sensilla. *Int. J. Insect Morphol. Embryol.* 26:217–28.

Leal, W.S. 2003. Proteins that make sense. In *Insect pheromone biochemistry and molecular biology*, ed. G.J. Blomquist and R.G. Vogt, 447–76. London: Elsevier Academic Press.

Leal, W.S. 2005. Pheromone reception. *Topics Curr. Chem.* 240:1–36.

Leal, W.S., A.M. Chen, Y. Ishida, et al. 2005. Kinetics and molecular properties of pheromone binding and release. *Proc. Natl. Acad. Sci. U.S.A.* 102:5386–91.

Linn, C. Jr., K. Poole, A. Zhang, and W. Roelofs. 1999. Pheromone-blend discrimination by European corn borer moths with inter-race and inter-sex antennal transplants. *J. Comp. Physiol. A* 184:273–78.

Linn, C. E. Jr., and W.L. Roelofs. 1995. Pheromone communication in moths and its role in the speciation process. In *Speciation and the recognition concept: Theory and application*, ed. D.M. Lambert and H.G. Spencer, 263–300. Baltimore: Johns Hopkins Univ. Press.

Linn, C.E., M.S. Young, M. Gendle, et al. 1997. Sex pheromone blend discrimination in two races and hybrids of the European corn borer moth, *Ostrinia nubilalis. Physiol. Entomol.* 22:212–23.

Liu, W., P.W.K. Ma, P. Marsella-Herrick, C.-L. Rosenfield, D.C. Knipple, and W.L. Roelofs. 1999. Cloning and functional expression of a cDNA encoding a metabolic acyl-CoA Δ9-desaturase of the cabbage looper moth, *Trichoplusia ni. Insect Biochem. Mol. Biol.* 29:435–43.

Löfstedt, C. 1990. Population variation and genetic control of pheromone communication systems in moths. *Entomol. Exp. Appl.* 54:199–218.

Löfstedt, C. 1993. Moth pheromone genetics and evolution. *Philos. Trans. R. Soc. Lond. B Biol. Sci.* 340:161–77.

Löfstedt, C., B.S. Hansson, W.L. Roelofs, and B.O. Bengtsson. 1989. No linkage between genes controlling female pheromone production and male pheromone response in the European corn borer, *Ostrinia nubilalis* Hübner (Lepidoptera: Pyralidae). *Genetics* 123:553–56.

Luxová, A., and A. Svatoš. 2006. Substrate specificity of membrane-bound alcohol oxidase from the tobacco hornworm moth (*Manduca sexta*) female pheromone glands. *J. Mol. Catal. B Enzymatic* 38:37–42.

Maïbèche-Coisne, M., E. Jacquin-Joly, M.C. François, and P. Nagnan-Le Meillour. 2002. cDNA cloning of biotransformation enzymes belonging to the cytochrome P450 family in the antennae of the noctuid moth *Mamestra brassicae*. *Insect Mol. Biol.* 11:273–81.

Maida, R., G. Ziegelberger, and K.E. Kaissling. 2003. Ligand binding to six recombinant pheromone-binding proteins of *Antheraea polyphemus* and *Antheraea pernyi*. *J. Comp. Physiol. B* 173:565–73.

Matoušková, P., I. Pichová, and A. Svatoš. 2007. Functional characterization of a desaturase from the tobacco hornworm moth (*Manduca sexta*) with bifunctional Z11- and 10,12-desaturase activity. *Insect Biochem. Mol. Biol.* 37:601–10.

Matsumoto, S., J.J. Hull, A. Ohnishi, K. Moto, and A. Fonagy. 2007. Molecular mechanisms underlying sex pheromone production in the silkmoth, *Bombyx mori*: Characterization of the molecular components involved in bombykol biosynthesis. *J. Insect Physiol.* 53:752–59.

Merlin, C., M.C. François, F. Bozzolan, J. Pelletier, E. Jacquin-Joly, and M. Maïbèche-Coisne. 2005. A new aldehyde oxidase selectively expressed in chemosensory organs of insects. *Biochem. Biophys. Res. Commun.* 332:4–10.

Metz, J.G., M.R. Pollard, L. Anderson, T.R. Hayes, and M.W. Lassner. 2000. Purification of a jojoba embryo fatty acyl-Coenzyme A reductase and expression of its cDNA in high erucic acid rapeseed. *Plant Physiol.* 122:635–44.

Mombaerts, P. 1999. Seven-transmembrane proteins as odorant and chemosensory receptors. *Science* 286:707–11.

Mombaerts, P. 2004. Odorant receptor gene choice in olfactory sensory neurons: The one receptor–one neuron hypothesis revisited. *Curr. Opin. Neurobiol.* 14:31–36.

Morse, D., and E. Meighen. 1986. Pheromone biosynthesis and the role of functional groups in pheromone specificity. *J. Chem. Ecol.*12:335–51.

Morse, D., and E. Meighen. 1987. Pheromone biosynthesis: Enzymatic studies in Lepidoptera. In *Pheromone biochemistry*, ed. G.D. Prestwich and G.J. Blomquist, 212–15. New York: Academic Press.

Moto, K., M.G. Suzuki, J.J. Hull, et al. 2004. Involvement of a bifunctional fatty-acyl desaturase in the biosynthesis of the silkmoth, *Bombyx mori*, sex pheromone. *Proc. Natl. Acad. Sci. U.S.A.* 101:8631–36.

Moto, K., T. Yoshiga, M. Yamamoto, et al. 2003. Pheromone gland-specific fatty-acyl reductase of the silkmoth, *Bombyx mori*. *Proc. Natl. Acad. Sci. U.S.A.* 100:9156–61.

Mustaparta, H. 1996. Central mechanisms of pheromone information processing. *Chem. Senses* 21:269–75.

Nakagawa, T., T. Sakurai, T. Nishioka, and K. Touhara. 2005. Insect sex-pheromone signals mediated by specific combinations of olfactory receptors. *Science* 307:1638–42.

Ohnishi, A., J.J. Hull, and S. Matsumoto. 2006. Targeted disruption of genes in the *Bombyx mori* sex pheromone biosynthetic pathway. *Proc. Natl. Acad. Sci. U.S.A.* 103:4398–4403.

Ozawa, R., and S. Matsumoto. 1996. Intracellular signal transduction of PBAN action in the silkworm, *Bombyx mori*: Involvement of acyl-CoA reductase. *Insect Biochem. Mol. Biol.* 26:259–65.

Pape, M.E., F. Lopez-Casillas, and K.-H. Kim. 1988. Physiological regulation of Acetyl-CoA carboxylase gene expression: Effects of diet, diabetes, and lactation on acetyl-CoA carboxylase mRNA. *Arch. Biochem. Biophys.* 267:104–09.

Pelletier, J., F. Bozzolan, M. Solvar, M.-C. François, E. Jacquin-Joly, and M. Maïbèche-Coisne. 2007. Identification of candidate aldehyde oxidases from the silkworm *Bombyx mori* potentially involved in antennal pheromone degradation. *Gene* 404:31–40.

Pelosi, P. 1996. Perireceptor events in olfaction. *J. Neurobiol.* 30:3–19.

Phelan, P.L. 1997. Genetic and phylogenetics in the evolution of sex pheromones. In *Insect pheromone research: New directions*, ed. R.T. Cardé and A.K. Minks, 563–79. New York: Chapman and Hall.

Prestwich, G.D., R.G. Vogt, and L.M. Riddiford. 1986. Binding and hydrolysis of radiolabeled pheromone and several analogs by male-specific antennal proteins of the moth *Antheraea polyphemus*. *J. Chem. Ecol.* 12:323–33.

Rafaeli, A. 2005. Mechanisms involved in the control of pheromone production in female moths: Recent developments. *Entomol. Exp. Appl.* 115:7–15.

Remington, D.L., R.W. Whetten, B.H. Liu, and D.M. O'Malley. 1999. Construction of genetic map with nearly complete genome coverage in *Pinus taeda*. *Theor. Appl. Genet.* 98:1279–92.

Rivière, S., A. Lartigue, B. Quennedey, et al. 2003. A pheromone-binding protein from the cockroach *Leucophaea maderae*: Cloning, expression and pheromone binding. *Biochem. J.* 371:573–79.

Robertson, H.M., R. Martos, C.R. Sears, E.Z. Todres, K.K. Walden, and J.B. Nardi. 1999. Diversity of odourant binding proteins revealed by an expressed sequence tag project on male *Manduca sexta* moth antennae. *Insect Mol. Biol.* 8:501–18.

Rodriguez, F., D.L. Hallahan, J.A. Pickett, and F. Camps. 1992. Characterization of the delta-11 palmitoyl-CoA-desaturase from *Spodoptera littoralis* (Lepidoptera, Noctuidae). *Insect Biochem. Mol. Biol.* 22:143–48.

Roelofs, W.L., and R.T. Cardé. 1977. Responses of Lepidoptera to synthetic sex-pheromone chemicals and their analogs. *Annu. Rev. Entomol.* 22:377–405.

Roelofs, W.L., T. Glover, X.H. Tang, et al. 1987. Sex-pheromone production and perception in European corn-borer moths is determined by both autosomal and sex-linked genes. *Proc. Natl. Acad. Sci. U.S.A.* 84:7585–89.

Roelofs, W.L., and R.A. Jurenka. 1996. Biosynthetic enzymes regulating ratios of sex pheromone components in female redbanded leafroller moths. *Bioorg. Med. Chem.* 4:461–66.

Roelofs, W.L., and A.P. Rooney. 2003. Molecular genetics and evolution of pheromone biosynthesis in Lepidoptera. *Proc. Natl. Acad. Sci. U.S.A.* 100:9179–84.

Roelofs, W.L., and W.A. Wolf. 1988. Pheromone biosynthesis in Lepidoptera. *J. Chem. Ecol.* 14:2019–31.

Rogers, M.E., J. Krieger, and R.G. Vogt. 2001. Antennal SNMPs (sensory neuron membrane proteins) of Lepidoptera define a unique family of invertebrate CD36-like proteins. *J. Neurobiol.* 49:47–61.

Rogers, M.E., M. Sun, M.R. Lerner, and R.G. Vogt. 1997. Snmp-1, a novel membrane protein of olfactory neurons of the silk moth *Antheraea polyphemus* with homology to the CD36 family of membrane proteins. *J. Biol. Chem.* 272:14792–99.

Rosenfield, C.-L., K.M. You, P. Marsella-Herrick, W.L. Roelofs, and D.C. Knipple. 2001. Structural and functional conservation and divergence among acyl-CoA desaturases of two noctuid species, the corn earworm, *Helicoverpa zea*, and the cabbage looper, *Trichoplusia ni*. *Insect Biochem. Mol. Biol.* 31:949–64.

Rützler, M., and L.J. Zwiebel. 2005. Molecular biology of insect olfaction: Recent progress and conceptual models. *J. Comp. Physiol. A* 191:777–90.

Rybczynski, R., J. Reagan, and M.R. Lerner. 1989. A pheromone-degrading aldehyde oxidase in the antennae of the moth *Manduca sexta*. *J. Neurosci.* 9:1341–53.

Rybczynski, R., R.G. Vogt, and M.R. Lerner. 1990. Antennal-specific pheromone-degrading aldehyde oxidases from the moths *Antheraea polyphemus* and *Bombyx mori*. *J. Biol. Chem.* 265:19712–15.

Sakurai, T., T. Nakagawa, H. Mitsuno, et al. 2004. Identification and functional characterization of a sex pheromone receptor in the silkmoth *Bombyx mori*. *Proc. Natl. Acad. Sci. U.S.A.* 101:16653–58.

Sambandan, D., A.Yamamoto, J.J. Fanara, T.F.C. Mackay, R.R.H. Anholt. 2006. Dynamic genetic interactions determine odor-guided behavior in *Drosophila melanogaster*. *Genetics* 174:1349–63.

Sandler, B.H., L. Nikonova, W.S. Leal, and J. Clardy. 2000. Sexual attraction in the silkworm moth: Structure of the pheromone-binding-protein-bombykol complex. *Chem. Biol.* 7:143–51.

Sato, K., M. Pellegrino, T. Nakagawa, T. Nakagawa, L.B. Vosshall, and K. Touhara. 2008. Insect olfactory receptors are heteromeric ligand-gated ion channels. *Nature* 452:1002–06.

Sheck, A.L., A.T. Groot, C.M. Ward, et al. 2006. Genetics of sex pheromone blend differences between *Heliothis virescens* and *Heliothis subflexa*: A chromosome mapping approach. *J. Evol. Biol.* 19:600–17.

Steinbrecht, R.A., M. Ozaki, and G. Ziegelberger. 1992. Immunocytochemical localization of pheromone-binding protein in moth antennae. *Cell Tissue Res.* 270:287–302.

Stengl, M., F. Zufall, H. Hatt, and J.G. Hildebrand. 1992. Olfactory receptor neurons from antennae of developing male *Manduca sexta* respond to components of the species-specific sex-pheromone *in vitro*. *J. Neurosci.* 12:2523–31.

Syed, Z., Y. Ishida, K. Taylor, D.A. Kimbrell, and W.S. Leal. 2006. Pheromone reception in fruit flies expressing a moth's odorant receptor. *Proc. Natl. Acad. Sci. U.S.A.* 103:16538–43.

Takanashi, T., Y.P. Huang, K.R. Takahasi, S. Hoshizaki, S. Tatsuki, and Y. Ishikaw. 2005. Genetic analysis and population survey of sex pheromone variation in the adzuki bean borer moth, *Ostrinia scapulalis*. *Biol. J. Linn. Soc.* 84:143–60.

Teal, P.E.A., and J.H. Tumlinson. 1987. The role of alcohols in pheromone biosynthesis by two noctuid moths that use acetate pheromone components. *Arch. Insect Biochem. Physiol.* 4:261–69.

Teal, P.E.A., and J.H. Tumlinson. 1997. Effects of interspecific hybridization between *Heliothis virescens* and *Heliothis subflexa* on the sex pheromone communication system. In *Insect pheromone research: New directions*, ed. R.T. Cardé and A.K. Minks, 535–47. New York: Chapman and Hall.

Tiku, P.E., A.Y. Gracey, A.I. Macartney, R.J. Beyton, and A.R. Crossins. 1996. Cold-induced expression of δ9-desaturase in carp by transcriptional and posttranslational mechanisms. *Science* 271:815–18.

Tsfadia, O., A. Azrielli, L. Falach, A. Zada, W. Roelofs, and A. Rafaeli. 2008. Pheromone biosynthetic pathways: PBAN-regulated rate-limiting steps and differential expression of desaturase genes in moth species. *Insect Biochem. Mol. Biol.* 38:552–67.

Tumlinson, J.H., and P.E.A. Teal. 1987. Relationship of structure and function to biochemistry in insect pheromone systems. In *Pheromone biochemistry*, ed. G.D. Prestwich and G.J. Blomquist, 3–26. New York: Academic Press.

van den Berg, M.J., and G. Zielgelberger. 1991. On the function of the pheromone binding protein in the olfactory hairs of *Antheraea polyphemus*. *J. Insect Physiol.* 37:79–85.

Vickers, N.J. 2006. Inheritance of olfactory preferences I. Pheromone-mediated behavioral responses of *Heliothis subflexa* x *Heliothis virescens* hybrid male moths. *Brain Behav. Evol.* 68:63–74.

Vickers, N.J., and T.A. Christensen. 2003. Functional divergence of spatially conserved olfactory glomeruli in two related moth species. *Chem. Senses* 28:325–38.

Vickers, N.J., K. Poole, and C.E. Linn. 2005. Plasticity in central olfactory processing and pheromone blend discrimination following interspecies antennal imaginal disc transplantation. *J. Comp. Neurol.* 491:141–56.

Vogt, R.G. 2003. Biochemical diversity of odor detection: OBPs, ODEs and SNMPs. In *Insect pheromone biochemistry and molecular biology*, ed. G.J. Blomquist and R.G. Vogt, 391–445. London: Elsevier Academic Press.

Vogt, R.G. 2005. Molecular basis of pheromone detection in insects. In *Comprehensive insect physiology, biochemistry, pharmacology and molecular biology, Vol. 3, Endocrinology*, ed. L.I. Gilbert, K. Iatrou, and S. Gill, 753–804. London: Elsevier.

Vogt, R.G., and L.M. Riddiford. 1981. Pheromone binding and inactivation by moth antennae. *Nature* 293:161–63.

Vogt, R.G., L.M. Riddiford, and G.D. Prestwich. 1985. Kinetic properties of a pheromone degrading enzyme: The sensillar esterase of *Antheraea polyphemus*. *Proc. Natl. Acad. Sci. U.S.A.* 82:8827–31.

Vosshall, L.B., H. Amrein, P.S. Morozov, A. Rzhetsky, and R. Axel. 1999. A spatial map of olfactory receptor expression in the *Drosophila* antennae. *Cell* 96:725–36.

Wade, M.J., and C.J. Goodnight. 1998. Perspective: The theories of Fisher and Wright in the context of metapopulations: When nature does many small experiments. *Evolution* 52:1537–53.

Wakil, S.J., J.K. Stoops, and V.C. Joshi. 1983. Fatty acid synthesis and its regulation. *Annu. Rev. Biochem.* 52:537–79.

Wanner, K.V., A.R. Anderson, S.C. Trowell, D.A. Theilmann, H.M. Robertson, and R.D. Newcomb. 2007. Female-biased expression of odourant receptor genes in the adult antennae of the silkworm, *Bombyx mori*. *Insect Mol. Biol.* 16:107–19.

Weber, K.E., R.J. Greenspan, D.R. Chicoine, K. Fiorentino, M.H. Thomas, and T.L. Knight. 2008. Microarray analysis of replicate populations selected against a wing-shape correlation in *Drosophila melanogaster*. *Genetics* 178:1093–1108.

Whitlock, M.C., and P.C. Phillips. 2000. The exquisite corpse: A shifting view of the shifting balance. *Trends Ecol. Evol.* 15:347–48.

Wicher, D., R. Schäfer, R. Bauernfeind, et al. 2008. *Drosophila* odorant receptors are both ligand-gated and cyclic-nucleotide-activated cation channels. *Nature* 452:1007–11.

Willett, C.S. 2000. Evidence for directional selection acting on pheromone-binding proteins in the genus *Choristoneura*. *Mol. Biol. Evol.* 17:553–62.

Wittkopp, P.J., B.K. Haerum, and A.G. Clark. 2004. Evolutionary changes in *cis* and *trans* regulation. *Nature* 430:85–88.

Wittkopp, P.J., B.K. Haerum, and A.G. Clark. 2008. Independent effects of cis- and trans-regulatory variation on gene expression in *Drosophila melanogaster*. *Genetics* 178:1831–35.

Witzgall, P., T. Lindblom, M. Bengtsson, and M. Tóth. 2004. The Pherolist. http://www-pherolist.slu.se (accessed September 12, 2008).

Wu, W-Q., J.-W. Zhu, J. Millar, and C. Löfstedt. 1998. A comparative study of sex pheromone biosynthesis in two strains of the turnip moth, *Agrotis segetum*, producing ratios of sex pheromone components. *Insect Biochem. Mol. Biol.* 28:895–900.

Xiu, W.-M., and S.-L. Dong. 2007. Molecular characterization of two pheromone binding proteins and quantitative analysis of their expression in the beet armyworm, *Spodoptera exigua* Hübner. *J. Chem. Ecol.* 33:947–61.

Xiu, W.-M., Y.-Z. Zhou, and S.-L. Dong. 2008. Molecular characterization and expression pattern of two pheromone-binding proteins from *Spodoptera litura* (Fabricius). *J. Chem. Ecol.* 34:487–98.

Xue, B., A.P. Rooney, M. Kajikawa, N. Okada, and W.L. Roelofs. 2007. Novel sex pheromone desaturases in the genomes of corn borers generated through gene duplication and retroposon fusion. *Proc. Natl. Acad. Sci. U.S.A.* 104:4467–72.

Zhu, J.W., B.B. Chastain, B.G. Spohn, and K.F. Haynes. 1997. Assortative mating in two pheromone strains of the cabbage looper moth, *Trichoplusia ni*. *J. Insect Behav.* 10:805–17.

Zhu, J.W., C.H. Zhao, F. Lu, M. Bengtsson, and C. Löfstedt. 1996. Reductase specificity and the ratio regulation of E/Z Isomers in pheromone biosynthesis of the European corn borer, *Ostrinia nubilalis* (Lepidoptera: Pyralidae). *Insect Biochem. Molec. Biol.* 26:171–76.

11 Genetics of Host Range in Lepidoptera

Sara J. Oppenheim and Keith R. Hopper

CONTENTS

INTRODUCTION

The genetic basis of complex, ecologically relevant traits is not well known for any organism, despite enormous interest in understanding how such traits evolve. The question is particularly compelling where closely related species have diverged radically in their adaptation to the environment. Differences in host plant use among moths and butterflies often provide such cases: Although close relatives tend to use similar hosts, there are many examples of congeneric species that differ widely in host range. In several systems, work is under way to identify the genetic changes that underlie shifts in host use. While such changes may or may not contribute to the well-documented speciosity of phytophagous insects, understanding the genetic architecture of host range is fundamental to understanding the evolution of Lepidoptera. Improved understanding of the genetics of host range is crucial for applied reasons as well: Both the safe practice of biological control and the breeding of plants with persistent resistance to pests demand greater understanding of the genetics of host range. Understanding the evolution of host range in Lepidoptera will require knowledge of its genetic architecture, that is, which genes are involved, how these genes interact, and how much change in each gene is needed for a change in host range.

Host range in phytophagous insects involves not one, but many, traits. To use a plant, an insect must find and lay eggs on it and feed and develop to adulthood on it. Thus host range is multifactorial, and the competing demands of each phase of host use must be integrated. (Although the term "host range" is used in several ways in the literature, we mean the list of host plant species on which an herbivore species will oviposit and on which its larvae have some chance of completing development.) Host range is dynamic because use of a given host depends on both external factors (e.g., local host availability, competition) and internal factors (e.g., female egg load, age, previous experience). As a result, a clear understanding of the genetic differences responsible for differences in host range is difficult to obtain because not only is host range a moving target but many genes may be involved in the numerous processes that determine host range.

The plant species on which larvae may feed are restricted by where their mothers lay eggs. Most neonate larvae can travel only a few meters in search of food before they starve, although ballooning neonates can travel greater distances. Even late instars have an ambit measured in meters to tens of meters, compared with the hundreds to tens of thousands of meters that lepidopteran adults can travel, either under their own power or carried by the wind. Thus adult females have a much greater opportunity than their progeny to choose suitable host plant species. Whether larvae themselves will have the opportunity to become adults and search for host plants for their progeny depends on larval feeding and performance on the host plant where they find themselves. As a result, host range involves genes underlying adult chemoreception and interneuronal processing, which lead to oviposition on one hand; and larval chemoreception, digestion, and nutritional metabolism, which determine larval feeding, growth, and survival on the other.

Large numbers of chromosomes and sex-limited recombination make Lepidoptera attractive models for investigations into the genetic basis of host range, as well as other complex traits. In Lepidoptera, within-chromosome recombination is restricted to males, so maternal-origin chromosomes are inherited intact (Suomalainen 1969 and references therein; Marec 1996; see Chapter 3 for details on absence of recombination in females). Genetic linkage mapping is simplified because maternal-origin "linkage groups" are actually chromosomes, and any putative recombination can be attributed to scoring error (in systems where recombination occurs, disentangling scoring error from true recombination can be a major challenge). The majority of Lepidoptera have between 28 and 32 small chromosomes of relatively uniform size (Suomalainen 1969; Robinson 1971), although chromosome numbers across Lepidoptera range from $n = 5$ to $n = 223$ (White 1973; De Prins and Saitoh 2003; see Chapter 3 for characteristics of lepidopteran chromosomes). Whereas the presence of numerous small chromosomes makes them difficult to distinguish cytologically, it also means that each chromosome comprises a relatively small fraction of the genome. If the distribution of chromosome sizes and the total number of genes in most Lepidoptera resemble estimates for the silkworm *Bombyx mori* (Xia et al. 2004; Yoshido et al. 2005), each chromosome will contain 2–5 percent of the genome or about 300 to 1,000 genes. This means that resolution to chromosome in Lepidoptera is at least as precise as in many systems having within-chromosome recombination. Finer-scale resolution can be achieved using a biphasic approach whereby one maps first to chromosome with female-informative markers and then within chromosome with male-informative markers (Heckel et al. 1999). This allows one to concentrate on chromosome(s) carrying genes of interest, a major advantage for fine-scale mapping and map-based (positional) cloning.

Over five thousand papers and books had been published on plant-insect interactions by 2002 (Scriber 2002), and the numbers continue to increase rapidly. Current knowledge on the evolutionary biology of herbivore-plant interactions has been recently and thoroughly reviewed, including phylogeny, biochemistry, behavior, and evolution (Tilmon 2008). Despite the volume of interesting research and the advantages of lepidopteran genetics discussed above, we do not know the detailed genetic architecture of host range for any species of moth or butterfly. In this chapter, we review what is known about the genetics of host range in Lepidoptera, discuss the biology of host range and its implications for genetic architecture, and suggest promising lines of research. Although we cover most thoroughly the system on which we work and thus know best—the generalist *Heliothis virescens* and the closely related specialist *Heliothis subflexa* (Noctuidae)—we treat several other systems in-depth as well. Many themes and questions that permeate the literature will become apparent in this review: adult oviposition preference versus larval performance; trade-offs in performance among host species; the pace of host range evolution; many versus few genes; genes on autosomes versus sex chromosomes; differences in the basis of interspecific versus intraspecific variation; expansion or contraction in host range versus shifts in host range; directionality of evolution from generalist to specialist or vice versa; and the role of host shifts in speciation.

THE GENETICS OF HOST SPECIFICITY

In this section, we summarize by genus the current evidence concerning the genetics of host range. There are various types of evidence: (1) phylogenetic patterns; (2) population and strain comparisons, especially in common-garden experiments; (3) responses to artificial and natural selection; (4) crosses between host races and species; (5) resemblance of relatives (parent-offspring regression, full-sib families, half-sib families); (6) marker-based mapping of quantitative trait loci; (7) differences in sequence and expression of proteins involved in chemoreception, detoxification, and assimilation of plant chemicals. Two additional types of evidence will soon become available: map-based (positional) cloning of genes, and silencing of candidate genes.

Heliothis

The *Heliothis virescens* complex comprises at least thirteen closely related species in North and South America that vary in host specificity and geographic range (Mitter, Poole, and Matthews 1993). Among the members of this complex, two are of particular interest: *H. virescens* and *H. subflexa*. *Heliothis virescens* is a major agricultural pest and has been the subject of much research (over thirteen hundred papers in refereed journals alone). *Heliothis subflexa* is not a pest but is closely related to *H. virescens*, with which it has 99 percent sequence similarity in the genes for which comparisons have been made (Cho et al. 1995; Fang et al. 1997). Their geographical ranges overlap broadly (Mitter, Poole, and Matthews 1993), and the two species are morphologically so similar that *H. subflexa* was only conclusively identified as a separate species in 1941 (McElvare 1941). In the laboratory, *H. virescens* and *H. subflexa* can be hybridized, producing fertile F_1 females and sterile F_1 males (male fertility is restored after several backcross generations; Karpenko and Proshold 1977). These two species are thought to have evolved quite recently from a shared, generalist ancestor (Mitter, Poole, and Matthews 1993; Poole, Mitter, and Huettel 1993; Fang et al. 1997). Despite the similarity between them, they differ greatly in host range. *Heliothis virescens* has a very broad host range, feeding on at least 37 species in 14 plant families, including *Nicotiana tabacum* (tobacco), *Gossypium hirsutum* (cotton), *Glycine max* (soybean) and other crops (Sheck and Gould 1993), whereas *H. subflexa* is narrowly specialized on the genus *Physalis* (e.g., ground cherry *P. pruinosa*; Laster, Pair, and Martin 1982). Interestingly, *H. virescens* is not known to feed on *Physalis* species in the field. Thus, the *H. virescens/H. subflexa* pair is an excellent model for studying the evolution of genetic differences responsible for divergence in host range, because genetic differentiation is likely to be concentrated in loci involved in host use (Sheck and Gould 1993) and mate recognition (Groot et al. 2004).

Several studies have examined the genetic basis of host range in *H. virescens* and *H. subflexa*. Sheck and Gould (1993) analyzed larval performance on four plant species by exposing *H. virescens*, *H. subflexa*, their F_1 hybrids, and a backcross to *H. subflexa* to cotton, soybean, tobacco (hosts of *H. virescens*), and *Physalis pubescens* (a host of *H. subflexa*). Each species survived and gained weight well on its own host(s) and poorly on nonhosts. Hybrid F_1 larvae survived well on all host plants, but had intermediate weight gain on all four plant species. In the backcross to *H. subflexa*, larval survival was lower on cotton, soybean, and tobacco than on *P. pubescens*, and larval weight gain was lower on cotton and tobacco than on soybean and *P. pubescens*. Analysis of the results from the four types of cross (within each species, F_1, and backcross), indicated that genes from *H. virescens* were partially dominant for larval survival and weight gain on cotton and tobacco, but additive for both traits on soybean. Genes from *H. subflexa* were overdominant for survival and dominant for weight gain on *P. pubescens*, so that backcross larvae survived better and gained weight as well as *H. subflexa*. However, epistatic or gene-environment interactions also appeared to be involved because additive and dominance effects alone did not explain the results. In a subsequent experiment, repeated backcrosses to *H. subflexa* with selection for larval performance on soybean were used to examine the genetic architecture for use of several plant species (Sheck and Gould 1996). After several generations of selection, larval preference and performance were tested on cotton, soybean, tobacco, and *P. pubescens*. Although performance on soybean had improved, no correlated changes occurred in performance on cotton, tobacco, or *P. pubescens*, indicating performance on these plants had an independent genetic basis. Interestingly, larval preference for soybean, though not selected on, had also increased, implying a common genetic basis for larval preference for and performance on soybean. Larval performance on *P. pubescens* did not differ from that of *H. subflexa*, showing that introgession of genes for using soybean into the *H. subflexa* background did not involve tradeoffs in ability to use *P. pubescens*.

Sheck and Gould (1995) also examined oviposition behavior of *H. virescens*, *H. subflexa*, and their reciprocal F_1 hybrids. Adult females were exposed to cotton, soybean, tobacco, and *Physalis angulata* (a favored host of *H. subflexa*) in laboratory assays. *Heliothis virescens* females oviposited mostly on tobacco and rarely on the other plant species; *H. subflexa* females oviposited mostly on *P. angulata*, but also oviposited occasionally on nonhosts; F_1 females from crosses in both directions oviposited preferentially on tobacco, indicating dominance of genes from *H. virescens*. Inheritance appeared to be autosomal with no indication of sex-linkage for genes affecting oviposition preference, larval performance, or larval preference (Sheck and Gould 1993, 1995).

In recent experiments with interspecific hybrids (unpublished collaborations between the authors and F. Gould), we have further explored the genetic basis of host range in *H. virescens* and *H. subflexa*. In laboratory experiments, we introgressed genes from each species into the background of the other species by backcrossing and assaying their backcross progeny on either cotton or *P. angulata*. Using amplified fragment length polymorphism (AFLP) markers and polymorphisms in published gene sequences, we made linkage maps covering the 31 chromosomes of *H. virescens* and *H. subflexa* (for chromosome numbers, see Chen and Graves 1970; Sheck et al. 2006) and used quantitative trait locus (QTL) analysis to determine the genetic architecture of variation in larval performance. In the experiments on cotton, we did five generations of backcrosses. For generations one to four, hybrid females were mated with *H. subflexa* males; these crosses generated female-informative markers that allowed us to identify introgressed chromosomes contributing to phenotypic variation. In generation five, hybrid males were backcrossed to *H. subflexa* females, giving us male-informative markers for within-chromosome mapping. In a preliminary analysis of first-generation backcross larvae, stepwise regression of larval feeding versus the presence/absence of *H. virescens*–origin chromosomes identified six chromosomes that together explained 39 percent of the variation in larval feeding on cotton (unpublished data). Four *H. virescens* chromosomes increased the amount of cotton consumed, and larvae with all four chromosomes had phenotypes indistinguishable from *H. virescens*. These chromosomes had additive effects with no interaction among chromosomes. Two of the introgressed *H. virescens* chromosomes had an unexpected effect: Their presence reduced, rather than increased, the amount of cotton eaten by backcross larvae. Perhaps these chromosomes carry genes for feeding on host plants other than cotton that interact epistatically with those for feeding on cotton. Because backcross larvae were either homozygous for *H. subflexa* alleles or heterozygous for *H. subflexa* and *H. virescens* alleles, the introgressed genes from *H. virescens* were at least additive and perhaps dominant. One of the sex chromosomes was among those that increased feeding on cotton, although its impact was no greater than that of autosomes. In backcrosses (BC) of hybrid females (W_vZ_s) to *H. subflexa* males (Z_sZ_s), all female progeny had their Z chromosome from *H. subflexa* and their W chromosome from *H. virescens*, but all male progeny had both sex chromosomes from *H. subflexa*. This means either that there were genes on the W_v chromosome that increased feeding on cotton, which seems unlikely given the paucity of expressed genes on the W chromosome (see Chapters 3 and 4 for candidate W-linked genes and molecular composition of the W chromosome), or there was an overall difference in feeding between the sexes. Although we have not yet mapped QTL in BC_5

larvae, the frequency of BC_5 larvae with *H. virescens*–like phenotypes indicates that a few genes explain much of the variance in feeding on cotton. Recent theory and evidence suggest that finding few QTL that explain most of the variation in quantitative traits is not surprising (Orr 2001, 2005; Remington, Ungerer, and Purugganan 2001).

To investigate the genetics of larval performance on *P. angulata*, we introgressed *H. subflexa* genes into the *H. virescens* background by backcrossing hybrids to *H. virescens*. When fed on the fruits of *P. angulata*, the assimilation efficiency (larval weight gain per gram of fruit consumed) of *H. subflexa* is thirty times greater than that of *H. virescens*, although *H. virescens* larvae feed readily on *P. angulata*. The phenotypes of backcross larvae ranged from *H. subflexa*–like to *H. virescens*–like. Five introgressed chromosomes affected the performance of backcross larvae on *P. angulata*, together explaining 45 percent of the variation in assimilation efficiency. Similar to the cotton results, three chromosomes increased assimilation efficiency, while two decreased assimilation efficiency. The presence of the three chromosomes that increased assimilation efficiency gave phenotypes equal to those of *H. subflexa* (unpublished data).

Much effort has been directed toward understanding how *H. virescens* detects and selects host plants and mates. Twenty-one genes coding for olfactory receptor proteins, each from a different group of olfactory neurons, have been sequenced in *H. virescens* (Krieger et al. 2002, 2004). Antennal lobe structure and patterns of innervation suggest that there at least thirty to sixty types of olfactory neurons and thus olfactory receptor proteins (Mustaparta 2002; Rostelien et al. 2005). Furthermore, sixteen types of olfactory neurons have been identified based on their electrophysiological response to plant odors, and all are finely tuned to specific plant odors (Rostelien et al. 2005). These receptor genes may provide candidates for explaining differences in host specificity between these two species.

HELICOVERPA

Like *Heliothis*, the genus *Helicoverpa* (see Chapter 12 for further information on this genus) includes species with broad host ranges such as *H. armigera*, recorded from over 150 host plant species in many families (Zalucki et al. 1994), and *H. zea*, recorded from at least 34 species of plants in 11 families (Sudbrink and Grant 1995), as well as species with narrow host ranges like *H. assulta*, recorded from only certain species in the Solanaceae (Fitt 1989). Laboratory experiments have been used to examine population variation and heritability of oviposition preference and larval performance in these species. Populations of *H. armigera* from various regions of Australia did not differ in ranking of plant species (maize, sorghum, tobacco, cotton, cowpea, lucerne) for oviposition, but females within populations did show heritable variation (parent-offspring regression) in ranking of these plants (Jallow and Zalucki 1996). Besides showing genetic variation in oviposition among plant species, female *H. armigera* also appear to learn: Females oviposit preferentially on plant species previously experienced (Cunningham et al. 1998). In another laboratory study on Australian *H. armigera*, a full-sib parent-offspring regression showed high heritability (60 percent) for oviposition on *Sonchus oleraceus* (Asteraceae), a preferred host plant from the indigenous geographical range of *H. armigera* (Gu and Walter 1999), versus *Gossypium hirsutum* (Malvaceae), a less-preferred host plant (Gu, Cao, and Walter 2001). Although *H. armigera* larvae survived better and gained more weight on *S. oleraceus* than on *G. hirsutum*, larval performance was not genetically correlated with oviposition preference (Gu, Cao, and Walter 2001). In a full-sib/half-sib experiment on Australian *H. armigera*, larvae gained more weight (73 percent for neonates and 23 percent for third instars) but did not differ in survival on resistant versus susceptible *Cicer arietinum* (chickpea; Cotter and Edwards 2006). Heritability was high for larval weight gain on both resistant and susceptible varieties, but heritability was zero for oviposition, and females did not distinguish between varieties in oviposition.

In laboratory experiments, populations of *Helicoverpa zea* from different regions of North America (where the moth is indigenous) differed in ranking of plant species and varieties (hairy vs. glabrous soybean and cotton) for oviposition, and within the one population tested oviposition preference was heritable (although with large variance; Ward et al. 1993).

In the most interesting experiment concerning the genetics of host range in *Helicoverpa*, interspecific crosses (F_1, F_2, and backcrosses) of the generalist *H. armigera* and specialist *H. assulta* indicated that at least one major autosomal gene was involved in larval feeding on cotton and that *H. armigera* alleles were partially dominant to *H. assulta* alleles (Tang et al. 2006).

Papilio

The genus *Papilio* broadly construed comprises about 205 species (which may, in fact, represent as many as six genera) whose ancestors appear to have fed on species in the Rutaceae; 80 percent of species still feed on plants in this family (Zakharov, Caterino, and Sperling 2004). However, several clades have diverged from Rutaceae use, including the *glaucus* complex (*Papilio* [*Pterourus*] *glaucus*, *P. canadensis*, and related species), which attack species in at least eight plant families (Bossart and Scriber 1995a), and the *machaon* complex (*Papilio machaon, P. zelicaon*, *P. oregonius*, and related species), which attack species in Apiaceae (Umbelliferae) and Asteraceae (Sperling and Harrison 1994).

Papilio zelicaon is reported from over sixty species of Apiaceae and Rutaceae (Wehling and Thompson 1997), but *P. oregonius* is reported only from a single species (*Artemisia dracunculus*) of Asteraceae (Thompson 1988). In laboratory experiments with *P. zelicaon*, *P. oregonius*, and their reciprocal interspecific F_1 hybrids, females of each species showed strong oviposition preference for the appropriate field hosts; but their hybrids showed preferences similar to that of their paternal source, indicating a major locus or loci on the Z sex chromosome, although genes on autosomes modified preferences (Thompson 1988). Larval survival of each species was high on the appropriate plant, but survival of hybrid larvae was intermediate on both host plants, indicating autosomal inheritance of genes with additive effects (Thompson, Wehling, and Podolsky 1990). On the other hand, hybrid pupal mass and to a lesser extent development time were closer to the maternal source, indicating maternal effects, but not sex linkage (Thompson, Wehling, and Podolsky 1990).

In laboratory analyses of oviposition preference hierarchies among five *machaon*-complex species for five species of Apiaceae and Asteraceae, the butterflies showed a range of preference hierarchies from narrow to broad (Thompson 1998). One pair of sister species (*P. machaon*/*P. oregonius*) differed strongly in ranking of plant species, with *P. machaon* laying eggs on most species and *P. oregonius* laying only on a plant barely used by *P. machaon*, while another pair of sister species (*P. polyxenes*/*P. zelicaon*) closely resembled one another in ranking of plant species (Thompson 1998). Populations within *P. machaon*, and to a lesser extent within *P. polyxenes* and *P. zelicaon*, differed somewhat in preference hierarchies, and these differences may provide the raw material for host range shifts (Thompson 1998). For example, a few females in some populations of *P. machaon* laid a few eggs on *A. dracunculus*, the only known host of *P. oregonius*. The shift by *P. oregonius* to ovipositing on this plant may have been easy because oviposition preference in *P. oregonius* appears to be sex-linked and may involve few loci (Thompson 1988, 1998). However, this does not explain why the shift to *A. dracunculus* by *P. oregonius* led to dropping other plants from its host range. Although there was a shift toward local plants in populations of *P. zelicaon*, butterflies did not strongly prefer local plants, despite genetic variation in preference within these populations, as determined by differences among full-sib families (Thompson 1993; Wehling and Thompson 1997). This lack of strong preference for local hosts may result from coadapted gene complexes involved in preference for certain plants, from gene flow among populations preventing a response to selection, or from a lack of strong selection for adaptation to local hosts (Thompson 1993; Wehling and Thompson 1997).

Bossart (1998, 2003) and Bossart and Scriber (1995a,b, 1999) conducted a series of laboratory experiments on differences in oviposition preference and larval performance on three tree species (*Liriodendron tulipifera*, *Magnolia virginiana*, and *Prunus serotina*) among geographical populations of *P. glaucus* with different field exposures to these plants. Females from Florida, where *M. virginiana* is common and *L. tulipifera* rare, oviposited more on *M. virginiana* than females from regions where *L. tulipifera* was common and *M. virginiana* rare (Georgia) or absent (Ohio; Bossart and Scriber 1995a). However, like *P. zelicaon* in California (Thompson 1993), Florida *P. glaucus* did not strongly prefer to oviposit on the local host. The Ohio population showed very high heritability (0.81) in oviposition preference between *L. tulipifera* and *M. virginiana*, with some families ovipositing on both trees and some on *L. tulipifera* only (Bossart and Scriber 1999). Larvae from Florida and Georgia performed better on *M. virginiana* (as measured by development time and pupal mass) than larvae from Ohio, indicating adaptation to a locally available host, although larvae from all three regions still did best on *L. tulipifera* (Bossart and Scriber 1995a; Bossart 2003). These geographical populations did not differ in allozyme frequencies, which suggests that gene flow between them is counteracted by local selection to maintain differences in oviposition preference and larval performance (Bossart and Scriber 1995a). Comparison of larval performance among full-sib families showed no heritability for larval performance on *M. virginiana* for the Florida population, but significant heritability for performance on this host for the other two populations, as well as for performance on *P. serotina* for all three populations (Bossart 1998). Performance on the three hosts was either genetically uncorrelated or positively correlated, indicating no trade-offs in host plant suitability (Bossart 1998). In the locally polyphagous Ohio population, larvae from mothers that oviposited preferentially on *L. tulipifera* did better than larvae from mothers that oviposited preferentially on *M. virginiana*, regardless of the host plant on which they were reared, revealing a negative correlation between preference versus performance on *M. virginiana* (Bossart 2003). This may not be surprising given that the Ohio population is not exposed to *M. virginiana*, so selection for a preference-performance correlation is lacking. In the locally monophagous Florida population, oviposition preference showed no correlation with three of the four measures of larval performance and a negative correlation with the fourth measure (Bossart 2003), which suggests that selection may have shifted both oviposition preference and larval performance (Bossart and Scriber 1995a), but not enough to have resulted in a positive relationship. Although the preference-performance relationship appears to have a genetic basis, it is not the relationship expected from optimal oviposition theory, perhaps because of constraints arising from coadaptation, pleiotropy, or epistasis among genes controlling both preference and performance (Bossart 2003).

Differences in regulation and activity of cytochrome P450 monooxygenases have been implicated in differences in host use among Lepidoptera in general and papilionids in particular (for review, see Berenbaum and Feeny 2008). *Papilio polyxenes* specializes on species of Apiaceae and Rutaceae with high levels of specific furanocumarins (xanthotoxin and angelicin) and has high activity of P450s specific for these allomones that are not very effective at metabolizing others; *P. glaucus* and *P. canadensis* have broader host ranges and have P450s that metabolize a variety of allomones with less efficiency but are highly inducible (Li, Schuler, and Berenbaum 2007). Differences in P450 regulation and activity between *P. glaucus* and *P. canadensis* may play a role in the differences in their host ranges (Li, Schuler, and Berenbaum 2003).

Euphydryas

The genus *Euphydryas* (Nymphalidae) in the broad sense comprises fourteen species (Zimmermann, Wahlberg, and Descimon 2000). Their larvae feed on plant species in five families that produce iridoids, and Neartic *Euphydryas* specialize on plants of the families Scrophulariaceae and Plantaginaceae that have iridoid glycosides that the butterflies sequester (Zimmermann, Wahlberg, and Descimon 2000, and references therein). These butterflies disperse little and show interpopulation variation in host plant use. These attributes, as well as oviposition behavior that can be manipulated and measured in the field, led to a series of studies on the genetics and evolution of host use, particularly in *E. editha*. Several rapid shifts in plant species by various populations of *E. editha* have been documented (for review, see Singer et al. 2008). One population shifted from most females ovipositing on a native plant, *Collinsia parviflora* (Scrophulariaceae), to most females ovipositing on an introduced plant, *Plantago lanceolata* (Plantagenaceae). This shift occurred quite rapidly, going from 5 percent to 53 percent of females showing postalightment preference for the exotic plant in eight generations (Singer, Thomas, and Parmesan 1993). Early in the shift (1983–1984), postalightment oviposition preference measured in the field and laboratory showed a heritability of 0.90, based on mother-daughter regression, although this may be an overestimate if there were maternal effects (Singer, Ng, and Thomas 1988). Apparently larvae were preadapted to doing well on the exotic species, so no genetic changes were required in larval performance (Thomas et al. 1987). Indeed, larvae did much better on the exotic host than on *C. parviflora* because the exotic host matched butterfly phenology better (Singer 1984). By 1985, there was an interaction between oviposition preference and host plant that explained 32 percent of variation in larval performance (measured as weight gain), with larvae doing better on the host plant their mothers preferred for oviposition (Singer, Ng, and Thomas 1988). Better larval performance on the exotic and high heritability of oviposition preference, together with the weaker correlation between preference and performance, explain why this shift was so rapid (Singer et al. 2008). Another population of *E. editha* shifted from most females preferring to oviposit on one native plant, *Pedicularis semibarbata* (Scrophulariaceae), reduced in abundance by logging, to most females preferring to oviposit on a different native plant, *Collinsia torreyi* (Scrophulariaceae), increased in suitability by logging (Singer, Thomas, and Parmesan 1993). As with the shift to an exotic, the change was in postalightment preference and occurred rapidly, in twelve generations for this population (Singer and Thomas 1996). However, these preferences differed between patch types (rocky outcrops with *P. semibarbata* vs. logged areas with *C. torreyi*), with females ovipositing preferentially on the plant most abundant and suitable in their patch type (Singer and Thomas 1996). Interestingly, the oviposition frequencies switched back to the starting point when succession occurred in the logged patches and *C. torreyi* ceased being so suitable. Two conclusions about genetic architecture of host range in *E. editha* can be drawn from the rapidity of evolution in these populations. First, selection was strong, and second, there was either substantial genetic variation in the starting populations, or mutations readily supplied such variation; if the latter was the case, it suggests few genes with simple interactions were involved.

Other Systems

In laboratory experiments with full-sib families, larvae of *Depressaria patinacella* (Oecophoridae) showed genetic variation in survival on diets with fruits from their original host, *Pastinaca sativa* (Apiaceae), and those of a novel host, *Heracleum lanatum* (Apiaceae) (Berenbaum and Zangerl 1991), as well as in metabolism of parsnip furanocoumarins at various concentrations (Berenbaum and Zangerl 1992). However, larvae showed no genetic variation in feeding preference, indicating that adaptation to plant allomones was physiological rather than behavioral (Berenbaum and Zangerl 1991, 1992).

In laboratory experiments on rice and corn strains of *Spodoptera frugiperda* (Noctuidae), larvae of both strains performed best on rice, with the rice strain performing poorly on corn but the corn strain performing well on both hosts (Prowell, McMichael, and Silvain 2004). Analysis of genotype by environment interactions of full-sib families within strains showed variation that would promote host-associated divergence (Pashley 1988).

Larch and pine host races of *Zeiraphera diniana* (Tortricidae) mate assortatively (Emelianov et al. 2003). Genome-wide variation in hybridization between these host races suggests selection for host use in small regions of the *Z. diniana* genome, implying that a limited number of genes are involved in using alternative hosts (Emelianov, Marec, and Mallet 2004). The two host races differed in oviposition on larch versus pine but gave the same electroantennogram response to their odors; however, the plants differed in the numbers and concentrations of stimuli that elicited responses (Syed, Guerin, and Baltensweiler 2003). Thus both host races could distinguish both plant species, but their decisions about what to do with this information differed.

Within cedar and cypress host races of *Mitoura* (Lycaenidae), oviposition preference was strongly correlated with larval performance (Forister 2004), but this correlation was lost in the F_1 progeny of reciprocal crosses between cedar and cypress races (Forister 2005). Survival of hybrid larvae on cypress was identical to the cypress race, but hybrid survival on cedar was 30 percent lower than the cedar race. Hybrid females preferred to oviposit on cedar, the same host that resulted in the reduced survival of hybrid larvae. Thus, oviposition preference for cedar was dominant, with hybrid preference indistinguishable from the cedar race, but larval performance on cedar was recessive, with hybrid performance indistinguishable from the cypress race.

Comparisons between host races may end up being comparisons between species. For example, the mugwort and maize host races of *Ostrinia nubilalis* are genetically isolated (Martel et al. 2003; Bethenod et al. 2005), mate assortatively (Malausa et al. 2005), and have different sex pheromones that attract essentially only males from the same host race (Pelozuelo et al. 2004). Recently these *O. nubilalis* host races have been determined to be different species (Frolov, Bourguet, and Ponsard 2007). Whether they are host races or cryptic species, studying the genetic basis of differences in host use will be useful; indeed, crosses between closely related species with different host ranges may prove to be the most useful approach to determining the genetic architecture of host range.

In laboratory experiments with F_1 hybrids and backcrosses of three closely related species of *Yponomeuta*, oviposition on *Euonymus europaeus* (Celastraceae), the normal host of *Y. cagnagellus*, was partially dominant to oviposition on *Prunus spinosa* (Rosaceae), a normal host of *Y. padellus*, and *Malus domestica* (Rosaceae), a normal host of *Y. malinellus* (Hora, Roessingh, and Menken 2005). Reciprocal crosses gave the same results, indicating that the genes involved were autosomal rather than sex-linked. In these experiments, both *Y. padellus* and *Y. malinellus* laid some eggs on their nonhost *E. europaeus*, perhaps retaining willingness to oviposit on this host because species of Celastraceae appear to be the ancestral hosts for *Yponomeuta* (Menken 1996; Hora, Roessingh, and Menken 2005).

Colias eurytheme and *C. philodice* (Pieridae) appear to be distinct species with diagnosable differences maintained by assortative mating. Nevertheless, they hybridize, which may account for a lack of differences in their adaptation to several novel, introduced host plant species (Porter and Levin 2007). However, differences in genetic correlations and heritabilities for fitness components among host plants for the two species suggest that the genetic architecture of host use may differ between them (Porter and Levin 2007).

Conclusions

The current knowledge of the genetic architecture of lepidopteran host ranges is limited mostly to heritability estimates, dominance relationships, and location of genes on autosomes versus sex chromosomes, although QTL mapping studies are in progress. Heritabilities for oviposition preference and larval performance can be high (e.g., 60–90 percent) but also can be zero. Dominance relationships run the gamut from additive to overdominant. Larval performance tends to be controlled by autosomal genes and oviposition preference by sex chromosome genes, but this trend is weak. Finally, evidence is accumulating that differences in host range between closely related species and host races appear to have a relatively simple architecture, involving few segregating factors (e.g., fewer than 10) that may interact epistatically.

INTEGRATION OF LARVAL AND ADULT TRAITS

Much attention has been devoted to the relationship between oviposition preference and larval performance because of its implications for host range evolution and speciation. With rare exceptions, lepidopteran adults suck nectar and sometimes eat pollen (if they feed at all), but their larvae chew on plant tissue. Adults choose their own food by sight, smell, and taste (Ramaswamy 1988; Fitt 1991) and may feed on nectar from a variety of host plants unsuitable for larval development. Where females oviposit is determined in part by visual appearance, but primarily by the smell and taste of the surfaces of intact plants. Larval feeding decisions are largely determined by the odor and taste of intact plant surfaces and macerated tissues, though larvae may also use visual cues when moving from one plant to another. Whether larvae thrive on a host plant and produce fit adults depends on the interaction between their digestive systems, including ability to detoxify phytochemicals (Berenbaum and Zangerl 1992; Berenbaum, Cohen, and Schuler 1992; Hung et al. 1995; Rose et al. 1997; Stevens et al. 2000; Li et al. 2002; Wittstock et al. 2004; Zagrobelny et al. 2004; Berenbaum and Feeny 2008), and the plant tissues they ingest, as well as their nutritional requirements (Lee, Behmer, and Simpson 2006), especially for essential nutrients, or defensive chemicals they cannot produce themselves (Engler-Chaouat and Gilbert 2007). Host use involves a balance between two sets of traits: those of adults (e.g., location of hosts over a relatively large area, recognition and acceptance of suitable oviposition sites, and success in finding mates) and those of larvae (e.g., feeding on suitable hosts, recolonization of the host plant if dislodged, location of a new host plant if one is eaten up, and the ability to cope with plant defense compounds). Historically, it has been assumed that oviposition choice and larval performance are linked, so that females will tend to oviposit on plants that maximize larval performance, and oviposition preference will be influenced by larval host (e.g., Darwin 1909). However, given that these traits may be under different selection regimes and may be controlled by different sets of genes, complete integration of preference and performance may not be possible (Scheirs and De Bruyn 2002; Quental, Patten, and Pierce 2007).

The observed correspondence between oviposition preference and larval performance ranges from excellent to poor (for review, see Thompson and Pellmyr 1991). Recent work on *H. subflexa* has revealed that even extreme specialists may not always oviposit on the hosts that maximize larval fitness. In a common garden experiment involving seven *Physalis* species, oviposition preference of wild *H. subflexa* females did not correlate with larval performance (Benda 2007). *Physalis pubescens* was the species most preferred for oviposition, but larval feeding was greatest on *P. angulata* and *P. philadelphica*, which were less preferred for oviposition. On seventeen naturally occurring *Physalis* species in Mexico (the center of *Physalis* diversity), larval densities of *H. subflexa* on *P. pubescens*, *P. angulata*, and *P. philadelphica* were indistinguishable, and far greater than on the other ten *Physalis* species infested by *H. subflexa* larvae (Bateman 2006). In laboratory bioassays, *H. subflexa* larvae survived best on *P. angulata* (46 percent of neonates survived to pupation) but less well on both *P. pubescens* (34 percent) and *P. philadelphica* (30 percent). Interestingly, poor decision making was not restricted to adults: Larval feeding also failed to reflect performance reliably. In assays on thirteen *Physalis* species, larval mortality from starvation (with no attempt to feed) was quite high, ranging from 25 percent to 83 percent among plant species (Bateman 2006). If larvae had failed to feed only on plant species where performance was poor, one might conclude that refusal to feed on these suboptimal hosts was adaptive. In fact, the relationship between willingness to feed and survival to pupation was not consistent: Only 52 percent of neonates attempted to feed on *P. angulata*, but 89 percent survived to pupation; in contrast, 75 percent of neonates attempted to feed on *P. philadelphica*, but only 43 percent survived to pupation. It is unclear why larvae would refuse to eat suitable plants, especially in the absence of other choices. When presented with artificial diet, 95 percent of neonates fed, and their strikingly lower willingness to feed on plant material may reflect variation in larval sensitivity to species-specific plant compounds (Bateman 2006). In any case, it appears that neither oviposition preference nor larval feeding is fine tuned to larval performance in the specialist *H. subflexa*.

Given the difference in selection between larvae and adults, it is perhaps not surprising that preference and performance appear to be controlled by different genes (Thompson, Wehling, and Podolsky 1990; Sheck and Gould 1993, 1995, 1996). Even in cases with a strong correlation between preference and performance, this correlation appears to reflect independent selection on these traits, rather than a shared genetic basis. If larval and adult host use traits were controlled by genes on the same chromosome, physical linkage might allow them to evolve in concert. However, genes affecting larval performance have consistently mapped to autosomes (Hagen 1990; Thompson, Wehling, and Podolsky 1990; Sheck and Gould 1996; Forister 2005), while genes affecting oviposition preference are less consistent, mapping sometimes to sex chromosomes and sometimes to autosomes (Sheck and Gould 1995; Forister 2005; Hora, Roessingh, and Menken 2005). Many traits associated with adult behavior (e.g., male response to pheromones: *Ostrinia* [Dopman et al. 2005]; female mate choice: *Colias* [Grula and Taylor 1980], Arctiidae [Iyengar, Reeve, and Eisner 2002]; female oviposition preference: *Papilio* [Thompson 1988; Scriber, Giebink, and Snider 1991], *Polygonia* [Nygren, Nylin, and Stefanescu 2006]) are sex linked, specifically to the male (Z) sex chromosome, suggesting that genes found on the Z chromosome may contribute disproportionately to the evolution of reproductive isolation and thus be important in speciation (Sperling 1994; Prowell 1998). However, sex linkage of oviposition preference may depend on the geographical scale of comparison: Janz (1998) found that variation between two populations of *Polygonia c-album* with different host specificity was sex linked, whereas Nylin et al. (2005) found strong variation in oviposition preference among females in a single population, but no evidence for sex linkage. Regardless of the autosomal versus sex chromosomal basis of oviposition preference, all research to date has suggested that oviposition preference and larval performance are controlled by genes on different chromosomes.

NEUROBIOLOGY OF HOST RANGE

Although some host-use genes (e.g., those involved in larval detoxification of plant allomones) may affect only one life stage, others probably act in both larvae and adults. Most notably, both larvae and adults use smell and taste to evaluate potential hosts, so genes involved in olfaction and gustation are likely to affect both egg laying and larval feeding. That the chemosensory systems of adults and larvae are often in harmony is demonstrated by females' generally ovipositing on plants where their larvae are willing to feed. In *Papilio*, adult oviposition and larval feeding are stimulated (or deterred) by the same chemicals, suggesting that the same chemosensory genes are responsible for host-use decisions in adults and larvae (Ono, Kuwahara, and Nishida 2004; Nishida 2005). Furthermore, P450s degrade odorants in adult *Papilio* as well as detoxify plant allomones in larvae (Ono, Ozaki, and Yoshikawa 2005) and may provide a link between adult oviposition and larval performance (Berenbaum and Feeny 2008).

Chemoreceptors are broadly classified as members of either the olfactory (Or) or gustatory (Gr) receptor subfamilies. Olfactory processing is mediated by olfactory binding proteins (OBPs) secreted into the aqueous lymph of sensilla and olfactory receptor proteins (ORPs) embedded in the membranes of olfactory receptor neurons (ORNs) that innervate sensilla. A thorough review of the neurobiology of insect olfaction is beyond the scope of this chapter (for recent reviews, see Mustaparta 2002; Chyb 2004; Rützler and Zwiebel 2005; Hallem, Dahanukar, and Carlson 2006; see Chapter 9 for a discussion of the phylogenetics of lepidopteran chemoreception genes). Briefly, odor molecules pass through the pores of olfactory sensilla on antennae and maxillary palps (the primary and secondary olfactory organs), are transported by an OBP to the membrane of an ORN, where they bind to an ORP, inducing an action potential that propagates along the axon of the ORN. While the dendrites of ORNs innervate the sensilla, their axons project into the glomeruli of the antennal lobe. Thus, stimulation at the periphery is quickly conveyed to the higher processing areas of the central nervous system.

ORPs are highly diverse, with many sharing less than 20 percent amino acid similarity. This diversity long delayed their discovery in insects, as homology-based similarity searches using known mammalian ORP sequences were unsuccessful. A combination of bioinformatic (Clyne et al. 1999) and genetic (Vosshall et al. 1999) approaches finally identified *Drosophila* ORPs, including sixty *Drosophila* Or genes. Identification of olfactory receptors in Lepidoptera has proved challenging because of low sequence similarity to Or genes in other organisms. Krieger et al. (2002) identified nine candidate ORPs by screening an *H. virescens* antennal cDNA library for proteins with partial sequence similarity to *Drosophila* ORPs and used in situ hybridization to find which candidates were expressed in ORN. These newly identified proteins had very low amino acid sequence similarity with any identified Or genes, and homology was restricted to small regions of *Drosophila* ORPs. More promising for identification of adult chemoreception genes in Lepidoptera are the findings of Wanner et al. (2007) with *Bombyx mori*. Once *H. virescens Or* genes were identified, they used traditional measures of sequence similarity to identify forty-one candidate ORPs in the newly released *B. mori* genome, many of which appear orthologous with *H. virescens* ORPs (for details, see Chapter 9).

It is unclear what role OBPs and ORPs play in determining the host range of lepidopterans. In *Drosophila*, some ORPs are narrowly tuned to a single odor and some are more broadly tuned. Each ORP is expressed in a subset of three to fifty ORNs, and the response properties of ORNs arise from differences in the ORPs they express (de Bruyne and Warr 2006). Many ORNs respond to the same odor, and thus one odor typically activates multiple receptors (Hallem, Ho, and Carlson 2004; Goldman et al. 2005). This result helps explain why flies with an engineered deletion of an odor receptor often show normal olfactory-mediated behavior (Elmore et al. 2003). In *Drosophila* and probably Lepidoptera, specificity of response to odors relies on combinatorial discrimination of odorants, which has been observed in the first center of neuronal integration, the antennal lobe (Ng et al. 2002; Wang et al. 2003). Behavioral changes can be induced by the inactivation of selected subsets of olfactory neurons (Suh et al. 2004), and one recent study demonstrated that overexpression of a single odor receptor resulted in reduced behavioral avoidance of benzaldehyde, a compound involved in host avoidance and attraction in some Lepidoptera (Stortkuhl et al. 2005).

OBPs were first discovered in the antenna of the moth *Antheraea polyphemus* (Vogt and Riddiford 1981) and have since been identified in *H. virescens* (Krieger et al. 2002), *S. exigua* (Xiu and Dong 2007), *O. nubialis* (Coates, Hellmich, and Lewis 2005), *M. sexta* (Vogt et al. 2002), and other species. A variety of biochemical roles have been proposed for OBPs, including the transport of odorants through sensillum lymph to ORPs and the deactivation of odorants following receptor activation (Park et al. 2000); whether OBPs will prove to be involved in multiple processes in Lepidoptera remains to be seen. In *Drosophila*, recent work by Matsuo et al. (2007) suggests that OBPs may be involved in host range. They examined the genetic basis of oviposition choice in *D. sechellia*, a specialist on *Morinda citrifolia*, which is toxic to the closely related *D. melanogaster*. *Drosophila sechellia* is preferentially attracted to *M. citrifolia* fruit, while *D. melanogaster* is deterred by its odor. Using targeted gene knockout and replacement, they replaced two *D. melanogaster* OBP genes (*Obp57d* and *Obp57e*) with the *D. sechellia* versions. In the resulting transformed flies, oviposition preference closely mirrored that of *D. sechellia*. While such manipulations are not yet possible for any lepidopteran species, OBPs and ORPs are attractive candidate genes for explaining differences in host plant use.

In discussions of host plant acceptance, the role of larval choice is frequently overlooked. Although larvae are less mobile than adults, they may also be more motivated to find optimal hosts. As with adults, progress is being made in understanding the neurophysiology of host plant recognition and acceptance by larvae. Larval host choice appears to be based on a small set of gustatory receptors on antennae, maxillary palps, and epipharynx (Hanson and Dethier 1973; de Boer 1993, 2006; Glendinning, Valcic, and Timmermann 1998; Schoonhoven and van Loon 2002; Schoonhoven 2005). Gustatory sensilla, which are also found in adults, are innervated by four gustatory receptor neurons (GRNs), each responding only to sweet, salt, or water stimuli (Dethier 1976). The axons of GRNs project into the subesophogeal ganglion of the central nervous system, the first relay center of taste processing in the brain. Gustatory sensilla have been studied in a wide variety of insects, including moths and butterflies (Zacharuk 1980). To date, Gr genes have only been identified in two Lepidoptera: *H. virescens* (Krieger et al. 2002) and *B. mori* (for details, see Chapter 9). In *Drosophila*, sixty Gr genes have been identified (Robertson, Warr, and Carlson 2003), but receptor specificity has been determined for only a few of these (i.e., sugar receptors, Dahanukar et al. 2007; carbon dioxide receptors, Jones et al. 2007; Kwon et al. 2007; and bitter receptors, Moon et al. 2006).

The proteins encoded by identified Gr genes are, like OBPs and ORPs, extremely divergent in sequence, sharing as little as 8 percent amino acid identity (Scott et al. 2001). As with Or genes, Gr genes show much higher levels of sequence homology within Lepidoptera than between, for example, Lepidoptera and Diptera. Thus the increasing availability of Lepidoptera-specific resources (e.g., ButterflyBase, Papanicolaou et al. 2008; and the *B. mori* genome project, Mita et al. 2004; Xia et al. 2004) should make it easier to identify these genes in a wide variety of lepidopterans.

Given that host selection/acceptance is mediated by a balance of phagostimulatory and deterrent inputs (Schoonhoven 1987), a simple (i.e., single-gene) explanation of host range is unlikely. In many cases, the experience of an insect with its environment interacts with its genome to produce the observed host range. The larvae of *Pieris rapae* and *Manduca sexta* are polyphagous at hatching and become oligophagous only after exposure to a host-specific compound (a glucosinolate for *P. rapae* [Renwick and Lopez 1999] and indioside D for *M. sexta* [del Campo et al. 2001]). Presumably, following this exposure, plants lacking the relevant compound are deterrent (alternatively, only plants with the compound are stimulatory). In *M. sexta*, changes in the activity of the peripheral nervous system are known to occur after exposure to indioside D (del Campo and Miles 2003), but the mechanism by which these changes affect larval behavior is unknown.

The possibility that changes in host range are caused by changes in sensitivity to deterrent or stimulatory compounds is supported at the behavioral level (Bernays and Chapman 1987; Bernays et al. 2000). For example, the presence of benzaldehyde had no effect on feeding in *Y. cagnagellus*, a species that retains the original ancestral association with Celastraceae (which do not contain benzaldehyde), but stimulated feeding in species in a more derived clade that has shifted to the benzaldehyde-containing Rosaceae (Roessingh, Xu, and Menken 2007). Interestingly, such changes in peripheral sensitivity to host-associated chemicals appear to be a consequence rather than a cause of shifts in host range. In studies of both larvae and adults, species with widely divergent host ranges appear to have similar receptor neuron sensitivities. In larvae of *H. subflexa* and *H. virescens*, interspecific behavioral differences could not be attributed to differences in sensory neuron responses to stimulatory or deterrent compounds (Bernays and Chapman 2000; Bernays et al. 2000). In the generalists *H. armigera* and *H. virescens* and the specialist *H. assulta*, four types of ORNs in adult females of each species responded to the same four volatile plant chemicals, and each type of neuron, though narrowly tuned to a single molecule, showed some response to closely related molecules (Stranden et al. 2003). In both larvae and adults, species-specific host acceptance appears to depend on differences in central processing of sensory input, so changes in host range may depend upon changes in the central nervous system (Bernays and Chapman 1987; Bernays et al. 2000; Chyb 2004).

Although the mechanisms causing Lepidoptera with different host ranges to produce very different behaviors from the same peripheral input have not yet been identified, such work is under way in *Drosophila*. Melcher and Pankratz (2005) have identified a neuropeptide (coded by the *hugin* gene) expressed in gustatory interneurons that link peripheral receptor neurons with motor neurons in the ventral nerve cord and the pharyngeal apparatus. Output from these *hugin*-expressing interneurons appears to integrate taste, the endocrine system, higher-order brain centers, and motor output to modify feeding. As with the olfactory and gustatory genes, identification of lepidopteran neuropeptides may depend on the development and exploitation of Lepidoptera-specific resources. In the search for pheromone biosynthesis activating neuropeptide receptor genes, for example, very low levels of similarity were found between lepidopteran and *Drosophila* sequences, but similarity within the Lepidoptera was high (Zheng et al. 2007).

DIRECTIONALITY OF HOST RANGE EVOLUTION

Most Lepidoptera have relatively narrow host ranges, feeding on a small fraction of available plants. This preponderance of specialists may reflect host-associated fitness trade-offs (Jaenike 1990), selection by natural enemies (Bernays and Graham 1988), or neural constraints (Bernays

2001). Neural constraints (i.e., limitations of insect nervous systems that restrict the rate of information processing; Dukas 1998) might be the proximate means by which many checks on host range operate because insects with broad host ranges may be less efficient at correctly accepting or rejecting plants (and choose plants on which their fitness is reduced) or may simply take longer to choose hosts (thus increasing their exposure to natural enemies). The ability of adult females to select the best oviposition sites depends on accurately assessing host quality, and generalists seem to perform poorly compared to specialists. In assays of three specialist and two generalist nymphalid species on nettles of different quality, specialist females oviposited preferentially on high-quality nettles, but generalist females did not (Janz and Nylin 1997). All larvae performed poorly on low-quality nettles, so the poor choices made by generalist females reduced their reproductive success.

Larval performance also varies between specialists and generalists in a manner consistent with the neural constraints hypothesis. Bernays et al. (2000) found that larvae of the specialist *H. subflexa* rejected toxic diets untasted or after a single bite, but larvae of the generalist *H. virescens* rejected such diets only after extensive feeding. Apparently, the specialist relied on swift sensory evaluation of the diet, whereas the generalist relied on negative postingestive effects. This "eat now, decide later" approach of *H. virescens* larvae greatly increased their risk of consuming fatal doses of toxins. Inefficient decision making can also lead to reduced feeding opportunities: In assays of larval foraging behavior of two specialist and two generalist arctiid species, generalist larvae took much longer to accept or reject a plant and rejected many suitable host plants (Bernays, Chapman, and Singer 2004).

Although Mayr (1963) considered host range evolution to be unidirectional and irreversible, with generalists giving rise to specialists and specialization a dead end, more recent work has shown this to be false (Nosil and Mooers 2005). Instead, transitions from specialist to generalist and generalist to specialist occur freely and are not constrained by phylogeny (Winkler and Mitter 2008). For example, optimization of host-use traits on the phylogeny of Nymphalini suggests that ancestral specialization on Urticales was followed by frequent expansions and contractions of host range (Janz, Nyblom, and Nylin 2001). Furthermore, larvae of many species could feed on plants outside their current host range, with a strong bias toward plants used as hosts by other species of Nymphalini (Janz, Nyblom, and Nylin 2001). Such a bias is consistent with the striking conservatism in host use in most Lepidoptera. The observation that closely related insects often use closely related plants is well supported by phylogenetic reconstructions (e.g., Mitter and Farrell 1991; Winkler and Mitter 2008) and probably results from retention of ancestral host-use genes. Novelty, however, is also a common theme for host-use evolution. In the Nymphalini, some "extreme" host shifts to plant families outside the ancestral host range of either the Nymphalini or their close relatives in the *Nymphalis-Polygonia* clade occurred. A similar phenomenon is found with the Troidini tribe of Papilionidae, in which host range reflects neither host plant phylogeny nor plant secondary chemistry (Silva-Brandao and Solferini 2007). Instead, Troidini host range is strictly opportunistic, and increases in geographical range are strongly correlated with increases in the number of plant species used.

The evolution of host range probably reflects both the constraints of phylogeny and the constructive effects of natural selection, changing in response to the availability and adaptive value of particular host plants. Interestingly, shifts to novel host plants are associated with increased speciosity in *Polygonia*, where clades that shifted to novel host plants were more speciose than sister clades that used only hosts from the ancestral group (Weingartner, Wahlberg, and Nylin 2006). Such patterns are consistent with expansions in host plant range driving the elevated diversification rates observed in Lepidoptera and other phytophagous insects. In one recent analysis of 145 phytophage speciation events, fully half of the events were accompanied by shifts to new host plant species (Winkler and Mitter 2008).

FUTURE PROSPECTS

Our current understanding of the genetic architecture of host range is limited. Although differences in host range between closely related species and host races appear to have a relatively simple architecture, this conclusion awaits corroboration by more detailed analysis of the actual genes involved. Differences in sequence and expression of detoxification enzymes and chemoreceptor proteins have been implicated in differences in host range, but their full roles remain to be determined. Further advances in our understanding will require either much larger experiments or new approaches—and probably both. Crosses between closely related species or races that differ in host range provide the strong phenotypic differences and distinct molecular markers that together greatly aid in the identification of the genes responsible for differences in host range. Furthermore, it is exactly these differences in host range between recently diverged populations and species that are most intriguing. The most promising systems for this approach have involved species and populations in the genera *Euphydryas*, *Helicoverpa*, *Heliothis*, and *Papilio.*

Three main strategies show great promise for delineating in more detail the genetic architecture of differences in host range: (1) more and finer-scale genetic mapping of QTL, including combined genetic and physical mapping, (2) analysis of sequence and expression differences between closely related species or races that differ in host range, and (3) the effects of targeted disruption of candidate genes.

Genetic mapping of QTL is a powerful technique for determining the number and interaction of loci affecting quantitative traits (for review, see Lynch and Walsh 1998). However, QTL mapping might be better named QTR (quantitative trait region) mapping, because the number of genes between markers flanking QTL may be large. Many researchers have argued forcefully for the need to go beyond QTL mapping to identify the specific genetic changes underlying adaptive divergence (e.g., Remington, Ungerer, and Purugganan 2001; Orr 2005). Going from QTL to candidate gene can be quite challenging, especially when several different QTL affect phenotypes. Even if a single QTL is strongly implicated, sequencing the region between flanking markers in a nonmodel organism can prove difficult.

The development of whole-genome integrated physical/genetic maps (Chang et al. 2001; Yamamoto et al. 2006) can significantly accelerate map-based (positional) cloning of genes underlying QTL. Furthermore, sequencing has recently become much easier and cheaper with the introduction of ultrahigh-throughput technologies such as the Genome Sequencer FLX System (454-Life Sciences/Roche Applied Science, Indianapolis, Indiana, USA), Illumina Genome Analyzer (Illumina, San Diego, California, USA), and the SOLiD System (Applied Biosystems, Foster City, California, USA). Vera et al. (2008) used 454 pyrosequencing to generate approximately half a million high-quality reads from the genome of the Glanville fritillary, *Melitaea cinxia* (Nymphalidae). BLAST searches against the *B. mori* genome resulted in about nine thousand hits. If, as has been estimated for *B. mori*, most Lepidoptera have about eighteen thousand genes, then at least half of all genes in the *M. cinxia* genome have high levels of homology with *B. mori*. Vera et al. (2008) were able to detect a large number of sequence polymorphisms, a valuable source of genetic markers for QTL mapping and population genetic analysis. In addition, the availability of large-scale transcriptome information will allow species-specific microarrays to be constructed. Assembly of the short reads from these new technologies is a problem, but approaches like paired-end sequencing and the use of scaffolds from related species should ease assembly (Goldsmith, Shimada, and Abe 2005).

Improved techniques for fluorescence in situ hybridization (FISH) using genomic bacterial artificial chromosomes (BACs) as probes promise a renaissance in the cytogenetics of Lepidoptera, transforming their small, uniformly sized, undifferentiated chromosomes into powerful tools for the analysis of genome organization (Yoshido et al. 2005; Yasukochi et al. 2006; Sahara et al. 2007). Already, genetic mapping combined with BAC-FISH has revealed synteny in a variety of lepidopterans (Jiggins et al. 2005; Kaplan et al. 2006; Lee and Heckel 2007; Sahara et al. 2007). Integrated genetic/physical maps and this synteny may make it possible to use the well-mapped and sequenced genome of *B. mori* to find candidate genes in chromosome segments delineated by common anchor loci.

For candidate genes, expression analysis combined with genetic mapping can determine whether the candidates map to the same region as QTL associated with phenotypic differences. The sequence differences among detoxification enzymes and olfactory receptor proteins can be quite large, so they should be readily distinguishable in expression analyses. An alternative method for testing gene function is to silence the expression of candidate genes using RNA interference (RNAi), which works by inducing intracellular enzymes that destroy native mRNA homologous to introduced double-stranded RNA (Bettencourt, Terenius, and Faye 2002). Several Lepidoptera species have been genetically transformed (Tamura et al. 2000; Thomas et al. 2002; Imamura et al. 2003; Marcus 2005), opening up the potential to develop RNAi constructs that express endogenously in the appropriate tissue or developmental stage.

The new and developing techniques in molecular genetics have the potential to move lepidopteran genetics beyond the limited world of model organisms and into one that better reflects the great diversity of lepidopteran biology. Unlike earlier insect model systems, existing research on Lepidoptera is deeply rooted in attempts to understand the evolution of traits that allow insects to adapt to a variety of environments. The combination of the vast knowledge of the behavior, ecology, and phylogeny of many species of Lepidoptera that has accumulated over the last half century with the rapidly expanding availability of cutting-edge genetic and genomic technologies promises exciting progress in our understanding of the genetic architecture of lepidopteran host ranges in the near future.

REFERENCES

Bateman, M.L. 2006. Impact of plant suitability, biogeography, and ecological factors on associations between the specialist herbivore *Heliothis subflexa* G. (Lepidoptera: Noctuidae) and the species in its host genus, *Physalis* L. (Solanaceae), in west-central Mexico. PhD diss., North Carolina State Univ.

Benda, N.D. 2007. Host location by adults and larvae of specialist herbivore *Heliothis subflexa* G. (Lepidoptera: Noctuidae). PhD diss., North Carolina State Univ.

Berenbaum, M.R., M.B. Cohen, and M.A. Schuler. 1992. Cytochrome-P450 monooxygenase genes in oligophagous Lepidoptera. *ACS Symp. Ser.* 505:114–24.

Berenbaum, M.R., and P.P. Feeny. 2008. Chemical mediation of host-plant specialization-the papillionid paradigm. In *Specialization, speciation and radiation: The evolutionary biology of herbivorous insects*, ed. K.J. Tilmon, pp. 3–19. Berkeley: Univ. California Press.

Berenbaum, M.R., and A.R. Zangerl. 1991. Acquisition of a native hostplant by an introduced oligophagous herbivore. *Oikos* 62:153–59.

Berenbaum, M.R., and A.R. Zangerl. 1992. Genetics of physiological and behavioral resistance to host furanocoumarins in the parsnip webworm. *Evolution* 46:1373–84.

Bernays, E.A. 2001. Neural limitations in phytophagous insects: Implications for diet breadth and evolution of host affiliation. *Annu. Rev. Entomol.* 46:703–27.

Bernays, E.A., and R.F. Chapman. 1987. The evolution of deterrent responses in plant-feeding insects. In *Perspectives in chemoreception and behavior*, ed. R.E. Chapman, E.A. Bernays, and J.G. Stoffolano, 159–74. New York: Springer.

Bernays, E.A., and R.F. Chapman. 2000. A neurophysiological study of sensitivity to a feeding deterrent in two sister species of *Heliothis* with different diet breadths. *J. Insect Physiol.* 46:905–12.

Bernays, E.A., R.F. Chapman, and M.S. Singer. 2004. Changes in taste receptor cell sensitivity in a polyphagous caterpillar reflect carbohydrate but not protein imbalance. *J. Comp. Physiol. A Neuroethol. Sens. Neural Behav. Physiol.* 190:39–48.

Bernays, E., and M. Graham. 1988. On the evolution of host specificity in phytophagous arthropods. *Ecology* 69:886–92.

Bernays, E.A., S. Oppenheim, R.F. Chapman, H. Kwon, and F. Gould. 2000. Taste sensitivity of insect herbivores to deterrents is greater in specialists than in generalists: A behavioral test of the hypothesis with two closely related caterpillars. *J. Chem. Ecol.* 26:547–63.

Bethenod, M.T., Y. Thomas, F. Rousset, et al. 2005. Genetic isolation between two sympatric host plant races of the European corn borer, *Ostrinia nubilalis* Hubner. II: Assortative mating and host-plant preferences for oviposition. *Heredity* 94:264–70.

Bettencourt, R., O. Terenius, and I. Faye. 2002. Hemolin gene silencing by ds-RNA injected into *Cecropia* pupae is lethal to next generation embryos. *Insect Mol. Biol.* 11:267–71.

Bossart, J.L. 1998. Genetic architecture of host use in a widely distributed, polyphagous butterfly (Lepidoptera: Papilionidae): Adaptive inferences based on comparison of spatio-temporal populations. *Biol. J. Linn. Soc. Lond.* 65:279–300.

Bossart, J.L. 2003. Covariance of preference and performance on normal and novel hosts in a locally monophagous and locally polyphagous butterfly population. *Oecologia* 135:477–86.

Bossart, J.L., and J.M. Scriber. 1995a. Maintenance of ecologically significant genetic variation in the tiger swallowtail butterfly through differential selection and gene flow. *Evolution* 49:1163–71.

Bossart, J.L., and J.M. Scriber. 1995b. Genetic variation in oviposition preference in tiger swallowtail butterflies: Interspecific, interpopulation and interindividual comparisons. In *Swallowtail butterflies: Their ecology and evolutionary biology*, ed. J.M. Scriber, Y. Tsubaki, and R.C. Lederhouse, 183–93. Gainesville: Scientific Publishers.

Bossart, J.L., and J.M. Scriber. 1999. Preference variation in the polyphagous tiger swallowtail butterfly (Lepidoptera: Papilionidae). *Environ. Entomol.* 28:628–37.

Chang, Y.-L., Q. Tao, C. Scheuring, K. Ding, K. Meksem, and H.-B. Zhang. 2001. An integrated map of *Arabidopsis thaliana* for functional analysis of its genome sequence. *Genetics* 159:1231–42.

Chen, G.T., and J.B. Graves 1970. Spermatogenesis of the tobacco budworm. *Ann. Entomol. Soc. Am.* 63:1095–104.

Cho, S.W., A. Mitchell, J.C. Regier, et al. 1995. A highly conserved nuclear gene for low-level phylogenetics: Elongation factor-1 alpha recovers morphology-based tree for heliothine moths. *Mol. Biol. Evol.* 12:650–56.

Chyb, S. 2004. *Drosophila* gustatory receptors: From gene identification to functional expression. *J. Insect Physiol,* 50:469–77.

Clyne, P.J., C.G. Warr, M.R. Freeman, D. Lessing, J.H. Kim, and J.R. Carlson. 1999. A novel family of divergent seven-transmembrane proteins: Candidate odorant receptors in *Drosophila. Neuron* 22:327–38.

Coates, B.S., R.L. Hellmich, and L.C. Lewis. 2005. Two differentially expressed ommochrome-binding protein-like genes (*obp1* and *obp2*) in larval fat body of the European corn borer, *Ostrinia nubilalis. J. Insect Sci.* 5:19.

Cotter, S.C., and O.R. Edwards. 2006. Quantitative genetics of preference and performance on chickpeas in the noctuid moth, *Helicoverpa armigera. Heredity* 96:396–402.

Cunningham, J.P., M.F.A. Jallow, D.J. Wright, and M.P. Zalucki. 1998. Learning in host selection in *Helicoverpa armigera* (Hubner) (Lepidoptera: Noctuidae). *Animal Behav.* 55:227–34.

Dahanukar A., Y.T. Lei, J.Y. Kwon, and J.R. Carlson. 2007. Two Gr genes underlie sugar reception in *Drosophila. Neuron* 56:503–16.

Darwin, C. 1909. Essay of 1844. In *The foundations of the origin of species*, ed. F. Darwin, 127. Cambridge: Cambridge Univ. Press.

de Boer, G. 1993. Plasticity in food preference and diet-induced differential weighting of chemosensory information in larval *Manduca sexta. J. Insect Physiol.* 39:17–24.

de Boer, G. 2006. The role of the antennae and maxillary palps in mediating food preference by larvae of the tobacco hornworm, *Manduca sexta. Entomol. Exp. Appl.* 119:29–38.

de Bruyne, M., and C.G. Warr. 2006. Molecular and cellular organization of insect chemosensory neurons. *Bioessays* 28:23–34.

De Prins, J., and K. Saitoh. 2003. Karyology and sex determination. In *Lepidoptera, moths and butterflies: Morphology, physiology, and development*, ed. N.P. Kristensen, 449–68. Berlin: Walter de Gruyter.

del Campo, M.L., and C.I. Miles. 2003. Chemosensory tuning to a host recognition cue in the facultative specialist larvae of the moth *Manduca sexta. J. Exp. Biol.* 206:3979–90.

del Campo, M.L., C.I. Miles, F.C. Schroeder, C. Mueller, R. Booker, and J.A. Renwick. 2001. Host recognition by the tobacco hornworm is mediated by a host plant compound. *Nature* 411:186–89.

Dethier, V.G. 1976. *The hungry fly: A physiological study of the behavior associated with feeding*. Cambridge: Harvard Univ. Press.

Dopman, E.B., L. Perez, S.M. Bogdanowicz, and R.G. Harrison. 2005. Consequences of reproductive barriers for genealogical discordance in the European corn borer. *Proc. Natl. Acad. Sci. U.S.A.* 102:14706–11.

Dukas, R. 1998. Constraints on information processing and their effects on behavior. In *Cognitive ecology*, ed. R. Dukas, 89–127. Chicago: Chicago Univ. Press.

Elmore, T., R. Ignell, J.R. Carlson, and D.P. Smith. 2003. Targeted mutation of a *Drosophila* odor receptor defines receptor requirement in a novel class of sensillum. *J. Neurosci.* 23:9906–12.

Emelianov, I., F. Marec, and J. Mallet. 2004. Genomic evidence for divergence with gene flow in host races of the larch budmoth. *Proc. Biol. Sci.* 271:97–105.

Emelianov, I., F. Simpson, P. Narang, and J. Mallet. 2003. Host choice promotes reproductive isolation between host races of the larch budmoth *Zeiraphera diniana*. *J. Evol. Biol.* 16:208–18.

Engler-Chaouat, H.S., and L.E. Gilbert. 2007. De novo synthesis vs. sequestration: Negatively correlated metabolic traits and the evolution of host plant specialization in cyanogenic butterflies. *J. Chem. Ecol.* 33:25–42.

Fang, Q.Q., S. Cho, J.C. Regier, et al. 1997. A new nuclear gene for insect phylogenetics: DOPA carboxylase is informative of relationships within Heliothinae (Lepidoptera: Noctuidae). *Syst. Biol.* 46:269–83.

Fitt, G.P. 1989. The ecology of *Heliothis* species in relation to agroecosystems. *Annu. Rev. Entomol.* 34:17–53.

Fitt, G.P. 1991. Host selection in Heliothinae. In *Reproductive behaviour of insects: Individuals and populations*, ed. W.J. Bailey and J. Ridsdill-Smith, 172–200. London: Chapman & Hall.

Forister, M.L. 2004. Oviposition preference and larval performance within a diverging lineage of lycaenid butterflies. *Ecol. Entomol.* 29:264–72.

Forister, M.L. 2005. Independent inheritance of preference and performance in hybrids between host races of *Mitoura* butterflies (Lepidoptera: Lycaenidae). *Evolution* 59:1149–55.

Frolov, A.N., D. Bourguet, and S. Ponsard. 2007. Reconsidering the taxomony of several *Ostrinia* species in the light of reproductive isolation: A tale for Ernst Mayr. *Biol. J. Linn. Soc. Lond.* 91:49–72.

Glendinning, J.I., S. Valcic, and B.N. Timmermann. 1998. Maxillary palps can mediate taste rejection of plant allelochemicals by caterpillars. *J. Comp. Physiol. A Neuroethol. Sens. Neural Behav. Physiol.* 183:35–43.

Goldman, A.L., W.V. van Naters, D. Lessing, C.G. Warr, and J.R. Carlson. 2005. Coexpression of two functional odor receptors in one neuron. *Neuron* 45:661–66.

Goldsmith, M.R., T. Shimada, and H. Abe. 2005. The genetics and genomics of the silkworm, *Bombyx mori*. *Annu. Rev. Entomol.* 50:71–100.

Groot, A.T., C. Ward, J. Wang, et al. 2004. Introgressing pheromone QTL between species: Towards an evolutionary understanding of differentiation in sexual communication. *J. Chem. Ecol.* 30:2495–2514.

Grula, J.W., and O.R. Taylor. 1980. The effect of X-chromosome inheritance on mate-selection behavior in the sulfur butterflies, *Colias eurytheme* and *Colias philodice*. *Evolution* 34:688–95.

Gu, H., A. Cao, and G.H. Walter. 2001. Host selection and utilisation of *Sonchus oleraceus* (Asteraceae) by *Helicoverpa armigera* (Lepidoptera: Noctuidae): A genetic analysis. *Ann. Appl. Biol.* 138:293–99.

Gu, H., and G.H. Walter. 1999. Is the common sowthistle (*Sonchus oleraceus*) a primary host plant of the cotton bollworm, *Helicoverpa armigera* (Lep., Noctuidae)? Oviposition and larval preference. *J. Appl. Entomol.* 123:99–105.

Hagen, R.H. 1990. Population structure and host use in hybridizing subspecies of *Papilio glaucus* (Lepidoptera, Papilionidae). *Evolution* 44:1914–30.

Hallem, E.A., A. Dahanukar, and J.R. Carlson. 2006. Insect odor and taste receptors. *Annu. Rev. Entomol.* 51:113–35.

Hallem, E.A., M.G. Ho, and J.R. Carlson. 2004. The molecular basis of odor coding in the *Drosophila* antenna. *Cell* 117:965–79.

Hanson, F.E., and V.G. Dethier. 1973. Role of gustation and olfaction in food plant discrimination in tobacco hornworm, *Manduca sexta*. *J. Insect Physiol.* 19:1019–34.

Heckel, D.G., L.J. Gahan, Y.-B. Liu, and B.E. Tabashnik. 1999. Genetic mapping of resistance to *Bacillus thuringiensis* toxins in diamondback moth using biphasic linkage analysis. *Proc. Natl. Acad. Sci. U.S.A.* 96:8373–77.

Hora, K.H., P. Roessingh, and S.B.J. Menken. 2005. Inheritance and plasticity of adult host acceptance in Yponomeuta species: Implications for host shifts in specialist herbivores. *Entomol. Exp. Appl.*115:271–81.

Hung, C.F., T.L. Harrison, M.R. Berenbaum, and M.A. Schuler. 1995. CYP6B3: A second furanocoumarin-inducible cytochrome P450 expressed in *Papilio polyxenes*. *Insect Mol. Biol.* 4:149–60.

Imamura, M., J. Nakai, S. Inoue, G.X. Quan, T. Kanda, and T. Tamura. 2003. Targeted gene expression using the GAL4/UAS system in the silkworm *Bombyx mori*. *Genetics* 165:1329–40.

Iyengar, V.K., H.K. Reeve, and T. Eisner. 2002. Paternal inheritance of a female moth's mating preference. *Nature* 419:830–32.

Jaenike, J. 1990. Host specialization in phytophagous insects. *Annu. Rev. Ecol. Syst.* 21:243–73.

Jallow, M.F.A., and M.P. Zalucki. 1996. Within- and between-population variation in host-plant preference and specificity in Australian *Helicoverpa armigera* (Hubner) (Lepidoptera: Noctuidae). *Aust. J. Zool.* 44:503–19.

Janz, N. 1998. Sex-linked inheritance of host-plant specialization in a polyphagous butterfly. *Proc. R. Soc. Lond. B. Biol. Sci.* 265:1675–78.

Janz, N., K. Nyblom, and S. Nylin. 2001. Evolutionary dynamics of host-plant specialization: A case study of the tribe Nymphalini. *Evolution* 55:783–96.

Janz, N., and S. Nylin. 1997. The role of female search behaviour in determining host plant range in plant feeding insects: A test of the information processing hypothesis. *Proc. R. Soc. Lond. B. Biol. Sci.* 264:701–07.

Jiggins, C.D., J. Mavarez, M. Beltrán, W.O. McMillan, J.S. Johnston, and E. Bermingham. 2005. A genetic linkage map of the mimetic butterfly *Heliconius melpomene*. *Genetics* 171:557–70.

Jones, W.D., P. Cayirlioglu, I.G. Kadow, and L.B. Vosshall. 2007. Two chemosensory receptors together mediate carbon dioxide detection in Drosophila. *Nature* 445:86–90.

Kaplan, D.D., N.S. Flanagan, A. Tobler, R. Papa, and R.D. Reed. 2006. Localization of Mullerian mimicry genes on a dense linkage map of *Heliconius erato*. *Genetics* 173:735–57.

Karpenko, C.P., and F.I. Proshold. 1977. Fertility and mating performance of interspecific crosses between *Heliothis virescens* and *H. subflexa* (Lepidoptera: Noctuidae) backcrossed for 3 generations to *H. subflexa*. *Ann. Entomol. Soc. Am.* 70:737–40.

Krieger, J., E. Grosse-Wilde, T. Gohl, Y.M.E. Dewer, K. Raming, and H. Breer. 2004. Genes encoding candidate pheromone receptors in a moth (*Heliothis virescens*). *Proc. Natl. Acad. Sci. U.S.A.* 101:11845–50.

Krieger, J., K. Raming, Y.M.E. Dewer, S. Bette, S. Conzelmann, and H. Breer. 2002. A divergent gene family encoding candidate olfactory receptors of the moth *Heliothis virescens*. *Eur. J. Neurosci.* 16:619–28.

Kwon, J.Y., A. Dahanukar, L.A. Weiss, and J.R. Carlson. 2007. The molecular basis of CO_2 reception in *Drosophila*. *Proc. Natl. Acad. Sci. U.S.A.* 104:3574–78.

Laster, M.L., S.D. Pair, and D.F. Martin. 1982. Acceptance and development of *Heliothis subflexa* and *Heliothis virescens* (Lepidoptera, Noctuidae), and their hybrid and backcross progeny on several plant species. *Environ. Entomol.* 11:979–80.

Lee, K.P., S.T. Behmer, and S.J. Simpson. 2006. Nutrient regulation in relation to diet breadth: A comparison of *Heliothis* sister species and a hybrid. *J. Exp. Biol.* 209:2076–84.

Lee, S.F., and D.G. Heckel. 2007. Chromosomal conservation in Lepidoptera: Synteny versus collinearity. In Iatrou, K., and P. Couble 2007. 7th International Workshop on the Molecular Biology and Genetics of the Lepidoptera. August 20–26, 2006. Orthodox Academy of Crete, Kolympari, Crete, Greece. *J. Insect Sci.* 7:29.

Li, W., R.A. Petersen, M.A. Schuler, and M. R. Berenbaum. 2002. CYP6B cytochrome P450 monooxygenases from *Papilio canadensis* and *Papilio glaucus*: Potential contributions of sequence divergence to host plant associations. *Insect Mol. Biol.* 11:543–51.

Li, W.M., M.A. Schuler, and M.R. Berenbaum. 2003. Diversification of furanocoumarin-metabolizing cytochrome P450 monooxygenases in two papilionids: Specificity and substrate encounter rate. *Proc. Natl. Acad. Sci. U.S.A.* 100:14593–98.

Li, X.C., M.A. Schuler, and M.R. Berenbaum. 2007. Molecular mechanisms of metabolic resistance to synthetic and natural xenobiotics. *Annu. Rev. Entomol.* 52:231–53.

Lynch, M., and B. Walsh. 1998. *Genetics and analysis of quantitative traits*. Sunderland: Sinauer.

Malausa, T., M.T. Bethenod, A. Bontemps, D. Bourguet, J.M. Cornuet, and S. Ponsard. 2005. Assortative mating in sympatric host races of the European corn borer. *Science* 308:258–60.

Marcus, J.M. 2005. Jumping genes and AFLP maps: Transforming lepidopteran color pattern genetics. *Evol. Dev.* 7:108–14.

Marec, F. 1996. Synaptonemal complexes in insects. *Int. J. Insect Morphol. Embryol.* 25:205–33.

Martel, C., A. Rejasse, F. Rousset, M.T. Bethenod, and D. Bourguet. 2003. Host-plant-associated genetic differentiation in Northern French populations of the European corn borer. *Heredity* 90:141–49.

Matsuo, T., S. Sugaya, J. Yasukawa, T. Aigaki, and Y. Fuyama. 2007. Odorant-binding proteins OBP57d and OBP57e affect taste perception and host-plant preference in *Drosophila sechellia*. *PLoS Biol.* 5:985–96.

Mayr, E. 1963. *Animal species and evolution*. Cambridge: Belknap Press of Harvard Univ. Press.

McElvare, R.R. 1941. Validity of the species *Heliothis subflexa*. *Bull. Brooklyn Entomol. Soc.* 36:29–30.

Melcher, C., and M.J. Pankratz. 2005. Candidate gustatory interneurons modulating feeding behavior in the *Drosophila* brain. *PLoS Biol.* 3:1618–29.

Menken, S.B.J. 1996. Pattern and process in the evolution of insect-plant associations: *Yponomeuta* as an example. *Entomol. Exp. Appl.* 80:297–305.

Mita, K., M. Kasahara, S. Sasaki, et al. 2004. The genome sequence of silkworm, *Bombyx mori*. *DNA Res.* 11:27–35.

Mitter, C., and B.D. Farrell. 1991. Macroevolutionary aspects of insect-plant relationships. In *Insect/plant interactions, vol. 3*, ed. E.A. Bernays, 35–78. Boca Raton: CRC Press.

Mitter, C., R.W. Poole, and M. Matthews. 1993. Biosystematics of the Heliothinae (Lepidoptera: Noctuidae). *Annu. Rev. Entomol.* 38:207–25.

Moon, S.J., M. Kottgen, Y.C. Jiao, H. Xu, and C. Montell. 2006. A taste receptor required for the caffeine response in vivo. *Curr. Biol.* 16:1812–17.

Mustaparta, H. 2002. Encoding of plant odour information in insects: Peripheral and central mechanisms. *Entomol. Exp. Appl.* 104:1–13.

Ng, M., R.D. Roorda, S.Q. Lima, B.V. Zemelman, P. Morcillo, and G. Miesenbock. 2002. Transmission of olfactory information between three populations of neurons in the antennal lobe of the fly. *Neuron* 36:463–74.

Nishida, R. 2005. Chemosensory basis of host recognition in butterflies—Multi-component system of oviposition stimulants and deterrents. *Chem. Senses* 30, Suppl. 1:i293–94.

Nosil, P., and A.O. Mooers. 2005. Testing hypotheses about ecological specialization using phylogenetic trees. *Evolution* 59:2256–63.

Nygren, G.H., S. Nylin, and C. Stefanescu. 2006. Genetics of host plant use and life history in the comma butterfly across Europe: Varying modes of inheritance as a potential reproductive barrier. *J. Evol. Biol.* 19:1882–93.

Nylin, S., G.H. Nygren, J.J. Windig, N. Janz, and A. Bergstrom. 2005. Genetics of host-plant preference in the comma butterfly *Polygonia c-album* (Nymphalidae), and evolutionary implications. *Biol. J. Linn. Soc. Lond.* 84:755–65.

Ono, H., Y. Kuwahara, and R. Nishida. 2004. Hydroxybenzoic acid derivatives in a nonhost rutaceous plant, *Orixa japonica*, deter both oviposition and larval feeding in a Rutaceae-feeding swallowtail butterfly, *Papilio xuthus* L. *J. Chem. Ecol.* 30:287–301.

Ono, H., K. Ozaki, and H. Yoshikawa. 2005. Identification of cytochrome P450 and glutathione-S-transferase genes preferentially expressed in chemosensory organs of the swallowtail butterfly, *Papilio xuthus* L. *Insect Biochem. Mol. Biol.* 35:837–46.

Orr, H.A. 2001. The genetics of species differences. *Trends Ecol. Evol.* 16:343–50.

Orr, H.A. 2005. Theories of adaptation: What they do and don't say. *Genetica* 123:3–13.

Papanicolaou, A., S. Gebauer-Jung, M.L. Blaxter, W.O. McMillan, and C.D. Jiggins. 2008. ButterflyBase: A platform for lepidopteran genomics. *Nucleic Acids Res.* 36:D582–87.

Park, S.K., S.R. Shanbhag, Q. Wang, G. Hasan, R.A. Steinbrecht, and C.W. Pikielny. 2000. Expression patterns of two putative odorant-binding proteins in the olfactory organs of *Drosophila melanogaster* have different implications for their functions. *Cell Tissue Res.* 300:181–92.

Pashley, D.P. 1988. Quantitative genetics, development and physiological adaptation in host strains of fall armyworm. *Evolution* 42:93–102.

Pelozuelo, L., C. Malosse, G. Genestier, H. Guenego, and B. Frerot. 2004. Host-plant specialization in pheromone strains of the European corn borer *Ostrinia nubilalis* in France. *J. Chem. Ecol.* 30:335–52.

Poole, R.W., C. Mitter, and M.D. Huettel. 1993. A revision and cladistic analysis of the *Heliothis virescens* species group (Lepidoptera: Noctuidae) with a preliminary morphometric analysis of *H. virescens*. In *Mississippi Agriculture and Forestry Experiment Station Technical Bulletin* 185:1–51.

Porter, A.H., and E.J. Levin. 2007. Parallel evolution in sympatric, hybridizing species: Performance of *Colias* butterflies on their introduced host plants. *Entomol. Exp. Appl.* 124:77–99.

Prowell, D.P. 1998. Sex linkage and speciation in Lepidoptera. In *Endless forms: Species and speciation*, ed. S. Berlocher and D. Howard, 309–19. New York: Oxford Univ. Press.

Prowell, D.P., M. McMichael, and J.F. Silvain. 2004. Multilocus genetic analysis of host use, introgression, and speciation in host strains of fall armyworm (Lepidoptera: Noctuidae). *Ann. Entomol. Soc. Am.* 97:1034–44.

Quental, T.B., M.M. Patten, and N.E. Pierce. 2007. Host plant specialization driven by sexual selection. *Amer. Nat.* 169:830–36.

Ramaswamy, S.B. 1988. Host finding by moths: Sensory modalities and behaviours. *J. Insect Physiol.* 34:235–49.

Remington, D.L., M.C. Ungerer, and M.D. Purugganan. 2001. Map-based cloning of quantitative trait loci: Progress and prospects. *Genet. Res.* 78:213–18.

Renwick, J.A.A., and K. Lopez. 1999. Experience-based food consumption by larvae of *Pieris rapae*: Addiction to glucosinolates? *Entomol. Exp. Appl.* 91:51–58.

Robertson, H.M., C.G. Warr, and J.R. Carlson. 2003. Molecular evolution of the insect chemoreceptor gene superfamily in *Drosophila melanogaster*. *Proc. Natl. Acad. Sci. U.S.A.* 100:14537–42.

Robinson, R. 1971. *Lepidoptera genetics*. New York: Pergamon Press.

Roessingh, P., S. Xu, and S.B.J. Menken. 2007. Olfactory receptors on the maxillary palps of small ermine moth larvae: Evolutionary history of benzaldehyde sensitivity. *J. Comp. Physiol. A Neuroethol. Sens. Neural Behav. Physiol.* 193:635–47.

Rose, R.L., D. Goh, D.M. Thompson, et al. 1997. Cytochrome P450 (CYP)9A1 in *Heliothis virescens*: The first member of a new CYP family. *Insect Biochem. Mol. Biol.* 27:605–15.

Rostelien, T., M. Stranden, A.K. Borg-Karlson, and H. Mustaparta. 2005. Olfactory receptor neurons in two heliothine moth species responding selectively to aliphatic green leaf volatiles, aromatic compounds, monoterpenes and sesquiterpenes of plant origin. *Chem. Senses* 30:443–61.

Rützler, M., and L.J. Zwiebel. 2005. Molecular biology of insect olfaction: Recent progress and conceptual models. *J. Comp. Physiol. A Neuroethol. Sens. Neural Behav. Physiol.* 191:777–90.

Sahara, K., A. Yoshido, F. Marec, et al. 2007. Conserved synteny of genes between chromosome 15 of *Bombyx mori* and a chromosome of *Manduca sexta* shown by five-color BAC-FISH. *Genome* 50:1061–65.

Scheirs, J., and L. De Bruyn. 2002. Temporal variability of top-down forces and their role in host choice evolution of phytophagous arthropods. *Oikos* 97:139–44.

Schoonhoven, L.M. 1987. What makes a caterpillar eat? The sensory codes underlying feeding behaviour. In *Advances in chemoreception and behavior*, ed. R. Chapman, E. Bernays, and J. Stoffolano, 69–97. New York: Springer.

Schoonhoven, L.M. 2005. Insect-plant relationships: The whole is more than the sum of its parts. *Entomol. Exp. Appl.* 115:5–6.

Schoonhoven, L.M., and J.J.A. van Loon. 2002. An inventory of taste in caterpillars: Each species its own key. *Acta Zool. Acad. Sci. Hung.* 48:215–63.

Scott, K., R. Brady, A. Cravchik, et al. 2001. A chemosensory gene family encoding candidate gustatory and olfactory receptors in *Drosophila*. *Cell* 104:661–73.

Scriber, J.M. 2002. Evolution of insect-plant relationships: Chemical constraints, coadaptation, and concordance of insect/plant traits. *Entomol. Exp. Appl.* 104:217–35.

Scriber, J.M., B.L. Giebink, and D. Snider. 1991. Reciprocal latitudinal clines in oviposition behavior of *Papilio glaucus* and *P. canadensis* across the Great Lakes hybrid zone: Possible sex-linkage of oviposition preferences. *Oecologia* 87:360–68.

Sheck, A.L., and F. Gould. 1993. The genetic basis of host range in *Heliothis virescens*: Larval survival and growth. *Entomol. Exp. Appl.* 69:157–72.

Sheck, A.L., and F. Gould. 1995. Genetic analysis of differences in oviposition preferences of *Heliothis virescens* and *Heliothis subflexa* (Lepidoptera, Noctuidae). *Environ. Entomol.* 24:341–47.

Sheck, A.L., and F. Gould. 1996. The genetic basis of differences in growth and behavior of specialist and generalist herbivore species: Selection on hybrids of *Heliothis virescens* and *Heliothis subflexa* (Lepidoptera). *Evolution* 50:831–41.

Sheck, A.L., A.T. Groot, C.M. Ward, et al. 2006. Genetics of sex pheromone blend differences between *Heliothis virescens* and *Heliothis subflexa*: A chromosome mapping approach. *J. Evol. Biol.* 19:600–17.

Silva-Brandao, K.L., and V.N. Solferini. 2007. Use of host plants by Troidini butterflies (Papilionidae, Papilioninae): Constraints on host shift. *Biol. J. Linn. Soc. Lond.* 90:247–61.

Singer, M.C. 1984. Butterfly-host plant relationships. In *The biology of butterflies, Symposium of the Royal Entomological Society XIII, London*, ed. R.L. Vane-Wright and P.R. Ackery, 81–88. London: Royal Entomological Society.

Singer, M.C., and C.D. Thomas. 1996. Evolutionary responses of a butterfly metapopulation to human- and climate-caused environmental variation. *Amer. Nat.* 148:S9–S39.

Singer, M.C., D. Ng, and C.D. Thomas. 1988. Heritability of oviposition preference and its relationship to offspring performance within a single insect population. *Evolution* 42:977–85.

Singer, M.C., C.D. Thomas, and C. Parmesan. 1993. Rapid human-induced evolution of insect host associations. *Nature* 366:681–83.

Singer, M.C., B. Wee, S. Hawkins, and M. Butcher. 2008. Rapid natural and antropogenic diet evolution: Three examples from checkerspot butterflies. In *Specialization, speciation and radiation: The evolutionary biology of herbivorous insects*, ed. K.J. Tilmon, pp. 311–24. Berkeley: Univ. California Press.

Sperling, F.A.H. 1994. Sex-linked genes and species differences in Lepidoptera. *Canad. Entomol.* 126:807–18.

Sperling, F.A.H., and R.G. Harrison. 1994. Mitochondrial DNA variation within and between species of the *Papilio machaon* group of swallowtail butterflies. *Evolution* 48:408–22.

Stevens, J.L., M.J. Snyder, J.F. Koener, and R. Feyereisen. 2000. Inducible P450s of the CYP9 family from larval *Manduca sexta* midgut. *Insect Biochem. Mol. Biol.* 30:559–68.

Stortkuhl, K.F., R. Kettler, S. Fischer, and B.T. Hovemann. 2005. An increased receptive field of olfactory receptor Or43a in the antennal lobe of Drosophila reduces benzaldehyde-driven avoidance behavior. *Chem. Senses* 30:81–87.

Stranden, M., T. Rostelien, I. Liblikas, T.J. Almaas, A.K. Borg-Karlson, and H. Mustaparta. 2003. Receptor neurones in three heliothine moths responding to floral and inducible plant volatiles. *Chemoecology* 13:143–54.

Sudbrink, D.L., and J.F. Grant. 1995. Wild host plants of *Helicoverpa zea* and *Heliothis virescens* (Lepidoptera, Noctuidae) in eastern Tennessee. *Environ. Entomol.* 24:1080–85.

Suh, G.S.B., A.M. Wong, A.C. Hergarden, et al. 2004. A single population of olfactory sensory neurons mediates an innate avoidance behaviour in *Drosophila. Nature* 431:854–59.

Suomalainen, E. 1969. Chromosome evolution in the Lepidoptera. *Chromosomes Today* 2:132–38.

Syed, Z., P.M. Guerin, and W. Baltensweiler. 2003. Antennal responses of the two host races of the larch bud moth, *Zeiraphera diniana*, to larch and cembran pine volatiles. *J. Chem. Ecol.* 29:1691–1708.

Tamura, T., C. Thibert, C. Royer, et al. 2000. Germline transformation of the silkworm *Bombyx mori* L. using a *piggyBac* transposon-derived vector. *Nature Biotechnol.* 18:81–84.

Tang, Q.B., J.W. Jiang, Y.H. Yan, J.J.A. van Loon, and C.Z. Wang. 2006. Genetic analysis of larval host-plant preference in two sibling species of *Helicoverpa*. *Entomol. Exp. Appl.* 118:221–28.

Thomas, J.L., M. Da Rocha, A. Besse, B. Mauchamp, and G. Chavancy. 2002. 3xP3-EGFP marker facilitates screening for transgenic silkworm *Bombyx mori* L. from the embryonic stage onwards. *Insect Biochem. Mol. Biol.* 32:247–53.

Thomas, C.D., D. Ng, M.C. Singer, J.L.B. Mallet, C. Parmesan, and H.L. Billington. 1987. Incorporation of a European weed into the diet of a North American herbivore. *Evolution* 41:892–901.

Thompson, J.N. 1988. Evolutionary genetics of oviposition preference in swallowtail butterflies. *Evolution* 42:1223–34.

Thompson, J.N. 1993. Preference hierarchies and the origin of geographic specialization in host use in swallowtail butterflies. *Evolution* 47:1585–94.

Thompson, J.N. 1998. The evolution of diet breadth: Monophagy and polyphagy in swallowtail butterflies. *J. Evol. Biol.* 11:563–78.

Thompson, J.N., and O. Pellmyr. 1991. Evolution of oviposition behavior and host preference in Lepidoptera. *Annu. Rev. Entomol.* 36:65–89.

Thompson, J.N., W. Wehling, and R. Podolsky. 1990. Evolutionary genetics of host use in swallowtail butterflies. *Nature* 344:148–50.

Tilmon, K.J. 2008. *Specialization, speciation, and radiation: The evolutionary biology of herbivorous insects*. Berkeley: Univ. California Press.

Vera, J.C., C.W. Wheat, H.W. Fescemyer, et al. 2008. Rapid transcriptome characterization for a nonmodel organism using 454 pyrosequencing. *Mol. Ecol.* 17:1636–47.

Vogt, R.G., and L.M. Riddiford. 1981. Pheromone binding and inactivation by moth antennae. *Nature* 293:161–63.

Vogt, R.G., M.E. Rogers, M.D. Franco, and M. Sun. 2002. A comparative study of odorant binding protein genes: differential expression of the PBP1-GOBP2 gene cluster in *Manduca sexta* (Lepidoptera) and the organization of OBP genes in *Drosophila melanogaster* (Diptera). *J. Exp. Biol.* 205:719–44.

Vosshall, L.B., H. Amrein, P.S. Morozov, A. Rzhetsky, and R. Axel. 1999. A spatial map of olfactory receptor expression in the *Drosophila* antenna. *Cell* 96:725–36.

Wang, J.W., A.M. Wong, J. Flores, L.B. Vosshall, and R. Axel. 2003. Two-photon calcium imaging reveals an odor-evoked map of activity in the fly brain. *Cell* 112:271–82.

Wanner, K.W., A.R. Anderson, S.C. Trowell, D.A. Theilmann, H.M. Robertson, and R.D. Newcomb. 2007. Female-biased expression of odourant receptor genes in the adult antennae of the silkworm, *Bombyx mori*. *Insect Mol. Biol.* 16:107–19.

Ward, K.E., J.L. Hayes, R.C. Navasero, and D.D. Hardee. 1993. Genetic variability in oviposition preference among and within populations of the cotton bollworm (Lepidoptera, Noctuidae). *Ann. Entomol. Soc. Am.* 86:103–10.

Wehling, W.F., and J.N. Thompson. 1997. Evolutionary conservatism of oviposition preference in a widespread polyphagous insect herbivore, *Papilio zelicaon*. *Oecologia* 111:209–15.

Weingartner, E., N. Wahlberg, and S. Nylin. 2006. Dynamics of host plant use and species diversity in *Polygonia* butterflies (Nymphalidae). *J. Evol. Biol.* 19:483–91.

White, M.J.D. 1973. *Animal cytology and evolution*. Cambridge: Cambridge Univ. Press.

Winkler, I.S., and C. Mitter. 2008. Phylogenetic dimension of insect-plant interactions. In *Specialization, speciation and radiation: The evolutionary biology of herbivorous insects*, ed. K.J. Tilmon, 240–63. Berkeley: Univ. California Press.

Wittstock, U., N. Agerbirk, E.J. Stauber, et al. 2004. Successful herbivore attack due to metabolic diversion of a plant chemical defense. *Proc. Natl. Acad. Sci. U.S.A.* 101:4859–64.

Xia, Q., Z. Zhou, C. Lu, et al. 2004. A draft sequence for the genome of the domesticated silkworm (*Bombyx mori*). *Science* 306:1937–40.

Xiu, W.M., and S.L. Dong. 2007. Molecular characterization of two pheromone binding proteins and quantitative analysis of their expression in the beet armyworm, *Spodoptera exigua* Hübner. *J. Chem. Ecol.* 33:947–61.

Yamamoto, K., J. Narukawa, K. Kadono-Okuda, et al. 2006. Construction of a single nucleotide polymorphism linkage map for the silkworm, *Bombyx mori*, based on bacterial artificial chromosome end sequences. *Genetics* 173:151–61.

Yasukochi, Y., L.A. Ashakumary, K. Baba, A. Yoshido, and K. Sahara. 2006. A second-generation integrated map of the silkworm reveals synteny and conserved gene order between lepidopteran insects. *Genetics* 173:1319–28.

Yoshido, A., H. Bando, Y. Yasukochi, and K. Sahara. 2005. The *Bombyx mori* karyotype and the assignment of linkage groups. *Genetics* 170:675–85.

Zacharuk, R.Y. 1980. Ultrastructure and function of insect chemosensilla. *Annu. Rev. Entomol.* 25:27–47.

Zagrobelny, M., S. Bak, A.V. Rasmussen, B. Jorgensen, C.M. Naumann, and B.L. Moller. 2004. Cyanogenic glucosides and plant-insect interactions. *Phytochemistry* 65:293–306.

Zakharov, E.V., M.S. Caterino, and F.A.H. Sperling. 2004. Molecular phylogeny, historical biogeography, and divergence time estimates for swallowtail butterflies of the genus *Papilio* (Lepidoptera: Papilionidae). *Syst. Biol.* 53:193–215.

Zalucki, M.P., D.A.H. Murray, P.C. Gregg, G.P. Fitt, P.H. Twine, and C. Jones. 1994. Ecology of *Helicoverpa armigera* (hubner) and *Heliothis punctigera* (wallengren) in the inland of Australia: Larval sampling and host-plant relationships during winter and spring. *Aust. J. Zool.* 42:329–46.

Zheng, L., C. Lytle, C.N. Njauw, M. Altstein, and M. Martins-Green. 2007. Cloning and characterization of the pheromone biosynthesis activating neuropeptide receptor gene in *Spodoptera littoralis* larvae. *Gene* 393:20–30.

Zimmermann, M., N. Wahlberg, and H. Descimon. 2000. Phylogeny of *Euphydryas* checkerspot butterflies (Lepidoptera: Nymphalidae) based on mitochondrial DNA sequence data. *Ann. Entomol. Soc. Am.* 93:347–55.

12 Genetics and Molecular Biology of the Major Crop Pest Genus *Helicoverpa*

Karl Gordon, Wee Tek Tay, Derek Collinge, Adam Williams, and Philip Batterham

CONTENTS

INTRODUCTION

Biology and Evolution of the Heliothinae

The noctuid genus *Helicoverpa* (Hardwick 1965) is one of the most widely studied groups in the Lepidoptera. This genus contains several of the world's major pests of agriculture. However, most of the species in the genus are characterized by limited geographic spread and narrow host plant ranges, making it a valuable model system for studies in the evolution of generalist phytophagous insects displaying the central characteristics of pests: fecundity, vagility, and polyphagicity. How the tools of modern molecular biology are being applied to species in this genus to address fundamental issues such as these will be the theme of this review.

Helicoverpa needs to be studied in the context of the subfamily to which it belongs, the Heliothinae, which are considered to have diversified in the second half of the Cenozoic era, that is, over the past approximately twenty million years (Cho et al. 1995; phylogenetic relationships of noctuids are discussed in Chapter 1). The 365 species in the subfamily are spread over twelve genera. The *Heliothis* group includes *Helicoverpa* (Matthews 1999; Cho et al. 2008) and its sister genus *Australothis*, represented by four species found only in Australia, Indonesia, and New Zealand (Matthews 1999). Within this group, the eponymous genus, *Heliothis* Oschenheimer, now represents a major anomaly in that it is paraphyletically distributed around the *Helicoverpa/Australothis* clade (Cho et al. 2008). However, a lineage containing two *Heliothis* species found only in the New World, the major pest *Heliothis virescens* and its sister species *H. subflexa*, is phylogenetically most closely related to *Helicoverpa/Australothis*, in what Cho et al. (2008) have termed the megapest lineage (Figure 12.1). The origins of this megapest lineage are unclear but were probably in Australia, since the basal *Helicoverpa* species *H. punctigera* and many of the earlier branching relatives in the *Heliocheilus* clade are found there (Matthews 1999). With the apparent disappearance of the *H. virescens* lineage from Australia, only the *Australothis/Helicoverpa* ancestor remained. It is striking that these major pest lineages appear to have diverged in Australia, possibly due to the arid nature of their predominant habitats, characterized by variable rainfall and diverse, unpredictable flora across large areas.

The above comprehensive phylogeny of the Heliothinae provides a solid foundation for studying the evolution of host-range specificity in this subfamily. Of the species in the subfamily, most (*ca.* 200), including the earliest diverging lineages, are host specialists, reflecting the inferred ancestral heliothine host range under parsimony (Cho et al. 2008). A single group, the *Heliothis* group, which includes the megapest lineage, also includes most of the polyphages (30 percent of heliothines). It is likely that this represents one of a few acquisitions of polyphagy in the subfamily (Cho et al. 2008). Within the *Heliothis* group, there has then been reversion to host specialization on a number of occasions (Mitter, Poole, and Matthews 1993; Cho 1997; Matthews 1999). Such reversions have occurred in the genus *Helicoverpa* and in several of the lineages comprising the genus *Heliothis* (Figure 12.1; see Cho et al. 2008).

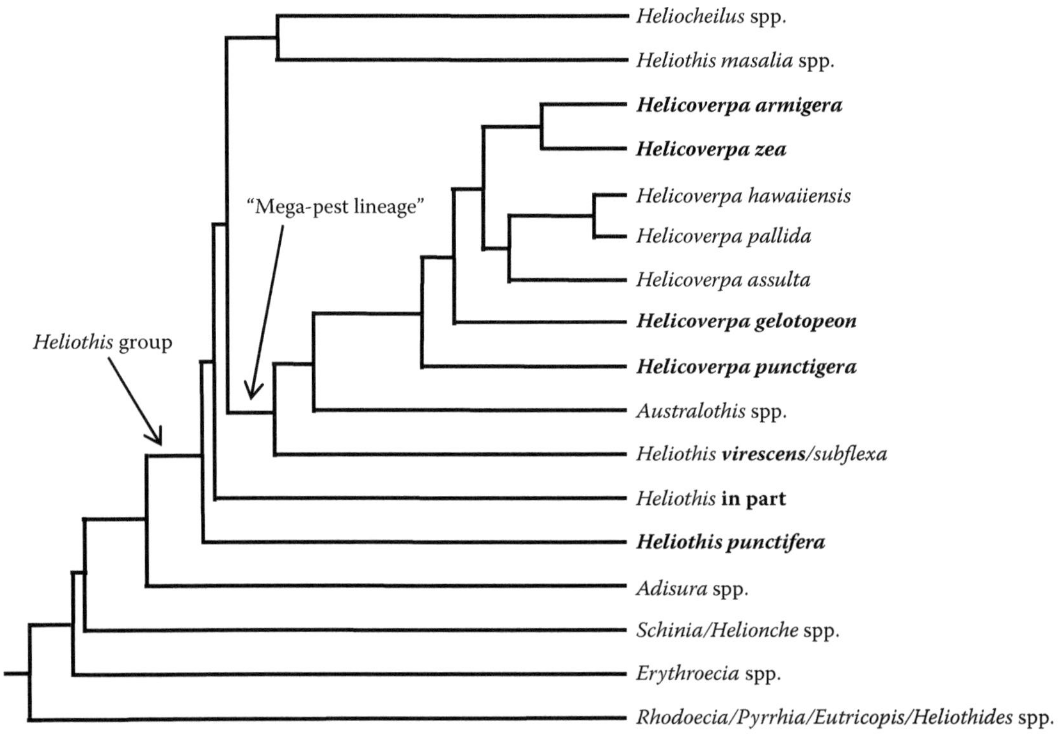

FIGURE 12.1 Phylogeny of the Heliothinae. The topology is based on Maximum Likelihood analysis of sequences of three genes (EF-1α, DDC, and COI) by Cho et al. (2008). Only representative lineages were selected from the complete set described by Cho et al. (2008). The *Heliothis* group comprises the genera *Heliothis* (paraphyletic), *Australothis*, *Helicoverpa*, and *Heliocheilus*. Polyphagous species are underlined, as are lineages that include some such species. See Cho et al. (2008) for details, and Table 12.1 for host plant families for the oligophages.

This genus *Helicoverpa* is currently recognized as having twenty described members (Matthews 1999). In establishing *Helicoverpa* as a new genus, Hardwick (1965) incorporated results from extensive morphological, rearing, and hybridization studies, and arranged these moths into a postulated evolutionary sequence of species groups. His conclusions on the genus *Helicoverpa* were later supported in an exhaustive study by Matthews (1999) using more sophisticated morphological methods he had developed. The species, listed in Table 12.1, include the most damaging insect pests of agriculture in the world, *H. armigera* (Hübner) and *H. zea* (Boddie), the dominant pest species in the Old World and New World, respectively, and first resolved as separate species by Hardwick (1965). These insects are the true megapests, enjoying a worldwide distribution as a result of a likely founder event by *H. armigera* that led to the establishment of *H. zea* in the New World (Mallet et al. 1993; Behere et al. 2007). A few other *Helicoverpa* species are pests of a range of crops, but they are either limited in host plant range or are geographically restricted (Cho et al. 2008). The majority of *Helicoverpa* species are oligophagous and are not considered as major agricultural pests. Indeed, a number of species are so rare that they are listed as endangered and are probably extinct.

Unfortunately the phylogeny of the genus is incompletely understood, with only eight of the species having been subjected to rigorous molecular phylogenetic analysis (Cho et al. 2008; Table 12.1).

TABLE 12.1
Species of the Genus *Helicoverpa*

	Distribution										
Species	Australia	Chr. Is.	NZ	Asia	Africa	Hawaii	South America	North America	Phagy	Hosts	Pest
H. armigera armigera	+		+	+	+				P		**
H. armigera conferta			+						P		**
H. zea							+	+	P		**
H. confusa						+					
H. helenae											
H. toddi					+						
H. tibetensis				+							
H. prepodes	+								O		
H. pacifica						+					
H. fletcheri					+				P		
H. assulta	+			+	+				O	Solan	*
H. hardwicki	+								O		
H. pauliana		+									
H. minuta						+					
H. hawaiiensis						+			O	Malv	
H. pallida						+			O	Cheno	
H. gelotopoeon							+		P		**
H. titicacae							+		P		
H. bracteae							+				
H. atacamae							+		O	Poac	
H. punctigera	+								P		**

Note: Compiled from data in Mitter, Poole, and Matthews 1993; Cho 1997; Matthews 1999; Cho et al. 2008; and M. Matthews, pers. comm. *H. conferta* and *H. confusa* may be extinct. *H. helenae* is restricted to St. Helena and may be a subspecies of *H. armigera*. *H. pauliana* is only found on Christmas Island (Chr. Is.). Oligophagous species are marked O, and polyphagous P, under Phagy. The following abbreviations apply to host plants, where given for the oliogophages: Solan: Solanaceae; Malv, Malvaceae; Cheno, Chenopodiaceae; Poac, Poaceae.

The basal species in this genus is *H. punctigera* (Wallengren), a polyphagous species widely distributed but restricted to Australia, and an occasional significant pest of cotton. The earlier morphological and allozyme-based assessment (Matthews 1999) of *H. punctigera* has been supported by molecular phylogeny (Cho et al. 2008), and it is unlikely that any of the other species, once subjected to phylogenetic analysis, will prove more basal. Subsequent diversification of the genus has been characterized by migration and the establishment of isolated species around the world, presumably from occasional founder events. The oldest branching species found outside Australia is *H. geloteon*, now established in South America, where it remains a specialized feeder of Poaceae. It has spawned a number of new species there that are all endemic to South America (Table 12.1). Later diversifications of the *Helicoverpa* lineage yielded *H. armigera* and *H. assulta* (Guenée), which both probably also diverged first in Australia (Matthews 1999). On the basis of morphological characters, *H. assulta* is closely related to *H. hardwickii* and *H. prepodes*, which are only found in Australia; confirmation of their relationship awaits characterization at the molecular level. Whereas *H. assulta* has achieved very wide distribution through the Orient, being endemic to Asia, Africa, and Australia (Cho et al. 1995), it remains an oligophage feeding on plants in the Solanaceae and is therefore not viewed as a major pest. It is most likely that *H. assulta* gave rise to another distinctive Hawaiian species (*H. pacifica*) now threatened or extinct as a result of the more recently introduced *H. virescens*.

The *Heliothis* group is an excellent model system for investigating the evolution of the key attributes of pests. With specialist and generalist feeders found in several heliothine genera, there have been a number of studies using the most closely related species pairs available. Such studies have been designed to address the genetics of oviposition, host preference, host feeding, and adaptation (genetics of host range is discussed in Chapter 11). One focus has been on the two species *H. armigera* and *H assulta*. Adaptation to novel hosts by the highly polyphagous *H. armigera* is in part due to the highly heritable ability of newly hatched larvae to feed and develop on novel hosts; feeding by older larvae and oviposition preference were much less heritable so probably have played a smaller role in the evolution of host plant adaptation (Cotter and Edwards 2006). *H. assulta* females show clear oviposition preference for Solanaceous hosts suitable for larval feeding (Cai, Konno, and Matsuda 2002). Interestingly, a host plant vulnerable to feeding by either species (tobacco) shows induction of the same response system (jasmonic acid) for both insect species (Zong and Wang 2007), although feeding by *H. assulta* resulted in higher levels of peroxidase and nicotine and lower levels of foliar polyphenol oxidase. The two species can be crossed to a limited degree: Crosses between males of *H. assulta* and females of *H. armigera* yield only fertile males; the remaining progeny are sterile and morphologically abnormal (Zhao et al. 2005). Although progeny of the reciprocal cross were initially reported as fertile, this has not been confirmed (S.C. Cotter and O.R. Edwards, unpubl.). These two species frequently occur on the same host plants in the field (Cai, Konno, and Matsuda 2002; Zong and Wang 2007), but a key role in their reproductive isolation is played by their respective sex pheromone blends (Ming, Yan, and Wang 2007).

The feeding characteristics of lines resulting from backcrosses of hybrid males to parental lines would allow definition of genes controlling host use. Such studies have begun but were limited by lack of detailed molecular and genetic knowledge for these species (Wang et al. 2004; Tang et al. 2006; S.C. Cotter and O.R. Edwards, unpubl.). The polyphage *H. virescens* and the specialist *H. subflexa* are also able to form hybrids and have been used in studies on host plant adaptation (described in Chapter 11). Interestingly, the specialist *H. subflexa* shows behavioral adaptation to its host plant (*Physalis*), which is also a host for *H. virescens*. *Heliothis subflexa* larvae "hide" within the inflated calyx of *Physalis* fruits while feeding, thus escaping attack by parasitoid wasps (Oppenheim and Gould 2002). In contrast, *H. virescens* feeds while exposed, resulting in survival rates on *Physalis* being much higher for *H. subflexa* than for *H. virescens*. Conceivably, the ability to feed on a wide range of plants may preclude host-specific adaptation.

The widely distributed *H. armigera* has increasingly been the subject of most of the genetic and molecular biology studies carried out on heliothines. Its voraciously polyphagous larvae inflict significant losses to agricultural crop production (Fitt 1989) and readily develop resistance to pesticides (Daly 1993; Heckel et al. 1998; Yang et al. 2004). Studies on the genetic variability of populations in Australia (Daly and Gregg 1985), western Africa (Nibouche et al. 1998), and the eastern Mediterranean region (Zhou et al. 2000) have shown them to be relatively unstructured, with comparable levels of heterozygosity. These studies provided evidence of gene flow across extensive distances, consistent with other evidence that adult moths are efficient flyers, capable of migrating long distances, especially under favorable weather conditions (Farrow 1984; Pedgley 1985; Gregg et al. 1993; Hmimina et al. 1993). Panmictic populations of *H. armigera* therefore cover large areas. Moreover, female *H. armigera* are highly fecund, laying up to three thousand eggs (Mitter, Poole, and Matthews 1993), whereas both female and male moths are capable of multiple mating. Under appropriate climatic conditions, *H. armigera* can have six to seven generations per year, and pupae can undergo diapause under extreme temperature. Indeed, Nibouche et al. (1998) found populations in southern Europe (France and Portugal) to show greater structure, presumably due to the greater relative effect of diapause under these climatic conditions. These characteristics give the species a very high adaptive capability.

DEVELOPMENT OF GENETIC AND GENOMIC RESOURCES FOR *HELICOVERPA ARMIGERA*

GENETIC MAPPING AND GENOMICS

Work in recent years has gradually grown to encompass a broad range of molecular approaches that have shaped our understanding of *Helicoverpa* species. A major step was the application of genetic linkage mapping to identify genes for pyrethroid resistance (Heckel et al. 1998). A similar approach was subsequently used to map *Bacillus thuringiensis* (Bt) resistance to the cadherin gene in *Heliothis* (Gahan, Gould, and Heckel 2001). This work has also indicated extensive microsynteny across the lepidopteran genomes (see abstracts by d'Alençon et al., and Lee and Heckel, in Iatrou and Couble 2007), making gene finding potentially easier in species for which no genomic data are available.

At the time of writing, very little genomic information is available for species in the genus *Helicoverpa*. It is restricted to a limited number of sequences of genomic DNA, predominantly covering genes identified as being implicated in Bt resistance and some expressed sequence tags (ESTs). However, a significant amount of unpublished material has been accumulated by a number of laboratories and is being contributed to the resources used for the proposed International *Helicoverpa* Genome Project. This material includes bacterial artificial chromosome (BAC) libraries and EST databases (see abstracts by d'Alençon et al. and Gordon et al. in Iatrou and Couble 2007). The BAC sequences are of particular interest because they include a significant proportion of the genes for key detoxification enzymes such as cytochrome P450s (Cyp450s), and glutathione S-transferases (GSTs; R. Feyereisen, P. Fournier et al., pers. comm.), considerably extending the data obtained from EST studies.

The study of the capabilities of the Cyp450s (see Feyereisen 2006 for an overview) has begun to shed light on the role of these genes in particular in enabling the host range of *Helicoverpa* species. Certain Cyp450s from *H. zea* show structural changes allowing them to metabolize diverse plant allelochemicals (Li, Berenbaum, and Schuler 2002; Rupasinghe et al. 2007). It is still unclear to what extent this diversification of metabolic capability has been accompanied by a diversification of the genes present in the polyphage genomes. Comparison of the available BAC sequences has given a more complete picture of the complement of detoxification genes in *H. armigera*, as well as in another noctuid pest, *Spodoptera frugiperda*. One finding emerging from the study of genome sequences and cDNA libraries is that gene duplication and divergence are common across the Noctuidae, with both *Spodoptera* and *Helicoverpa* showing different gene duplications. In contrast, the genome of the monophage *Bombyx mori* shows it to have fewer genes for these key enzymes involved in host plant allelochemical detoxification (Mita et al. 2004; Xia et al. 2004).

The identification of genes encoding receptors for the δ-endotoxin of Bt has led to the analysis of significant portions of the heliothine genome. These include the genes for cadherin and the aminopeptidases. A number of deletion mutations of the cadherin gene have been identified in *H. armigera* resistant to Bt Cry1Ac (Xu, Yu, and Wu 2005). A number of aminopeptidases binding to Bt Cry1Ac have also been identified as part of a complete characterization of this family of proteins from *H. armigera* (Angelucci et al. 2008). Comparison of aminopeptidase 3 sequences from *H. armigera* strains susceptible and resistant to Bt Cry1Ac disclosed a small number of mutations that may be associated with the resistance (Wang et al. 2005). The aminopeptidases are among the proteins identified on the peritrophic membrane of *H. armigera* (Campbell et al. 2008); other major classes of proteins comprising this protective membrane include the insect intestinal mucins and the chitin deacetylases.

New Markers for Population Biology and Genetics

The cotton bollworm *H. armigera* represents a major and unique challenge to pest control. Efforts in controlling *H. armigera* infestation have relied on both insecticides (e.g., organophosphates, carbamates, spinosyn, synthetic pyrethroids) and transgenic crops such as Bt cotton. The widespread use of insecticides in controlling *H. armigera* has led to development of insecticide-resistant populations in Asia (e.g., India, Armes et al. 1992; Kranthi et al. 2001; China, Yang et al. 2004); nevertheless, a concerted effort in integrated pest management strategies involving Bt cotton in Australia has so far delayed the development of Bt-resistant populations.

Polyphagy and the ability of larvae to develop insecticide resistance rapidly have resulted in the urgent need to understand the genetic structure of *H. armigera* populations. Despite its significant pest status and socioeconomic costs associated with pesticide usage (i.e., insecticide-related health problems, environmental pollution, and negative impact on the ecology of beneficial insects and parasitoids; Fitt 1989), the population genetics of *H. armigera* remain poorly understood in many affected countries. This is in part due to the limited availability of effective molecular genetic markers such as microsatellites, and has highlighted the need for better markers to be identified and more research in this area.

Development of microsatellite DNA as genetic markers (Tautz 1989) has rapidly replaced allozyme markers in population and evolutionary genetic studies. Unlike for most organisms, efforts in developing lepidopteran simple sequence repeat (SSR) markers have met with limited success (e.g., Butcher, Wright, and Cook 2001; Meglécz et al. 2004; Perera et al. 2007). Although the frequencies of SSR DNA were initially considered low in the Lepidoptera (Nève and Meglécz 2000), significant numbers of SSR DNA markers have since been reported in the silkworm *Bombyx mori* (Reddy, Abraham, and Nagaraju 1999; Prasad et al. 2005a,b), and used to develop a genetic linkage map with more than five hundred markers (Miao et al. 2005; see also Chapter 2). To date, twenty SSR markers have been developed in *H. armigera* (Tan et al. 2001; Ji et al. 2003; Scott et al. 2004; Ji, Wu, and Zhang 2005), and thirteen markers were also reported by Perera et al. (2007) in the closely related (Mitter, Poole, and Matthews 1993; Behere et al. 2007) New World pest *H. zea*. Compared with the results of allozyme, mitochondrial DNA, random amplification of polymorphic DNA (RAPD) markers, and the sodium channel gene as markers in *H. armigera* population genetic studies (Daly and Gregg 1985; McKechnie et al. 1993; Stokes, McKechnie, and Forrester 1997; Nibouche et al. 1998; Zhou et al. 2000; Behere et al. 2007), SSR DNA markers have produced conflicting findings in Australian populations (Scott et al. 2003, 2005a,b, 2006; Endersby et al. 2007). With careful treatment of microsatellite DNA data, Endersby et al. (2007) found *H. armigera* populations from the east coast of Australia to exhibit no significant population substructure patterns, similar to findings from previous population genetic studies of Australian *H. armigera* based on other DNA marker systems (e.g., Daly and Gregg 1985; McKechnie et al. 1993; Stokes, McKechnie, and Forrester 1997). In contrast, Scott et al. (2003, 2005a,b, 2006) consistently detected patterns of population substructure in *H. armigera* from Australian cotton-growing regions based on SSR DNA markers that were suspected to be problematic (i.e., deficient in heterozygosity; Scott et al. 2005b).

To understand factors underlying the conflicting results between Scott et al. (2003, 2005a,b, 2006) and Endersby et al. (2007), it is necessary first to consider the problems encountered and identified in the great majority of lepidopteran SSR DNA markers developed to date. Although low frequencies of SSR DNA in the Lepidoptera can be greatly increased through microsatellite enrichment isolation protocols (e.g., Keyghobadi, Roland, and Strobeck 1999; Scott et al. 2004), a more important factor that prevented the widespread use of lepidopteran SSR DNA markers has been the occurrence of bilateral and unilateral microsatellite DNA families (Ji and Zhang 2004; Meglécz et al. 2004, 2007; van't Hof et al. 2007), loci that share significant homology at the 5′ and/or 3′ flanking sequences. A given set of primers in these flanking sequences may therefore amplify from several loci in the genome, producing multiple DNA banding patterns (Meglécz et al. 2004). Other factors typically affecting lepidopteran microsatellite loci, such as excess of homozygosity, high frequencies of null alleles and allele dropouts, and the lack of Hardy-Weinberg equilibrium may share common underlying causes; their clarification will require detailed investigation of the genesis of microsatellite DNA families. For example, the 3′ termini of non-LTR (long terminal repeat) retrotransposable elements of the RTE (retrotransposable element) clade were recently identified by Tay and colleagues (W.T. Tay, pers. comm.) as being responsible for generating unilateral microsatellite DNA families in over 190 microsatellite DNA loci from at least 22 lepidopteran species, including *H. armigera* (e.g., HarSSR3, Ji et al. 2003; HaD47, Scott et al. 2004; HarSSR7, Ji, Wu, and Zhang 2005) and *H. zea* (HzMS1–6, Perera et al. 2007). These findings confirmed a previous suspicion of association between unilateral microsatellite DNA families and mobile elements (Ji and Zhang 2004; Meglécz et al. 2004, 2007; van't Hof et al. 2007), and contrasts with the mechanism proposed by van't Hof et al. (2007) of unequal crossing-over during DNA recombination. Mechanisms leading to the genesis of new RTE-associated SSR units may involve inaccurate and non-site-specific insertion events of complete and partial copies of these retrotransposable elements at low complexity regions within the host genome (W.T. Tay, pers. comm.). The association of microsatellite DNA loci with non-LTR RTE transposable elements can explain why markers such as HaD47 in *H. armigera* regularly showed high frequencies of null alleles in population genetics analysis (e.g., Endersby et al. 2007), or the significant deficit in observed levels of heterozygosity (e.g., *H. armigera* HarSSR7 marker, Ji, Wu, and Zhang 2005; *H. zea* HzMS1–6 marker, Perera et al. 2007).

Some *H. armigera* SSR markers are known to also occur as bilateral microsatellite DNA families (e.g., Figure 1 in Ji and Zhang 2004), whereby 5' and 3' sequences flanking the SSR units shared high levels of sequence homologies between loci but are nonhomologous within individual loci. Closer examination of the published *H. armigera* HarSSR1 (Ji et al. 2003) and HarSSR8 (Ji, Wu, and Zhang 2005) DNA markers found that these were present in unsuspected families (W.T. Tay, unpubl.), thereby potentially indicating similar difficulties in their application as those identified to be associated with non-LTR RTE1 retroelements (W.T. Tay, pers. comm.). Factors leading to the genesis of bilateral microsatellite DNA families in the Lepidoptera have not been identified and are likely to involve associations with novel transposable elements. Application of published lepidopteran microsatellite DNA markers in population genetic studies will therefore require careful treatment (e.g., Endersby et al. 2007).

The presence of microsatellite DNA families and associated marker problems in the great majority of lepidopteran species may limit their effectiveness in evolutionary and population genetic studies. Alternative lepidopteran nuclear DNA markers are therefore needed. Tay et al. (2008) developed exon-primed intron-crossing (EPIC) DNA markers based on intron length polymorphisms (Palumbi and Baker 1994; Palumbi 1996) of the *Dopa decarboxylase* (*Ddc*) and *ribosomal protein* (*rp*) genes in *H. armigera.* In EPIC DNA markers, the primer annealing sites are located within conserved coding sequences, therefore substantially reducing the likelihood occurrences of allele dropouts and null alleles due to random mutations such as INDELs (insertions and/or deletions) and SNPs (single-nucleotide polymorphisms) at noncoding genome regions. Careful design of EPIC markers to avoid duplicated genes, genes under selection pressure, and known pseudogenes will also greatly minimize undesirable effects such as multiple banding patterns typically associated with lepidopteran SSR DNA markers. Family studies of these EPIC DNA markers in *H. armigera* detected no allele dropouts or null alleles; furthermore, consistent PCR amplification of these EPIC markers was demonstrated in Australian, China, and Indian populations of *H. armigera*. The *H. armigera* EPIC DNA markers also exhibited similar levels of observed heterozygosity as expected for SSR DNA markers. Based on conserved genes that can serve as anchor loci, the *H. armigera* EPIC DNA markers will be an important tool for studying the evolutionary and population genetic structures of this important agricultural pest species, as well as related pest species such as *H. assulta*, *H. zea*, and *H. punctigera*, while also offering a way forward in lepidopteran population and evolutionary genetic studies.

HELIOTHINE TRANSFORMATION: CHALLENGES AND PROSPECTS

There are several benefits of developing germ line transformation in *H. armigera*. Transgenic overexpression can be used to confirm candidate insecticide resistance genes, while transposon-based mutagenesis can rapidly identify resistance genes *de novo*. Such knowledge allows more rational insecticide design as well as prediction and monitoring of resistance alleles in natural populations (Vreysen et al. 2007). The sterile insect technique (SIT), which controls pink bollworm in California annually, would be greatly assisted by the transgenic generation of sterile male-only lines. Transgenic markers such as the green and red fluorescent proteins EGFP and DsRed would also allow the monitoring of released insects to optimize SIT strategies (Alphey 2002; Vreysen et al. 2007).

The vast majority of insect transformations are performed by delivery of class II "cut-and-paste" transposons into eggs by microinjection, targeting progenitor germ line "pole cells" in the developing embryo (G_0; Handler and James 2000). Any G_0 germ line cells transformed will produce insects in the following (G_1) generation that have their whole genome transformed.

The lepidopteran-derived transposon *piggyBac* has been the focus of recent heliothine transformation efforts due to its successful transformation of four insect orders, including Lepidoptera (Handler and James 2000). The transposon *piggyBac* consists of an intronless transposase gene flanked by inverted terminal repeats (ITRs) that are recognized by the transposase and allow precise genome excision and integration (Fraser et al. 1996). For transformation the element is cloned into a bacterial plasmid for propagation, and then its transposase gene is replaced by the DNA of interest to form an "integrator" plasmid. A second "helper" plasmid containing the transposase gene with deleted ITRs is coinjected with the integrator to provide transposase activity, but cannot itself integrate (Handler and James 2000).

Several Lepidoptera have been transformed with *piggyBac*: the pink bollworm, *Pectinophora gossypiella* (Peloquin et al. 2000), a butterfly, *Bicyclus anynana* (Marcus, Ramos, and Monteiro 2004), and the codling moth, *Cydia pomonella* (Neven 2007). The silkworm *B. mori* has been transformed in over fifteen studies (e.g., Tamura et al. 2000), and the transformation frequency (TF: the proportion of injected fertile survivors that produce transformed progeny) is now routinely around 5–10 percent. The other lepidopteran TFs were 3.5–5 percent suggesting all these species could be routinely transformed, and that the vector should be suitable for *H. armigera*. However, attempts to transform *H. armigera*, *H. zea*, and *H. virescens* have proven fruitless with *piggyBac* and other vectors (A.M. Handler, T.A. Miller, pers. comm.; A. Williams, D. Collinge, unpubl.). PBLEs (*piggyBac*-like elements) have recently been identified in these species (Wang et al. 2006; Zimowska and Handler 2006; Sun et al. 2008). Many of them appear to be full-length and therefore could be active, which may prevent element integration or force remobilization. However, multiple PBLEs exist in *P. gossypiella* (T.A. Miller, pers. comm.) and *B. mori* (Xu et al. 2006), which have both been transformed. Although *piggyBac* transformation frequencies in *B. mori* vary significantly between three strains from about 1.5 percent to 14 percent, it is unknown whether this is related to PBLE copy number (Zhong et al. 2007). Hence, in some Lepidoptera multiple homologous elements do not hinder stable integration, while in the heliothines they may, and success or efficiency may be strain-specific.

Promoter Activity

Apart from a good vector, strong promoters are needed to drive expression of the transposase and selectable marker. The first is important to integrate the vector, and the second subsequently to screen transformants. Sometimes these promoters serve both roles (e.g., *Actin A3*, Tamura et al. 2000). The *hsp70* "heat shock" promoter functions in the absence of heat shock when injected into eggs (Tamura et al. 2000), although when transgenically incorporated in *B. mori* it functions as a normal heat-inducible driver (Uhlirova et al. 2002).

Both the *Actin A3* and *hsp70 piggyBac* helpers have been tested in *H. armigera* in conjunction with fluorescent reporters with no successful G_1 transformants isolated (D. Collinge, A. Williams, unpubl.). As mentioned above, this failure may be due to repression from PBLEs, although visualization of the reporter signal may also be a problem as discussed below. Two main promoters, *Actin A3* and *3xP3*, are used to drive the most commonly used selectable markers, enhanced GFP (EGFP) and DsRed in lepidopteran transformations. When injected into 15 to 30 minute-old *H. armigera* embryos, EGFP transcripts are detectable as early as 4 to 6 hours later. Both promoters show EGFP fluorescence in about 80 percent of injected eggs within 16 hours, peaking at 48 hours and continuing up to early first instar stage on day 5 (A. Williams, unpubl.).

Most EGFP in *H. armigera* eggs is excreted in the first frass, and no embryonic tissue fluoresces after that, although EGFP transcripts can be detected weakly in late first instars by RT-PCR (A. Williams, unpubl.). This is consistent with *B. mori* where the bulk of transient expression in eggs is limited to extra-embryonic vitellophage cells that participate in yolk protein metabolism and are subsequently partitioned into the mesenteron (gut) and excreted after hatching (Coulon-Bublex et al. 1993). Whereas the *Actin A3* promoter expresses strongly from plasmids in *B. mori* eggs, hence its usefulness as a helper promoter, it has very low expression in eggs when integrated, but strong expression in the larval gut (Tamura et al. 2000). *Actin A3* driving EGFP in the gut allowed transformants to be identified readily through the larval epidermis in the silkworm (e.g., Tamura et al. 2000) and the pink bollworm, *P. gossypiella* (e.g., Peloquin et al. 2000).

3xP3, which is an eye and nervous system promoter (Berghammer, Klingler, and Wimmer 1999), has been used to drive EGFP in the silkworm (Thomas et al. 2002) and *Bicyclus* (Marcus, Ramos, and Monteiro 2004). An artificial tandem binding site for the metazoan conserved *Pax6* (*eyeless*) transcription factor, *3xP3* expresses strongly in *B. mori* larval stemmata (Thomas et al. 2000); however, the stemmata of *H. armigera* larvae that are expressing large amounts of EGFP upon hatching after transient transfection are not fluorescent (A. Williams, unpubl.), and several germ line screens using *3xP3-EGFP* have failed to detect any transgenics based on stemmata or adult eye fluorescence. It is not known whether this is due to blocking of signal by eye pigmentation or whether this promoter is not a faithful eye enhancer in heliothines. The former seems more likely given that *3xP3* expresses even in the primitive planarian (flatworm) eye (Gonzalez-Estevez et al. 2003), indicating its deep conservation. The role of depigmented mutants in assisting *3xP3* screening is discussed below.

The yeast binary GAL4-UAS system allows for modular tissue-specific targeting of genes (reviewed in Handler and James 2000). In *B. mori* GAL4 has been driven by the *Actin A3* or *3xP3* promoter (Imamura et al. 2003) to drive expression of EGFP or juvenile hormone esterase (Tan et al. 2005). In *H. armigera* eggs the Imamura vectors (Imamura et al. 2003) were able specifically to drive EGFP in the presence but not the absence of the GAL4 drivers, suggesting they would be very useful for ectopic driving of, for example, candidate insecticide resistance genes (A. Williams, unpubl.).

A *piggyBac* vector containing a *D. melanogaster polyubiquitin* (*PUb*) promoter driving EGFP with a nuclear localization signal (Handler and Harrell 1999) was injected into *H. armigera* embryos and produced fluorescence localized to the nuclei of blastoderm cells just beneath the chorion (D. Collinge, unpubl.). Fluorescence peaked after two days but was not visible before hatching. Interestingly, after mating the G_0 survivors, two putative G_1 transgenic larvae were identified based on first instar larval fluorescence, which faded over time. Unfortunately, these putative transformants, which were generated using the *Actin A3* helper, failed to reproduce as adults. Another potential candidate is the lepidopteran viral *hr5-ie1* promoter, which is a good driver in certain Lepidoptera (Mohammed and Coates 2004). This vector expresses EGFP in *H. armigera* eggs, although its strength initially seemed weak (A. Williams, unpubl.).

Fluorescent Screening

A range of fluorescent selectable markers is now in common use in insect transformation systems due to improved sensitivity over previous pigmentation gene complementation systems such as *white*$^+$ and the fact that more than one reporter can be combined in the one insect. Additionally a preexisting pigmentation mutant is not absolutely required, allowing the transformation of a wide range of species (Berghammer, Klingler, and Wimmer 1999; Handler and Harrell 1999; Horn, Jaunich, and Wimmer 2000; Horn et al. 2003).

In all lepidopteran transformations to date, only EGFP and DsRed have been used, and several labs have noted spectral overlap with EGFP and embryonic/larval autofluorescence. *H. armigera* first instar larvae often consume their eggshells after hatching, and this reflective material can easily be mistaken for EGFP fluorescence, leading to false positive G_1 scoring (A. Williams and D. Collinge, pers. obs.). Conversely, the DsRed.T3 variant in conjunction with a Texas Red filter has very low autofluorescence and is the preferable of the two reporters for *H. armigera* transformation.

Although EGFP fluorescence driven by *Actin A3* is easily seen through the pigmented epidermis of wild-type transformed larvae, *3xP3* screens are often performed on G_1 adult eyes where fluorescence can be heavily obscured by wild-type eye pigments (e.g., Marcus, Ramos, and Monteiro 2004). *3xP3* is best visualized in pale-eyed mutant backgrounds (Berghammer, Klingler, and Wimmer 1999; Handler and Harrell 1999; Horn Jaunich, and Wimmer 2000; Thomas et al. 2002). This may in part explain the inability to "see" potential *H. armigera* and other heliothine G_1 transformants using *3xP3-EGFP.*

The *yellow* mutant, isolated from the Australian *H. armigera* Toowoomba strain, lacks ommochrome pigment, converting the adult eye from green to yellow, and the larval stemmata from brown to white (A. Williams, pers. obs.). Adult eye pigmentation can be reduced to almost zero by making *yellow* homozygous with a second Toowoomba dark-eyed mutant (A. Williams, pers. obs.). The depigmented stemmata could be particularly useful in late embryo stage fluorescent stemmata screening as is performed in *B. mori* lines transformed with *3xP3* (e.g., Thomas et al. 2002), which would save extensive G_1 rearing and adult screening.

DNA Delivery Methods

By far the most common method of delivering DNA into insects is egg microinjection (Handler and James 2000; Handler 2002). Unlike *D. melanogaster*, dechorionation is rarely if ever used in lepidopteran microinjection due to its rendering the eggs extreme fragile. Various groups have employed mechanical and micromanipulation techniques to optimize embryo survival and needle longevity (Peloquin et al. 1997; Tamura et al. 2000). In *H. armigera* up to three hundred eggs can routinely be injected with a single sharp silicon borate needle, and survival is generally greater than 50 percent, comparable to *Drosophila* microinjections (A. Williams, unpubl.). If *H. armigera* embryos are injected 15 to 30 minutes after deposition, vector dispersal is improved, consistent with reports in other Lepidoptera (Coulon-Bublex et al. 1993; Bossin et al. 2007).

Early injection should allow spread of vector into all regions of the egg including the crucial pole cells.

Alternative methods for DNA delivery into lepidopteran embryos include biolistics, electroporation, and viral mediated, the latter two having successfully produced transformants. Electroporation was used to transform *H. zea* using *hobo* (DeVault et al. 1996), and remains the only report of a heliothine germ line transformation. The transformation frequency was 1.7 percent, and the reporter, *lacZ*, was detected for five generations by Southern blot. Also notable was that PCR detection of the *lacZ* gene in the pupal cases of surviving G_0 individuals predicted in every instance that the individuals would produce germ line transformants. Although this technical feat has not been repeated for heliothines, electroporation would theoretically be a far more efficient method of DNA delivery than microinjection, and has been used recently to deliver DNA to silkworm embryos (Guo et al. 2004).

The first germ line transformations of *B. mori* were achieved using baculovirus vectors (Mori et al. 1995; Yamao et al. 1999). In Yamao et al. (1999), less than three hundred fifth instar larvae were injected with the virus, which spread systemically into the germ cells and went on to produce transgenics with a TF of 2.7 percent. This paper was also remarkable in that it achieved homologous recombination, a first for insects, fusing *EGFP* into exon 7 of the fibroin light chain gene in the vector, which then homologously recombined with the endogenous fibroin gene in a small proportion of transformants. Like DeVault et al. (1996) the researchers also used a PCR-based screening strategy, supported by Southern and Western blots. Although *B. mori* can survive infection by this recombinant baculovirus (specifically Baculogold AcNPV), it is generally assumed to be lethal for most other Lepidoptera.

Biolistic delivery (e.g., bombardment with DNA-coated particles) has been used to deliver DNA into *B. mori* embryos and tissue (Thomas et al. 2001) and dissected silk glands (Takahashi et al. 2003). Biolistic delivery was tested on *H. armigera* embryos using gold particles coated with *piggyBac* DNA (D. Collinge, unpubl.). The outer chorion formed an impassable barrier, requiring its partial removal to allow penetration. As a consequence the structural integrity of the embryo was lost, resulting in high mortality rates. None of the attempts using biolistics produced a transgenic *H. armigera* line.

More sophisticated transformation methods to improve integration frequencies and provide routine and efficient site-specific and consistent genomic targeting are currently in use in *D. melanogaster*, such as phiC31 integrase, but still rely on the basic type II transposon system to integrate these robust recombinatory systems into the genome (Bischof et al. 2007). Hence unfortunately one must have a basic transformation system in place before one can improve upon it.

Future Prospects in *H. armigera* Transformation

Many "mysterious" failures in transforming heliothines are starting to be understood in the broader light of lepidopteran transformation research. Challenges to address in the development of a functional transgenic system for *H. armigera* include (1) screening and exclusion of strains carrying endogenous transposases such as PBLEs prior to transformation, (2) maximizing transposase and reporter expression by isolating native promoters, and (3) improving visibility and screening by using depigmented mutants.

Southern blotting or PCR screening for PBLE-free *yellow* and dark-eyed mutant lines may identify a line potentially permissive to *piggyBac 3xP3*, and DsRed would be the reporter of choice. Strong candidate native egg-specific promoters include the egg pigmentation gene *white3* (A. Williams, unpubl.), the germ line–specific gene *vasa*, or a constitutively expressed ribosomal protein gene like *RpS3A*. A return to preliminary PCR screens of G_0 or G_1 batches followed by Southern blot confirmation of transformants (DeVault et al. 1996; Yamao et al. 1999) may also bypass any reporter visualization problems and automate the screening process to a degree.

It may be advantageous to revisit the use of *hobo* because it is still the only vector demonstrated to transform a heliothine (DeVault et al. 1996). Despite a *Hermes* screen failing in *H. armigera* with *3xP3-EGFP* (A. Williams, unpubl.), it was successful in *B. anynana* (Marcus, Ramos, and Monteiro 2004) so may also be worth trying after a "rebuild" with native promoters. Success of *Minos* in *B. mori* (Uchino et al. 2007) emphasizes its potential, while *Mariner* has also transformed numerous phyla and may prove useful (Hartl et al. 1997; Wang, Swevers, and Iatrou 2000). Transposition assays (Handler and James 2000) will be useful to test new vectors and strains in light of potential negative effects of PBLEs and related transposon copies/relics, as well as the effectiveness of coinjection of transposase RNA, which may bypass any driver limitations.

Despite setbacks the future seems positive for *H. armigera* transformation. Many of the technical basics are in place: Microinjection efficiencies and survival are good; EGFP, DsRed and various promoters are active; and GAL4-UAS is functional. Utilizing strong egg-specific native promoters and transposon-free and depigmented recipient strains may assist in transgenic screening and boost transformation efficiencies. Overcoming these hurdles should bring heliothine transformation on par with *B. mori* and provide greater tools for the resistance and SIT communities.

RNA INTERFERENCE AND ITS STUDY IN INSECTS

RNAi describes the process by which dsRNA triggers the sequence-specific destruction of a cellular mRNA, thereby silencing expression of that gene. First identified in plants, where it was known as posttranscriptional gene silencing (PTGS), it has also been observed in some animals, and in particular in *C. elegans* (Guo and Kemphues 1995; Fire et al. 1998). In these systems cellular dicerlike enzymes cleave the dsRNA into short fragments; the short RNAs are then able to trigger cleavage of cellular mRNAs through binding to an RNA-induced silencing complex that includes the RNase Argonaute (see Meister and Tuschl 2004, for a review). In plants and nematodes, amplification of the dsRNA that triggers further silencing is enabled by cellular RNA-dependent RNA polymerases (RdRps) that are able to recognize short RNAs as primers on the mRNA, resulting in the systemic spread of the gene silencing (for overviews, see Baulcombe 2007; Gordon and Waterhouse 2007). The RNAi machinery appears to play a central role in defense against viruses (Ding and Voinnet 2007). While this was originally perceived to be its main function, the RNAi machinery has now been implicated in many other aspects of cell biology, including transposon silencing, pairing-sensitive silencing, telomere function, chromatin insulator activity, nucleolar stability, and heterochromatin formation; it is also involved in the regulation of gene expression by miRNAs (for reviews, see Nilsen 2007; Filipowicz, Bhattacharyya, and Sonenberg 2008; Kavi et al. 2008). Genome-wide screens using RNAi have been widely used on model organisms such as *C. elegans* (Kim et al. 2005) and *D. melanogaster* (Mathey-Prevot and Perrimon 2006). For researchers of nonmodel organisms, RNAi has provided a unique opportunity to study gene function through the specific knockdown of gene expression. This does not require complete knowledge of the sequence of the gene of interest: generally, a gene fragment of 200–500 bp is all that is required to generate dsRNA for the initiation of an RNAi response.

Attempts to achieve RNAi in insects, including Lepidoptera, have usually involved injection with dsRNA. Systemic RNAi, in which this gene silencing spreads throughout the insect, has been observed in coleopteran insects such as *Tribolium* (Ober and Jokusch 2006; Tomoyasu et al. 2008); parental RNAi (i.e., affecting the next generation) has also been demonstrated in these insects through the silencing of three developmental genes (Bucher, Scholten, and Klingler 2002; Schröder 2003). Similar observations have been made in the wasp *Nasonia* (Lynch and Desplan 2006). In the grasshopper *Schistocerca americana*, RNAi targeting the eye-color gene vermilion in first instar nymphs triggered suppression of ommochrome formation in the eye lasting through two instars (Dong and Friedrich 2005). Although attempts to achieve systemic RNAi in dipterans have been unsuccessful (Roignant et al. 2003), tissue- and stage-specific RNAi are now routine in *Drosophila* by transgenesis (Kennerdell and Carthew 2000), allowing detailed studies on the RNAi pathway

(Kavi et al. 2008). Comparative genomic studies have confirmed that most insects possess almost all genes required for RNAi, including (outside the Diptera) candidate dsRNA transport proteins (SID-1); they lack the RdRp required for amplification (Honeybee Genome Sequencing Consortium 2006; Gordon and Waterhouse 2007; Tomoyasu et al. 2008).

RNAi has been demonstrated in a variety of Lepidoptera, including *B. mori* (Quan, Kanda, and Tamura 2002; Uhlirova et al. 2003; Huang et al. 2007), *Plodia interpunctella* (Fabrick, Kanost, and Baker 2004), *Spodoptera litura* (Rajagopal et al. 2002), *Spodoptera exigua* (Herrero et al. 2005), *Hyalophora cecropia* (Bettencourt, Terenius, and Faye 2002), and in cultured *Manduca sexta* neuronal cells (Vermehren, Qazi, and Trimmer 2001). In *B. mori*, silencing of the *white* gene yielded a clear phenotype (white eggs, translucent larval cuticle) upon injection of eggs (Quan, Kanda, and Tamura 2002). Another phenotype-altering RNAi experiment in *B. mori* involved use of a Sindbis virus vector to deliver dsRNA silencing the Broad-Complex transcription factor (Br-C; Uhlirova et al. 2003); larvae showed a range of clear morphogenetic defects, suggesting a conserved role for Br-C in metamorphosis. In an experiment asking whether a phenotypic effect could be observed in adults, silencing of the *bursicon* gene by pupal injection of dsRNA was shown to affect wing expansion (Huang et al. 2007). For the most part, these studies have focused on the knockdown of genes for functional analysis rather than for control purposes. However, they have provided encouragement for the development of an RNAi-based control method for other Lepidoptera.

Recent work on *H. armigera* embryos has demonstrated functional RNAi machinery and the possibility of achieving at least a transient RNAi effect. In the absence of any transgenic line, a transient silencing assay was developed by injecting an EGFP reporter construct into the preblastoderm embryos. For silencing, different forms of dsRNA were coinjected with the expression plasmid and resulting levels of EGFP expression compared. This approach showed the transient expression of EGFP could be silenced by coinjection of dsRNA (D. Collinge, unpubl.). Extending this work, A. Williams (unpubl.) showed that the *white* gene could be transiently silenced by injection of dsRNA into preblastoderm embryos, resulting in larval phenotypes similar to those seen for *B. mori* (Quan, Kanda, and Tamura 2002).

Additionally, strong mRNA knockdown associated with a depigmented phenotype was observed after embryo injection of either of two 22-bp duplex short interfering RNAs (siRNAs) against the *H. armigera white* gene, a phenotype extending to the end of first instar. Similar knockdown of about 90 percent was observed with EGFP and the pyrethroid target *para* using siRNAs. Mixing and matching of siRNAs and two vectors provided unambiguous phenotypes from up to four components at a time in the same egg (A. Williams, pers. obs.). This is, therefore, a promising method to analyze efficiently the functions of complex redundant combinatorial systems such as the subunit composition of the nicotinic acetylcholine receptor (a target of neonicotinoid insecticides), which has remained elusive in insects. Such studies could facilitate improved insecticide design.

RNAi for Pest Control

The development of an effective, orally delivered RNAi system for a major pest insect like *H. armigera* would provide a powerful tool, not only for reverse genetics but also as a potential species-specific method of controlling this pest. Transgene-encoded RNAi has long been known to be feasible in plants (Waterhouse, Graham, and Wang 1998), as oral delivery of dsRNA can trigger RNAi in nematodes (Timmons and Fire 1998). Although these findings have led to efforts to ask whether plants might be protected from herbivorous insects by engineering them to express dsRNAs targeting insect genes, the absence of reports of success from this approach seemed to suggest that the expression of hairpin RNA in plant cells did not provide sufficient intact dsRNA to trigger potent RNAi upon ingestion by the insect. Indeed, until recently, the only report of orally delivered systemic RNAi in a lepidopteran showed only partial silencing of a gene in the light brown apple moth, *Epiphyas postvittana*, in response to the feeding of large amounts of dsRNA (Turner et al. 2006).

Two recent papers have dramatically shown the potential to use RNAi from transgene-encoded hairpin RNAs for defense for crops against insect pests. The first report that *H. armigera* may be susceptible to transgene-encoded ingestible dsRNA has come from work by Mao et al. (2007), who initially identified a cytochrome P450 monooxygenase (Cyp6AE12) induced in *H. armigera* by exposure to gossypol, the main defense allelochemical produced in cotton (*Gossypium hirsutum*) plants. They then generated transgenic *Arabidopsis thaliana* plants expressing dsRNA hairpins corresponding to this gene's sequence. This dsRNA was able to silence the gene in larvae feeding on the plants, so that the larvae were susceptible to gossypol added to the leaves. Mao et al. (2007) were further able to use mutant lines of *A. thaliana* available to ask whether the dsRNA molecules inducing the RNAi in the feeding larvae were long hairpin RNAs rather than siRNAs. These siRNAs are normally generated by the plant's dicer-like (dcl) enzymes: *Arabidopsis* plants with mutations in *Dcl2, Dcl3*, and *Dcl4* produced more intact hairpin RNA and caused more pronounced silencing of the P450 gene. Some siRNAs were still detected in the plants, probably due to the presence of another, unmutated dicer gene—*Arabidopsis* has four such genes (Margis et al. 2006). These observations suggest that the efficacy of insect gene silencing by plant-produced long dsRNA may depend to a significant extent on the hairpin dsRNAs being produced more quickly in transgenic plants than they can be processed by the plant's dicer enzymes (Gordon and Waterhouse 2007), although the importance of the siRNAs produced in the plant itself cannot be ruled out. These observations on silencing of an *H. armigera* gene were corroborated in a paper published at the same time by Baum et al. (2007), who showed that corn plants were effectively protected against coleopteran pests such as western corn rootworm (*Diabrotica virgifera*), southern corn rootworm (*Diabrotica undecimpunctata*), and Colorado potato beetle (*Leptinotarsa decemlineata*).

Although preliminary, this work has raised exciting new possibilities in pest control, particularly as more comprehensive genetic information and even complete genome data become available for pests in the genus *Helicoverpa*. Nevertheless, the approach faces many important questions, not only about the mechanism but also the challenges facing adoption of this control strategy in the field (Gordon and Waterhouse 2007). In particular for the Heliothinae pests, an outstanding question is whether the plant-mediated provision of ingestible RNAi will be circumvented by sequence polymorphisms in pest populations. It is likely that for the coleopteran pests the successes seen in the lab will translate into effective pest control in the field, as the assay involved measuring root damage on the crop itself. For *Helicoverpa*, the question is how effectively crops like cotton could be protected.

CONCLUSIONS AND PROSPECTS

The gradual application of the tools of molecular biology, genetics, and genomics to insects in the genus *Helicoverpa* is bringing closer the day when their full potential for research into fundamental questions of insect biology and host range evolution can be explored. While some genes and gene families have already been well studied, it will be the completion of the genome analysis of *H. armigera* in the near future that has the most potential to greatly increase our understanding of this important pest genus and transform our ability to study it. Researchers would have the capability to study population genomics and map resistance genes, and QTL for such traits as pesticide resistance, diapause proclivity, and host choice. Evolution of host plant adaptation (e.g., through detoxification of plant allelochemicals) could be studied using selected host plants, including those for which genomes are available. Comparisons to closely related species and those for which interspecific crosses are possible (e.g., *H. assulta*) would allow detailed analyses of the difference between generalist and specialist feeders (host specificity is discussed in Chapter 11). Migration patterns could be studied with greater accuracy, allowing the rigorous assessment of population suppression options and the value of refuges for resistance management.

REFERENCES

Alphey, L. 2002. Re-engineering the sterile insect technique. *Insect Biochem. Mol. Biol.* 32:1243–47.

Angelucci, C., G.A. Barrett-Wilt, D.F. Hunt, et al. 2008. Diversity of aminopeptidases, derived from four lepidopteran gene duplications, and polycalins expressed in the midgut of *Helicoverpa armigera*: Identification of proteins binding the δ-endotoxin, Cry1Ac of *Bacillus thuringiensis*. *Insect Biochem. Mol. Biol.* 38:685–96.

Armes, N.J., D.R. Jadhav, G.S. Bond, and A.B.S. King. 1992. Insecticide resistance in *Helicoverpa armigera* in South India. *Pestic. Sci.* 34:355–64.

Baulcombe, D.C. 2007. Amplified silencing. *Science* 315:199–200.

Baum, J.A., T. Bogaert, W. Clinton, et al. 2007. Control of coleopteran insect pests through RNA interference. *Nat. Biotechnol.* 25:1322–26.

Behere, G.T., W.T. Tay, D.A. Russell, et al. 2007. Mitochondrial DNA analysis of field populations of *Helicoverpa armigera* (Lepidoptera: Noctuidae) and of its relationship to *H. zea*. *BMC Evol. Biol.* 7:17.

Berghammer, A.J., M. Klingler, and E.A. Wimmer. 1999. A universal marker for transgenic insects. *Nature* 402:370–71.

Bettencourt, R., O. Terenius, and I. Faye. 2002. Hemolin gene silencing by ds-RNA injected into *Cecropia* pupae is lethal to next generation embryos. *Insect Mol. Biol.* 11:267–71.

Bischof, J., R.K. Maeda, M. Hediger, F. Karch, and K. Basler. 2007. An optimized transgenesis system for *Drosophila* using germ-line-specific phiC31 integrases. *Proc. Natl. Acad. Sci. U.S.A.* 104:3312–17.

Bossin, H., R.B. Furlong, J.L. Gillett, M. Bergoin, and P.D. Shirk. 2007. Somatic transformation efficiencies and expression patterns using the JcDNV and *piggyBac* transposon gene vectors in insects. *Insect Mol. Biol.* 16:37–47.

Bucher, G., J. Scholten, and M. Klingler. 2002. Parental RNAi in *Tribolium* (Coleoptera). *Curr. Biol.* 12: R85–86.

Butcher, R.D.J., D. J. Wright, and J.M. Cook. 2001. Development and assessment of microsatellites and AFLPs for *Plutella xylostella*. In *The management of diamondback moth and other crucifer pests*, Proc. 4th Int. Workshop, ed. N.M. Endersby and P.M. Ridland, 87–93. Melbourne: The Regional Institute.

Cai, C.Y., Y. Konno, and K. Matsuda. 2002. Studies on the ovipositional preferences and feeding preferences in *Helicoverpa assulta* and *Helicoverpa armigera*. *Tohoku J. Agric. Res.* 53:11–24.

Campbell, P.M., A.T. Cao, E.R. Hines, P.D. East, and K.H.J. Gordon. 2008. Proteomic analysis of the peritrophic matrix from the gut of the caterpillar, *Helicoverpa armigera*. *Insect Bioch. Mol. Biol.* 38:950–8.

Cho, S. 1997. Molecular phylogenetics of the Heliothinae (Lepidoptera: Noctuidae) based on the nuclear genes for elongation factor-1α and dopa decarboxylase. PhD diss., Univ. Maryland.

Cho, S., A. Mitchell, C. Mitter, J. Regier, M. Matthews, and R. Robertson. 2008. Molecular phylogenetics of heliothine moths (Lepidoptera: Noctuidae: Heliothinae), with comments on the evolution of host range and pest status. *Syst. Entomol.* 33:581–94.

Cho, S., A. Mitchell, J.C. Regier, et al. 1995. A highly conserved nuclear gene for low-level phylogenetics: Elongation factor-1α recovers morphology-based tree for heliothine moths. *Mol. Biol. Evol.* 12:650–56.

Cotter, S.C., and O.R. Edwards. 2006. Quantitative genetics of preference and performance on chickpeas in the noctuid moth, *Helicoverpa armigera*. *Heredity* 96:396–402.

Coulon-Bublex, M., N. Mounier, P. Couble, and J.C. Prudhomme. 1993. Cytoplasmic *Actin A3* gene promoter injected as supercoiled plasmid is transiently active in *Bombyx mori* embryonic vitellophages. *Roux's Arch. Dev. Biol.* 203:123–27.

Daly, J.C. 1993. Ecology and genetics of insecticide resistance in *Helicoverpa armigera*: Interactions between selection and gene flow. *Genetica* 90:217–26.

Daly, J.C., and P. Gregg. 1985. Genetic variation in *Heliothis* in Australia: Species identification and gene flow in the two species *H. armigera* (Hübner) and *H. punctigera* (Wallengren) (Lepidoptera: Noctuidae). *Bull. Entomol. Res.* 75:169–84.

DeVault, J.D., K.J. Hughes, R.A. Leopold, O.A. Johnson, and S.K. Narang. 1996. Gene transfer into corn earworm (*Helicoverpa zea*) embryos. *Genome Res.* 6:571–79.

Ding, S.W., and O. Voinnet. 2007. Antiviral immunity directed by small RNAs. *Cell* 130:413–26.

Dong, Y., and M. Friedrich. 2005. Nymphal RNAi: Systemic RNAi mediated gene knockdown in juvenile grasshopper. *BMC Biotechnol.* 5:25.

Endersby, N.M., A.A. Hoffmann, S.W. Mckechnie, and A.R. Weeks. 2007. Is there genetic structure in populations of *Helicoverpa armigera* from Australia? *Entomol. Exp. Appl.* 122:253–63.

Fabrick, J.A., M.R. Kanost, and J.E. Baker. 2004. RNAi-induced silencing of embryonic tryptophan oxygenase in the pyralid moth, *Plodia interpunctella*. *J. Insect Sci.* 4:15.

Farrow, R.A. 1984. Detection of transoceanic migration of insects to a remote island in the Coral Sea, Willis Island. *Aust. J. Ecol.* 9:253–72.

Feyereisen, R. 2006. Evolution of insect P450. *Biochem. Soc. Trans.* 34:1252–55.

Filipowicz, W., S.N. Bhattacharyya, and N. Sonenberg. 2008. Mechanisms of post-transcriptional regulation by microRNAs: Are the answers in sight? *Nat. Rev. Genet.* 9:102–14.

Fire, A., S. Xu, M.K. Montgomery, S.A. Kostas, S.E. Driver, and C.C. Mello. 1998. Potent and specific genetic interference by double-stranded RNA in *Caenorhabditis elegans*. *Nature* 391:806–11.

Fitt, G.P. 1989. The ecology of *Heliothis* species in relation to agroecosystems. *Annu. Rev. Entomol.* 34:17–52.

Fraser, M.J., T. Ciszczon, T. Elick, and C. Bauser. 1996. Precise excision of TTAA-specific lepidopteran transposons *piggyBac* (IFP2) and tagalong (TFP3) from the baculovirus genome in cell lines from two species of Lepidoptera. *Insect Mol. Biol.* 5:141–51.

Gahan, L.J., F. Gould, and D.G. Heckel. 2001 Identification of a gene associated with Bt resistance in *Heliothis virescens*. *Science* 293:857–60.

Gonzalez-Estevez, C., T. Momose, W.J. Gehring, and E. Salo. 2003. Transgenic planarian lines obtained by electroporation using transposon-derived vectors and an eye-specific GFP marker. *Proc. Natl. Acad. Sci. U.S.A.* 100:14046–51.

Gordon, K.H.J., and P.M. Waterhouse. 2007. RNAi for insect-proof plants. *Nat. Biotechnol.* 25:1231–32.

Gregg, P.C., G.P. Fitt, M. Coombs, and G.S. Henderson. 1993. Migrating moths (Lepidoptera) collected in tower-mounted light traps in northern New South Wales, Australia: Species composition and seasonal abundance. *Bull. Entomol. Res.* 83:53–78.

Guo, X.Y., L. Dong, S.P. Wang, T.Q. Guo, J.Y. Wang, and C.D. Lu. 2004. Introduction of foreign genes into silkworm eggs by electroporation and its application in transgenic vector test. *Acta Biochim. Biophys. Sin. (Shanghai)* 36:323–30.

Guo, S., and K.J. Kemphues. 1995. *par-1*, a gene required for establishing polarity in *C. elegans* embryos, encodes a putative Ser/Thr kinase that is asymmetrically distributed. *Cell* 81:611–20.

Handler, A.M. 2002. Use of the *piggyBac* transposon for germ-line transformation of insects. *Insect Biochem. Mol. Biol.* 32:1211–20.

Handler, A.M., and R.A. Harrell II. 1999. Germline transformation of *Drosophila melanogaster* with the *piggyBac* transposon vector. *Insect Mol. Biol.* 8:449–57.

Handler, A.M., and A.A. James, Eds. 2000. *Insect transgenesis: Methods and applications*. Boca Raton: CRC Press.

Hardwick, D.F. 1965. The corn earworm complex. *Mem. Entomol. Soc. Canada* 40:1–248.

Hartl, D.L., E.R. Lozovskaya, D.I. Nurminsky, and A.R. Lohe. 1997. What restricts the activity of mariner-like transposable elements? *Trends Genet.* 13:197–201.

Heckel, D.G., L.J. Gahan, J.C. Daly, and S. Trowell. 1998. A genomic approach to understanding *Heliothis* and *Helicoverpa* resistance to chemical and biological insecticides. *Phil. Trans. R. Soc. Lond. B* 353:1713–22.

Herrero, S., T. Gechev, P.L. Bakker, W.J. Moar, and R.A. de Maagd. 2005. *Bacillus thuringiensis* Cry1Ca-resistant *Spodoptera exigua* lacks expression of one of four Aminopeptidase N genes. *BMC Genomics* 6:96.

Hmimina, M., S. Poitout, and R. Buès. 1993. Variabilité des potentialités diapausantes intra et interpopulations chez *Heliothis armigera* Hb (Lep: Noctuidae). *J. Appl. Entomol.* 116:273–83.

Honeybee Genome Sequencing Consortium. 2006. Insights into social insects from the genome sequence of the honeybee *Apis mellifera*. *Nature* 443:931–49.

Horn, C., B. Jaunich, and E.A. Wimmer. 2000. Highly sensitive, fluorescent transformation marker for *Drosophila* transgenesis. *Dev. Genes Evol.* 210:623–29.

Horn, C., N. Offen, S. Nystedt, U. Hacker, and E.A. Wimmer. 2003. *piggyBac*-based insertional mutagenesis and enhancer detection as a tool for functional insect genomics. *Genetics* 163:647–61.

Huang, J., Y. Zhang, M.L.S. Wang, et al. 2007. RNA interference-mediated silencing of the bursicon gene induces defects in wing expansion of silkworm. *FEBS Lett.* 581:697–701.

Iatrou, K., and P. Couble. 2007. 7th International Workshop on the Molecular Biology and Genetics of the Lepidoptera. August 20–26, 2006. Orthodox Academy of Crete, Kolympari, Crete, Greece. 52 pp. *J. Insect Sci.* 7:29.

Imamura, M., J. Nakai, S. Inoue, G.X. Quan, T. Kanda, and T. Tamura. 2003. Targeted gene expression using the GAL4/UAS system in the silkworm *Bombyx mori*. *Genetics* 165:1329–40.

Ji, Y.-J., and D.-X. Zhang. 2004. Characteristics of microsatellite DNA in lepidopteran genomes and implications for their isolation. *Acta Zool. Sin.* 50:608.

Ji, Y.-J., Y.-C. Wu, and D.-X. Zhang. 2005. Novel polymorphic microsatellite markers developed in the cotton bollworm *Helicoverpa armigera* (Lepidoptera: Noctuidae). *Insect Sci.* 12:331–34.

Ji, Y.-J., D.-X. Zhang, G.M. Hewitt, L. Kang, and D.-M. Li. 2003. Polymorphic microsatellite loci for the cotton bollworm *Helicoverpa armigera* (Lepidoptera: Noctuidae) and some remarks on their isolation. *Mol. Ecol. Notes* 3:102–04.

Kavi, H.H., H. Fernandez, W. Xie, and J.A. Birchler. 2008. Genetics and biochemistry of RNAi in *Drosophila*. *Curr. Top. Microbiol. Immunol.* 320:37–75.

Kennerdell, J.R., and R.W. Carthew. 2000. Heritable gene silencing in *Drosophila* using double-stranded RNA. *Nat. Biotechnol.* 18:896–98.

Keyghobadi, N., J. Roland, and C. Strobeck. 1999. Influence of landscape on the population genetic structure of the alpine butterfly *Parnassius smintheus* (Papilionidae). *Mol. Ecol.* 8:1481–95.

Kim, J.K., H.W. Gabel, R.S. Kamath, et al. 2005. Functional genomic analysis of RNA interference in *C. elegans*. *Science* 308:1164–67.

Kranthi, K.R., D. Jadhav, R. Wanjari, S. Kranthi, and D. Russell. 2001. Pyrethroid resistance and mechanisms of resistance in field strains of *Helicoverpa armigera* (Lepidoptera: Noctuidae). *J. Econ. Entomol.* 94:253–63.

Li, X., M.R. Berenbaum, and M.A. Schuler. 2002. Plant allelochemicals differentially regulate *Helicoverpa zea* cytochrome p450 genes. *Insect Mol. Biol.* 11:343–51.

Lynch, J.A., and C. Desplan. 2006. A method for parental RNA interference in the wasp *Nasonia vitripennis*. *Nat. Protoc.* 1:486–94.

Mallet, J., A. Korman, D. Heckel, and P. King. 1993. Biochemical genetics of *Heliothis* and *Helicoverpa* (Lepidoptera: Noctuidae) and evidence for a founder event in *Helicoverpa zea*. *Ann. Entomol. Soc. Am.* 86:189–97.

Mao, Y.-B., W.-J. Cai, J.-W. Wang, et al. 2007. Silencing a cotton bollworm P450 monooxygenase gene by plant-mediated RNAi impairs larval tolerance of gossypol. *Nat. Biotechnol.* 25:1307–13.

Marcus, J.M., D.M. Ramos, and A. Monteiro. 2004. Germline transformation of the butterfly *Bicyclus anynana*. *Proc. Biol. Sci.* 27:S263–65.

Margis, R., A.F. Fusaro, N.A. Smith, et al. 2006. The evolution and diversification of Dicers in plants. *FEBS Lett.* 580: 2442–50.

Mathey-Prevot, B., and N. Perrimon. 2006. *Drosophila* genome-wide RNAi screens: Are they delivering the promise? *Cold Spring Harb. Symp. Quant. Biol.* 71:141–48.

Matthews, M. 1999. *Heliothine moths of Australia: a guide to pest bollworms and related noctuid groups*. Melbourne: CSIRO Publishing.

McKechnie, S.W., M.E. Spackman, N.E. Naughton, I.V. Kovacs, M. Ghosn, and A.A. Hoffman. 1993. Assessing budworm population structure in Australia using the A-T rich region of mitochondrial DNA. In *Proc. Beltwide Cotton Conf.*, ed. D.J. Herber and D.J. Richter, 838–840. New Orleans, LA. January 10–14, 1993. Natl. Cotton Council of Am., Memphis, TN.

Meglécz, E., S.J. Anderson, D. Bourguet, et al. 2007. Microsatellite flanking region similarities among different loci within insect species. *Insect Mol. Biol.* 16:175–85.

Meglécz, E., F. Petenian, E. Danchin, A.C. D'Acier, J.-Y. Rasplus, and E. Faure. 2004. High similarity between flanking regions of different microsatellites detected within each of two species of Lepidoptera: *Parnassius apollo* and *Euphydryas aurinia*. *Mol. Ecol.* 13:1693–700.

Meister, G., and T. Tuschl. 2004. Mechanisms of gene silencing by double-stranded RNA. *Nature* 431:343–49.

Miao, X.-X., S.-J. Xu, M.-H. Li, et al. 2005. Simple sequence repeat-based consensus linkage map of *Bombyx mori*. *Proc. Natl. Acad. Sci. U.S.A.* 102:16303–08.

Ming, Q.L., Y.H. Yan, and C.Z. Wang. 2007. Mechanisms of premating isolation between *Helicoverpa armigera* (Hübner) and *Helicoverpa assulta* (Guenee) (Lepidoptera: Noctuidae). *J. Insect Physiol.* 53:170–78.

Mita, K., M. Kasahara, S. Sasaki, et al. 2004. The genome sequence of silkworm, *Bombyx mori*. *DNA Res.* 11:27–35.

Mitter, C., R.W. Poole, and M. Matthews. 1993. Biosystematics of the Heliothinae (Lepidoptera: Noctuidae). *Annu. Rev. Entomol.* 38:207–25.

Mohammed, A., and C.J. Coates. 2004. Promoter and *piggyBac* activities within embryos of the potato tuber moth, *Phthorimaea operculella*, Zeller (Lepidoptera: Gelechiidae). *Gene* 342:293–301.

Mori, H., M. Yamao, H. Nakazawa, et al. 1995. Transovarian transmission of a foreign gene in the silkworm, *Bombyx mori*, by *Autographa californica* nuclear polyhedrosis virus. *Biotechnology (NY)* 13:1005–07.

Nève, G., and E. Meglécz. 2000. Microsatellite frequencies in different taxa. *Trends Ecol. Evol.* 15:376–77.

Neven, L. 2007. Development of transgenic codling moth for ABC and SIT. *Entomol. Res.* 37:A54.

Nibouche, S., R. Bues, J.-F. Toubon, and S. Poitout. 1998. Allozyme polymorphism in the cotton bollworm *Helicoverpa armigera* (Lepidoptera: Noctuidae): Comparison of African and European populations. *Heredity* 80:438–45.

Nilsen, T.W. 2007. Mechanisms of microRNA-mediated gene regulation in animal cells. *Trends Genet.* 23:243–49.

Ober, K.A., and E.L. Jockusch. 2006. The roles of wingless and decapentaplegic in axis and appendage development in the red flour beetle, *Tribolium castaneum. Dev. Biol.* 294:391–405.

Oppenheim, S.J., and F. Gould. 2002. Behavioral adaptations increase the value of enemy-free space for *Heliothis subflexa*, a specialist herbivore. *Evolution* 56:679–89.

Palumbi, S.R. 1996. Nucleic acids II: The polymerase chain reaction. In *Molecular systematics*, ed. D.M. Hillis, C. Moritz, and B.K. Mable, 205–47. Sunderland, MA: Sinauer Associates.

Palumbi, S.R., and C.S. Baker. 1994. Contrasting population structure from nuclear intron sequences and mtDNA of humpback whales. *Mol. Biol. Evol.* 11:426–35.

Pedgley, D.E. 1985. Windborne migration of *Heliothis armigera* (Hübner) (Lepidoptera: Noctuidae) to the British Isles. *Entomol. Gaz.* 36:15–20.

Peloquin, J.J., S.T. Thibault, L.P. Schouest, Jr., and T.A. Miller. 1997. Electromechanical microinjection of pink bollworm *Pectinophora gossypiella* embryos increases survival. *Biotechniques* 22:496–99.

Peloquin, J.J., S.T. Thibault, R. Staten, and T.A. Miller. 2000. Germ-line transformation of pink bollworm (Lepidoptera: Gelechiidae) mediated by the *piggyBac* transposable element. *Insect Mol. Biol.* 9:323–33.

Perera, O.P., C.A. Blanco, B.E. Scheffler, and C.A. Abel. 2007. Characteristics of 13 polymorphic microsatellite markers in the corn earworm, *Helicoverpa zea* (Lepidoptera: Noctuidae). *Mol. Ecol. Notes* 7:1132–34.

Prasad, M.D., M. Muthulakshmi, K.P. Arunkumar, et al. 2005a. SilkSatDb: A microsatellite database of the silkworm, *Bombyx mori. Nucleic Acids Res.* 33:D403–06.

Prasad, M.D., M. Muthulakshmi, M. Madhu, S. Archak, K. Mita, and J. Nagaraju. 2005b. Survey and analysis of microsatellites in the silkworm, *Bombyx mori*: frequency, distribution, mutations, marker potential and their conservation in heterologous species. *Genetics* 169:197–214.

Quan, G.X., T. Kanda, and T. Tamura. 2002. Induction of the white egg 3 mutant phenotype by injection of the double-stranded RNA of the silkworm white gene. *Insect Mol. Biol.* 11:217–22.

Rajagopal, R., S. Sivakumar, N. Agrawal, P. Malhotra, and R.K. Bhatnagar. 2002. Silencing of midgut aminopeptidase N of *Spodoptera litura* by double-stranded RNA establishes its role as *Bacillus thuringiensis* toxin receptor. *J. Biol. Chem.* 277:46849–51.

Reddy, K.D., E.G. Abraham, and J. Nagaraju. 1999. Microsatellites in the silkworm, *Bombyx mori*: Abundance, polymorphism, and strain characterization. *Genome* 42:1057–65.

Roignant, J.Y., C. Carré, B. Mugat, D. Szymczak, J.A. Lepesant, and C. Antoniewski. 2003. Absence of transitive and systemic pathways allows cell-specific and isoform-specific RNAi in *Drosophila. RNA* 9:299–308.

Rupasinghe, S.G., Z. Wen, T.-L. Chiu, and M.A. Schuler. 2007. *Helicoverpa zea* CYP6B8 and CYP321A1: Different molecular solutions to the problem of metabolizing plant toxins and insecticides. *Protein Eng. Des. Sel.* 20:615–24.

Schröder, R. 2003. The genes orthodenticle and hunchback substitute for bicoid in the beetle *Tribolium. Nature* 422:621–25.

Scott, K.D., C.L. Lange, L.J. Scott, and L.J. Gahan. 2004. Isolation and characterization of microsatellite loci from *Helicoverpa armigera* Hübner (Lepidoptera: Noctuidae). *Mol. Ecol. Notes* 4:204–05.

Scott, K.D., N. Lawrence, C.L. Lange, et al. 2005a. Assessing moth migration and population structuring in *Helicoverpa armigera* (Lepidoptera: Noctuidae) at the regional scale: Example from the Darling Downs, Australia. *J. Econ. Entomol.* 98:2210–19.

Scott, L.J., N. Lawrence, C.L. Lange, et al. 2006. Population dynamics and gene glow of *Helicoverpa armigera* (Lepidoptera: Noctuidae) on cotton and grain crops in the Murrumbidgee Valley, Australia. *J. Econ. Entomol.* 99:155–63.

Scott, K.D., K.S. Wilkinson, N. Lawrence, et al. 2005b. Gene-flow between populations of cotton bollworm *Helicoverpa armigera* (Lepidoptera: Noctuidae) is highly variable between years. *Bull. Entomol. Res.* 95:381–92.

Scott, K.D., K.S. Wilkinson, M.A. Merritt, et al. 2003. Genetic shifts in *Helicoverpa armigera* Hübner (Lepidoptera: Noctuidae) over a year in the Dawson/Callide Valleys. *Aust. J. Agric. Res.* 54:739–44.

Stokes, N.H., S.W. McKechnie, and N.W. Forrester. 1997. Multiple allelic variation in sodium channel gene from populations of Australian *Helicoverpa armigera* (Hübner) (Lepidoptera: Noctuidae) detected via temperature gel electrophoresis. *Aust. J. Entomol.* 36:191–96.

Sun, Z.C., M. Wu, T.A. Miller, and Z.J. Han. 2008. *piggyBac*-like elements in cotton bollworm, *Helicoverpa armigera* (Hubner). *Insect Mol. Biol.* 17:9–18.

Takahashi, M., K. Kikuchi, S. Tomita, et al. 2003. Transient in vivo reporter gene assay for ecdysteroid action in the *Bombyx mori* silk gland. *Comp. Biochem. Physiol. B, Biochem. Mol. Biol.* 135:431–37.

Tamura, T., C. Thibert, C. Royer, et al. 2000. Germline transformation of the silkworm *Bombyx mori* L. using a *piggyBac* transposon-derived vector. *Nat. Biotechnol.* 18:81–84.

Tan, S., X. Chen, A. Zhang, and D.-M. Li. 2001. Isolation and characterization of DNA microsatellite from cotton bollworm (*Helicoverpa armigera*, Hubner). *Mol. Ecol. Notes* 1:243–44.

Tan, A., H. Tanaka, T. Tamura, and T. Shiotsuki. 2005. Precocious metamorphosis in transgenic silkworms overexpressing juvenile hormone esterase. *Proc. Natl. Acad. Sci. U.S.A.* 102:11751–56.

Tang, Q.B., J.W. Jiang, Y.H. Yan, J.J.A. Van Loon, and C.Z. Wang. 2006. Genetic analysis of larval host-plant preference in two sibling species of *Helicoverpa*. *Entomol. Exp. Appl.* 118:221–28.

Tautz, D. 1989. Hypervariability of simple sequences as a general source for polymorphic DNA markers. *Nucleic Acids Res.* 17:6463–71.

Tay, W.T., G.T. Behere, D.G. Heckel, S.F. Lee, and P. Batterham. 2008. Exon-primed intron-crossing (EPIC) PCR markers of *Helicoverpa armigera* (Lepidoptera: Noctuidae). *Bull. Entomol. Res.* 98:509–18.

Thomas, J.L., J. Bardou, S. L'Hoste, B. Mauchamp, and G. Chavancy. 2001. A helium burst biolistic device adapted to penetrate fragile insect tissues. *J. Insect Sci.* 1:9.

Thomas, J.L., M. Da Rocha, A. Besse, B. Mauchamp, and G. Chavancy. 2002. 3xP3-EGFP marker facilitates screening for transgenic silkworm *Bombyx mori* L. from the embryonic stage onwards. *Insect Biochem. Mol. Biol.* 32:247–53.

Thomas, D.D., C.A. Donnelly, R.J. Wood, and L.S. Alphey. 2000. Insect population control using a dominant, repressible, lethal genetic system. *Science* 287:2474–6.

Timmons, L., and A. Fire. 1998. Specific interference by ingested dsRNA. *Nature* 395:854.

Tomoyasu, Y., S.C. Miller, S. Tomita, M. Schoppmeier, D. Grossmann, and G. Bucher. 2008. Exploring systemic RNA interference in insects: A genome-wide survey for RNAi genes in *Tribolium*. *Genome Biol.* 9:R10.

Turner, C.T., M.W. Davy, R.M. MacDiarmid, K.M. Plummer, N.P. Birch, and R.D. Newcomb. 2006. RNA interference in the light brown apple moth, *Epiphyas postvittana* (Walker) induced by double-stranded RNA feeding. *Insect Mol. Biol.* 15:383–91.

Uchino, K., M. Imamura, K. Shimizu, T. Kanda, and T. Tamura. 2007. Germ line transformation of the silkworm, *Bombyx mori*, using the transposable element *Minos*. *Mol. Genet. Genomics* 277:213–20.

Uhlirova, M., M. Asahina, L.M. Riddiford, and M. Jindra. 2002. Heat-inducible transgenic expression in the silkmoth *Bombyx mori*. *Dev. Genes Evol.* 212:145–51.

Uhlirova, M., B.D. Foy, B.J. Beaty, K.E. Olson, L.M. Riddiford, and M. Jindra. 2003. Use of Sindbis virus-mediated RNA interference to demonstrate a conserved role of Broad-Complex in insect metamorphosis. *Proc. Natl. Acad. Sci. U.S.A.* 100:15607–12.

van't Hof, A.E., P.M. Brakefield, I.J. Saccheri, and B.J. Zwaan. 2007. Evolutionary dynamics of multi-locus microsatellite arrangements in the genome of the butterfly *Bicyclus anynana*, with implications for other Lepidoptera. *Heredity* 95:320–28.

Vermehren, A., S. Qazi, and B.A. Trimmer. 2001. The nicotinic α subunit MARA1 is necessary for cholinergic evoked calcium transients in *Manduca* neurons. *Neurosci. Lett.* 313:113–16.

Vreysen, M.J.B., A.S. Robinson, and J. Hendrichs, Eds. 2007. *Area-wide control of insect pests: From research to field implementation*. Dordrecht: Springer.

Wang, C.Z., J.F. Dong, D.L. Tang, J.H. Zhang, W. Li, and J. Qin. 2004. Host selection of *Helicoverpa armigera* and *H. assulta* and its inheritance. *Prog. Nat. Sci.* 14:880–84.

Wang, G.R., G.M. Liang, K.M. Wu, and Y.Y. Guo. 2005. Gene cloning and sequencing of aminopeptidase N3, a putative receptor for *Bacillus thuringiensis* insecticidal Cry1Ac toxin in *Helicoverpa armigera* (Lepidoptera: Noctuidae). *Eur. J. Entomol.* 102:13–19.

Wang, J., X. Ren, T.A. Miller, and Y. Park. 2006. *piggyBac*-like elements in the tobacco budworm, *Heliothis virescens* (Fabricius). *Insect Mol. Biol.* 15:435–43.

Wang, W., L. Swevers, and K. Iatrou. 2000. *Mariner* (*Mos*1) transposase and genomic integration of foreign gene sequences in *Bombyx mori* cells. *Insect Mol. Biol.* 9:145–55.

Waterhouse, P.M., M.W. Graham, and M.B.Wang. 1998. Virus resistance and gene silencing in plants is induced by double-stranded RNA. *Proc. Natl. Acad. Sci. U.S.A.* 95:13959–63.

Xia, Q., Z. Zhou, C. Lu, et al. 2004. A draft sequence for the genome of the domesticated silkworm (*Bombyx mori*). *Science* 306:1937–40.

Xu, H.F., Q.Y. Xia, C. Liu, et al. 2006. Identification and characterization of *piggyBac*-like elements in the genome of domesticated silkworm, *Bombyx mori*. *Mol. Genet. Genomics* 276:31–40.

Xu, X., L. Yu, and Y. Wu. 2005. Disruption of a cadherin gene associated with resistance to Cry1Ac δ-endotoxin of *Bacillus thuringiensis* in *Helicoverpa armigera*. *Appl. Environ. Microbiol.* 71:948–54.

Yamao, M., N. Katayama, H. Nakazawa, et al. 1999. Gene targeting in the silkworm by use of a baculovirus. *Genes Dev.* 13:511–16.

Yang, Y., Y. Wu, S. Chen, et al. 2004. The involvement of microsomal oxidases in pyrethroid resistance in *Helicoverpa armigera* from Asia. *Insect Biochem. Mol. Biol.* 34:763–73.

Zhao X.C., J.F. Dong, Q.B. Tang, et al. 2005. Hybridization between *Helicoverpa armigera* and *Helicoverpa assulta* (Lepidoptera: Noctuidae): Development and morphological characterization of F1 hybrids. *Bull. Entomol. Res.* 95:409–16.

Zhong, B., J. Li, J. Chen, J. Ye, and S. Yu. 2007. Comparison of transformation efficiency of piggyBac transposon among three different silkworm *Bombyx mori* Strains. *Acta Biochim. Biophys. Sin. (Shanghai)* 39:117–22.

Zhou, X., O. Faktor, S.W. Applebaum, and M. Coll. 2000. Population structure of the pestiferous moth *Helicoverpa armigera* in the eastern Mediterranean using RAPD analysis. *Heredity* 85:251–56.

Zimowska, G.J., and A.M. Handler. 2006. Highly conserved *piggyBac* elements in noctuid species of Lepidoptera. *Insect Biochem. Mol. Biol.* 36:421–28.

Zong, N., and C.Z. Wang. 2007. Larval feeding induced defensive responses in tobacco: Comparison of two sibling species of *Helicoverpa* with different diet breadths. *Planta* 226:215–24.

13 Molecular Genetics of Insecticide Resistance in Lepidoptera

David G. Heckel

CONTENTS

INTRODUCTION

RESISTANCE AS AN EVOLUTIONARY PHENOMENON

Chemical and biological insecticides are indispensable in modern agriculture for controlling damage to food and fiber crops by insect pests. One of the many problems resulting from this dependency is the frequent and widespread occurrence of insecticide resistance. Resistance is defined as a genetically based decrease in the susceptibility of a population to a toxin over time, in response to long-term exposure. The mechanism of this decrease is an evolutionary response of the population to differential mortality from the toxin, causing the frequencies of alleles controlling resistance mechanisms to increase over several generations. Insecticide resistance is a global problem, with more than 450 pest species displaying resistance to at least one insecticide (Mota-Sanchez, Bills, and Whalon 2002).

A striking example is provided by the lepidopteran pest of crucifers, *Plutella xylostella* L. (diamondback moth). It evolved resistance to the widely used insecticide DDT very early (Ankersmit 1953) and has continued to develop resistance to insecticides, often two to three years after their introduction to the market (Talekar and Shelton 1993). In addition to the popular organophosphorus insecticides, carbamates, cyclodienes, and pyrethroids with targets in the nervous system, it has pioneered the development of resistance to less commonly used insect growth regulators such as chitin synthesis inhibitors (Morishita 1998). In the early 1990s, after years of successful control by sprays based on crystal/spore preparations of *Bacillus thuringiensis* (Bt), *P. xylostella* developed resistance to them as well (Tabashnik et al. 1990).

Molecular genetic approaches have been applied to insecticide resistance only recently, but the progress has been impressive. Knowledge of the structure, function, and resistance-causing mutations of the genes underlying decreased toxin susceptibility in different species has enabled the beginnings of a conceptual unification of the field that was not possible with previous phenomenological approaches. This is because many of the genes are the same, that is, homologous, and different species exposed to the same selection thus provide independent evolutionary replicates of the selective response, resulting in resistance. This will have the practical benefit of pointing to common strategies based on this evolutionary perspective that could be effective in a wide range of species.

Overview of Resistance Mechanisms

Resistance to chemical insecticides classically has been described as caused by one or more of the following mechanisms: behavioral avoidance, reduced penetration through the cuticle, increased metabolic detoxification, or decreased sensitivity of the target. The latter two are the most commonly encountered in practice. Increased detoxification is usually due to mutation or upregulation of one or more members of certain gene families: P450 enzymes, carboxylesterases, or glutathione transferases. This resistance can often be broad-spectrum because such enzymes may detoxify insecticides belonging to different chemical classes. This is one basis for cross-resistance, a decrease in sensitivity to an insecticide different from the one to which the population was exposed. Target-site mutations resulting in decreased sensitivity are usually specific for insecticides within a single chemical class that share the same target. However, the recent trend to develop novel insecticides directed against specific subsites of the same target may result in target-site cross-resistance among different chemical classes.

By the time resistance develops in the field to detectable levels, two or more different mechanisms of dealing with the same insecticide may coexist in the same population, for example, increased carboxylesterase levels as well as insensitive acetylcholinesterase. These may interact in additive, multiplicative, or other ways (Raymond, Heckel, and Scott 1989), greatly complicating biochemical and physiological analysis. Identification of the genetic and molecular basis of each mechanism is ultimately necessary to account for the observed resistance level, a goal still rarely attained in practice. Moreover, a population may be exposed to different insecticides, simultaneously or successively, and thus develop "multiple resistance" to many of them.

The Role of Lepidoptera in Resistance Studies

Most chemical insecticides are poisonous to a broad range of species because their targets in the nervous system are highly conserved evolutionarily and essential to life in all insects. Thus, resistance mechanisms arising in different insect orders as well as other arthropods have important similarities, and scientific results obtained in one group are useful in illuminating general principles and facilitating their investigation in other groups. For historical reasons, molecular genetic tools have been employed much longer in the study of Diptera than Lepidoptera. Most resistance genes currently known in Lepidoptera were first cloned in *Drosophila* and extensively studied in Diptera.

In addition, several important mechanisms are known from other insect orders but have not been identified at the molecular level in Lepidoptera yet.

Therefore, a review of the molecular genetics of chemical insecticide resistance in Lepidoptera must present background information gleaned from studies of other orders for a minimal understanding of the mechanisms. And in attaining a more complete understanding of these mechanisms, examples limited to Lepidoptera give an incomplete picture. These factors may explain why, although there are many reviews of molecular aspects of insecticide resistance, few are focused on Lepidoptera. This review attempts to illustrate how this order has added to our understanding of the general mechanisms and to highlight areas where Lepidoptera may make a unique contribution. By contrast, resistance to insecticidal toxins from *Bacillus thuringienesis*, where studies on Lepidoptera have played a preeminent role, have been extensively reviewed elsewhere (Ferré and Van Rie 2002; Griffitts and Aroian 2005; Bravo, Gill, and Soberon 2007; Heckel et al. 2007; Pigott and Ellar 2007) and will not be covered here.

TARGET SITE RESISTANCE TO CHEMICAL INSECTICIDES

Compared with herbicides and fungicides, insecticide development has focused on a relatively small number of distinct molecular targets, the majority playing key roles in the nervous system. These targets have the advantage of producing a very rapid behavioral response or death due to high-affinity binding of a toxin that may be present at a very low concentration. Their normal function in the nervous system often has stringent structural requirements that are preserved in the face of evolutionary diversification, so that the target in phylogenetically diverse insects still presents the same vulnerabilities to the toxin. Most structural variants of the target are thus rare in the pre-exposed population; however, some are eventually selected by one of the wide array of structurally related pesticides employed, and the favored forms may also be resistant to a more or less broad spectrum from the same chemical class of insecticides. I illustrate this tension between structural conservation and mutability with the four main targets of chemical insecticides: one enzyme and three ion channels.

ACETYLCHOLINESTERASE

The enzyme acetylcholinesterase (AChE) is localized at cholinergic synapses in the insect central nervous system, where it hydrolyzes the neurotransmitter acetylcholine after it has diffused across the synaptic cleft and activated the acetylcholine receptor. Hydrolysis normally ensures that the same population of neurotransmitter molecules does not continually activate the postsynaptic neurons. Organophosphorus (OP) and carbamate insecticides inhibit the enzyme and prevent this hydrolysis, eventually leading to death of the insect. They also inhibit vertebrate AChE functioning in the central nervous system and at neuromuscular junctions, and thus are also toxic to humans.

Decreased inhibition by OPs and carbamates can be conferred by amino acid substitutions in AChE, and this type of target-site insecticide resistance has evolved repeatedly in arthropods. Direct in vitro biochemical characterization of this resistance mechanism is straightforward since the target is an enzyme. Activity of partially purified preparations is measured using an indicator substrate such as acetylthiocholine in the presence or absence of inhibitors. In the absence of interfering proteins that hydrolyze or bind to the insecticides, this provides a direct and quantitative measurement of the degree of target-site insensitivity of a caliber that is not achievable with the targets of other insecticides. This is particularly useful in insect strains with multiple resistance mechanisms.

Acetycholinesterase Resistance and Inhibition Phenotypes

In a survey of insensitive AChE in a number of insect species, Russell et al. (2004) discerned two major patterns of target-site resistance. In Pattern I, insensitivity to carbamates is much greater than insensitivity to OPs; in Pattern II, they are about the same. Both types have been observed in

TABLE 13.1
Acetylcholinesterase Insensitivity Patterns in Lepidoptera

Species	Insensitivity Pattern[a]	Location	Comments	References
Spodoptera frugiperda	II	Florida	2 isozymes	Yu 1992; Yu, Nguyen, and Abo-Elghar 1993; Yu 2006
Spodoptera exigua	I	Netherlands; California		Vanlaecke, Smagghe, and Degheele 1995; Byrne and Toscano 2001
Spodoptera litura	II	China		Huang and Han 2007
Heliothis virescens	I	Southern United States	AceIn, autosomal, chromosome 2	Brown and Bryson 1992; Heckel, Bryson, and Brown 1998
Helicoverpa armigera	I, II	Australia, India	2 or 3 different R alleles	Gunning, Moores, and Devonshire 1996, 1998; Srinivas et al. 2004
Grapholita molesta	I	Ontario, Canada	Sex-linked	Kanga et al. 1997
Plutella xylostella	?	Korea	Ace-1 substitutions A201S, G227A[b]	Baek et al. 2005; Lee et al. 2007
Cydia pomonella	I	Spain	Ace-1 substitution F290V[b]	Cassanelli et al. 2006

[a] Pattern I, Acetylcholinesterase (AChE) insensitivity to carbamates is greater than to organophosphates; Pattern II, AChE sensitivity to carbamates and organophosphates is equivalent.

[b] Amino acid substitutions numbered according to mature *Torpedo californica* enzyme.

Lepidoptera, although in many cases not enough data on different insecticides are available for classification according to this scheme (Table 13.1). Several cases will be described where comparative data are adequate.

Fall armyworm, *Spodoptera frugiperda* (J.E. Smith), collected from corn in Florida in 1990 was resistant to many insecticides, including the OP methyl parathion (517-fold) and the carbamate carbaryl (507-fold). In addition to increased activity of a number of detoxicative enzymes, AChE inhibition by dichlorvos was 4- to 8-fold lower (Yu 1992). A strain collected 10 years later exhibited similarly high resistance, and decreases in sensitivity measured on crude homogenates from adult heads were 9-fold for methyl paraoxon and 85-fold for carbaryl (Yu, Nguyen, and Abo-Elghar 2003). These differences were confirmed on AChE purified by gel filtration and affinity chromatography, in which comparison of bimolecular rate constants indicated that the resistant enzyme was 218-fold less sensitive to methyl paraoxon and 345-fold less sensitive to carbaryl (Yu 2006). Two isozymes could be resolved by nondenaturing polyacrylamide gel electrophoresis, the major form at 66.1 kDa and a minor form at 63.7 kDa. Susceptible and field-resistant strains showed the same electrophoretic patterns, and all inhibition curves of the purified preparation containing both forms were linear, suggesting that they had the same biochemical properties. This example appears to fall into Pattern II.

A strain of the beet armyworm, *Spodoptera exigua* (Hübner), from greenhouses in the Netherlands, in addition to detoxicative resistance mechanisms, showed 1.9-fold lower inhibition of AChE by methomyl and 2.7-fold lower by dichlorvos (Vanlaecke, Smagghe, and Degheele 1995). A strain from cotton in California was 68-fold more resistant to the carbamate methomyl but not significantly cross-resistant to chlorpyrifos; moreover, its AChE was 30-fold less sensitive to inhibition by methomyl but only 6-fold less sensitive to chlorpyrifos-oxon (Byrne and Toscano 2001). These properties are more characteristic of Pattern I.

Strains of common cutworm, *Spodoptera litura* (Fabricius), from soybean or cabbage in China showed 5.7- to 26-fold resistance to carbamates and OPs, as well as higher resistance to pyrethroids.

Cross-resistance to these differing classes of insecticides could be accounted for by cytochrome P450 and esterases, but AChE was also two to three times less sensitive to the organothiophosphate phoxim and the carbamate methomyl (Pattern II; Huang and Han 2007).

During the 1960s and 1970s, the tobacco budworm, *Heliothis virescens* (Fabricius), developed resistance to methyl parathion control on cotton in the southern United States. AChE of adults or larvae from a South Carolina resistant strain showed 20- to 25-fold decreased sensitivity to methyl paraoxon, the activated, toxic form of methyl parathion (Brown and Bryson 1992). Bimolecular reaction constants for propoxur were 10- to 100-fold lower than for several OPs, leading Russell et al. (2004) to classify this as Pattern I; however, inhibition constants for other carbamates were comparable to OPs. Some "antiresistant" compounds like the OP insecticides monocrotophos and dichrotophos actually inhibited the methyl paraoxon–resistant enzyme more. Three distinct phenotypes could be detected in field samples of tobacco budworm by dividing homogenate from the heads of individual moths into separate portions and testing them with methyl paraoxon or monocrotophos. Crossing experiments showed that these phenotypes corresponded to three genotypes, *RR*, *RS*, and *SS*, at a biallelic autosomal locus named *AceIn*, with *R* encoding the methyl paraoxon–resistant, monocrotophos-susceptible form of AChE (Brown and Bryson 1992). This genotypic characterization was used to demonstrate genetic linkage to the phenotype of methyl parathion resistance (Gilbert, Bryson, and Brown 1996) and to map the *AceIn* gene to LG (linkage group) 2 (Heckel, Bryson, and Brown 1998). Because the allozyme marker, IDH-2, occurring on this autosome in the noctuid, *H. virescens*, is sex linked in many tortricid moths, Heckel, Bryson, and Brown (1998) predicted that insensitive AChE in tortricids would be sex linked. Moreover, the ability to discern individual genotypes of field-collected individuals also formed the basis of a unique surveillance program in which the *R* allele frequency was monitored over several years (Brown et al. 1996).

Carbamate resistance in an Australian population of the cotton bollworm, *Helicoverpa armigera* (Hübner), led to the characterization of a form of insensitive AChE that was resistant to methomyl and thiodicarb but not OPs (Gunning, Moores, and Devonshire 1996), another case of Pattern I. However, a later study on a strain resistant to the OPs profenofos and methyl parathion, but not carbamates, described an AChE 100-fold less sensitive to inhibition by methyl paraoxon, yet no resistance to chlorpyrifos oxon (Gunning, Moores, and Devonshire 1998). The second type conforms to neither pattern, and it was suggested that there were two different types of insensitive AChE in Australian populations of this species. Srinivas et al. (2004) documented 3- to 4-fold insensitivity of AChE to monocrotophos and methyl paraoxon in a population of *H. armigera* from India, suggestive of Pattern II and possibly indicating a third type.

The oriental fruit moth, *Grapholita molesta* (Busck), from a population in Ontario, Canada, showed resistance to both carbamates and OPs, with 3- to 6-fold less sensitivity to the OP gluthoxon and 3- to 44-fold less sensitivity to carbamates, following Pattern I (Kanga et al. 1997). Resistance to carbofuran as determined by bioassay differed among reciprocal F_1 crosses of susceptible and resistant strains, and also differed among male and female offspring in the backcross and F_2. Decreased inhibition of AChE by carbaryl and propoxur showed a similar pattern, indicating that a single sex-linked locus could account for insensitive AChE in this species. This confirms the previous prediction (Heckel, Bryson, and Brown 1998) for at least one tortricid; whether other tortricid pests such as codling moth, *Cydia pomonella* (L.), also have sex-linked insensitive AChE has yet to be reported.

Additional lepidopteran species have shown some evidence of altered AChE involved in resistance. A strain of the rice stem borer, *Chilo suppressalis* (Walker), collected from China that was greater than 700-fold resistant to triazophos had AChE with a 32 percent lower V_{max}, 65 percent lower K_m, and 2.5-fold decreased sensitivity to triazophos relative to a susceptible strain (Qu et al. 2003). Two strains of diamondback moth, *P. xylostella* (L.), selected with phenthoate showed about 10-fold decreased sensitivity of AChE to acephate and phenothoate (Noppun, Miyata, and Saito 1987). A strain of codling moth, *C. pomonella*, from Spain that was selected with azinphosmethyl had AChE with decreased sensitivity by 1.7-fold to azinphosmethyl and 14-fold to carbaryl (Cassanelli et al. 2006).

It is evident that a wide variety of AChE-insensitive resistance phenotypes has been found in Lepidoptera. Interspecific comparisons are complicated by the diverse set of insecticides studied in different species, and so far it has not been possible to relate the Pattern I/Pattern II classification to other biological characteristics. Mechanistic explanations will depend on an understanding of the specific amino acid substitutions involved and how these relate to the structure of the enzymes; and this information is just now becoming available for some lepidopteran species.

Acetylcholinesterase Structure, Sequence, and Resistance Genotypes

Much is known about the structure and function of acetylcholinesterase from extensive studies on vertebrates, including 3-D structural determination (Sussman et al. 1991) based on isolation of the protein from the electric organ of the ray *Torpedo californica*, where it is highly abundant. The sequence of the AChE gene in *Drosophila melanogaster* was one of the first determined by the technique of chromosomal walking (Hall and Spierer 1986). The structure of the *Drosophila* enzyme has also been solved (Harel et al. 2000) and has provided the basis for modeling target-site mutations on insects. Like most esterases, acetylcholinesterase belongs to the structural superfamily termed the α/β hydrolase fold (Ollis et al. 1992). The core of the structure is a series of 11 nearly parallel β-sheets arranged in a slightly twisted configuration so that the first and last cross at an angle of about 90 degrees. There are additionally 14 α-helices occurring in the loops connecting the β-strands. The active site is buried deep within the enzyme, connected to the outside by the catalytic gorge, a channel 20 Å long and a few Å wide, lined with aromatic residues facilitating the efficient entry of the substrate.

The active site comprises four subsites, three of which accommodate different regions of the substrate. The P1 or anionic site cradles the positively charged choline moiety, the P2 subsite or acyl binding pocket holds the methyl group, and the oxyanion hole faces the carbonyl oxygen and stabilizes the negative charge displaced onto it during the progress of the reaction. The fourth region is the so-called catalytic triad consisting of Serine-238, Histidine-480, and Aspartate-367. Although distant in the primary sequence, these three residues are brought together by the folding of the protein into a characteristic orientation that is crucial for catalytic activity. The serine projects into the active site from its position in a sharp turn between a β-strand and an α-helix termed the nucleophilic elbow. The nucleophilicity of the serine is increased by interaction with the histidine as oriented by the aspartate. A transitory covalent bond to this serine oxygen produces the acyl-enzyme intermediate in the reaction cycle, and a longer-lasting bond to the same oxygen results in inhibition by the insecticide (Oakeshott et al. 2005a).

Hydrolysis of acetylcholine occurs in two steps. First, the serine oxygen makes a nucleophilic attack on the substrate's carbonyl carbon, forming a covalent bond and displacing the alcohol group. The bonds to this carbon adopt a tetrahedral configuration stabilized by the oxyanion hole. In the second step, the oxygen of a water molecule makes a nucleophilic attack on the same carbon, hydrolyzing the acyl-serine bond, releasing the free acid, and restoring the enzyme to its original condition.

Insecticides inhibit acetylcholinesterase by blocking progress of the reaction after the first step. The serine's oxygen nucleophilic attack on the phosphate of the oxon form of OP insecticides, or the carbonyl carbon of carbamate insecticides, results in a very stable covalent bond that cannot be hydrolyzed in the second step. Thus, the enzyme is irreversibly trapped in an inactive form by covalent linkage to the "suicide inhibitor." Target-site AChE insecticide resistance, therefore, should be eventually interpretable with respect to how specific amino acid substitutions interfere with the production of this inactive phosphoryl- or carbamyl-enzyme. No cases of target-site resistance due to hydrolysis of the enzyme-inhibitor complex are known for AChE, but a mechanistically similar type of resistance has been shown for certain carboxylesterases in the dipterans *Lucilia cuprina* (Newcomb et al. 1997) and *Musca domestica* (Claudianos, Russell, and Oakeshott 1999).

The pattern of mutations in *Drosophila* and housefly revealed that substitutions to larger amino acids in any of six sites near the base of the catalytic gorge could hinder inhibitor access to the active site (Oakeshott et al. 2005a). Attempts to extend these results to mosquitoes were initially met with frustration, as the AChE cloned from these species on the basis of sequence similarity to the *Drosophila* enzyme failed to show the same pattern of substitutions in resistant strains, and even failed to map to the same chromosomal location as the phenotype of AChE insensitivity (Malcolm et al. 1998). These inconsistencies were resolved when it was discovered that mosquitoes and indeed most insect species possess two similar but distinct genes, now usually called *Ace-1* and *Ace-2* (Weill et al. 2002), each encoding a different protein with AChE activity. In most insects, it appears that the product of *Ace-1* functions at the cholinergic synapse and mutations in *Ace-1* may directly affect AChE insensitivity; the role of *Ace-2* is unknown. However, *Ace-1* has been lost from higher Diptera (Huchard et al. 2006), including *Drosophila* and *Musca*, where the product of *Ace-2* functions at the cholinergic synapse instead. Thus the growing list of mutations in *Ace-1* (but not *Ace-2*) in other insects (Fournier 2005; Oakeshott et al. 2005b) illustrates the same principles as found in *Ace-2* in higher Diptera: Substitutions in many parts of the molecule can slow down the formation of a specific enzyme-inhibitor complex, but this effect is insecticide specific and can often speed it up for other inhibitors.

The first AChE genes cloned in Lepidoptera were *Ace-2* homologues. Ni et al. (2003) cloned a full-length *Ace-2* from susceptible *P. xylostella*. Ren, Han, and Wang (2002) compared the entire *Ace-2* sequence of *H. armigera* from five individuals from a Chinese monocrotophos-resistant strain and six susceptible individuals, and found extensive within- and among-strain polymorphism showing no clear association with resistance except for Ala585Thr; however, a threonine occurs in susceptible strains of other species and this position is not highly conserved across Lepidoptera.

More recently, both *Ace-1* and *Ace-2* genes from several lepidopteran species have been cloned and compared, including *Helicoverpa assulta* (Guenée); (Lee et al. 2006), *C. pomonella* (Cassanelli et al. 2006), *P. xylostella* (Lee et al. 2007), and *Bombyx mori* L. (Seino et al. 2007; Shang et al. 2007). In the latter study both proteins from insecticide-susceptible *B. mori* were expressed in cell culture and shown to exhibit AChE activity, with Ace-2 protein exhibiting more sensitivity to inhibition by eserine and paraoxon than Ace-1. Heterologously expressed Ace-1 protein from a Korean prothiofos-resistant strain of *P. xylostella* was less sensitive to paraoxon than Ace-1 from a susceptible strain, which is the first such comparison for Lepidoptera (Lee et al. 2006).

Mutations in *Ace-1* likely to cause target-site insensitivity to carbamates and OPs have been identified in two lepidopteran species so far. Baek et al. (2005) compared partial sequences of *Ace-1* from the Korean prothiofos-resistant strain of *P. xylostella* with a susceptible strain from Japan, and found three amino acid substitutions. Two of these were also found by Lee et al. (2007) in their comparison of a full-length *Ace-1* sequence from the same resistant strain with a different susceptible strain from Korea. The first of these occurs at position A201 (Tc; numbering according to sequence of the active form of *Torpedo californica*), one of the three residues in the oxyanion hole that immediately follows the catalytic serine in the primary sequence, in the motif GESAG. Lee et al. (2007) found an S in this position in the resistant sequence (thus GESSG); a sequence chromatograph generated from pooled DNA from resistant-strain individuals showed that it was polymorphic at this site with some alleles encoding A and others S (Baek et al. 2005). A serine in this position has also been found in *Ace-1* from resistant strains of the aphid *Aphis gossypii* (Andrews et al. 2004; Li and Han 2004; Toda et al. 2004); all other known sequences from arthropods have an alanine. The second substitution found by both groups was at the highly conserved glycine at residue 227 (Tc), which was replaced by an alanine fixed in the resistant strain. The same substitution occurs in *Ace-2* in resistant populations of *D. melanogaster* (Mutero et al. 1994) and *M. domestica* (Kozaki et al. 2001; Walsh et al. 2001), and confers AChE insensitivity to many insecticides in heterologously expressed recombinant enzymes from several dipteran species (reviewed in Fournier 2005). A third mutation reported by Lee et al. (2007) at position 131 (Tc; D229G in the *Plutella Ace-1* sequence) was considered irrelevant on the basis of structural modeling; moreover, this change was not reported by Baek et al. (2005). The third mutation reported by Baek et al. (2005) occurred at the residue just after the histidine in the catalytic triad; an A was found in the susceptible strain, and the resistant strain was polymorphic for A or G. Lee et al. (2007) reports a G in both strains studied; a survey of insect sequences shows an A present in most lepidopterans and a G present in most nonlepidopteran sequences here. Thus, two of the substitutions found in the prothiofos-resistant strain of diamondback moth are homologous to resistance-conferring mutations in other species.

Cassanelli et al. (2006) studied the Raz strain of codling moth, *C. pomonella*, from Spain that was 6.7-fold resistant to azinphos-methyl and 130-fold resistant to carbaryl, which conforms to Pattern I and is mirrored by a decrease in sensitivity of AChE to azinphos-methyl oxon and carbaryl by factors of 1.7 and 14, respectively. The *Ace-2* gene from Raz showed no amino acid differences to that from a susceptible strain, but the *Ace-1* gene showed a single amino acid substitution. The authors developed a PCR-based assay to detect this mutation in DNA from individuals for population studies. The substitution was F290V (Tc), a residue in the P2 subsite (acyl binding pocket), F399V in the *Cydia* sequence. A substitution to tyrosine in the same position is associated with Pattern II resistance in *Drosophila* (Mutero et al. 1994) and *Musca* (Walsh et al. 2001), which could potentially restrict access by larger inhibitors by steric hindrance while still allowing acetylcholine to the active site. Villatte et al. (2000) replaced this F290 residue in the *Drosophila* sequence with in vitro mutagenesis and tested nine such substitutions for inhibition by 19 compounds. The valine substitution F290V (F368V in their nomenclature based on the *Drosophila* precursor sequence) was 1.9-fold less sensitive to azinphos-methyl oxon and 3.7-fold less sensitive to carbaryl, and even less so for most other carbamates, which is also reminiscent of Pattern I.

The recent progress seen with *Plutella* and *Cydia* bodes well for future work in decoding the molecular basis of AChE target-site resistance in Lepidoptera. Examining sequence variation within *Ace-1* in the rich array of resistant strains already biochemically characterized will likely yield many more mutations, and their investigation will lead to a deeper understanding of structure-function relationships. This may enable exploitation of the increase in sensitivity to certain insecticides that often follows when resistance evolves to others, in discovering or developing "antiresistant" OPs and carbamates. Although studies on *Ace-2* of *Drosophila* have provided the foundation for this understanding, the loss of *Ace-1* from its lineage may have paradoxically delayed progress by temporarily leading to a focus on the "wrong" enzyme in other insects. So far, no 3-D structure has been determined for any insect *Ace-1* protein, and no site-directed mutagenesis studies have systematically examined the effects on inhibition by specific amino acid substitutions found in resistant lepidopteran sequences compared with susceptible sequences from the same species, which, after all, is the variation immediately acted upon by insecticide selection.

Voltage-Gated Sodium Channel

Pharmacological and toxicological evidence has identified the voltage-gated sodium channel as the primary target for DDT and pyrethroid insecticides (Narahashi 1996). The vulnerability of this target is well illustrated by the diverse evolutionary appearance of neurotoxins involved in the rapid chemical immobilization and killing of prey by predators (reviewed in Soderlund 2005). It is still being exploited in new ways in agriculture, as the target of the pyrazoline insecticide indoxacarb (Wing et al. 2005) and scorpion venom toxins that have been engineered into baculoviruses as biopesticides (Zlotkin et al. 1995).

Although sodium channel mutations may well have played a role in DDT resistance, the heyday of DDT use occurred well before the availability of molecular genetic tools to study them. Environmental and safety problems as well as resistance led to a general cessation of DDT applications in the 1970s; however use is increasing in Africa for mosquito control in the face of increasing difficulties with malaria. Most current research on the role of the sodium channel in insecticide resistance has been motivated by widespread resistance to the synthetic pyrethroids. Development of this class of insecticides based on the structures of naturally occurring insecticidal pyrethrins was one of the major advances in chemical control of insects, for crop protection and control of diseases of humans and animals that are spread by insects (Elliott 1996). The enthusiastic use of pyrethroids in many sectors worldwide was followed by development of resistance in a huge variety of arthropods.

Pyrethroid resistance is often multifactorial, with detoxicative mechanisms playing a role as well as target-site resistance. Teasing these mechanisms apart has required indirect methods. One approach has been to focus on symptoms of insecticide poisoning at the behavioral or neurophysiological levels, changes that might indicate target-site, rather than metabolic resistance. Another approach has been to employ insecticide synergists, chemicals that inhibit a large class of detoxicative enzymes, to suppress the metabolic defenses and to use any residual resistance remaining as an estimate of the target-site resistance. A picture began to emerge of the DDT and pyrethroid target-site resistance typified by the kdr (knockdown resistant) and super-kdr strains of the housefly. Similar resistance syndromes occurring in other species were called "kdr-like." The best justification for this classification as target-site resistance came from neurophysiological studies, including early work on pest Lepidoptera.

Sodium Channel Resistance and Neurophysiological Phenotypes

Early evidence of intrinsic nerve insensitivity as a mechanism of pyrethroid resistance in Lepidoptera was provided by Gammon (1980) in a strain of *Spodoptera littoralis* (Boisduval) resistant to permethrin but not cypermethrin. Isolated larval nerve cords exhibited a burst of increased firing after a single electrical stimulus, after incubation in a threshold concentration of permethrin for a certain time. Nerve preparations from resistant larvae required a significantly longer incubation with permethrin than susceptible larvae, but this effect was not seen with cypermethrin.

Fenvalerate-resistant *H. armigera* from cotton were detected in eastern Australia in 1983 (Gunning et al. 1984), and multiple mechanisms were investigated (Gunning et al. 1991). Nerve insensitivity was quantified by the onset time of spontaneous repetitive firing with extracellular measurements of peripheral nerves partially dissected from the ventral body muscle, after exposure to a diagnostic concentration of 10 μM fenvalerate. Responding larvae of a resistant strain collected in 1983 showed a greater than 5.1-fold increase in this time of onset than a susceptible strain showed; one-third of the resistant larvae did not respond at all to this concentration (Gunning et al. 1991). F_1 hybrid larvae showed a 3.7-fold increase on average. However, larvae from a resistant strain collected in 1987 averaged only a 2.8-fold increase, and the frequency of insensitive individuals in the resistant sample decreased through 1989 to less than 20 percent. The apparent decline in the potency and frequency of the nerve insensitivity mechanism was countered by an increase in metabolic resistance that could be suppressed by the cytochrome P450 inhibitor piperonyl butoxide (PBO; Gunning et al. 1991). Unlike most cotton-growing regions of the world during this period, the early detection of pyrethroid resistance led to a resistance management strategy with highly restricted use of pyrethroids on cotton in Australia (Forrester et al. 1993). This enabled sustainable use over several years even as pyrethroid resistance steadily increased, and the restrained selection pressure could thus partially explain the increase and then decrease of the nerve insensitivity mechanism, which may have a fitness cost in the absence of insecticide.

Using similar neurophysiological methods, nerve insensitivity has been shown to be a component of pyrethroid resistance in *H. armigera* infesting heavily sprayed cotton in Thailand (Ahmad, Gladwell, and McCaffery 1989); Andhra Pradesh, India (McCaffery et al. 1997); Jiangsu Province, China (Tan and McCaffery 1999); and Hebei Province, China (Ru et al. 1998). In most of the strains there was also significant metabolic resistance, as treatment of larvae with PBO decreased resistance levels considerably.

Another cotton pest, *H. virescens*, developed nerve insensitivity as one component of multifactorial pyrethroid resistance in the United States (Nicholson and Miller 1985). Church and Knowles (1993) studied crude preparations of neural membranes from adult heads and measured binding by radiolabeled batrachotoxin, a sodium channel–selective neurotoxin, as a function of increasing pyrethroid concentrations. Although differences between resistant and susceptible strains were found, they eluded a mechanistic explanation. By exposing larval peripheral nerve preparations to increasing concentrations of permethrin, a cumulative dose–response assay was developed to construct a population frequency distribution of the degree of insensitivity as measured by extracellular recordings of multiunit activity from larvae (McCaffery, Holloway, and Gladwell 1995) or adults (Holloway and McCaffery 1996). This was correlated with the degree of residual resistance after PBO treatment, and increased during the cotton growing season. Ottea and Holloway (1998) found similar results in the same species with allethrin. Lee et al. (1999) cultured neurons from adult thoracic and abdominal ganglia and used the whole-cell patch-clamp technique to compare properties of sodium channels from susceptible individuals and those from the resistant Pyr-R strain. Pyr-R channels had different gating properties, with a 13 mV positive shift in voltage-dependent activation, and were less excitable with an 11 mV elevation in the action potential threshold. They were also 21-fold less sensitive to the effects of permethrin in prolonging the tail current. The Pyr-R strain was homozygous for a V410M substitution in the sodium channel (see below). Park et al. (2000) correlated similar single-cell patch-clamp measurements with adult knockdown bioassays and extracellular recordings of larval neuromuscular junctions in progeny from a series of crosses that were segregating for a different mutation in the sodium channel, L1014H (see below). Although the three measures were correlated across the three genotypes defined by this mutation, there was still a wide range of variation in measures of nerve insensitivity within each genotype, suggesting that mutations elsewhere in the sodium channel, or in other genes, affected this trait.

Nerve insensitivity was measured in *P. xylostella* with intracellular recordings of EPSPs (excitatory postsynaptic potentials) from cells of the ventral internal longitudinal muscle of larvae, either of spontaneous mini-EPSPs or EPSPs evoked by electrical stimulation of the segmental nerve (Schuler et al. 1998). Mini-EPSP activity increased and EPSP amplitude declined upon exposure to deltamethrin, and the EC50 value based on a composite measure of these two responses was about 390,000-fold greater for the FEN resistant strain from Taiwan.

Sodium Channel Structure, Sequence, and Resistance Genotypes

Following the cloning of a voltage-sensitive sodium channel from the electric eel (Noda et al. 1984) and extensive study of other vertebrate sodium channels, two homologues were found in *Drosophila*. The first to be identified was *dsc1* (Salkoff et al. 1987), which does not seem to be involved in pyrethroid resistance but functions in processing olfactory information (Kulkarni et al. 2002) and may actually be a calcium channel (Zhou et al. 2004). The second was found at the *para*^{ts} locus, defined by a temperature-sensitive paralytic phenotype (Loughney, Kreber, and Ganetzky 1989). *Vssc1*, the ortholog of *para* in housefly *M. domestica*, was found to harbor mutations in the kdr and super-kdr strains, and subsequently additional mutations have been found in *para* homologues from at least twenty-six arthropod species (reviewed in Davies et al. 2007).

The sodium channel is formed by a large (approximately 2,100 residues) protein consisting of four internally homologous domains (I–IV) connected in series, each of which is composed of six transmembrane helices (S1–S6) and intracellular and extracellular loops. A low-resolution 3-D structure has been determined using the channel from electric eel (Sato et al. 2001). These four domains are assembled in a tetramer with S5 and S6 of each domain oriented centrally, lining the ion pore. The short "P-loops" between S5 and S6 dip into the extracellular side of the membrane and provide selectivity for the sodium ion, which enters the cell through the pore when the channel is in the open state. From the resting, closed state, opening is caused by membrane depolarization, which is sensed by structures in the S1–S4 segments. Opening of the voltage-dependent activation "m-gate" at the extracellular mouth of the pore allows influx of sodium ions for a few milliseconds, and is followed by a coupled, delayed closure of the inactivation "h-gate" at the intracellular end of the pore. At this point, the channel is in an inactivated state. Following repolarization of the membrane, the m-gate closes, the h-gate reopens, and the channel returns to the resting configuration, ready to be activated by the next membrane depolarization event (Davies et al. 2007).

Radioligand and functional studies have defined at least ten distinct sites on the channel at which binding of neurotoxins and insecticides affects different aspects of channel activation and inactivation (reviewed in Soderlund 2005). Binding by pyrethroids stabilizes the open configuration, inhibits inactivation, and thus causes a large tail current during repolarization. Type II pyrethroids possessing an α-cyano group at the phenylbenzyl alcohol position have a much stronger effect than Type I pyrethroids lacking this group. Information about the binding site for DDT and pyrethroids has been greatly augmented by the study of resistance mutants.

Full-length transcripts of *Drosophila para* compared with the genomic sequence reveal a complex pattern of alternative splicing, including seven optional exons occurring in intracellular loops and two pairs of mutually exclusive exons in transmembrane regions (reviewed in Dong 2007). Many of these are conserved in other insects and are likely to be functionally important. In addition, there is evidence for mRNA editing, that is, posttranscriptional conversion of A to I or U to C in the mRNA, resulting in at least eight amino acid substitutions in transmembrane or intracellular region. Both features complicate the relationship between functional protein and detection of mutations at the gene level.

Nevertheless, a number of DNA mutations resulting in amino acid substitutions conferring resistance have been identified, beginning with the mapping and sequencing of the *Vsscl* mutations responsible for knockdown resistance phenotypes in housefly. (In the following, substitutions will be numbered by their position in the *Drosophila para* amino acid sequence.) The *kdr* mutation, conferring low but broad-spectrum resistance to all classes of pyrethroids, is caused by the substitution L1029F in domain IIS6 (L1014F in the *Musca Vsscl* sequence; Ingles et al. 1996; Miyazaki et al. 1996; Williamson et al. 1996). This is the most common site of resistance mutation in arthropods, with the leucine substituted by a phenylalanine, histidine, or serine in eleven species including cockroach, aphids, mosquitoes, beetle, thrips, flea, and moths. The super-kdr strain of housefly showing even higher levels of resistance mainly to Type II pyrethroids possesses the additional substitution M933T in the linker between IIS4 and S5 (M918T in *Vsscl*; Williamson et al. 1996). Mutations in the same position have been found in hornfly, aphids, leafminer, and whitefly. An additional eight sites occurring in the domain IIS4–S6 region and eighteen sites outside of it are mutated in one or more species of arthropod resistant to pyrethroids (reviewed in Davies et al. 2007). Five have been found so far in Lepidoptera, and three of these uniquely so. This sample of mutations may be biased, as few studies present the entire sequence of the gene from resistant and susceptible strains. Moreover, many studies have been restricted to testing for the presence of mutations already found in other species.

Pyrethroid resistance in *H. virescens* was first mapped to the *para*-homologous sodium channel locus *hscp* using polymorphisms within an intron (Taylor et al. 1993). Further sequencing in this species revealed an L1029H substitution in the same location as the domain IIS6 housefly *kdr* mutation (Park and Taylor 1997). This polymorphism was used to determine genotypes of individuals used in bioassays and nerve insensitivity assays, further strengthening the association with resistance. In another resistant individual retaining the susceptible L1029, a V421M change was found, at a position within domain IS6 analogous to the domain IIS6 *kdr* mutation (Park, Taylor, and Feyereisen 1997). The V421M substitution has not been reported in any other arthropod species to date. Park, Taylor, and Feyereisen (1999) subsequently sequenced a genomic clone covering most of the *hscp* gene, revealing some similarities as well as differences to the pattern of alternate splicing earlier found in *Drosophila para*. Head, McCaffery, and Callaghan (1998) looked specifically for mutations in the intracellular loop between domains III and IV (DIII-DIV linker) and found two: D1561V and E1565G. Both substitutions were found in a pyrethroid-resistant strain of *H. virescens* exhibiting nerve insensitivity, and also in a nerve-insensitive strain of *H. armigera* from Jiangsu Province in China, but not in susceptible strains of either species (Head, McCaffery, and Callaghan 1998). Neither mutation has been found in any other pyrethroid-resistant arthropod species to date, but mutations in another linker DI-DII have been found in cockroaches along with *kdr*. Unfortunately the status of L1029 and V421 was not monitored in this study.

The FEN strain of *P. xylostella* exhibiting marked nerve insensitivity to deltamethrin was screened for substitutions in a 350bp fragment amplified from cDNA covering the region of the previously discovered *kdr* mutation in housefly (Schuler et al. 1998). The expected L1029F substitution was found, and additionally the novel substitution T944I, both homozygous in twelve sequenced larvae showing insensitivity in the neurophysiological assay. The latter is close to but distinct from the *super-kdr* mutation site M933T, and also appears to greatly increase nerve insensitivity when occurring in combination with *kdr*. Substitutions at T944 have subsequently been found in flower thrips, cat flea, tobacco whitefly, and head louse, to a different amino acid in each case.

Because of the large size and high intron number of the sodium channel gene, and the frequent occurrence and close spacing of the *kdr* and *super-kdr* mutations occurring in some resistant species, many other studies have screened only that region in comparing resistant and susceptible strains. Such a study revealed the presence of L1029F, but no changes at M933 and T944, in a deltamethrin-resistant strain of *C. pomonella* (Brun-Barale et al. 2005).

Functional Tests of Resistance-Associated Sodium Channel Genotypes

To examine the effect of amino acid substitutions on the function of the sodium channel protein, functional expression along with the TipE protein is performed in *Xenopus* oocytes. These proteins are translated from microinjected mRNA and incorporated into the extracellular membrane of these large cells, and electrophysiological recordings are made of channel properties in the presence and absence of pyrethroids and other ligands. As full-length cDNA clones are often not available from the species of interest, homologous mutations are introduced into *Drosophila*, housefly, or cockroach cDNAs instead. These three are well-characterized in the *Xenopus* system, facilitating comparison with the introduced mutations. On the other hand, it is not clear whether the mutations would have exactly the same effects when introduced into the wild-type background of the species of interest; for example, *H. virescens* shows overall sequence divergence of 15–20 percent compared with these three species.

Lee and Soderlund (2001) introduced the V421M substitution found in *H. virescens* into the housefly *Vssc1* cDNA and compared its properties to wild-type and the L1029F *kdr* mutation from housefly. Similar to the effect of L1029F, the V421M modified housefly channel showed a shift of the midpoint potential for activation upward 9 mV compared with wild type. It was also about 20-fold less sensitive than wild type to the effect of cismethrin in prolonging the tail current. These differences compared with wild type are similar to those found by Lee et al. (1999) in cultured neurons from *H. virescens* homozygous for V421M. Compared to the L1029F mutation, *Vssc1* carrying V421M was virtually identical in the degree and pattern of increased cismethrin insensitivity in the *Xenopus* expression system, suggesting that it could confer the same level of resistance in vivo as the classical *kdr* mutation.

Zhao, Park, and Adams (2000) introduced the V421M or the L1029H mutations found in *H. virescens* into the wild-type *para* channel from *D. melanogaster* for expression in *Xenopus* oocytes. Several differences in the mutants were observed. The upward shift in midpoint activation potential for V421M was 6.6 mV, twice that for L1029H. The steady-state inactivation potential was also shifted upward, 4.5 mV for V421M and 2.1mV for L1029H. Both mutants had a 45-fold faster rate of deactivation in the presence of 10 μM permethrin than wild type, a beneficial effect that reduced the time allowed by permethrin-modified channels for sodium influx into the cell leading to depolarization. And with respect to the ratio of tail current conductance to peak current conductance corrected for driving force, L1029H had a greater protective effect in 10 μM permethrin. The authors suggested that the differences measured in the absence of permethrin would translate into lower fitness of individuals carrying them, and more so for the V421M mutation. There is some evidence that nerve-insensitive pyrethroid-resistant moths have lower fecundity, lower production of the sex pheromone in females, and lower rate of attraction by the sex pheromone in males (Campanhola et al. 1991), although the sodium channel genotypes were not directly determined in that study. From the *Xenopus* expression results, the authors predicted that L1029H would have a competitive advantage over V421M since it confers more protection in the presence of insecticide, and has a lower fitness cost in its absence.

The V421M mutation has also been tested in combination with additional mutations occurring in other species. In the cockroach *Blattella germanica*, the *kdr* mutation L1029F occurs as well as two additional mutations in the intracellular DI-DII linker loop (denoted as E434K and C764R in the cockroach protein sequence). L1029F alone confers about a 5-fold decrease in sensitivity to deltamethrin in the *Xenopus* system, and each of the other two alone has no effect. When L1029F is combined with either one of the other two substitutions a 100-fold decrease in sensitivity results, and when combined with both the effect is 500-fold (Tan et al. 2002). To test whether the same enhancement effect operated on the V421M mutation occurring in the IS6 position analogous to the IIS6 *kdr*, the cockroach sequence was modified to contain V421M, producing a 10-fold decrease in sensitivity, and this was tested in combination with the other two. Here there was no enhancement effect unless both of the other two mutations were present, reducing the sensitivity by 100-fold (Liu et al. 2002). If these findings can be extrapolated to the background of the *Heliothis* wild-type channel, they would suggest that sequential accumulation of such enhancing mutations would be more likely in *Blattella* than in *Heliothis*. In this context, it would be interesting to test the DIII-DIV linker mutations D1549V and E1553G occurring in *Heliothis* for single and pairwise enhancing effects on V421M and L1029H.

Population Genetics of Sodium Channel Resistance Genotypes

DNA-based genotyping of individuals for sequence variation at the sodium channel has been used to detect selection and track the dynamics of allele frequency change in time. Even before the identity of the *kdr* mutation was known, Taylor, Shen, and Kreitman (1995) used the population frequencies of haplotypes defined by noncoding sequence variation in the *hscp* gene to test hypotheses about insecticide selection on *H. virescens* in the U.S. cotton belt. Haplotypes were identified by mobility differences in denaturing gradient gel electrophoresis, which can detect most nucleotide substitutions in a ~500 bp PCR product. A total of 660 adult males were collected in pheromone traps in four different states in 1990 and tested for *kdr*. The sample with the highest resistance had the lowest haplotype diversity, consistent with selective removal of less common haplotypes. Frequencies of the more common *hscp* haplotypes showed substantial geographic variation, in contrast to allozymes and another neutral genetic marker *Hejs*. Samples taken in 1995 still showed significant geographic heterogeneity in *hscp* haplotype frequencies but not *Hejs* (Taylor, Park, and Shen 1996). One haplotype increased significantly after pyrethroid selection in both 1990 and 1995 samples, suggesting that disequilibrium between that haplotype and the actual site under selection persisted over this time. After discovery of the resistance-causing mutations, retrospective analysis of field samples showed that L1029H and V421M changed in frequency over time. In 1990, both occurred at a frequency of 20 percent in Louisiana populations; by 1997, L1029H had increased to 78 percent, while V421M was not detected (Zhao, Park, and Adams 2000). This pattern of "evolutionary succession of mutations" was consistent with the predictions made on the basis of properties seen in the *Xenopus* expression system.

Similar approaches in other species include the use of temperature-gradient gel electrophoresis to detect haplotype variation in Australian *H. armigera* (Stokes, McKechnie, and Forrester 1997). Although extensive geographic variation was found, there was no correlation between haplotype variation and fenvalerate survivorship, which is consistent with the previously reported decline in the frequency of the nerve insensitivity mechanism in Australia eight years previously (Gunning et al. 1991). Population surveys of codling moth, on the other hand, did show a correlation between the frequency of the *kdr* mutation L1014F determined with a PCR assay for specific amplification of alternative alleles (PASA) and the frequency of pyrethroid treatments in French orchards, but no such pattern for neutral microsatellite markers (Franck et al. 2007).

The occurrence of two mutations in *Plutella* provided another opportunity for comparison of selection dynamics. The L1029F *kdr* mutation, but not V421M, D1561V, or E1565G, was found in a resistant Japanese strain (Tsukahara et al. 2003). The T944I substitution previously found by Schuler et al. (1998) was also found in this strain. This region is alternatively spliced such that only one of two possible alternative exons, A1 and A2, appear in the final mRNA, and the T944I replacement occurs in A2 in the resistant strain. Thus alternative splicing may affect the level of resistance (Sonoda et al. 2006).

By developing PASA assays for the L1029F and T944I substitutions, Kwon, Clark, and Lee (2004) have gained insight into the historical sequence of these two mutations in Korea by comparing present-day allele frequencies with those in a series of specimens collected from 1974 to 1995. Surprisingly, the L1029F mutation was present in 1974, even before the use of pyrethroids to control DBM, and was probably selected by the much earlier use of DDT. The T944I mutation appeared in 1995 after pyrethroids had been used extensively. Both mutations are currently at a high frequency in all Korean populations sampled. The PASA technique could be useful for assessing resistance mutation frequencies in field populations to predict the effectiveness of pyrethroid sprays.

The wide diversity of point mutations found in other insect orders in contrast to Lepidoptera suggests that the true diversity among the latter has been underestimated. Even common mutations may behave in unpredictable, species-specific manners in aspects important enough to influence their evolutionary dynamics. Thus the spectrum of resistance mutations continues to be explored, yielding fundamental information about mode of action that may be used to improve efficacy and possibly to combat resistance. This should also lead to a better understanding of the fitness costs of *kdr* and other mutations that seem to have led to a documented succession of mutations in at least one case, and could be responsible for the decline and eventual disappearance of target-site resistance at the sodium channel in another.

Nicotinic Acetylcholine Receptor

Two types of acetylcholine receptors (AChRs) are distinguished, nicotinic and muscarinic. Muscarinic AChRs are members of the superfamily of heterotrimeric seven-transmembrane-domain GPCRs and will not be considered further. The nicotinic acetylcholine receptors (nAChRs) belong to the Cys-loop ligand-gated ion channel (LGIC) superfamily, and are the targets of nicotine, of the relatively new neonicotinoid insecticides (imidacloprid, nitenpyram), and of the spinosyns derived from fermentation of the actinomycete bacterium *Saccharopolyspora spinosa*. They function as LGICs mediating fast synaptic transmission activated by the neurotransmitter acetylcholine (ACh) at cholinergic synapses.

The functional receptor is made up of five homologous subunits, each with an N-terminal extracellular domain containing the so-called Cys loop of thirteen amino acid residues and four transmembrane domains (TM1–TM4), the second of which lines the ion channel (Corringer, Le Novere, and Changeux 2000). An additional intracellular loop between TM3 and TM4 functions in receptor localization and modulation. Each subunit possesses two faces contacting the adjacent subunits when the pentamer is assembled. Six loops (A–F) form the ACh binding site, A–C on the "principal" face and D–F on the "complementary" face (Arias 2000). If loop C contains the two adjacent cysteines which are required for ACh binding (Kao and Karlin 1986), the subunit is termed an α-subunit. The neurotransmitter ACh binds at the interface of two subunits, between A and C of an α-subunit and between D and F of the adjacent subunit. At least two α-subunits in a pentamer are required for receptor function.

The genome of *D. melanogaster* contains genes for seven α-subunits (Dα1–Dα7) and three β subunits (Dβ1–Dβ3; Sattelle et al. 2005). Although Dβ2 lacks the adjacent cysteines that define α-subunits, it appears to be homologous to α-subunits from honeybee and mosquito. Alternative splicing of Dα4 and Dα6 and mRNA editing of Dα6 (Grauso et al. 2002), Dα5, Dβ1, and Dβ2 (Hoopengardner et al. 2003) greatly increases the potential number of isoforms. Heterologous expression in *Xenopus* oocytes has been used to probe the function of several of the α-subunits from *Drosophila* and other insect species; proper assembly requires coexpression of a vertebrate β-subunit.

In contrast to the sodium channel or AChE, very few examples of target-site resistance of nAChRs are known in insects. An imidacloprid-resistant strain of brown planthopper *Nilaparvata lugens* (Homoptera: Delphacidae) displayed reduced binding of tritiated imidacloprid to membranes. Cloning and sequencing of five nAChR subunit genes revealed a substitution Y151S (numbering according to *Torpedo californica* mature nAChR α1) in the highly conserved loop B in both α1 and α3 in resistant insects (Liu et al. 2005). Using allele-specific PCR to detect the mutation in α1 revealed a positive correlation between imidacloprid resistance and the Y151S allele frequency. When coexpressed in *Drosophila* S2 cells with rat β2, the mutated form had a lower specific binding to tritiated imidacloprid (Liu et al. 2005). Functional studies in *Xenopus* oocytes showed a reduced sensitivity to suppression of the maximal current by imidacloprid and rightward shifts in agonist dose-response curves for other neonicotinoids (Liu et al. 2006). Thus an amino acid substitution in a highly conserved, critical region for ACh binding conferred higher resistance when it occurred in two distinct α-subunits.

To identify which *Drosophila* subunits might be targets for neonicotinoids, Perry et al. (2008) conducted EMS mutagenesis and selection for nitenpyram resistance on hybrids carrying a deficiency in the cytological region 96A containing a cluster of three nAchR subunit genes. A mutation causing the loss of TM4 of Dα1 conferred resistance. So did mutants in Dβ2, including one in loop A and others, some distant from the ACh binding site that might interfere with subunit assembly or channel activation. By screening deficiency strains hemizygous for known chromosomal regions containing nAChR genes, a mutation disrupting the Dα6 gene was found that conferred 1,180-fold resistance to spinosad (Perry, McKenzie, and Batterham 2007). Thus not only point mutations but also major disruptions of subunit genes can result in target-site resistance, possibly by enabling the substitution of a less sensitive target among a partially redundant class of subunits.

Studies in Lepidoptera have proceeded in ignorance of the full spectrum of nAChR subunits in the insect until recently. Using the genome sequence of *Bombyx mori*, Shao, Dong, and Zhang (2007) identified nine α-type and three β-type subunits. Bmα1–Bmα4, Bmα6, Bmα7, and Bmβ1 are homologous to the *Drosophila* subunits of the same designation. Bmα8 appears to be homologous to Dβ2 (and α8 from honeybee and mosquito); relations of the other subunits are less clear. As in *Drosophila*, Bmα6 shows evidence of RNA editing; and Bmα4, Bmα6, and Bmα8 show evidence of alternate splicing.

Eastham et al. (1998) cloned an α-subunit (homologous to Bmα3) from *Manduca sexta*, which feeds on tobacco and is tolerant of nicotine. Residues in the α-bungaratoxin (α-btx) and nicotine-binding sites (including Y151) were conserved with respect to other α-btx sensitive nAChRs. Total larval or adult brain membranes were assayed for binding to radiolabelled α-btx and its competitive displacement by cholinergic ligands imidacloprid, nicotine, and ACh. The results were similar to those from nicotine-sensitive insects. Tritiated epibatidine was used to probe for α-btx insensitive binding sites, but showed no specific binding. The authors concluded that nicotine tolerance of this species is probably not due to a nicotinic receptor with reduced binding.

The lepidopteran family Sphingidae, to which *M. sexta* belongs, contains both nicotine-tolerant and nicotine-sensitive species. Wink and Theile (2002) quantified tolerance by injection of nicotine into the hemocoel and found a range of responses, with *M. sexta* most tolerant. A phylogenetic reconstruction of sphingids based on 16S rRNA sequences was used to map the evolution of nicotine tolerance. All Sphinginae and Macroglossinae tested were tolerant, all Smerenthinae tested were sensitive. Portions of two subunits, Msα1 (homologous to Bmα1) and Msα2 (Bmα3), were cloned from *M. sexta*, and a region of thirty-one residues encompassing loop C was amplified from the fourteen other sphingids that had been assayed for nicotine sensitivity. No coding differences among the species were found, ruling out sequence changes in this region of these two subunits as the basis for nicotine tolerance in this group.

Adamczewski, Oellers, and Schulte (2005) cloned two subunits from *H. virescens*, α7-1 (homologous to Bmα7) and α7-2 (Bmα6). Their function was assayed by creating chimeric constructs encoding the N-terminus of the *Heliothis* AChR fused to the C-terminus of the mouse 5-hydroxytryptamine receptor, transfecting HEK293 cells, and measuring calcium uptake with Fura-2 dye upon exposure to 500 μM nicotine. Sequences of four other subunits from *H. virescens* were deposited in GenBank but not reported. Grauso et al. (2002) examined patterns of RNA editing in the two α7 subunits from *H. virescens*. In *Drosophila* Dα6, seven positions were found where A-to-I editing could take place, five near loop E of the ACh binding site, some of which altered the amino acid. Four of these are also edited in the *H. virescens* homolog α7-2, but no evidence of RNA editing was found for α7-1.

Little evidence exists at present for target-site-mediated resistance to neonicotinoids in Lepidoptera, but there are emerging cases of spinosad resistance. Moulton, Pepper, and Dennehy (2000) reported up to 85-fold resistance to spinosad in a population of *Spodoptera exigua* from Thailand, but the mechanism was not investigated. To test the potential for resistance development, a susceptible strain of *H. virescens* was selected with spinosad in the laboratory and developed 1,068-fold resistance to topical application, 314-fold resistance to dietary spinosad, and 163-fold resistance to hemocoelic injection (Young, Bailey, and Roe 2003). Crosses provided evidence for a single autosomal partly recessive gene for resistance (Wyss et al. 2003). Resistance did not appear to be associated with increased metabolism or excretion, or decreased penetration (Young et al. 2000), but the resistant strain did show a decreased neural sensitivity at low concentrations of spinosad (Young et al. 2001), suggesting that target-site resistance might be responsible.

Significant resistance to spinosad has appeared in field populations of *P. xylostella*. The CH1 strain from Malaysia is more than 20,000-fold resistant to spinosad, and crosses have shown that resistance is autosomal, probably due to a single locus, completely recessive at high concentrations, and codominant at low concentrations (Sayyed, Omar, and Wright 2004). As a consequence of nearly continuous applications of spinosad over two years on *P. xylostella* in Hawaii, field control failures occurred in 2000 and variable levels of resistance were found in several populations (Mau and Gusukuma-Minuto 2004). The Pearl-Sel strain attained resistance levels of 13,100-fold after one generation of selection in the lab. Resistance was autosomally inherited and not affected by the synergists S,S,S-tributyl phosphorotrithioate and PBO (Zhao et al. 2002). A major gene affecting this resistance was mapped to AFLP LG 5, within 4.2 cM of the *Plutella* homologue of Bmα6 (Baxter 2005). RNA editing was observed at four of the same sites in the *Plutella* gene as previously found in *Heliothis* and *Drosophila*. These results are suggestive of target-site resistance, but the mutation responsible has not yet been identified.

We have seen that target-site resistance to these relatively new insecticides is starting to appear in subunits of the AChR, possibly even in Lepidoptera. The multiplicity of nAChR subunits with possibly different intrinsic sensitivities to an insecticide provides a level of redundancy not seen for AChE or the sodium channel, which could greatly widen the scope for resistance-conferring mutations. Single amino acid substitutions within a sensitive subunit could reduce its binding to the insecticide. In addition, more disruptive mutations in a sensitive subunit that interferes with its synthesis or assembly into receptors could also confer resistance, provided that other less sensitive but functionally similar subunits can take its place. The fitness costs of these "subunit substitutions" would then determine their significance in field populations.

GABA-Gated Chloride Channel

Insect GABA-gated chloride channels ($GABA_A$ receptors) are the targets of long-used cyclodiene insecticides such as dieldrin and endosulfan, hexachlorocyclohexanes such as lindane, and the more recently developed phenylpyrazole insecticide fipronil. They mediate rapid inhibitory synaptic transmission by enabling a chloride ion flux in response to binding by the neurotransmitter GABA (γ-aminobutyric acid). GABA-gated chloride channels are pentamers, with each subunit comprising four transmembrane domains, the second of which lines the pore, and a large N-terminal extracellular domain containing the GABA binding site. Many subunit types are known from vertebrates; fewer are known from insects, and their characterization is still ongoing. Like the nAChR family, they belong to the Cys-loop LGIC superfamily.

Twenty years ago, up to 60 percent of the documented cases of pest resistance to insecticides involved dieldrin and its relatives. In spite of this, the molecular basis of resistance was unknown until the *Rdl* gene was cloned from field-collected dieldrin-resistant *D. melanogaster* (ffrench-Constant et al. 1991). *Rdl* encodes a $GABA_A$ receptor subunit, and a single amino acid substitution named A302S (actually occurring in position 301) in the second transmembrane domain was found to confer dieldrin resistance (ffrench-Constant et al. 1993). This was established by site-directed mutagenesis of the wild-type cDNA converting the alanine in this position to a serine, and expression in *Xenopus* oocytes demonstrating reduced sensitivity to dieldrin and the $GABA_A$ agonist picrotoxin. Subsequent analysis showed that there is a single *Rdl* gene in the *Drosophila* genome, but that alternative splicing involving two pairs of alternative exons can produce four different protein isoforms from the same gene (ffrench-Constant and Rocheleau 1993).

Since those pioneering studies, the A302S (or A302G) substitution has been looked for and found in a diverse range of species, including cyclodiene-resistant coffee berry borer *Hypothenemus hampei*, the whitefly *Bemesia tabaci*, the aphids *Nasonovia ribisnigri* and *Myzus persicae*, and the flour beetle *Tribolium castaneum* (reviewed in Buckingham and Sattelle 2005). Lepidoptera have been notably absent from this list.

A full-length homologue of *Rdl* denoted as HVRDL was cloned from a strain of *H. virescens* susceptible to cyclodienes and fipronil (Wolff and Wingate 1998). This sequence exhibited about 95 percent sequence identity to *Drosophila Rdl* except for the intracellular loop between transmembrane domains 3 and 4. Curiously, the *Heliothis* sequence encoded a serine in position 285 corresponding to the A302S substitution in dieldrin-resistant *Drosophila*. By site-directed mutagenesis, this site was converted to encode an alanine; and the resulting HVRDL-Ala285 construct as well as the original HVRDL-Ser285 were separately injected into *Xenopus* oocytes for electrophysiological studies of the homopentamers. *Drosophila* Rdl-Ala302 and Rdl-Ser302 were also examined. The alanine form of each protein was more sensitive to chloride ion flux blockage by dieldrin and by picrotoxinin than the corresponding serine form. HVRDL-Ala285 was about fifteen times more sensitive to fipronil than was HVRDL-Ser285; however, the two forms of the *Drosophila* protein did not differ in this respect, both showing a fipronil sensitivity intermediate to the two *Heliothis* proteins at comparable GABA concentrations. Thus the effect of the A302S substitution on dieldrin resistance may be fairly general, while its effect on fipronil resistance depends on the sequence and structure of the rest of the subunit.

The unexpected serine in position 285 in dieldrin-susceptible *Heliothis* hints at greater complexity than expected from the *Drosophila* model. Indeed, each of three isoforms (GenBank accessions a1: AF006190; a2: AF006189; and a3: AF006192) from the same susceptible strain of *H. virescens* (Halling and Yuhas 2001) has a different residue at this site: glutamine, serine, or alanine, respectively. A glutamine was found at this position from both endosulfan susceptible and resistant strains of *H. armigera* (ffrench-Constant et al. 1996). Evidence that these isoforms are encoded by three distinct genes emerges from the *Bombyx mori* genome sequence where partially assembled contigs from the Dazao strain (never exposed to dieldrin or fipronil) can be found containing Gln285 (GenBank AADK01004540), Ser285 (AADK01022486), or Ala285 (AADK01003208). However, searches of the whole genome sequences of insects in other orders yield evidence for only a single gene: *D. melanogaster* and eleven additional sequenced *Drosophila* species, *Culex quinquefasciatus*, *Aedes aegypti*, the hymenopterans *Apis mellifera* and *Nasonia vitripennis*, and the coleopteran *T. castaneum*. All of these have an alanine in the corresponding 285 position. The aphid *Acyrthosiphon pisum* appears to have two genes, one (ABLF01022071) with S and the other

(ABLF01036703) with A. Anthony et al. (1998) provided evidence for two genes in the aphid *M. persicae*, one with a serine and the other polymorphic with either alanine or glycine, the latter correlating with endosulfan resistance.

Plutella xylostella shows a similar correlation. The SZ-F strain selected with fipronil developed 300-fold resistance to that compound, but only 3.5-fold resistance to dieldrin and 6.5-fold to endosulfan (Li et al. 2006). PCR was used to amplify a product from genomic DNA encoding 54 amino acids identical to the *D. melanogaster Rdl* gene in the region encompassing A302. Only an alanine was found in this position in two susceptible strains, but the resistant strain was polymorphic A/S. Sequencing PCR products from individual SZ-F larvae showed that the frequency of the allele encoding a serine in this position was 30 percent, which increased to 57 percent among survivors of a fipronil concentration that killed 25 percent of the SZ-F strain. The protein sequence of the 54-aa region surveyed is identical to all three *Bombyx* genes except for the polymorphic position, so the identity of the *Bombyx* ortholog is unknown. Variation at other positions in the gene was not explored in this study.

When fipronil was brought onto the market, concerns were raised about its sustainability in light of the widespread resistance that had appeared decades earlier to cyclodienes acting at the same molecular target. However, early studies suggested that the two insecticides acted at different subsites and that cross-resistance to fipronil in existing dieldrin-resistant strains was low (reviewed in Bloomquist 2001). Additional mutations in cyclodiene-resistant species may be required for significant fipronil resistance. A second substitution T351M in TM3 co-occurring with A302G in a lab-selected strain of *Drososophila simulans* does appear to greatly enhance the resistance to fipronil, and comparison of channels expressed in *Xenopus* oocytes showed that the double mutant was less sensitive to fipronil than either single mutant, although not as insensitive as expected from the bioassay results (Le Goff et al. 2005).

With the likely occurrence of three *Rdl* paralogs in Lepidoptera, functional redundancy in the gene family could have consequences for the mode of fipronil resistance evolution, compared with other insect groups with a single *Rdl* gene. If the paralog possessing an alanine at site 302 is intrinsically most sensitive, any mutation reducing its expression could decrease the insect's sensitivity providing that other channel paralogs could compensate. Should a substitution such as T351M in *D. simulans* occur in the more dieldrin- and fipronil-tolerant lepidopteran subunit already possessing a serine, then the preexisting functional redundancy would appear to have facilitated fipronil evolution in comparison to insects with a single *Rdl* gene. On the other hand, the sensitivity of the form encoding a glutamine has apparently never been assayed in *Xenopus* oocytes. If the glutamine form is even more sensitive than the alanine form, mutations in both genes would be required to "hide" them from the insecticide and so functional redundancy would appear to delay resistance development. In any event, the evolutionary response in Lepidoptera to dieldrin or fipronil selection is likely to be more complicated than extrapolations from dieldrin resistance in *Drosophila* would suggest.

METABOLIC RESISTANCE TO CHEMICAL INSECTICIDES

In Lepidoptera, less progress has been made to date on the molecular genetics of metabolic resistance than target-site resistance. Advantages enjoyed by workers on the latter include a relatively small number of targets that are highly enough conserved across insect orders, so that findings in different species can all be related back to the same basic molecular structure. Detoxicative enzymes, on the other hand, occur in huge, constantly diversifying gene families. Their great number and rapid evolutionary change makes it nearly impossible to make close comparisons with predictive value, except among closely related species in the best of cases. So far these have not included Lepidoptera. We illustrate this with brief reviews of the three major classes of detoxicative enzymes important in insecticide resistance: glutathione transferases, carboxylesterases, and cytochrome P450 enzymes.

Glutathione Transferases

Glutathione transferases (GSTs) conjugate glutathione to electrophilic compounds in Phase II detoxification reactions, protecting the cell against them and facilitating their excretion. Glutathione is the tripeptide γ-L-glutamyl-L-cysteinyl-glycine, abbreviated GSH to emphasize the importance of the reactive thiol group -SH. It is abundant in cells where it plays an important role in maintaining redox homeostasis and in scavenging electrophiles to which it is conjugated. The majority of GSTs are dimeric cytosolic proteins. These are classified according to sequence similarity, and six of the named classes occur in insects: Delta, Epsilon, Omega, Sigma, Theta, and Zeta. Some microsomal GSTs (MAPEG enzymes) also occur in insects; these have a different structure to the soluble forms. GSTs can detoxify OP insecticides by O-dealkylation or O-dearylation; the enzyme DDT dehydrochlorinase is a GST. GSTs may also sequester pyrethroids, but have been suggested to be more important in resistance by detoxifying lipid peroxidation products that are induced by pyrethroids (Vontas, Small, and Hemingway 2001).

Despite a number of studies where biochemical evidence implicates GSTs in resistance (Ranson and Hemingway 2005), there is but a single such case leading to the cloning of a GST in a lepidopteran. Chiang and Sun (1993) isolated three GST isozymes from a Taiwanese strain of *P. xylostella* resistant to methyl parathion and the chitin synthesis inhibitor teflubenzuron by affinity chromatography on a glutathione-agarose column. Compared to the other two isozymes, GST-3 had a lower activity against the model substrate CDNB but a higher activity against DCNB and the OP insecticides parathion, paraoxon, and methyl parathion, for which the conjugation product was S-(4-nitrophenyl) glutathione (Chiang and Sun 1993). A fourth isozyme, GST-4, was later isolated from the same strain, with even higher specific activity against DCNB and the OP insecticides (Ku et al. 1994). A polyclonal antibody raised against GST-3 cross-reacted with GST-4 and showed higher protein content in the Taiwanese resistant strains as well as a selected methyl parathion resistant strain from France.

The gene encoding GST-3 was subsequently cloned and expressed in *E. coli*, exhibiting similar conjugation properties as the purified GST-3 isozyme (Huang et al. 1998). The levels of mRNA were much greater in the resistant strains than in susceptible strains, but as shown by Southern blots this was not due to gene amplification. Sonoda and Tsumuki (2005) studied expression of the same gene in Japanese strains resistant to permethrin and the chitin synthesis inhibitor chlorfluazuron, where it also showed higher mRNA levels and no gene amplification. No protein sequence differences could be found between resistant and susceptible strains, and genomic flanking regions of the gene were nearly identical. The gene contains a single intron four nucleotides upstream of the start codon (Sonoda, Ashfaq, and Tsumuki 2006). Enayati, Ranson, and Hemingway (2005) classified it as an Epsilon GST. The four isozymes appear to be present in all strains examined, but GST-3 is produced in higher amounts in resistant strains. In the absence of linkage mapping it is not known whether *cis-* or *trans-*mediated upregulation is responsible.

Carboxylesterases

Carboxylesterases are hydrolytic enzymes belonging to the α/β-hydrolase fold superfamily that also contains AChE. They cleave carboxylic acid esters by a two-step mechanism similar to the action of AChE, with the oxygen of the active-site serine first making a nucleophilic attack on the carbonyl carbon of the substrate, forming an acyl-enzyme linkage while the alcohol product is displaced. A water molecule then makes a similar nucleophilic attack, displacing the serine, releasing the acid product and regenerating the free enzyme. These enzymes and their noncatalytic relatives form a large superfamily for which nomenclature is not yet standardized; however, in a comprehensive review of insect esterases, Oakeshott et al. (2005a) have defined fourteen major clades on the basis of sequence similarity.

Many esterases involved in OP resistance in Diptera and Hemiptera have been cloned, and a recurrent finding is that gene amplification leads to higher expression levels that protect the insect by sequestering rather than hydrolyzing the insecticide (reviewed in Oakeshott et al. 2005a). Thus the enzyme phosphorylated at the active site serine, trapped in the first stage of the reaction cycle, manages to keep one molecule of the insecticide away from the target AChE, at the expense of one molecule of the protein. Esterase genes in Clade E are amplified in Hemiptera, including the brown rice planthopper, *N. lugens*, and the aphid *M. persicae*, where the amplified exterase can account for up to 1 percent of the soluble protein of the aphid. Esterase genes in Clade C are likewise amplified in culicine and anopheline mosquitoes. An enzyme in Clade B, αE7 in housefly *M. domestica* and sheep blowfly *L. cuprina*, is not amplified but instead has point mutations that increase its OP hydrolase activity. Hydrolysis of the OP occurs slowly but at a sufficient rate to confer some level of resistance. Numerous biochemical studies provide evidence of esterase involvement in OP and pyrethroid resistance in Lepidoptera, including many cases of overexpressed enzymes. However, reports of the cloning of these genes from Lepidoptera are still lacking.

Cytochrome P450 Enzymes

Cytochrome P450 enzymes are a large and ubiquitous class of heme-thiolate proteins that collectively catalyze a huge range of reactions, many of which play an important role in detoxification of xenobiotics such as insecticides. The protein holds a heme group with an iron atom at its center, changes in the oxidation state of which are essential for the reaction cycle in which one atom of molecular oxygen is transferred to a substrate and the other reduced to water. The fidelity of this cycle can vary considerably, and depending on the enzyme and substrate, a large variety of other reactions can result. The membrane-associated P450s require a supply of electrons that are transferred from NADPH or NADH by membrane-associated redox partners, including NADPH cytochrome P450 reductase. These features pose special challenges for in vitro studies of P450 enzyme activity, whether in membrane fractions from the organism of interest or by heterologous expression. The biochemistry and molecular biology of P450s is an enormous subject and the literature relevant to insects has been reviewed (Scott and Wen 2001; Feyereisen 2005; Li, Schuler, and Berenbaum 2007).

P450s play many roles in insects, including synthesis of hormones and pheromones, fatty acid metabolism, and detoxification of xenobiotic compounds such as hostplant toxins and insecticides. Many of the detoxifying type are inducible by xenobiotics; that is, transcription rates are normally regulated at a low level but increase in response to the appearance of a foreign or endogenous compound. Although this may function adaptively to detoxify harmful compounds from the environment, the inducers are not necessarily the same molecules as the toxins that need to be eliminated. The induction process is best understood in mammals, as exemplified by the role of the aryl hydrocarbon receptor (AHR) in the upregulation of CYP1A1 by polycylic aromatic hydrocarbons. A protein complex containing AHR and molecular chaperones resides in the cytoplasm. On binding to a suitable inducer, AHR moves to the nucleus, forms a heterodimer with the related ARNT (aryl hydrocarbon receptor nuclear translocator) protein, binds to DNA at specific enhancer sequences near the promoters of CYP1A1 and other genes, and interacts with transcription factors and other coactivators to promote transcription. Homologues of these proteins exist in insects, but their roles in the regulation of insect P450s are not understood.

Biochemical approaches to studying the role of P450s include the use of model substrates to monitor the progress of a particular type of reaction such as O-demethylation, P450 enzyme inhibitors such as PBO, and inducers such as phenobarbital. Many studies have compared resistant and susceptible strains with such methods, but their lack of specificity for a single enzyme, combined with the multiplicity of P450 genes in insect genomes (e.g., 160 in *A. aegypti*, Strode et al. 2008; 87 in *B. mori*, Kozaki et al. 2008) has made the implication of a specific P450 gene very difficult. Lepidopteran species for which such studies provide evidence that P450s are a component of metabolic insecticide resistance include *H. armigera* (Forrester et al. 1993; Kranthi et al. 2001; Yang et al. 2004), *S. frugiperda* (Yu 1992), *S. littoralis* (Huang and Han 2007), *P. xylostella* (Yu and Nguyen 1992), *H. virescens* (Rose et al. 1995), and *C. pomonella* (Bouvier et al. 2002).

From studies that have successfully followed the trail of such evidence all the way to the gene, a general paradigm is emerging. One or more of the many P450s already being expressed by the susceptible insect may be capable of detoxifying an insecticide, but not fast enough for protection against lethal effects. The induction mechanism as well may respond too slowly or not at all. Thus, any of a great variety of mutations that could increase the mRNA and protein level of this P450 will be selected for by insecticide exposure. These could include loss of function of regulatory proteins that repress P450 transcription, disruption of P450 gene promoter binding sites for such repressor proteins, creation of new enhancer sites in the vicinity of the gene, or an increase in the gene copy number. Some of these changes could disrupt the feedback control required by an inducible system; thus, loss of inducibility often accompanies high constitutive expression. Many of these mutations may occur in genes other than the P450, increasing the difficulty of their identification. There is evidence that amino acid substitutions in the P450 protein itself can enhance activity against insecticides (*Drosophila* Cyp6a2, Amichot et al. 2004) but the increase in expression of a preexisting minimally competent enzyme seems to be more important in the majority of cases studied so far.

The first insect P450 gene to be cloned was CYP6A1 from housefly, and key findings in this system illustrate the paradigm outlined above (reviewed in Feyereisen 2005). CYP6A1 expression is low and inducible by phenobarbital in susceptible strains, and high and constitutive in the diazinon-resistant Rutgers strain and some others with P450-based resistance. Elevated mRNA levels are reflected in immunologically detectable increases in the protein. When expressed in *E. coli* along with its redox partners, CYP6A1 metabolizes diazinon with a high turnover. The CYP6A1 gene maps to chromosome 5; however, the factor increasing the expression level maps to chromosome 2. This shows that the factor differing between resistant and susceptible strains is *trans*-acting, not a part of the CYP6A1 gene itself. In another case, *cis*-acting overexpression of CYP6G1 in *D. melanogaster* is due to mutations in the gene itself; transposon-mediated insertions upstream of the coding sequence result in greater levels of expression (Daborn et al. 2002) of a P450 that is capable of metabolizing DDT and imidacloprid (Joussen et al. 2008).

Reactions catalyzed by P450s also activate proinsecticides to insecticides. For example, the oxidative desulfuration converting P = S in the inactive phosphorothioate forms of OP insecticides (e.g. triazophos) to P = O in the AChE-inhibiting oxon forms (e.g., triazophos-oxon) by insect P450s is responsible for their effectiveness as insecticides. This phenomenon may provide the basis for "negative cross-resistance," in which a strain resistant to one insecticide due to increased detoxification is more susceptible to another due to increased activation by the same enzyme system. Among West African populations of *H. armigera*, such negative cross-resistance was observed between pyrethroids and triazophos, and using the inhibition of purified recombinant *Drosophila* AChE as an assay, higher levels of triazophos-oxon were found in triazophos-treated larvae of the pyrethroid-resistant strain (Martin et al. 2003). A strain of *H. virescens* resistant to pyrethroids partly due to P450 activity had a genetically linked higher susceptibility to the pyrrole insecticide chlorfenapyr, which must be bioactivated before it is active in uncoupling oxidative phosphorylation (Pimprale et al. 1997).

The first P450s implicated in insecticide resistance to be cloned from Lepidoptera belong to the CYP6B subfamily. CYP6B2 (Wang and Hobbs 1995) and CYP6B6 and CYP6B7 (Ranasinghe and Hobbs 1998) were cloned from Australian *H. armigera* in studies of pyrethroid resistance. CYP6B7 mRNA levels were elevated in many field-collected pyrethroid-resistant larvae relative to a susceptible strain (Ranasinghe, Campbell, and Hobbs 1998) and could be induced by exposure to phenobarbital and pyrethroids in fat body organ culture (Ranasinghe and Hobbs 1999). These authors hypothesized that overexpression of CYP6B7 was the predominant cause of resistance in Australian field populations. In support of this idea is the fact that a similar gene CYP6B8 from the closely related species *H. zea* has been shown to be inducible by phenobarbital (Li, Berenbaum, and Schuler 2002) and is capable of metabolizing the pyrethroid cypermethrin (Li et al. 2004).

Grubor and Heckel (2007) tested this hypothesis by linkage mapping in the AN02 strain of *H. armigera*, which showed 50-fold PBO-suppressible fenvalerate resistance in larvae and adults due to a single semidominant resistance gene *RFen1*. Sequencing of a genomic BAC clone revealed that the three genes were arranged in a cluster, in the order CYP6B7-CYP6B6-CYP6B2. This cluster was mapped to LG 14, whereas the *RFen1* gene mapped to LG 13 (Heckel et al. 1998); thus, changes in the coding sequences or *cis*-regulatory sequences of the three CYP6B genes could not be responsible for the resistance controlled by *RFen1*. To test whether *RFen1* could encode a *trans*-acting factor regulating the constitutive expression of either of the three CYP6B genes, their mRNA levels were measured by quantitative real-time RT-PCR in individuals that had not been exposed to pyrethroids, but no correlation between mRNA level and *RFen1* genotype was found. Thus, in the AN02 strain, CYP6B levels did not vary with resistance.

In a screen for genes whose expression did covary with pyrethroid resistance in the AN02 strain, Wee et al. (2008) used the cDNA-AFLP technique as a differential display method to search for mRNAs that were up- or downregulated in individuals carrying the resistant allele at *RFen1*. One of the genes showing a significant upregulation in resistant individuals was a novel P450, CYP337B1, that also mapped to within 1 centimorgan of RFen1 on LG 13. The metabolic competency of CYP337B1 against pyrethroids remains to be tested.

Yang et al. (2006) cloned two P450 genes, CYP9A12 and CYP9A14, from the YGF strain from Shandong Province, China, of *H. armigera* which developed 1,690-fold resistance to fenvalerate after fourteen generations of selection. Quantitative real-time RT-PCR showed that YGF had elevated levels for CYP9A12 of 19-fold in midgut and 433-fold in fat body, and for CYP9A14, 4-fold and 59-fold, respectively, compared with the unselected parent strain. The resistance gene or genes were not mapped in this study.

Additional P450s from Lepidoptera showing higher expression levels in resistant strains include the first member of the CYP9 family, cloned from *H. virescens* (Rose et al. 1997). CYP9A1 showed 29-fold higher expression in RNA blots from a thiodicarb-resistant strain from North Carolina than in neigboring susceptible strains. Bautista, Tanaka, and Miyata (2007) cloned four P450 genes from a Japanese strain of *P. xylostella* 200-fold resistant to permethrin that could be partially suppressed by PBO. CYP6BG1 and CYP6BG2 mRNA levels were 5- and 4-fold higher, respectively, in the resistant strain. These and the other two, CYP6AE13 and CYPBF1v4, were induced by low levels of permethrin in the susceptible strain. An additional p450 cloned from a Chinese *Plutella* strain 30-fold resistant to permethrin is CYP9G2 (Shen et al. 2004).

The paradigm of selectively driven upregulation of individual P450s from a preexisting, metabolically diverse arsenal of enzymes remains to be established in Lepidoptera. As additional P450s are cloned, the opportunities to test this idea will increase. One important feature of Lepidoptera is its long coevolutionary history of insect-plant interactions. P450s are employed in the detoxification of many chemicals that plants synthesize in order to avoid herbivory, and the induction patterns and metabolic capabilities of several of these have been investigated. To cite just one example, CYP6B enzymes in specialist *Papilio polyxenes* are induced by and metabolize furanocoumarins produced by their host plants; other CYP6B enzymes in generalist *Papilio glaucus* have a broader range of substrates, and *Papilio multicaudatus*, intermediate in specialization with one furanocoumarin-containing host, has CYP6B enzymes with features indicative of the transition from oligophagy to polyphagy (Mao, Schuler, and Berenbaum 2007). These and other signs of ancient chemical interactions with plants may serve as preadaptations to surviving the modern challenge of insecticides (reviewed in Despres, David, and Gallet 2007). Are host-plant generalists endowed with a set of P450s with a wider metabolic range as hypothesized by Krieger, Feeny, and Wilkinson (1971), and if so, does that predispose them to develop insecticide resistance sooner than host-plant specialists? Greater integration of classically oriented resistance studies with chemical ecology will be required to answer this question.

CONCLUSIONS AND FUTURE PROSPECTS

For most chemical insecticides, there are enough similarities in the targets and detoxicative systems to ensure that results from studies of other insect orders are broadly applicable to Lepidoptera. Yet many species differences and lepidopteran-specific aspects have been overlooked. Resistance-conferring substitutions in insecticide targets that seem to recur repeatedly in different insect orders may still have different effects depending on the rest of the protein in which they occur. Directed mutational studies starting with sequences from the species of interest are becoming increasingly feasible and should be pursued more often. Studies on newly occurring resistant strains should characterize the entire target rather than focusing only on mutations that have been found before in other groups.

The resources necessary for a more systematic study of lepidopteran resistance mechanisms are rapidly improving, largely due to the emerging information on the genome sequence of *B. mori* (International Silkworm Genome Consortium 2008). This and other lepidopteran genome sequences (including *H. armigera*) to appear in the next few years will help enormously in understanding the diversity of targets and detoxicative enzymes, most of which occur in gene families more diverse than *Drosophila* would lead us to believe. As well as filling in the blanks in our partial lists of known targets and their paralogs, comparative genomics will illuminate other gene families similar enough for broad-spectrum control of pest Lepidoptera but different enough from vertebrates and other insect orders to minimize off-target effects.

A more complete understanding of insect-plant coevolutionary interactions will also be useful in understanding the techniques that plants have developed to reduce herbivory and attractiveness, and how insects have coevolved to overcome those defenses (for genetics of host range, see Chapter 11). Consideration of future resistance development should pay attention to the "ghosts of selection past" that have shaped interactions we see today. One example of this is the high diversity of proteinase inhibitors produced by plants (Ryan 1990) as opposed to the adaptive responses of proteases produced by insects (Jongsma et al. 1995).

There has been a shift away from research on chemical insecticide resistance in developed countries with stable or declining populations. Developing countries should start to take the lead as demands on food production increase, providing that scientific infrastructure develops rapidly enough. Developing countries bore the brunt of past resistance problems, which must be better understood to successfully exploit new subtargets of the same targets, as illustrated by insecticides such as fipronil, indoxacarb, and neonicotinoids. In adopting the new generation of improved chemistries, the lingering effects of insecticides from the past that have fallen into disuse because of resistance, environmental concerns, and economic disadvantages should not be overlooked. Although impossible to quantify absolutely, the ghost of selection past from DDT resistance undoubtedly accelerated the selective response to pyrethroids. More emphasis is needed on finding improved compounds that are also "anti-resistant" to previously evolved resistance mechanisms.

Molecular genetics should also contribute to prolonging the effectiveness of existing insecticides by providing information that can be used in insecticide resistance management strategies for sustainable and safe crop protection. This would be most useful if there were a manageable spectrum of resistance mutations that could be monitored in pest populations at the gene level when resistance allele frequencies are low, rather than detected at the level of a control failure when allele frequencies are high. It would also be useful in support of resistance management plans such as the still successful high-dose/refuge strategy for transgenic Bt cotton implemented in the United States and Australia. Most developing countries do not implement such plans and thus may provide an unfortunate "control" for the development of resistance in their absence, unless an increase in allele frequencies can be detected in time for corrective action.

As already seen in the case of *B. thuringiensis* toxins, many new opportunities for resistance research will be opened as crop plants are transformed with a greater variety of toxic or bioactive proteins: novel bacterial toxins, proteins from spider and scorpion venoms, protease and other digestive enzyme inhibitors, and lectins coupled to neuropeptides, to name a few. Manipulation of pest gene expression by RNA interference is just now being explored and appears to be very promising for some insect orders like Coleoptera (Baum et al. 2007). Experimental utilization of RNA interference has been less successful in Lepidoptera, for reasons that are still not understood, but the potential has been illustrated by the suppression of a P450 in *H. armigera* using double-stranded RNA produced by its food plant (Mao et al. 2007). Concerns about resistance are usually lost in the promise of new technologies and the excitements of early success. But evolution never stands still. A sobering reminder of this is the first report of field-evolved resistance to an insecticidal granulovirus used to control codling moth (Asser-Kaiser et al. 2007), a strategy widely held to be resistance-proof. Under the "right" conditions, resistance isn't impossible—it's inevitable.

REFERENCES

Adamczewski, M., N. Oellers, and T. Schulte. 2005. Nucleic acids encoding insect acetylcholine receptor subunits. US Patent No. 6933131.

Ahmad, M., R.T. Gladwell, and A.R. McCaffery. 1989. Decreased nerve sensitivity is a mechanism of resistance in a pyrethroid resistant strain of *Heliothis armigera* from Thailand. *Pestic. Biochem. Physiol.* 35:165–71.

Amichot, M., S. Tares, A. Brun-Barale, et al. 2004. Point mutations associated with insecticide resistance in the *Drosophila* cytochrome P450 Cyp6a2 enable DDT metabolism. *Eur. J. Biochem.* 271:1250–57.

Andrews, M.C., A. Callaghan, L.M. Field, et al. 2004. Identification of mutations conferring insecticide-insensitive AChE in the cotton-melon aphid, *Aphis gossypii* Glover. *Insect Mol. Biol.* 13:555–61.

Ankersmit, G.W. 1953. DDT-resistance in *Plutella maculipennis* (Curt.) (Lep.) in Java. *Bull. Entomol. Res.* 44:421–23.

Anthony, N., T. Unruh, D. Ganser, et al. 1998. Duplication of the *Rdl* GABA receptor subunit gene in an insecticide-resistant aphid, *Myzus persicae*. *Mol. Gen. Genet.* 260:165–75.

Arias, H.R. 2000. Localization of agonist and competitive agonist binding sites on nicotinic acetylcholine receptors. *Neurochem. Int.* 36:595–645.

Asser-Kaiser, S., E. Fritsch, K. Undorf-Spahn, et al. 2007. Rapid emergence of baculovirus resistance in codling moth due to dominant, sex-linked inheritance. *Science* 317:1916–18.

Baek, J.H., J.I. Kim, D.W. Lee, et al. 2005. Identification and characterization of *Ace1*-type acetylcholinesterase likely associated with organophosphate resistance in *Plutella xylostella*. *Pestic. Biochem. Physiol.* 81:164–75.

Baum, J.A., T. Bogaert, W. Clinton, et al. 2007. Control of coleopteran insect pests through RNA interference. *Nat. Biotechnol.* 25:1322–26.

Bautista, M.A.M., T. Tanaka, and T. Miyata. 2007. Identification of permethrin-inducible cytochrome P450s from the diamondback moth, *Plutella xylostella* (L.) and the possibility of involvement in permethrin resistance. *Pestic. Biochem. Physiol.* 87:85–93.

Baxter, S.W. 2005. Molecular and genetic analysis of Bt and spinosad resistance in diamondback moth, *Plutella xylostella*. PhD diss. Univ. Melbourne.

Bloomquist, J. 2001. GABA and glutamate receptors as biochemical sites for insecticide action. In *Biochemical sites of insecticide action and resistance*, ed. I. Ishaaya, 17–41. Berlin: Springer-Verlag.

Bouvier, J.C., T. Boivin, D. Beslay, et al. 2002. Age-dependent response to insecticides and enzymatic variation in susceptible and resistant codling moth larvae. *Arch. Insect Biochem. Physiol.* 51:55–66.

Bravo, A., S.S. Gill, and M. Soberon. 2007. Mode of action of *Bacillus thuringiensis* Cry and Cyt toxins and their potential for insect control. *Toxicon* 49:423–35.

Brown, T.M., and P.K. Bryson. 1992. Selective inhibitors of methyl parathion-resistant acetylcholinesterase from *Heliothis virescens*. *Pestic. Biochem. Physiol.* 44:155–64.

Brown, T.M., P.K. Bryson, F. Arnette, et al. 1996. Surveillance of resistant acetylcholinesterase in *Heliothis virescens*. In *Molecular genetics and evolution of pesticide resistance*, ed. T.M. Brown, 149–57. Washington, D.C.: American Chemical Society.

Brun-Barale, A., J.C. Bouvier, D. Pauron, et al. 2005. Involvement of a sodium channel mutation in pyrethroid resistance in *Cydia pomonella* L, and development of a diagnostic test. *Pest Manag. Sci.* 61:549–54.

Buckingham, S.D., and D.B. Sattelle 2005. GABA receptors of insects. In *Comprehensive molecular insect science*, Vol. 5, ed. L. Gilbert, K. Iatrou, and S. Gill, 107–42. Oxford: Elsevier.

Byrne, F.J., and N.C. Toscano. 2001. An insensitive acetylcholinesterase confers resistance to methomyl in the beet armyworm *Spodoptera exigua* (Lepidoptera: Noctuidae). *J. Econ. Entomol.* 94:524–28.

Campanhola, C., B.F. McCutchen, E.H. Baehrecke, et al. 1991. Biological constraints associated with resistance to pyrethroids in the tobacco budworm (Lepidoptera, Noctuidae). *J. Econ. Entomol.* 84:1404–11.

Cassanelli, S., M. Reyes, M. Rault, et al. 2006. Acetylcholinesterase mutation in an insecticide-resistant population of the codling moth *Cydia pomonella* (L.). *Insect Biochem. Mol. Biol.* 36:642–53.

Chiang, F.M., and C.N. Sun. 1993. Glutathione transferase isozymes of diamondback moth larvae and their role in the degradation of some organophosphorus insecticides. *Pestic. Biochem. Physiol.* 45:7–14.

Church, C.J., and C.O. Knowles. 1993. Relationship between pyrethroid enhanced batrachotoxinin A 20-alpha-benzoate binding and pyrethroid toxicity to susceptible and resistant tobacco budworm moths *Heliothis virescens*. *Comp. Biochem. Physiol. C Pharmacol. Toxicol. Endocrinol.* 104:279–87.

Claudianos, C., R.J. Russell, and J.G. Oakeshott. 1999. The same amino acid substitution in orthologous esterases confers organophosphate resistance on the house fly and a blowfly. *Insect Biochem. Mol. Biol.* 29:675–86.

Corringer, P.J., N. Le Novere, and J.P. Changeux. 2000. Nicotinic receptors at the amino acid level. *Annu. Rev. Pharmacol. Toxicol.* 40:431–58.

Daborn, P.J., J.L. Yen, M.R. Bogwitz, et al. 2002. A single P450 allele associated with insecticide resistance in *Drosophila*. *Science* 297:2253–56.

Davies, T.G.E., L.M. Field, P.N.R. Usherwood, et al. 2007. DDT, pyrethrins, pyrethroids and insect sodium channels. *IUBMB Life* 59:151–62.

Despres, L., J.P. David, and C. Gallet. 2007. The evolutionary ecology of insect resistance to plant chemicals. *Trends Ecol. Evol.* 22:298–307.

Dong, K. 2007. Insect sodium channels and insecticide resistance. *Invertebr. Neurosci.* 7:17–30.

Eastham, H.M., R.J. Lind, J.L. Eastlake, et al. 1998. Characterization of a nicotinic acetylcholine receptor from the insect *Manduca sexta*. *Eur. J. Neurosci.* 10:879–89.

Elliott, M. 1996. Synthetic insecticides related to natural pyrethrins. In *Crop protection agents from nature: Natural products and analogues*, ed. L. Copping, 254–300. Cambridge: Royal Society of Chemistry.

Enayati, A.A., H. Ranson, and J. Hemingway. 2005. Insect glutathione transferases and insecticide resistance. *Insect Mol. Biol.* 14:3–8.

Ferré, J., and J. Van Rie. 2002. Biochemistry and genetics of insect resistance to *Bacillus thuringiensis*. *Annu. Rev. Entomol.* 47:501–33.

Feyereisen, R. 2005. Insect cytochrome P450. In *Comprehensive molecular insect science*, Vol. 4, ed. L. Gilbert, K. Iatrou, and S. Gill, 1–77. Oxford: Elsevier.

ffrench-Constant, R.H., N. Anthony, D. Andreev, et al. 1996. Single versus multiple origins of insecticide resistance: Inferences from the cyclodiene resistance gene *Rdl*. In *Molecular genetics and evolution of pesticide resistance. ACS Symposium Series 645*, ed. T.M. Brown, 106–11. Washington, D.C.: American Chemical Society.

ffrench-Constant, R.H., D.P. Mortlock, C.D. Shaffer, et al. 1991. Molecular cloning and transformation of cyclodiene resistance in *Drosophila*: An invertebrate γ-aminobutyric acid subtype A receptor locus. *Proc. Natl. Acad. Sci. U.S.A.* 88:7209–13.

ffrench-Constant, R.H., and T.A. Rocheleau. 1993. *Drosophila* γ-aminobutyric acid receptor gene *Rdl* shows extensive alternative splicing. *J. Neurochem.* 60:2323–26.

ffrench-Constant, R.H., T.A. Rocheleau, J.C. Steichen, et al. 1993. A point mutation in a *Drosophila* GABA receptor confers insecticide resistance. *Nature* 363:449–51.

Forrester, N.W., M. Cahill, L.J. Bird, et al. 1993. Management of pyrethroid and endosulfan resistance in *Helicoverpa armigera* (Lepidoptera, Noctuidae) in Australia. *Bull. Entomol. Res.*, Suppl. 1:R1–132.

Fournier, D. 2005. Mutations of acetylcholinesterase which confer insecticide resistance in insect populations. *Chem. Biol. Interact.* 157:257–61.

Franck, P., M. Reyes, J. Olivares, et al. 2007. Genetic architecture in codling moth populations: Comparison between microsatellite and insecticide resistance markers. *Mol. Ecol.* 16:3554–64.

Gammon, D.W. 1980. Pyrethroid resistance in a strain of *Spodoptera littoralis* is correlated with decreased sensitivity of the CNS *in vitro*. *Pestic. Biochem. Physiol.* 13:53–62.

Gilbert, R.D., P.K. Bryson, and T.M. Brown. 1996. Linkage of acetylcholinesterase insensitivity to methyl parathion resistance in *Heliothis virescens*. *Biochem. Genet.* 34:297–312.

Grauso, M., R.A. Reenan, E. Culetto, et al. 2002. Novel putative nicotinic acetylcholine receptor subunit genes, Dα5, Dα6 and Dα7 in *Drosophila melanogaster* identify a new and highly conserved target of adenosine deaminase acting on RNA-mediated A-to-I pre-mRNA editing. *Genetics* 160:1519–33.

Griffitts, J.S., and R.V. Aroian. 2005. Many roads to resistance: How invertebrates adapt to Bt toxins. *Bioessays* 27:614–24.

Grubor, V.D., and D.G. Heckel. 2007. Evaluation of the role of CYP6B cytochrome P450s in pyrethroid resistant Australian *Helicoverpa armigera. Insect Mol. Biol.* 16:15–23.

Gunning, R.V., C.S. Easton, M.E. Balfe, et al. 1991. Pyrethroid resistance mechanisms in Australian *Helicoverpa armigera. Pestic. Sci.* 33:473–90.

Gunning, R.V., C.S. Easton, L.R. Greenup, et al. 1984. Pyrethroid resistance in *Heliothis armigera* (Hübner) (Lepidoptera, Noctuidae) in Australia. *J. Econ. Entomol.* 77:1283–87.

Gunning, R.V., G.D. Moores, and A.L. Devonshire. 1996. Insensitive acetylcholinesterase and resistance to thiodicarb in Australian *Helicoverpa armigera* Hübner (Lepidoptera: Noctuidae). *Pestic. Biochem. Physiol.* 55:21–28.

Gunning, R.V., G.D. Moores, and A.L. Devonshire. 1998. Insensitive acetylcholinesterase and resistance to organophosphates in Australian *Helicoverpa armigera. Pestic. Biochem. Physiol.* 62:147–51.

Hall, L.M.C., and P. Spierer. 1986. The *Ace* locus of *Drosophila melanogaster*: Structural gene for acetylcholinesterase with an unusual 5' leader. *EMBO J.* 5:2949–54.

Halling, B., and D. Yuhas. 2001. Lepidopteran GABA-gated chloride channels. US Patent No. 6329516.

Harel, M., G. Kryger, T.L. Rosenberry, et al. 2000. Three-dimensional structures of *Drosophila melanogaster* acetylcholinesterase and of its complexes with two potent inhibitors. *Protein Sci.* 9:1063–72.

Head, D.J., A.R. McCaffery, and A. Callaghan. 1998. Novel mutations in the *para*-homologous sodium channel gene associated with phenotypic expression of nerve insensitivity resistance to pyrethroids in Heliothine Lepidoptera. *Insect Mol. Biol.* 7:191–96.

Heckel, D.G., P.K. Bryson, and T.M. Brown. 1998. Linkage analysis of insecticide-resistant acetylcholinesterase in *Heliothis virescens. J. Hered.* 89:71–78.

Heckel, D.G., L.J. Gahan, S.W. Baxter, et al. 2007. The diversity of Bt resistance genes in species of Lepidoptera. *J. Invertebr. Pathol.* 95:192–97.

Heckel, D.G., L.J. Gahan, J.C. Daly, et al. 1998. A genomic approach to understanding *Heliothis* and *Helicoverpa* resistance to chemical and biological insecticides. *Philos. Trans. R. Soc. Lond. B Biol. Sci.* 353:1713–22.

Holloway, J.W., and A.R. McCaffery. 1996. Nerve insensitivity to cis-cypermethrin is expressed in adult *Heliothis virescens. Pestic. Sci.* 47:205–11.

Hoopengardner, B., T. Bhalla, C. Staber, et al. 2003. Nervous system targets of RNA editing identified by comparative genomics. *Science* 301:832–36.

Huang, H.S., N.T. Hu, Y.E. Yao, et al. 1998. Molecular cloning and heterologous expression of a glutathione S-transferase involved in insecticide resistance from the diamondback moth, *Plutella xylostella. Insect Biochem. Mol. Biol.* 28:651–658.

Huang, S.J., and Z.J. Han. 2007. Mechanisms for multiple resistances in field populations of common cutworm, *Spodoptera litura* (Fabricius) in China. *Pestic. Biochem. Physiol.* 87:14–22.

Huchard, E., M. Martinez, H. Alout, et al. 2006. Acetylcholinesterase genes within the Diptera: Takeover and loss in true flies. *Proc. Biol. Sci.* 273:2595–2604.

Ingles, P.J., P.M. Adams, D.C. Knipple, et al. 1996. Characterization of voltage-sensitive sodium channel gene coding sequences from insecticide-susceptible and knockdown-resistant house fly strains. *Insect Biochem. Mol. Biol.* 26:319–26.

International Silkworm Genome Consortium. 2008. The genome of a lepidopteran model insect, the silkworm *Bombyx mori. Insect Biochem. Mol. Biol.* 38:1036–45.

Jongsma, M.A., P.L. Bakker, J. Peters, et al. 1995. Adaptation of *Spodoptera exigua* larvae to plant proteinase-inhibitors by induction of gut proteinase activity insensitive to inhibition. *Proc. Nat. Acad. Sci. U.S.A.* 92:8041–45.

Joussen, N., D.G. Heckel, M. Haas, et al. 2008. Metabolism of imidacloprid and DDT by P450 GYP6G1 expressed in cell cultures of *Nicotiana tabacum* suggests detoxification of these insecticides in Cyp6g1-overexpressing strains of *Drosophila melanogaster*, leading to resistance. *Pest Manag. Sci.* 64:65–73.

Kanga, L.H.B., D.J. Pree, J.L. van Lier, et al. 1997. Mechanisms of resistance to organophosphorus and carbamate insecticides in oriental fruit moth populations (*Grapholita molesta* Busck). *Pestic. Biochem. Physiol.* 59:11–23.

Kao, P.N., and A. Karlin. 1986. Acetylcholine receptor binding site contains a disulfide cross-link between adjacent half-cystinyl residues. *J. Biol. Chem.* 261:8085–88.

Kozaki, T., H. Sezutsu, R. Feyereisen, et al. 2008. The *Bombyx mori* P450s. In *Ninth international symposium on cytochrome P450 biodiversity and biotechnology*, ed. R. Feyereisen, 43. Nice, France: Institut National de la Recherche Agronomique.

Kozaki, T., T. Shono, T. Tomita, et al. 2001. Fenitroxon insensitive acetylcholinesterases of the housefly, *Musca domestica* associated with point mutations. *Insect Biochem. Mol. Biol.* 31:991–97.

Kranthi, K.R., D. Jadhav, R. Wanjari, et al. 2001. Pyrethroid resistance and mechanisms of resistance in field strains of *Helicoverpa armigera* (Lepidoptera: Noctuidae). *J. Econ. Entomol.* 94:253–63.

Krieger, R.I., P.P. Feeny, and C.F. Wilkinson. 1971. Detoxication enzymes in the guts of caterpillars: An evolutionary answer to plant defenses? *Science* 172:579–81.

Ku, C.C., F.M. Chiang, C.Y. Hsin, et al. 1994. Glutathione transferase isozymes involved in insecticide resistance of diamondback moth larvae. *Pestic. Biochem. Physiol.* 50:191–97.

Kulkarni, N.H., A.H. Yamamoto, K.O. Robinson, et al. 2002. The DSC1 channel, encoded by the smi60E locus, contributes to odor-guided behavior in *Drosophila melanogaster*. *Genetics* 161:1507–16.

Kwon, D.H., J.M. Clark, and S.H. Lee. 2004. Estimation of knockdown resistance in diamondback moth using real-time PASA. *Pestic. Biochem. Physiol.* 78:39–48.

Le Goff, G., A. Hamon, J.B. Bergé, and M. Amichot. 2005. Resistance to fipronil in *Drosophila simulans*: Influence of two point mutations in the RDL GABA receptor subunit. *J. Neurochem.* 92:1295–1305.

Lee, D.W., J.Y. Choi, W.T. Kim, et al. 2007. Mutations of acetylcholinesterase1 contribute to prothiofos-resistance in *Plutella xylostella* (L.). *Biochem. Biophys. Res. Commun.* 353:591–97.

Lee, D.W., S.S. Kim, S.W. Shin, et al. 2006. Molecular characterization of two acetylcholinesterase genes from the oriental tobacco budworm, *Helicoverpa assulta* (Guenee). *Biochim. Biophys. Acta* 1760:125–33.

Lee, D., Y. Park, T.M. Brown, et al. 1999. Altered properties of neuronal sodium channels associated with genetic resistance to pyrethroids. *Mol. Pharmacol.* 55:584–93.

Lee, S.H., and D.M. Soderlund. 2001. The V410M mutation associated with pyrethroid resistance in *Heliothis virescens* reduces the pyrethroid sensitivity of housefly sodium channels expressed in *Xenopus* oocytes. *Insect Biochem. Mol. Biol.* 31:19–29.

Li, X.C., J. Baudry, M.R. Berenbaum, et al. 2004. Structural and functional divergence of insect CYP6B proteins: From specialist to generalist cytochrome P450. *Proc. Natl. Acad. Sci. U.S.A.* 101:2939–2944.

Li, X.C., M.R. Berenbaum, and M.A. Schuler. 2002. Plant allelochemicals differentially regulate *Helicoverpa zea* cytochrome P450 genes. *Insect Mol. Biol.* 11:343–51.

Li, F., and Z.J. Han. 2004. Mutations in acetylcholinesterase associated with insecticide resistance in the cotton aphid, *Aphis gossypii* Glover. *Insect Biochem. Mol. Biol.* 34:397–405.

Li, X.C., M.A. Schuler, and M.R. Berenbaum. 2007. Molecular mechanisms of metabolic resistance to synthetic and natural xenobiotics. *Annu. Rev. Entomol.* 52:231–53.

Li, A.G., Y.H. Yang, S.W. Wu, et al. 2006. Investigation of resistance mechanisms to fipronil in diamondback moth (Lepidoptera: Plutellidae). *J. Econ. Entomol.* 99:914–19.

Liu, Z.Q., J.G. Tan, S.M. Valles, et al. 2002. Synergistic interaction between two cockroach sodium channel mutations and a tobacco budworm sodium channel mutation in reducing channel sensitivity to a pyrethroid insecticide. *Insect Biochem. Mol. Biol.* 32:397–404.

Liu, Z.W., M.S. Williamson, S.J. Lansdell, et al. 2005. A nicotinic acetylcholine receptor mutation conferring target-site resistance to imidacloprid in *Nilaparvata lugens* (brown planthopper). *Proc. Natl. Acad. Sci. U.S.A.* 102:8420–25.

Liu, Z.W., M.S. Williamson, S.J. Lansdell, et al. 2006. A nicotinic acetylcholine receptor mutation (Y151S) causes reduced agonist potency to a range of neonicotinoid insecticides. *J. Neurochem.* 99:1273–81.

Loughney, K., R. Kreber, and B. Ganetzky. 1989. Molecular analysis of the *para* locus, a sodium-channel gene in *Drosophila*. *Cell* 58:1143–54.

Malcolm, C.A., D. Bourguet, A. Ascolillo, et al. 1998. A sex-linked *Ace* gene, not linked to insensitive acetylcholinesterase-mediated insecticide resistance in *Culex pipiens*. *Insect Mol. Biol.* 7:107–20.

Mao, Y.B., W.J. Cai, J.W. Wang, et al. 2007. Silencing a cotton bollworm P450 monooxygenase gene by plant-mediated RNAi impairs larval tolerance of gossypol. *Nat. Biotechnol.* 25:1307–13.

Mao, W., M.A. Schuler, and M.R. Berenbaum. 2007. Cytochrome P450s in *Papilio multicaudatus* and the transition from oligophagy to polyphagy in the Papilionidae. *Insect Mol. Biol.* 16:481–90.

Martin, T., O.G. Ochou, M. Vaissayre, et al. 2003. Oxidases responsible for resistance to pyrethroids sensitize *Helicoverpa armigera* (Hübner) to triazophos in West Africa. *Insect Biochem. Mol. Biol.* 33:883–87.

Mau, R.F.L., and L. Gusukuma-Minuto. 2004. Diamondback moth, *Plutella xylostella* (L.), resistance management in Hawaii. In *The management of diamondback moth and other crucifer pests: Proceedings of the 4th international workshop*, ed. N. Endersby and P. Ridland, 307–11. Melbourne: Victorian Department of Primary Industries.

McCaffery, A.R., D.J. Head, T. Jianguo, et al. 1997. Nerve insensitivity resistance to pyrethroids in heliothine Lepidoptera. *Pestic. Sci.* 51:315–20.

McCaffery, A.R., J.W. Holloway, and R.T. Gladwell. 1995. Nerve insensitivity resistance to cypermethrin in larvae of the tobacco budworm *Heliothis virescens* from USA cotton field populations. *Pestic. Sci.* 44:237–47.

Miyazaki, M., K. Ohyama, D.Y. Dunlap, et al. 1996. Cloning and sequencing of the *para*-type sodium channel gene from susceptible and *kdr*-resistant German cockroaches (*Blattella germanica*) and house fly (*Musca domestica*). *Mol. Gen. Genet.* 252:61–68.

Morishita, M. 1998. Changes in susceptibility to various pesticides of diamondback moth (*Plutella xylostella* L.) in Gobo, Wakayama Prefecture. *Jap. J. Appl. Entomol. Zool.* 42:209–13.

Mota-Sanchez, D., P.S. Bills, and M.E. Whalon. 2002. Arthropod resistance to pesticides: Status and overview. In *Pesticides in agriculture and the environment*, ed. W.B. Wheeler, 241–72. New York: Marcel Dekker Inc.

Moulton, J.K., D.A. Pepper, and T.J. Dennehy. 2000. Beet armyworm (*Spodoptera exigua*) resistance to spinosad. *Pest Manag. Sci.* 56:842–48.

Mutero, A., M. Pralavorio, J.M. Bride, et al. 1994. Resistance-associated point mutations in insecticide-insensitive acetylcholinesterase. *Proc. Nat. Acad. Sci. U.S.A.* 91:5922–26.

Narahashi, T. 1996. Neuronal ion channels as the target sites of insecticides. *Pharmacol. Toxicol.* 79:1–14.

Newcomb, R.D., P.M. Campbell, D.L. Ollis, et al. 1997. A single amino acid substitution converts a carboxylesterase to an organophosphorus hydrolase and confers insecticide resistance on a blowfly. *Proc. Nat. Acad. Sci. U.S.A.* 94:7464–68.

Ni, X.Y., T. Tomita, S. Kasai, et al. 2003. cDNA and deduced protein sequence of acetylcholinesterase from the diamondback moth, *Plutella xylostella* (L.) (Lepidoptera: Plutellidae). *Appl. Entomol. Zool.* 38:49–56.

Nicholson, R.A., and T.A. Miller. 1985. Multifactorial resistance to transpermethrin in field-collected strains of the tobacco budworm *Heliothis virescens* F. *Pestic. Sci.* 16:561–70.

Noda, M., S. Shimizu, T. Tanabe, et al. 1984. Primary structure of *Electrophorus electricus* sodium channel deduced from cDNA sequence. *Nature* 312:121–27.

Noppun, V., T. Miyata, and T. Saito. 1987. Insensitivity of acetylcholinesterase in phenthoate resistant diamondback moth, *Plutella xylostella*. *Appl. Entomol. Zool.* 22:116–18.

Oakeshott, J., C. Claudianos, P. Campbell, et al. 2005a. Biochemical genetics and genomics of insect esterases. In *Comprehensive molecular insect science*, Vol. 5, ed. L. Gilbert, K. Iatrou, and S. Gill, 309–81. Oxford: Elsevier.

Oakeshott, J.G., A.L. Devonshire, C. Claudianos, et al. 2005b. Comparing the organophosphorus and carbamate insecticide resistance mutations in cholin- and carboxyl-esterases. *Chem. Biol. Interact.* 157:269–75.

Ollis, D.L., E. Cheah, M. Cygler, et al. 1992. The alpha/beta hydrolase fold. *Protein Eng.* 5:197–211.

Ottea, J.A., and J.W. Holloway. 1998. Target-site resistance to pyrethroids in *Heliothis virescens* (F.) and *Helicoverpa zea* (Boddie). *Pestic. Biochem. Physiol.* 61:155–67.

Park, Y.S., D.W. Lee, M.F.J. Taylor, et al. 2000. A mutation Leu1029 to His in *Heliothis virescens* F.: *hscp* sodium channel gene associated with a nerve-insensitivity mechanism of resistance to pyrethroid insecticides. *Pestic. Biochem. Physiol.* 66:1–8.

Park, Y., and M.F.J. Taylor. 1997. A novel mutation L1029H in sodium channel gene *hscp* associated with pyrethroid resistance for *Heliothis virescens* (Lepidoptera: Noctuidae). *Insect Biochem. Mol. Biol.* 27:9–13.

Park, Y., M.F.J. Taylor, and R. Feyereisen. 1997. A valine421 to methionine mutation in IS6 of the *hscp* voltage-gated sodium channel associated with pyrethroid resistance in *Heliothis virescens* F. *Biochem. Biophys. Res. Commun.* 239:688–91.

Park, Y., M.F.J. Taylor, and R. Feyereisen. 1999. Voltage-gated sodium channel genes *hscp* and *hDSC1* of *Heliothis virescens* F.: Genomic organization. *Insect Mol. Biol.* 8:161–70.

Perry, T., D.G. Heckel, J.A. McKenzie, et al. 2008. Mutations in Dα1 or Dβ2 nicotinic acetylcholine receptor subunits can confer resistance to neonicotinoids in *Drosophila melanogaster*. *Insect Biochem. Mol. Biol.* 38:520–28.

Perry, T., J.A. McKenzie, and P. Batterham. 2007. A Dα6 knockout strain of *Drosophila melanogaster* confers a high level of resistance to spinosad. *Insect Biochem. Mol. Biol.* 37:184–88.

Pigott, C.R., and D.J. Ellar. 2007. Role of receptors in *Bacillus thuringiensis* crystal toxin activity. *Microbiol. Mol. Biol. Rev.* 71:255–81.

Pimprale, S.S., C.L. Besco, P.K. Bryson, et al. 1997. Increased susceptibility of pyrethroid-resistant tobacco budworm (Lepidoptera: Noctuidae) to chlorfenapyr. *J. Econ. Entomol.* 90:49–54.

Qu, M.J., Z.J. Han, X.J. Xu, et al. 2003. Triazophos resistance mechanisms in the rice stem borer (*Chilo suppressalis* Walker). *Pestic. Biochem. Physiol.* 77:99–105.

Ranasinghe, C., B. Campbell, and A.A. Hobbs. 1998. Over-expression of cytochrome P450 CYP6B7 mRNA and pyrethroid resistance in Australian populations of *Helicoverpa armigera* (Hübner). *Pestic. Sci.* 54:195–202.

Ranasinghe, C., and A.A. Hobbs. 1998. Isolation and characterization of two cytochrome P450 cDNA clones for CYP6B6 and CYP6B7 from *Helicoverpa armigera* (Hübner): Possible involvement of CYP6B7 in pyrethroid resistance. *Insect Biochem. Mol. Biol.* 28:571–80.

Ranasinghe, C., and A.A. Hobbs. 1999. Induction of cytochrome P450 CYP6B7 and cytochrome b5 mRNAs from *Helicoverpa armigera* (Hübner) by pyrethroid insecticides in organ culture. *Insect Mol. Biol.* 8:443–47.

Ranson, H., and J. Hemingway. 2005. Glutathione transferases. In *Comprehensive molecular insect science*, Vol. 5, ed. L. Gilbert, K. Iatrou, and S. Gill, 383–402. Oxford: Elsevier.

Raymond, M., D.G. Heckel, and J.G. Scott. 1989. Interactions between pesticide resistance genes: Model and experiment. *Genetics* 123:543–51.

Ren, X.X., Z.J. Han, and Y.C. Wang. 2002. Mechanisms of monocrotophos resistance in cotton bollworm, *Helicoverpa armigera* (Hübner). *Arch. Insect Biochem. Physiol.* 51:103–10.

Rose, R.L., L. Barbhaiya, R.M. Roe, et al. 1995. Cytochrome P450-associated insecticide resistance and the development of biochemical diagnostic assays in *Heliothis virescens*. *Pestic. Biochem. Physiol.* 51:178–91.

Rose, R.L., D. Goh, D.M. Thompson, et al. 1997. Cytochrome P450 CYP9A1 in *Heliothis virescens*: The first member of a new CYP family. *Insect Biochem. Mol. Biol.* 27:605–15.

Ru, L.J., C. Wei, J.Z. Zhao, et al. 1998. Differences in resistance to fenvalerate and cyhalothrin and inheritance of knockdown resistance to fenvalerate in *Helicoverpa armigera*. *Pestic. Biochem. Physiol.* 61:79–85.

Russell, R.J., C. Claudianos, P.M. Campbell, et al. 2004. Two major classes of target site insensitivity mutations confer resistance to organophosphate and carbamate insecticides. *Pestic. Biochem. Physiol.* 79:84–93.

Ryan, C.A. 1990. Protease inhibitors in plants: Genes for improving defenses against insects and pathogens. *Ann. Rev. Phytopathol.* 28:425–49.

Salkoff, L., A. Butler, A. Wei, et al. 1987. Genomic organization and deduced amino acid sequence of a putative sodium channel gene in *Drosophila*. *Science* 237:744–49.

Sato, C., Y. Ueno, K. Asai, et al. 2001. The voltage-sensitive sodium channel is a bell-shaped molecule with several cavities. *Nature* 409:1047–51.

Sattelle, D.B., A.K. Jones, B.M. Sattelle, et al. 2005. Edit, cut and paste in the nicotinic acetylcholine receptor gene family of *Drosophila melanogaster*. *Bioessays* 27:366–76.

Sayyed, A.H., D. Omar, and D.J. Wright. 2004. Genetics of spinosad resistance in a multi-resistant field-selected population of *Plutella xylostella*. *Pest Manage. Sci.* 60:827–32.

Schuler, T.H., D. Martinez-Torres, A.J. Thompson, et al. 1998. Toxicological, electrophysiological, and molecular characterisation of knockdown resistance to pyrethroid insecticides in the diamondback moth, *Plutella xylostella* (L.). *Pestic. Biochem. Physiol.* 59:169–82.

Scott, J.G., and Z.M. Wen. 2001. Cytochromes P450 of insects: The tip of the iceberg. *Pest Manag. Sci.* 57:958–67.

Seino, A., T. Kazuma, A.J. Tan, et al. 2007. Analysis of two acetylcholinesterase genes in *Bombyx mori*. *Pestic. Biochem. Physiol.* 88:92–101.

Shang, J.Y., Y.M. Shao, G.J. Lang, et al. 2007. Expression of two types of acetylcholinesterase gene from the silkworm, *Bombyx mori*, in insect cells. *Insect Sci.* 14:443–49.

Shao, Y.M., K. Dong, and C.X. Zhang. 2007. The nicotinic acetylcholine receptor gene family of the silkworm, *Bombyx mori*. *BMC Genomics* 8:324–34.

Shen, B.C., D.X. Zhao, C.L. Qiao, et al. 2004. Cloning of CYP9G2 from the diamondback moth, *Plutella xylostella* (Lepidoptera: Yponomeutidae). *DNA Sequence* 15:228–33.

Soderlund, D. 2005. Sodium channels. In *Comprehensive molecular insect science*, Vol 5. ed. L. Gilbert, K. Iatrou, and S. Gill, 1–24. Oxford: Elsevier.

Sonoda, S., M. Ashfaq, and H. Tsumuki. 2006. Genomic organization and developmental expression of glutathione S-transferase genes of the diamondback moth, *Plutella xylostella*. *J. Insect Sci.* 6:1–9.

Sonoda, S., C. Igaki, M. Ashfaq, et al. 2006. Pyrethroid-resistant diamondback moth expresses alternatively spliced sodium channel transcripts with and without T929I mutation. *Insect Biochem. Mol. Biol.* 36:904–10.

Sonoda, S., and H. Tsumuki. 2005. Studies on glutathione S-transferase gene involved in chlorfluazuron resistance of the diamondback moth, *Plutella xylostella* L. (Lepidoptera: Yponomeutidae). *Pestic. Biochem. Physiol.* 82:94–101.

Srinivas, R., S.S. Udikeri, S.K. Jayalakshmi, et al. 2004. Identification of factors responsible for insecticide resistance in *Helicoverpa armigera*. *Comp. Biochem. Physiol. C Toxicol. Pharmacol.* 137:261–69.

Stokes, N.H., S.W. McKechnie, and N.W. Forrester. 1997. Multiple allelic variation in a sodium channel gene from populations of Australian *Helicoverpa armigera* (Hübner) (Lepidoptera: Noctuidae) detected via temperature gradient gel electrophoresis. *Aust. J. Entomol.* 36:191–96.

Strode, C., C.S. Wondji, J.P. David, et al. 2008. Genomic analysis of detoxification genes in the mosquito *Aedes aegypti*. *Insect Biochem. Mol. Biol.* 38:113–23.

Sussman, J.L., M. Harel, F. Frolow, et al. 1991. Atomic structure of acetylcholinesterase from *Torpedo californica*: A prototypic acetylcholine-binding protein. *Science* 253:872–79.

Tabashnik, B.E., N.L. Cushing, N. Finson, and M.W. Johnson. 1990. Field development of resistance to *Bacillus thuringiensis* in diamondback moth (Lepidoptera, Plutellidae). *J. Econ. Entomol.* 83:1671–76.

Talekar, N.S., and A.M. Shelton. 1993. Biology, ecology, and management of the diamondback moth. *Annu. Rev. Entomol.* 38:275–301.

Tan, J., Z. Liu, T.D. Tsai, et al. 2002. Novel sodium channel gene mutations in *Blattella germanica* reduce the sensitivity of expressed channels to deltamethrin. *Insect Biochem. Mol. Biol.* 32:445–54.

Tan, J.G., and A.R. McCaffery. 1999. Expression and inheritance of nerve insensitivity resistance in larvae of *Helicoverpa armigera* (Lepidoptera: Noctuidae) from China. *Pestic. Sci.* 55:617–25.

Taylor, M.F.J., D.G. Heckel, T.M. Brown, et al. 1993. Linkage of pyrethroid insecticide resistance to a sodium channel locus in the tobacco budworm. *Insect Biochem. Mol. Biol.* 23:763–75.

Taylor, M.F.J., P. Park, and Y. Shen. 1996. Molecular population genetics of sodium channel and juvenile hormone esterase markers in relation to pyrethroid resistance in *Heliothis virescens* (Lepidoptera: Noctuidae). *Ann. Entomol. Soc. Am.* 89:728–38.

Taylor, M.F.J., Y. Shen, and M.E. Kreitman. 1995. A population genetic test of selection at the molecular level. *Science* 270:1497–99.

Toda, S., S. Komazaki, T. Tomita, et al. 2004. Two amino acid substitutions in acetylcholinesterase associated with pirimicarb and organophosphorous insecticide resistance in the cotton aphid, *Aphis gossypii* Glover (Homoptera: Aphididae). *Insect Mol. Biol.* 13:549–53.

Tsukahara, Y., S. Sonoda, Y. Fujiwara, et al. 2003. Molecular analysis of the *para*-sodium channel gene in the pyrethroid-resistant diamondback moth, *Plutella xylostella* (Lepidoptera: Yponomeutidae). *Appl. Entomol. Zool.* 38:23–29.

Vanlaecke, K., G. Smagghe, and D. Degheele. 1995. Detoxifying enzymes in greenhouse and laboratory strains of beet armyworm (Lepidoptera, Noctuidae). *J. Econ. Entomol.* 88:777–81.

Villatte, F., P. Ziliani, V. Marcel, et al. 2000. A high number of mutations in insect acetylcholinesterase may provide insecticide resistance. *Pestic. Biochem. Physiol.* 67:95–102.

Vontas, J.G., G.J. Small, and J. Hemingway. 2001. Glutathione S-transferases as antioxidant defence agents confer pyrethroid resistance in *Nilaparvata lugens*. *Biochem. J.* 357:65–72.

Walsh, S.B., T.A. Dolden, G.D. Moores, et al. 2001. Identification and characterization of mutations in housefly (*Musca domestica*) acetylcholinesterase involved in insecticide resistance. *Biochem. J.* 359:175–81.

Wang, X.P., and A.A. Hobbs. 1995. Isolation and sequence analysis of a cDNA clone for a pyrethroid inducible cytochrome P450 from *Helicoverpa armigera*. *Insect Biochem. Mol. Biol.* 25:1001–09.

Wee, C.W., S.F. Lee, C. Robin, et al. 2008. Identification of candidate genes for fenvalerate resistance in *Helicoverpa armigera* using cDNA-AFLP. *Insect Mol. Biol.* 17:351–60.

Weill, M., P. Fort, A. Berthomieu, et al. 2002. A novel acetylcholinesterase gene in mosquitoes codes for the insecticide target and is non-homologous to the *Ace* gene in *Drosophila*. *Proc. Biol. Sci.* 269:2007–16.

Williamson, M.S., D. MartinezTorres, C.A. Hick, et al. 1996. Identification of mutations in the housefly *para*-type sodium channel gene associated with knockdown resistance (*kdr*) to pyrethroid insecticides. *Mol. Gen. Genet.* 252:51–60.

Wing, K., J. Andaloro, S. McCann, et al. 2005. Indoxacarb and the sodium channel blocker insecticides: Chemistry, physiology, and biology in insects. In *Comprehensive molecular insect science*, Vol. 5., ed. L. Gilbert, K. Iatrou, and S. Gill, 31–53. Oxford: Elsevier.

Wink, M., and V. Theile. 2002. Alkaloid tolerance in *Manduca sexta* and phylogenetically related sphingids (Lepidoptera: Sphingidae). *Chemoecology* 12:29–46.

Wolff, M.A., and V.P.M. Wingate. 1998. Characterization and comparative pharmacological studies of a functional γ-aminobutyric acid (GABA) receptor cloned from the tobacco budworm, *Heliothis virescens* (Noctuidae: Lepidoptera). *Invertebr. Neurosci.* 3:305–15.

Wyss, C.F., H.P. Young, J. Shukla, et al. 2003. Biology and genetics of a laboratory strain of the tobacco budworm, *Heliothis virescens* (Lepidoptera: Noctuidae), highly resistant to spinosad. *Crop Protect.* 22:307–14.

Yang, Y.H., S. Chen, S.W. Wu, et al. 2006. Constitutive overexpression of multiple cytochrome P450 genes associated with pyrethroid resistance in *Helicoverpa armigera*. *J. Econ. Entomol.* 99:1784–89.

Yang, Y., Y. Wu, S. Chen, et al. 2004. The involvement of microsomal oxidases in pyrethroid resistance in *Helicoverpa armigera* from Asia. *Insect Biochem. Mol. Biol.* 34:763–73.

Young, H.P., W.D. Bailey, and R.M. Roe. 2003. Spinosad selection of a laboratory strain of the tobacco budworm, *Heliothis virescens* (Lepidoptera: Noctuidae), and characterization of resistance. *Crop Protect.* 22:265–73.

Young, H.P., W.D. Bailey, R.M. Roe, et al. 2001. Mechanism of resistance and cross-resistance in a laboratory, spinosad-selected strain of the tobacco budworm and resistance in laboratory-selected cotton bollworms. In *Proceedings of the 2001 Beltwide Cotton Conference*, 1167–71. Memphis: National Cotton Council of America.

Young, H.P., W.D. Bailey, C.F. Wyss, et al. 2000. Studies on the mechanisms of tobacco budworm resistance to spinosad (Tracer). In *Proceedings of the 2000 Beltwide Cotton Conference*, 1197–1201. Memphis: National Cotton Council of America.

Yu, S.J. 1992. Detection and biochemical characterization of insecticide resistance in fall armyworm (Lepidoptera, Noctuidae). *J. Econ. Entomol.* 85:675–82.

Yu, S.J. 2006. Insensitivity of acetylcholinesterase in a field strain of the fall armyworm, *Spodoptera frugiperda* (J.E. Smith). *Pestic. Biochem. Physiol.* 84:135–42.

Yu, S.J., and S.N. Nguyen. 1992. Detection and biochemical characterization of insecticide resistance in diamondback moth. *Pestic. Biochem. Physiol.* 44:74–81

Yu, S.J., S.N. Nguyen, and G.E. Abo-Elghar. 2003. Biochemical characteristics of insecticide resistance in the fall armyworm, *Spodoptera frugiperda* (J.E. Smith). *Pestic. Biochem. Physiol.* 77:1–11.

Zhao, J.Z., Y.X. Li, H.L. Collins, et al. 2002. Monitoring and characterization of diamondback moth (Lepidoptera: Plutellidae) resistance to spinosad. *J. Econ. Entomol.* 95:430–36.

Zhao, Y., Y. Park, and M.E. Adams. 2000. Functional and evolutionary consequences of pyrethroid resistance mutations in S6 transmembrane segments of a voltage-gated sodium channel. *Biochem. Biophys. Res. Commun.* 278:516–21.

Zhou, W., I.B. Chung, Z.Q. Liu, et al. 2004. A voltage-gated calcium-selective channel encoded by a sodium channel-like gene. *Neuron* 42:101–12.

Zlotkin, E., H. Moskowitz, R. Herrmann, et al. 1995. Insect sodium channel as the target for insect selective neurotoxins from scorpion venom. In *Molecular action of insecticides on ion channels*, ed. J.M. Clark, 56–85. Washington, D.C.: American Chemical Society.

14 Innate Immune Responses of *Manduca sexta*

Michael R. Kanost and James B. Nardi

CONTENTS

INTRODUCTION

Insect immune systems face a great diversity of challenges from parasites, parasitoids, and pathogens. Parasitoid insects alone are estimated to comprise about 15 percent of the approximately one million known species of insects. Pathogens and parasites include vast numbers of viruses, bacteria, microsporidia, fungi, nematodes, and protozoa such as sporozoa, gregarines, and coccidia; the number of described species of these organisms grows each year. The success of these parasitoids and pathogens has depended to a great extent on their ability to either physically avoid encounters with cells of the insect immune system (Kathirithamby, Ross, and Johnston 2003; Manfredini et al. 2007) or to produce substances that can circumvent the defenses mounted by a host's blood cells. These substances can be (1) viruses or molecules that suppress the immune response, (2) molecules that mimic host antigens and are mistaken for *self* antigens by the host, or (3) molecules that mask the *nonself* antigens of the intruders and result in the pathogens' or parasitoids' not being recognized as foreign or nonself by the immune cells of the host (e.g., Salt 1970; Rizki and Rizki 1984; Davies and Vinson 1986; Schmidt et al. 1990; Strand and Noda 1991; Strand and Wong 1991; Hoek et al. 1996; Lavine and Beckage 1996; Asgari et al. 1998; Galibert et al. 2003; Beck and Strand 2005; Wang and St. Leger 2006).

Understanding how pathogens and parasitoids prevent cells of the innate immune system from being activated as part of a humoral or cell-mediated response to intruders requires addressing the basic question of what exogenous or endogenous signals are recognized by receptors in plasma or on surfaces of hemocytes and how these signals activate innate immune responses. Extensive progress has been made in understanding genetics and pathways of immune responses in *Drosophila melanogaster* (Ferrandon et al. 2007; Lemaitre and Hoffmann 2007), and genomic information is providing a strong basis for experimental characterization of immune processes in mosquitoes that vector human diseases (Michel and Kafatos 2005; Waterhouse et al. 2007). For different reasons, lepidopteran insects have proven to be important experimental species for investigations of insect immunity. The large hemolymph volume and hemocyte number of caterpillars provide sufficient experimental material for conducting detailed biochemical experiments on plasma proteins and use of modern cell biological techniques for investigations of hemocyte function. Studies of silk moths, particularly *Bombyx mori* (Ashida and Brey 1997; Ponnuvel and Yamakawa 2002) and *Hyalophora cecropia* (Boman et al. 1991; Su et al. 1998), the wax moth *Galleria mellonella*, and noctuid species, including *Pseudoplusia includens* (Lavine and Strand 2002), have produced important discoveries of cellular and humoral immune mechanisms that were subsequently found to be widely conserved in innate immune systems of insects and even of vertebrates. We focus in this chapter on recent advances in our understanding of the immune system of the tobacco hornworm, *Manduca sexta*, which has become one of the best characterized insect model systems (Kanost, Jiang, and Yu 2004; Jiang 2008) with regard to function of plasma proteins and hemocyte adhesion molecules (Table 14.1).

PATTERN RECOGNITION RECEPTORS TRIGGER INNATE IMMUNE RESPONSES OF INSECTS AND MAMMALS

Innate immune responses of insects include humoral responses such as melanization and induced synthesis of antimicrobial proteins and other plasma proteins with immune function (Gillespie, Kanost, and Trenczek 1997; Nappi and Christensen 2005; Ferrandon et al. 2007) and cell-mediated responses such as encapsulation, phagocytosis, and nodulation (Lavine and Strand 2002; Jiravanichpaisal, Lee, and Söderhäll 2006). Humoral and cell-mediated responses of immune systems are intertwined and are triggered by the same or similar phenomena. These triggering phenomena have been defined by the strategies used to recognize normal self. As outlined by Medzhitov and Janeway (2002), strategies employed to discriminate self from nonself involve the recognition of (a) *microbial nonself*, (b) *missing self*, and (c) *altered self*.

In discriminating microbial pathogens from self, the innate immune system uses proteins referred to as pattern-recognition receptors (PRRs), which recognize conserved molecular patterns of nonself, or pathogen-associated molecular patterns (PAMPs) representing highly conserved motifs. The latter include such microbial structures as lipopolysaccharide (LPS), lipoteichoic acid (LTA), peptidoglycan, and ß-1,3-glucan found on surfaces of microbes but not on cells of nonmicrobial eukaryotes (Janeway and Medzhitov 2002; Sansonetti 2006). However, the activation of resting, nonadherent hemocytes of insects can occur in the absence of any of these PAMPs. Discrimination between self and nonself by the innate immune system encompasses not only the recognition of PAMPs or (a) microbial nonself but also recognition of (b) missing self and (c) altered self (Medzhitov and Janeway 2002). These latter two recognition strategies of the innate immune response include recognition of wounding or recognition of the many apoptotic cells produced during development (altered self) as well as foreign surfaces such as abiotic and parasitoid surfaces that lack microbial PAMPs (missing self). Furthermore, these three different recognition strategies (a–c) of the innate immune system can be subsumed by a model of immunity, hypothesizing that the immune system is actually responding to endogenous danger signals or substances released by stressed, wounded, or dying cells in addition to the well-studied exogenous danger signals from pathogens (Aderem and Ulevitch 2000; Matzinger 2002).

in the gene for *M. sexta* serpin-1, which contains twelve copies of its 9th exon. Each version of exon 9 encodes a different reactive site loop sequence and inhibits a different spectrum of proteinases (Jiang, Wang, and Kanost 1994; Jiang et al. 1996; Jiang and Kanost 1997). Structures of two of the *Manduca* serpin-1 variants have been determined by X-ray crystallography (Li et al. 1999; Ye et al. 2001). Serpin genes with alternate exons in the same position as in *M. sexta* serpin-1 have been identified in several other insect species (Kanost and Clarke 2005).

Three serpins from *Manduca* hemolymph (serpin-1J, serpin-3, and serpin-6) can inhibit PAPs to regulate their activity (Jiang et al. 2003b; Zhu et al. 2003b; Wang and Jiang 2004b). Recombinant serpins-4 and -5 block proPO activation, but they do not inhibit the PAPs, suggesting that they inhibit proteinases upstream of the PAPs (Tong and Kanost 2005). Serpin-4 inhibits HP1, HP6, and HP21. Serpin-5 inhibits HP1 and HP6 (Tong, Jiang, and Kanost 2005). As serpin-4 and serpin-5 can each block proPO activation in plasma, HP1, HP6, and HP21 are candidates for components of the proPO activation cascade. This is consistent with the observation that HP21 can activate proPAP-2 and proPAP-3 (Gorman, An, and Kanost 2007; Wang and Jiang 2007).

IMMUNE RESPONSES MEDIATED BY HEMOCYTES: A SWITCH FROM A NONADHERENT TO AN ADHERENT STATE

Hemocytes take part in phagocytosis, nodule formation, and encapsulation-immune responses that involve binding of these cells to the surface of pathogens or adherence of hemocytes to one another. Nodules are aggregations of hemocytes that trap large numbers of bacteria. In encapsulation, multiple layers of hemocytes adhere to a foreign object to form an organized structure that encases and kills the parasite. Nodules and capsules are often melanized, indicating coordination of cellular and humoral responses (Stanley, Miller, and Howard 1998; Yu and Kanost 2004; Rodríguez-Pérez et al. 2005). Two classes of hemocytes—granular cells and plasmatocytes—are involved in the cell-mediated response of *M. sexta*. They can be readily distinguished with specific lectin and antibody markers (Nardi et al. 2003, 2005, 2006; Willott et al. 1994) as well as by their different ploidy levels (granular cells are diploid, plasmatocytes are polyploid; Figure 14.2). During a cell-mediated response, circulating, nonadherent hemocytes are transformed to adherent, spreading cells (Figure 14.3). Either an exogenous signal (PAMP) from a foreign surface induces the cell-mediated response or simply an endogenous danger signal released by wounded or stressed cells, as hypothesized by Matzinger (2002), activates cells of a caterpillar's immune system. Eicosanoids are one such danger signal that promotes hemocyte aggregation and nodule formation (Stanley 2006). In addition, proteins released by granular hemocytes or produced by processing of plasma proteins elicit cell-mediated responses of plasmatocytes: phagocytosis, encapsulation, or nodule formation (Ratcliffe and Gagen 1977; Schmit and Ratcliffe 1977; Gillespie, Kanost, and Trenczek 1997; Loret and Strand 1998; Lavine and Strand 2002; Dean et al. 2004). A cytokine-like peptide called plasmatocyte spreading peptide can also stimulate plasmatocyte adhesion and spreading in *Manduca* and other lepidopterans (Clark, Pech, and Strand 1997; Wang, Jiang, and Kanost 1999). This peptide is present in plasma as a precursor that is activated by specific proteolysis, perhaps as a response to altered self produced by damage to wounded cells, and functions through binding to a receptor on the plasmatocyte surface (Clark et al. 2004).

ADHESION PROTEINS ON HEMOCYTE SURFACES AND THEIR INTERACTION WITH PROTEINS RELEASED FROM GRANULAR HEMOCYTES

Several superfamilies of adhesion proteins control cell-mediated responses of insect and mammalian immune cells. Cell adhesion molecules control interactions of mammalian immune cells not only during an inflammatory response when leukocytes and endothelial cells interact but also during interaction of naive T cells with antigen-presenting cells (i.e., macrophages, dendritic cells)

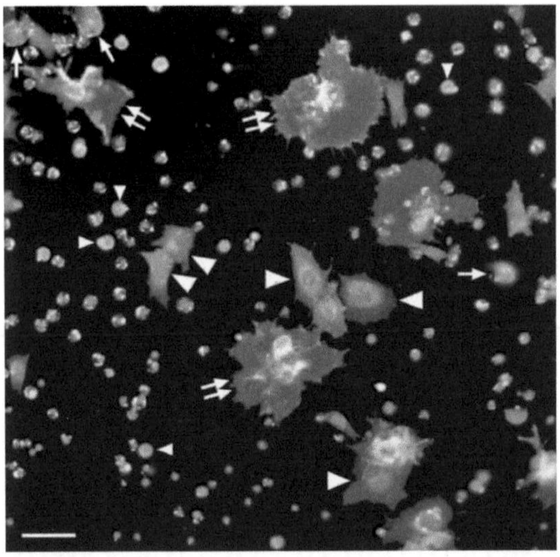

FIGURE 14.2 ***A color version of this figure follows page 176.*** *Manduca sexta* hemocytes cultured for 60 minutes in Grace's medium. Some of the more important features: DAPI (blue) distinguishes larger polyploid nuclei of plasmatocytes from smaller diploid nuclei of granular cells. PNA (peanut agglutinin)-rhodamine (orange) labels hemocytin. Note the capping of rhodamine and beta1-integrin on small, unspread plasmatocytes (small arrowheads) and the loss of rhodamine caps from plasmatocytes that have spread on the substrate. Note universal labeling of granular cells with rhodamine. Alexa Fluor 488 (green) labels beta1-integrin of large, spread plasmatocytes (double arrows) as well as small, spread plasmatocytes (large arrowheads). Alexa Fluor 647 (magenta) labels neuroglian of large, spread plasmatocytes as well as large, spherical unspread plasmatocytes (small arrows). Bar represents 20 μm.

that activate T lymphocytes and initiate an adaptive immune response (Aderem and Ulevitch 2000; Janeway et al. 2001). Below we describe the characteristics of three families of cell adhesion molecules found in lepidopterans.

Integrins

Functions of integrins on insect hemocytes are just beginning to be uncovered (Lavine and Strand 2003; Irving et al. 2005; Levin et al. 2005; Moita et al. 2006). Multiple integrin subunits have been described for hemocytes of the noctuid *P. includens*: three α subunits (αPi1–3) and one β subunit (βPi1; Lavine and Strand 2003). Expression levels for αPi2 and βPi1 of *P. includens* increase during the encapsulation response, suggesting that one or more integrins play an important role in regulating hemocyte adhesion. Two monoclonal antibodies (MS13, MS34) that label *Manduca* plasmatocytes and block spreading and encapsulation (Willott et al. 1994; Wiegand et al. 2000) were subsequently shown to recognize the ligand-binding site of a β-integrin designated ß1 (Levin et al. 2005). Antibody MS13, which binds to integrin ß1, was used to purify the corresponding integrin heterodimer from hemocyte membranes, and yielded an α-subunit designated α1. This integrin (α1ß1), referred to as hemocyte-specific (HS) integrin, is expressed exclusively in hemocytes (Zhuang et al. 2007b). The importance of HS integrin for cell-mediated immune responses has been demonstrated by disruption of the encapsulation response following treatment of larval hemocytes not only with the specific antibodies to *Manduca* ß1-integrin mentioned above but also using ß1-integrin siRNA (Levin et al. 2005). Sequences for two more α-subunits were obtained by using degenerate primers designed against the conserved domain of α-integrin subunits to amplify PCR products from total RNA of larval hemocytes (Zhuang et al. 2008).

Expression of different integrins on different populations of hemocytes may reflect a division of labor among hemocytes. Striking differences in spatio-temporal expression of ß1-integrin on three

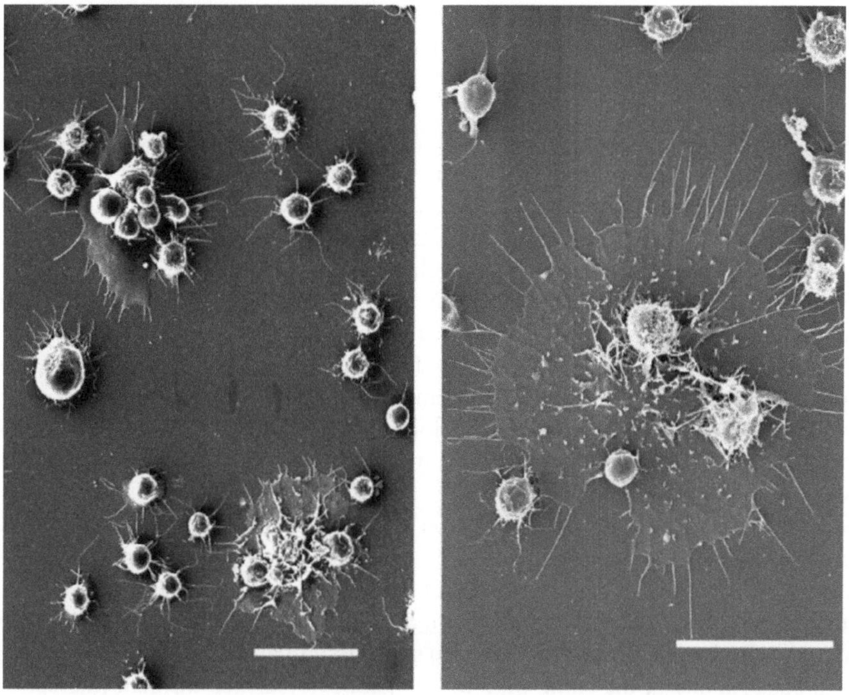

FIGURE 14.3 Adhesion and spreading of *Manduca sexta* hemocytes demonstrated by scanning electron microscopy. Large plasmatocytes spread extensively on a foreign surface and form a focus of adhesion, to which other plasmatocytes and granular cells attach and subsequently extend filopodia and spread. Bars represent 20 μm.

different populations of hemocytes may be a function of there being three different α-subunits of integrin present among *Manduca* hemocytes. Different hemocytes show directional adhesion that is related to spatio-temporal patterns of integrin expression. (1) Small plasmatocytes simultaneously adhere to other hemocytes and spread on substrates; these hemocytes label with anti-ß1-integrin as nonadherent hemocytes as well as adherent hemocytes. (2) Granular cells adhere to substrates and to plasmatocytes, but not to other granular cells; these hemocytes label with anti-ß1-integrin as adherent hemocytes but not as circulating, nonadherent hemocytes. (3) Large neuroglian-positive plasmatocytes first adhere to substrate before spreading and adhering to other plasmatocytes and granular cells; only as they start to spread on substrates do these plasmatocytes label with anti-ß1-integrin. Immunolabeling of integrins is thus a function of cell activation that accompanies conformational change(s) in integrins (Hynes 1992).

Tetraspanins

In vertebrates and insects, tetraspanins are mediators of cell interactions not only during development but also during immune responses (Hemler, Mannion, and Berditchevski 1996; Kopczynski, Davis, and Goodman 1996; Maecker, Todd, and Levy 1997; Berditchevski 2001; Boucheix and Rubinstein 2001; Zhuang et al. 2007b). Based on an EST project, four members of the tetraspanin superfamily of proteins were identified from *Manduca* (Todres, Nardi, and Robertson 2000); based on the *Drosophila melanogaster* genome project, thirty-seven members of this superfamily of integral membrane proteins are known to exist (Todres, Nardi, and Robertson 2000). A functional analysis of one of the four described *Manduca* tetraspanins (D76) that is expressed on hemocyte surfaces revealed that the large extracellular loop (LEL) of this tetraspanin binds to HS integrin (Zhuang et al. 2007b). Blocking of function with monoclonal antibodies to tetraspanin and HS integrin as well as dsRNA and binding assays for these two proteins imply a *trans* interaction of integrin and tetraspanin on hemocyte surfaces. This *trans* interaction of tetraspanin with HS integrin on lepidopteran hemocytes contrasts with the known *cis* interactions of mammalian tetraspanins with α-subunits of integrins (Zhuang et al. 2007b).

Neuroglian

In *Manduca*, the neural cell adhesion protein neuroglian, which contains six Ig domains, is expressed not only in the nervous system (Chen et al. 1997) but also by a subset of hemocytes that act as founder cells for hemocyte aggregation (Nardi et al. 2006). Neuroglian of these hemocyte surfaces binds in a homophilic fashion to neuroglian on opposing hemocytes and also heterophilically to HS integrin. Like other members of the Ig superfamily, neuroglian can function as a ligand for integrins. Integrins of mammalian T cells, for example, can bind to at least three members of the Ig superfamily: ICAM-1, ICAM-2, ICAM-3 (Janeway et al. 2001). Blocking these homophilic and heterophilic interactions of neuroglian on hemocyte surfaces with a monoclonal antibody to its Ig domains or with dsRNA disrupts encapsulation responses of *Manduca* hemocytes (Zhuang et al. 2007a), indicating that neuroglian takes part in cell adhesion events occurring in formation of hemocyte capsules.

We speculate that hemolin may also function as a ligand for neuroglian and/or HS integrin. This would explain the ability of hemolin to disrupt hemocyte adhesion (Ladendorff and Kanost 1991) by competing for homophilic interactions of neuroglian on neighboring cells or by interfering with binding between neuroglian and HS integrin and thus preventing hemocyte aggregation. Such interaction could also be responsible for hemolin's opsonizing activity in promoting phagocytosis, by binding to bacteria and at the same time to the hemocyte surface through interaction with HS integrin or neuroglian.

Proteins Released by Granular Hemocytes That Promote Plasmatocyte Adhesion

Proteins released from granules of granular hemocytes appear to promote nodule formation and encapsulation (Ratcliffe and Gagen 1977; Schmit and Ratcliffe 1977; Pech and Strand 1996; Loret and Strand 1998). We are attempting to identify such proteins from *M. sexta* granular cells and study their function in hemocyte adhesion and have identified two likely candidates. Lacunin is a large extracellular matrix protein produced by granular cells (Nardi, Gao, and Kanost 2001). Its synthesis is induced by inoculation of larva with bacterial PAMPs (M.R. Kanost, unpubl.). Lacunin is also released by granular cells and binds to surfaces of plasmatocytes (Nardi et al. 2005). Lacunin is localized at interfaces between substratum and adherent plasmatocyte surfaces (Nardi et al. 2005). The patchy distribution of this surface protein on surfaces of adherent plasmatocytes in an aggregate follows outlines of cells revealed by the uniform cell surface labeling with anti-ß1-integrin. Thus, lacunin released by granular hemocytes may promote plasmatocyte adhesion.

Hemocytin is a high-molecular-weight (~340 kDa) protein produced by granular cells. It was first identified in *B. mori* (Kotani et al. 1995) and is also present in *M. sexta* (Scholz 2002; J.B. Nardi, unpubl. results). Hemocytin binds to hemocytes and is involved in hemolymph clotting in moths and flies. The protein is strongly induced simply by wounding, is prevalent in hemolymph clots (Karlsson et al. 2004; Scherfer et al. 2004), and acts as a signal for immune activation. Clotting fails to occur in fly larvae in which hemolectin (the *Drosophila* counterpart of hemocytin) has been eliminated by RNAi suppression (Goto et al. 2003).

In response to wounding, hemocytin and lacunin are released by hemocytes and bind to surfaces of plasmatocytes (Li et al. 2002; Theopold et al. 2004; J.B. Nardi, unpubl.; Figure 14.4). Knowledge of the domain architecture for these two proteins released from granular hemocytes offers some predictions about their possible receptors on hemocyte surfaces. They each have at least one and as many as three of the following domains: thrombospondin type 1, immunoglobulin, von Willebrand factor, and Arg-Gly-Asp motifs. Each of these has the potential to function as a ligand for an integrin (Humphries, Byron, and Humphries 2006).

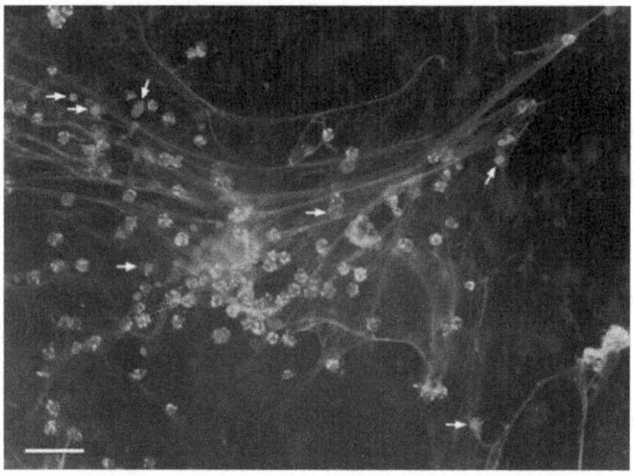

FIGURE 14.4 ***A color version of this figure follows page 176.*** Labeling of adhesion molecules in aggregation of *Manduca sexta* hemocytes. Hemocytes from a last instar larva were exposed for 5 minutes in situ to a sterile cover glass before being fixed and immunolabeled. Note the green labeling (with peanut agglutinin-FITC, which labels hemocytin) of granular cells as well as labeling of extracellular fibers and matrix that surround each of the granular cells. Small plasmatocytes are labeled red with TRITC-MS13 (anti-ß1-integrin). Nuclei are labeled blue with DAPI. The larger nuclei (arrows) belong to plasmatocytes. The two obvious foci of aggregation (two red masses in center of image) contain granular cells (with green granules, diploid nuclei), small plasmatocytes (red membranes, polyploid nuclei), and large, neuroglian-positive plasmatocytes (slight green label, no red label, polyploid nuclei). Bar represents 20 μm.

Certain proteins released by granular cells and/or fat body cells are postulated to act as multivalent ligands that cross-link and cluster receptors on hemocyte surfaces. A multivalent protein binding to surface receptors such as integrin and neuroglian (Nardi et al. 2005) would induce their aggregation in the plane of the membrane. Clustering of receptors would promote binding capacity (avidity) of surface receptors such as integrin and neuroglian, transforming hemocytes from nonadherent to adherent cells (Pech and Strand 1996; Xavier et al. 1998; Corinti et al. 1999; van Kooyk and Figdor 2000; Fahmy et al. 2001; Hynes 2003; Lavine and Strand 2003; Nardi et al. 2005). Clustering of receptors would also activate multiple signaling pathways that influence the activity of multiple cellular proteins: various protein kinases, cytoskeletal organization, ion channels, and growth receptors (Giancotti and Ruoslahti 1999).

Adhesion receptors on circulating, nonadherent platelets and leukocytes of vertebrates often are not activated until they bind ligands. This activation is accompanied by conformational change(s) in integrin molecules (Hynes 1992). A similar phenomenon probably occurs on nonadherent hemocytes of *Manduca*. Although ß1-integrin is involved in the encapsulation response (Wiegand et al. 2000; Levin et al. 2005), it does not behave like an adhesion receptor until hemocytes are exposed to nonself surfaces. Presumably, clustering of integrins on surfaces of hemocytes by multivalent PRR ligands switches nonadherent cells with nonactivated integrin receptors to adherent cells with activated receptors.

During this adhesive transformation of hemocytes, proteins are released by granular cells and bind to surfaces of all hemocytes. Binding of proteins to cell surfaces is detected on plasmatocyte surfaces within a few minutes of culture in the nonself environment of a culture dish. Bound protein is observed to arrange in ring-shaped patches on surfaces of granular cells (Nardi et al. 2005). These patches presumably represent an evolutionary antecedent of the mammalian immunological synapse (Davis 2002).

CONCLUSIONS AND FUTURE PROSPECTS

The immune system of *M. sexta* and other insects is complex, with multiple responses that protect the insect from infection. The actions of hemocytes are rapid, and decrease pathogen number dramatically within the first hour of infection. Phenoloxidase activation is also quite fast, as it makes use of constitutively expressed plasma proteins and amplifies an initial interaction through the proteinase cascade pathway to begin producing reactive quinones and melanin at the surface of a pathogen or parasite within minutes. Induced synthesis of antimicrobial proteins and peptides, as well as other immune response proteins, occurs within hours and can clear microbial infections that have escaped hemocytes, and then provides protection from further infection for several days. We are beginning to understand the molecules involved in hemocyte adhesion to pathogens, extracellular matrix, other hemocytes, and the many plasma proteins that function in immune responses of *M. sexta*. However, this knowledge, especially information about regulation of the immune response, is far from complete.

Important topics for future research include more thorough identification of the set of genes that function in immunity, which will require an *M. sexta* genome sequence and extensive experimental work. Although pattern recognition receptors that trigger responses to bacteria and fungi have been identified, we know little about how responses to other eukaryotic parasites are initiated or about immune responses to viral infections. The changes in the immune system that occur with development, particularly embryonic development and metamorphosis, should be rich areas for future research with *Manduca* as a model system, about which there has been a small but intriguing start (Russell and Dunn 1996; Nardi et al. 2003; Nardi 2004; Gorman, Kankanala, and Kanost 2004; Beetz et al. 2008). In addition, future study is needed to learn how immune responses are regulated (e.g., the roles of cytokines) and to investigate the degree of cross-talk between immune pathways.

ACKNOWLEDGMENTS

Research in the authors' laboratories was supported by National Institutes of Health Grants GM41247 and HL64657. Charles Mark Bee of the Imaging Technology Group at the Beckman Institute (University of Illinois) helped to prepare the hemocyte images.

REFERENCES

Aderem, A. and R.J. Ulevitch. 2000. Toll-like receptors in the induction of the innate immuine response. *Nature* 406:782–787.

Ao, J.Q., E. Ling, and X.Q. Yu. 2008. A Toll receptor from *Manduca sexta* is in response to *Escherichia coli* infection. *Mol. Immunol.* 45:543–552.

Asgari, S., U. Theopold, C. Wellby, and O. Schmidt. 1998. A protein with protective properties against the cellular defense reactions in insects. *Proc. Natl. Acad. Sci. U.S.A.* 95:3690–3695.

Ashida, M. and P. Brey. 1997. Recent advances in research on the insect prophenoloxidase cascade. In *Molecular mechanisms of immune responses in insects,* ed. P. Brey and D. Hultmark, 135–171. New York: Chapman and Hall.

Beck, M., and M.R. Strand. 2005. Glc1.8 from *Microplitis demolitor* bracovirus induces a loss of adhesion and phagocytosis in insect High Five and S2 cells. *J. Virol.* 79:1861–1870.

Beetz, S., T.K. Holthusen, J. Koolman, and T. Trenczek. 2008. Correlation of hemocyte counts with different developmental parameters during the last larval instar of the tobacco hornworm, *Manduca sexta. Arch. Insect Biochem. Physiol.* 67:63–75.

Berditchevski, F. 2001. Complexes of tetraspanins with integrins: More than meets the eye. *J. Cell Sci.* 114:4143–4151.

Boman, H.G., I. Faye, G.H. Gudmundsson, J.Y. Lee, and D.A. Lidholm. 1991. Cell-Free immunity in Cecropia—A model system for antibacterial proteins. *Eur. J. Biochem.* 201:23–31.

Boucheix, C., and E. Rubinstein. 2001. Tetraspanins. *Cell. Mol. Life Sci.* 58:1189–1205.

Chen, C.L., D.J. Lampe, H.M. Robertson, and J.B. Nardi. 1997. Neuroglian is expressed on cells destined to form the prothoracic glands of *Manduca* embryos as they segregate from surrounding cells and rearrange during morphogenesis. *Dev. Biol.* 181:1–13.

Clark K.D., S.F Garczynski, A. Arora, J.W. Crim, and M.R. Strand. 2004. Specific residues in plasmatocyte-spreading peptide are required for receptor binding and functional antagonism of insect immune cells. *J. Biol. Chem.* 279:33246–33252.

Clark, K.D., L.L. Pech, and M.R. Strand. 1997. Isolation and identification of a plasmatocyte-spreading peptide from the hemolymph of the lepidopteran insect *Pseudoplusia includens*. *J. Biol. Chem.* 272:23440–23447.

Corinti, S., E. Fanales-Belasio, C. Albanesi, A. Cavani, P. Angelisova, and G. Girolomoni. 1999. Cross-linking of membrane CD43 mediates dendritic cell maturation. *J. Immunol.* 162:6331–6336.

Davies, D.H., and S.B Vinson. 1986. Passive evasion by eggs of braconid parasitoid *Cardiochiles nigriceps* of encapsulation in vitro by haemocytes of host *Heliothis virescens*. Possible role for fibrous layer in immunity. *J. Insect Physiol.* 32:1003–1010.

Davis, D.M. 2002. Assembly of the immunological synapse for T cells and NK cells. *Trends Immunol.* 23:356–363.

Dean, P., U. Potter, E.H. Richards, J.P Edwards, A.K. Charnley, and S.E. Reynolds. 2004. Hyperphagocytic haemocytes in *Manduca sexta*. *J. Insect Physiol.* 50:1027–1036.

Dickinson, L., V.W. Russell, and P.E. Dunn. 1988. A family of bacteria-regulated cecropin D-like peptides from *Manduca sexta*. *J. Biol. Chem.* 263:19424–19429.

Eleftherianos, I., F. Gokcen, G. Felfoldi, et al. 2007. The immunoglobulin family protein Hemolin mediates cellular immune responses to bacteria in the insect *Manduca sexta*. *Cellular Microbiol.* 9:1137–1147.

Eleftherianos, I., P.J. Millichap, R.H. ffrench-Constant, and S.E. Reynolds. 2006. RNAi suppression of recognition protein mediated immune responses in the tobacco hornworm *Manduca sexta* causes increased susceptibility to the insect pathogen *Photorhabdus*. *Dev. Comp. Immunol.* 30:1099–1107.

Fabrick, J.A., J.E. Baker, and M.R. Kanost. 2004. Innate immunity in a pyralid moth: Functional evaluation of domains from a β-1,3-glucan recognition protein. *J. Biol. Chem.* 279:26605–26611.

Fahmy, T.M., J.G. Bieler, M. Edidin, and J.P. Schneck. 2001. Increased TCR Avidity after T cell activation: A mechanism for sensing low-density antigen. *Immunity* 14:135–143.

Faye, I., and M.R. Kanost. 1998. Function and regulation of hemolin. In *Molecular mechanisms of immune responses in insects*, ed. P.T. Brey and D.D. Hultmark, 173–188. New York: Chapman and Hall.

Ferrandon, D., J.-L. Imler, C. Hetru, and J.A. Hoffmann. 2007. The *Drosophila* systemic immune response: Sensing and signaling during bacterial and fungal infections. *Nature Rev. Immunol.* 7:862–874.

Finnerty, C.M., and R.R. Granados. 1997. The plasma protein scolexin from *Manduca sexta* is induced by Baculovirus infection and other immune challenges. *Insect Biochem. Mol. Biol.* 27:1–7.

Finnerty, C.M., P.A. Karplus, and R.R. Granados. 1999. The insect immune protein scolexin is a novel serine proteinase homolog. *Protein Sci.* 8:242–248.

Galibert, L., J. Rocher, M. Ravallec, M. Duonon-Cerutti, B.A. Webb, and A.N. Volkoff. 2003. Two *Hyposoter didymator* ichnovirus genes expressed in the lepidopteran host encode secreted or membrane-associated serine-and threonine-rich proteins in segments that may be nested. *J. Insect Physiol.* 49:441–451.

Gan, H., Y. Wang, H.B. Jiang, K. Mita, and M.R. Kanost. 2001. A bacteria-induced, intracellular serpin in granular hemocytes of *Manduca sexta*. *Insect Biochem. Mol. Biol.* 31:887–898.

Gandhe, A.S., S.H. John, and J. Nagaraju. 2007. Noduler, a novel immune up-regulated protein mediates nodulation response in insects. *J. Immunol.* 179:6943–6951.

Giancotti, F.G., and E. Ruoslahti. 1999. Integrin signaling. *Science* 285:1028–1032.

Gillespie, J.P., M.R Kanost, and T. Trenczek. 1997. Biological mediators of insect immunity. *Annu. Rev. Entomol.* 42:611–643.

Gorman, M. J., C.J. An, and M.R. Kanost. 2007. Characterization of tyrosine hydroxylase from *Manduca sexta*. *Insect Biochem. Mol. Biol.* 37:1327–1337.

Gorman, M.J., P. Kankanala, and M.R. Kanost. 2004. Bacterial challenge stimulates innate immune responses in extra-embryonic tissues of tobacco hornworm eggs. *Insect Mol. Biol.* 13:19–24.

Gorman, M.J., Y. Wang, H.B. Jiang, and M.R. Kanost. 2007. *Manduca sexta* hemolymph proteinase 21 activates prophenoloxidase-activating proteinase 3 in an insect innate immune response proteinase cascade. *J. Biol. Chem.* 282:11742–11749.

Goto, A., T. Kumagai, C. Kumagai, et al. 2003. *Drosophila* hemolectin gene is expressed in embryonic and larval hemocytes and its knockdown causes bleeding defects. *Dev. Biol.* 264:582–591.

Gupta, S., Y. Wang, and H.B. Jiang. 2005a. Purification and characterization of *Manduca sexta* prophenoloxidase-activating proteinase-1, an enzyme involved in insect immune responses. *Protein Express. Purif.* 39:261–268.

Gupta, S., Y. Wang, and H.B. Jiang. 2005b. *Manduca sexta* prophenoloxidase (proPO) activation requires proPO-activating proteinase (PAP) and serine proteinase homologs (SPHs) simultaneously. *Insect Biochem. Mol. Biol.* 35:241–248.

Hall, M., T. Scott, M. Sugumaran, K. Söderhäll, and J.H. Law. 1995. Proenzyme of *Manduca sexta* phenol oxidase: Purification, activation, substrate specificity of the active enzyme, and molecular cloning. *Proc. Natl. Acad. Sci. U.S.A.* 92:7764–7768.

Hemler, M., B.A. Mannion, and F. Berditchevski. 1996. Association of TM4SF proteins with integrins: Relevance to cancer. *Biochim. Biophys. Acta* 1287:67–71.

Hiruma, K., M.S. Carter, and L.M. Riddiford. 1995. Characterization of the dopa decarboxylase gene of *Manduca sexta* and its suppression by 20-hydroxyecdysone. *Dev. Biol.* 169:195–209.

Hoek, R.M., A.B. Smit, H. Frings, J.M. Vink, M. deJong-Brink, and W.P.M. Geraerts. 1996. A new Ig-superfamily member, molluscan defence molecule (MDM) from *Lymnaea stagnalis*, is down-regulated during parasitosis. *Eur. J. Immunol.* 26:939–944.

Huang, R.D., Z.Q. Lu, H.E. Dai, D.V. Velde, O. Prakash, and H.B. Jiang. 2007. The solution structure of clip domains from *Manduca sexta* prophenoloxidase activating proteinase-2. *Biochemistry* 46:11431–11439.

Humphries, J.D., A. Byron, M.J. Humphries. 2006. Integrin ligands at a glance. *J. Cell Sci.* 119: 3901–3903.

Hynes, R.O. 1992. Integrins: Versatility, modulation, and signaling in cell adhesion. *Cell* 69:11–25.

Hynes, R.O. 2003. Changing partners. *Science* 300:755–756.

Irving, P., J.M. Ubeda, D. Doucet, et al. 2005. New insights into *Drosophila* larval haemocyte functions through genome-wide analysis. *Cell. Microbiol.* 7:335–350.

Janeway, C.A., and R. Medzhitov. 2002. Innate immune recognition. *Annu. Rev. Immunol.* 20:197–216.

Janeway, C.A., P. Travers, M. Walport, and M.J. Shlomchik. 2001. *Immunobiology: The immune system in health and disease.* New York: Garland Publishing.

Ji, C.Y., Y. Wang, X.P. Guo, S. Hartson, and H.B. Jiang. 2004. A pattern recognition serine proteinase triggers the prophenoloxidase activation cascade in the tobacco hornworm, *Manduca sexta. J. Biol. Chem.* 279:34101–34106.

Jiang, H. 2008. The biochemical basis of antimicrobial responses in *Manduca sexta. Insect Science.* 15:53–66.

Jiang H., and M.R. Kanost. 1997. Characterization and functional analysis of twelve naturally occurring reactive site variants of serpin-1 from *Manduca sexta. J. Biol. Chem.* 272:1082–1087.

Jiang, H., and M.R. Kanost. 2000. The clip-domain family of serine proteinases in arthropods. *Insect Biochem. Molec. Biol.* 30:95–105.

Jiang, H.B., C.C. Ma, Z.Q. Lu, and M.R. Kanost. 2004. β-1,3-Glucan recognition protein-2 (β GRP-2) from *Manduca sexta*: An acute-phase protein that binds β-1,3-glucan and lipoteichoic acid to aggregate fungi and bacteria and stimulate prophenoloxidase activation. *Insect Biochem. Mol. Biol.* 34:89–100.

Jiang H.B., A.B. Mulnix, and M.R. Kanost. 1995. Expression and characterization of recombinant *Manduca sexta* serpin-1B and site-directed mutants that change its inhibitory selectivity. *Insect Biochem. Molec. Biol.* 25:1093–1100.

Jiang, H.B., Y. Wang, Y.L. Gu, et al. 2005. Molecular identification of a bevy of serine proteinases in *Manduca sexta* hemolymph. *Insect Biochem. Mol. Biol.* 35:931–943.

Jiang H., Y. Wang, Y. Huang, et al. 1996. Organization of serpin gene-1 from *Manduca sexta*: Evolution of a family of alternate exons encoding the reactive site loop. *J. Biol. Chem.* 271:28017–28023.

Jiang H, Y. Wang, and M.R. Kanost. 1994. Mutually exclusive exon use and reactive center diversity in insect serpins. *J. Biol. Chem.* 269:55–58.

Jiang, H., Y. Wang, and M.R. Kanost. 1998. Pro-phenol oxidase activating proteinase from an insect, *Manduca sexta*: A bacteria-inducible protein similar to *Drosophila* easter. *Proc. Natl. Acad. Sci. U.S.A.* 95:12220–12225.

Jiang, H., Y. Wang, C. Ma, and M.R. Kanost. 1997. Subunit composition of pro-phenol oxidase from *Manduca sexta*: Molecular cloning of subunit proPO-p1. *Insect Biochem. Molec. Biol.* 27:835–850.

Jiang, H.B., Y. Wang, X.Q. Yu, and M.R. Kanost. 2003a. Prophenoloxidase-activating proteinase-2 from hemolymph of *Manduca sexta. J. Biol. Chem.* 278:3552–3561.

Jiang, H.B., Y. Wang, X.Q. Yu, Y.F. Zhu, and M. Kanost. 2003b. Prophenoloxidase-activating proteinase-3 (PAP-3) from *Manduca sexta* hemolymph: A clip-domain serine proteinase regulated by serpin-1J and serine proteinase homologs. *Insect Biochem. Mol. Biol.* 33:1049–1060.

Jiravanichpaisal, P., B.L. Lee, and K. Söderhäll. 2006. Cell-mediated immunity in arthropods: Hematopoiesis, coagulation, melanization and opsonization. *Immunobiol.* 211:213–236.

Kang, D.W., G. Liu, A. Lundstrom, E. Gelius, and H. Steiner. 1998. A peptidoglycan recognition protein in innate immunity conserved from insects to humans. *Proc. Nat. Acad. Sci. U.S.A.* 95:10078–10082.

Kanost, M.R. 1999. Serine proteinase inhibitors in arthropod immunity. *Dev. Comp. Immunol.* 23:291–301.

Kanost, M.R. 2007. Serpins in a Lepidopteran insect, *Manduca sexta*. In *The serpinopathies: Molecular and cellular aspects of serpins and their disorders*, ed. G.A. Silverman and D.A. Lomas, 229–242. Singapore: World Scientific Publishing Co.

Kanost, M.R., and T. Clarke. 2005. Proteases. In *Comprehensive molecular insect science*, Vol. 4, ed. L.I. Gilbert, K. Iatrou, and S. Gill, 247–266. San Diego: Elsevier.

Kanost, M.R., W. Dai, and P.E. Dunn. 1988. Peptidoglycan fragments elicit antibacterial protein synthesis in larvae of *Manduca sexta. Arch. Insect Biochem. Physiol.* 8:147–164.

Kanost, M.R., and M.G. Gorman. 2008. Phenoloxidases in insect immunity. In *Insect immunology*, ed. N. Beckage, 69–96. San Diego: Academic Press/Elsevier.

Kanost, M.R., H. Jiang, and X.-Q. Yu. 2004. Innate immune responses of a lepidopteran insect, *Manduca sexta. Immunol. Rev.* 198:97–105.

Kanost, M.R., J.K. Kawooya, J.H. Law, R.O. Ryan, M.C. Van Heusden, and R. Ziegler. 1990. Insect Haemolymph proteins. *Adv. Insect Physiol.* 22:299–396.

Kanost, M.R., S.V. Prasad, Y. Huang, and E. Willott. 1995. Regulation of serpin gene-1 in *Manduca sexta. Insect Biochem. Mol. Biol.* 25:285–291.

Kanost M.R., S.V. Prasad, and M.A. Wells. 1989. Primary structure of a member of the serpin superfamily of proteinase inhibitors from an insect, *Manduca sexta. J. Biol. Chem.* 264:965–972.

Kanost, M.R., M.K. Zepp, N.E. Ladendorff, and L.A. Andersson. 1994. Isolation and characterization of a hemocyte aggregation inhibitor from hemolymph of *Manduca sexta* larvae. *Arch. Insect Biochem. Physiol.* 27:123–136.

Karlsson, C., A.M. Korayem, C. Scherfer, O. Loseva, M.S. Dushay, and U. Theopold. 2004. Proteomic analysis of the *Drosophila* larval hemolymph clot. *J.Biol. Chem.* 279:52033–52041.

Kathirithamby, J., L.D. Ross, and J.S. Johnston. 2003. Masquerading as self? Endoparasitic Strepsiptera (Insecta) enclose themselves in host-derived epidermal bag. *Proc. Natl. Acad. Sci. U.S.A.* 100:7655–7659.

Koizumi, N., M. Imamura, T. Kadotani, K. Yaoi, H. Iwahana, and R. Sato. 1999. The lipopolysaccharide-binding protein participating in hemocyte nodule formation in the silkworm *Bombyx mori* is a novel member of the C-type lectin superfamily with two different tandem carbohydrate-recognition domains. *FEBS Letters* 443:139–143.

Kopczynski, C.C., G.W. Davis, and C.S. Goodman. 1996. A neural tetraspanin, encoded by late bloomer, that facilitates synapse formation. *Science* 271:1867–1870.

Kotani, E., M. Yamakawa, S. Iwamoto, et al. 1995. Cloning and expression of the gene of hemocytin, an insect humoral lectin which is homologous with the mammalian von Willebrand factor. *Biochim. Biophys. Acta* 1260:245–258.

Kyriakides, T.R., J.L. McKillip, and K.D. Spence. 1995. Biochemical characterization, developmental expression, and induction of the immune protein scolexin from *Manduca sexta. Arch. Insect Biochem. Physiol.* 29:269–80.

Ladendorff, N.E., and M.R. Kanost. 1991. Bacteria-induced protein P4 (hemolin) from *Manduca sexta*: A member of the immunoglobulin superfamily which can inhibit hemocyte aggregation. *Arch. Insect Biochem. Physiol.* 18:285–300.

Lavine, M.D., and N.E. Beckage 1996. Temporal pattern of parasitism-induced immunosuppression in *Manduca sexta* larvae parasitized by *Cotesia congregata. J. Insect Physiol.* 42:41–51.

Lavine, M.D., and M.R. Strand. 2002. Insect hemocytes and their role in immunity. *Insect Biochem. Mol. Biol.* 32:1295–1309.

Lavine, M.D., and M.R. Strand. 2003. Haemocytes from *Pseudoplusia includens* express multiple α and β integrin subunits. *Insect Molec. Biol.* 12:441–452.

Lemaitre, B., and J. Hoffmann. 2007. The host defense of *Drosophila melanogaster. Annu. Rev. Immunol.* 25:697–743.

Levin, D., L.N. Breuer, S. Zhuang, S.A. Anderson, J.B. Nardi, and M.R. Kanost. 2005. A hemocyte-specific integrin required for hemocytic encapsulation in the tobacco hornworm, *Manduca sexta. Insect Biochem. Mol. Biol.* 35:369–380.

Li, D., C. Scherfer, A.M. Korayem, Z. Zhao, O. Schmidt, and U. Theopold. 2002. Insect hemolymph clotting: Evidence for interaction between the coagulation system and the prophenoloxidase activating cascade. *Insect Biochem. Mol. Biol.* 32:919–928.

Li J., Z. Wang, B. Canagarajah, H. Jiang, M.R. Kanost, and E.J. Goldsmith. 1999. The structure of active serpin K from *Manduca sexta* and a model for serpin-protease complex formation. *Structure* 7:103–109.

Ling, E., and X.-Q. Yu. 2006. Cellular encapsulation and melanization are enhanced by immulectins, pattern recognition receptors from the tobacco hornworm *Manduca sexta. Dev. Comp. Immunol.* 30:289–299.

Loret, S.M., and M.R. Strand. 1998. Follow-up of protein release from *Pseudoplusia includens* hemocytes: A first step toward identification of factors mediating encapsulation in insects. *Eur. J. Cell Biol.* 76:146–155.

Lu, Z., and H. Jiang. 2007. Regulation of phenoloxidase activity by high- and low-molecular-weight inhibitors from the larval hemolymph of *Manduca sexta. Insect Biochem. Mol. Biol.* 37:478–485.

Lu, Z., and H. Jiang. 2008. Expression of *Manduca sexta* serine proteinase homolog precursors in insect cells and their proteolytic activation. *Insect Biochem. Mol. Biol.* 38:89–98.

Ma, C., and M.R. Kanost. 2000. A beta 1,3-glucan recognition protein from an insect, *Manduca sexta*, agglutinates microorganisms and activates the phenoloxidase cascade. *J. Biol. Chem.* 275:7505–7514.

Maecker, H.T., S.C. Todd, and S. Levy. 1997. The tetraspanin superfamily: Molecular facilitators. *FASEB J.* 11:428–442.

Manfredini, F., F. Giusti, L. Beani, and R. Dallai. 2007. Developmental strategy of the endoparasite *Xenos vesparum* (Strepsiptera, Insecta): Host invasion and elusion of its defense reactions. *J. Morph.* 268:588–601.

Matzinger, P. 2002. The danger model: A renewed sense of self. *Science* 296:301–305.

Medzhitov, R., and C.A. Janeway. 2002. Decoding the patterns of self and nonself by the innate immune system. *Science* 296:298–300.

Michel, K., and F.C. Kafatos. 2005. Mosquito immunity against *Plasmodium. Insect Biochem. Mol. Biol.* 35:677–89.

Moita, L.F., G. Vriend, V. Mahairaki, C. Louis, and F.C. Kafatos. 2006. Integrins of *Anopheles gambiae* and a putative role of a new integrin, BINT2, in phagocytosis of *E. coli. Insect Biochem. Mol. Biol.* 36:282–290.

Mulnix, A.B., and P.E. Dunn. 1994. Structure and induction of a lysozyme gene from the tobacco hornworm, *Manduca sexta. Insect Biochem. Mol. Biol.* 24:271–281.

Mulnix, A.B., and P.E. Dunn. 1995. Molecular biology of the immune response. In *Molecular model systems in the lepidoptera*, ed. M.R. Goldsmith and A.S. Wilkins, 369–395. New York: Cambridge University Press.

Nappi, A.J., and B.M. Christensen. 2005. Melanogenesis and associated cytotoxic reactions: Applications to insect innate immunity. *Insect Biochem. Mol. Biol.* 35:443–59.

Nardi, J.B. 2004. Embryonic origins of the two main classes of hemocytes—granular cells and plasmatocytes—in *Manduca sexta. Dev. Genes Evol.* 214:19–28.

Nardi, J.B., C. Gao, and M.R. Kanost. 2001. The extracellular matrix protein lacunin is dynamically expressed by a subset of the hemocytes involved in basal lamina morphogenesis. *J. Insect Physiol.* 47:997–1006.

Nardi, J.B., B. Pilas, C.M. Bee, K. Garsha, S. Zhuang, and M.R. Kanost. 2006. Neuroglian-positive plasmatocytes and the initiation of hemocyte attachment to foreign surfaces. *Dev. Comp. Immunol.* 30:447–462.

Nardi, J.B., E. Ujhelyi, B. Pilas, K. Garsha, and M.R. Kanost. 2003. Hematopoietic organs of *Manduca sexta* and hemocyte lineages. *Dev. Genes Evol.* 213:477–491.

Nardi, J.B., S. Zhuang, B. Pilas, C.M. Bee, and M.R. Kanost. 2005. Clustering of adhesion receptors following exposure of insect blood cells to foreign surfaces. *J. Insect Physiol.* 51:555–564.

Ochiai, M., and M. Ashida. 1988. Purification of a β-1,3-glucan recognition protein in the prophenoloxidase activating system from hemolymph of the silkworm, *Bombyx mori. J. Biol. Chem.* 263:12056–12062.

Ochiai, M., and M. Ashida. 1999. A pattern recognition protein for peptidoglycan—Cloning the cDNA and the gene of the silkworm, *Bombyx mori. J. Biol. Chem.* 274:11854–11858.

Ochiai, M., and M. Ashida. 2000. A pattern recognition protein for β-1,3-glucan. *J. Biol. Chem.* 275:4995–5002.

Pech, L.L., and M.R. Strand. 1996. Granular cells are required for encapsulation of foreign targets by insect hemocytes. *J. Cell Sci.* 109:2053–2060.

Ponnuvel, K.M., and M. Yamakawa. 2002. Immune responses against bacterial infection in *Bombyx mori* and regulation of host gene expression. *Curr. Sci.* 83:447–454.

Ratcliffe, N.A., and S.J. Gagen. 1977. Studies on the in vivo cellular reactions of insects: An ultrastructural analysis of nodule formation in *Galleria mellonella. Tissue Cell* 9:73–85.

Rizki, R.M., and T.M. Rizki. 1984. The cellular defense system of *Drosophila melanogaster*. In *Insect ultrastructure*, Vol. 2, ed. R.C. King and H. Akai, 579–604. New York: Plenum.

Rodríguez-Pérez, M.A., R.F. Dumpit, J.M. Lenz, E.N. Powell, S.Y. Tam, and N.E. Beckage. 2005. Host refractoriness of the tobacco hornworm, *Manduca sexta*, to the braconid endoparasitoid *Cotesia flavipes. Arch. Insect Biochem. Physiol.* 60:159–71.

Royet, J., and R. Dziarski. 2007. Peptidoglycan recognition proteins: Pleiotropic sensors and effectors of antimicrobial defences. *Nat. Rev. Microbiol.* 5:264–277.

Russell, V., and P.E. Dunn 1996. Antibacterial proteins in the midgut of *Manduca sexta* during metamorphosis. *J. Insect Physiol.* 42:65–71.

Salt, G., 1970. *The cellular defence reactions of insects.* Cambridge, UK: Cambridge Univ. Press.

Sansonetti, P.J. 2006. The innate signaling of dangers and the dangers of innate signaling. *Nature Immunol.* 7:1237–1242.

Scherfer, C., C. Karlsson, O. Loseva, et al. 2004. Isolation and characterization of hemolymph clotting factors in *Drosophila melanogaster* by a pullout method. *Curr. Biol.* 14:625–629.

Schmidt, O., K. Andersson, A. Will, and I. Schuchmann-Feddersen. 1990. Viruslike particle proteins from a hymenopteran endoparasitoid are related to a protein component of the immune system in the lepidopteran host. *Arch. Insect Biochem. Physiol.* 13:107–115.

Schmit, A.R., and N.A. Ratcliffe. 1977. The encapsulation of foreign tissue implants in *Galleria mellonella* larvae. *J. Insect Physiol.* 23:175–184.

Scholz, F. 2002. Aktivierung von Hämozyten des Tabakschwärmers *Manduca sexta* nach bakteriellen Infektionen. PhD dissertation, Justus-Liebig Universität, Giessen.

Shin, S.W., S.S. Park, D.S. Park, et al. 1998. Isolation and characterization of immune-related genes from the fall webworm, *Hyphantria cunea*, using PCR-based differential display and subtractive cloning. *Insect Biochem. Mol. Biol.* 28:827–837.

Shrestha, S., and Y. Kim. 2008. Eicosanoids mediate prophenoloxidase release from oenocytoids in the beet armyworm *Spodoptera exigua. Insect Biochem. Mol. Biol.* 38:99–112.

Silverman, G.A., P.I. Bird, R.W. Carrell, et al. 2001. The serpins are an expanding superfamily of structurally similar but functionally diverse proteins: Evolution, mechanism of inhibition, novel functions, and a revised nomenclature. *J Biol. Chem.* 276:33293–33296.

Stanley, D. 2006. Prostaglandins and other eicosanoids in insects: Biological significance. *Annu. Rev. Entomol.* 51:25–44.

Stanley, D.W., J.S. Miller, and R.W. Howard. 1998. The influence of bacterial species and intensity of infections on nodule formation in insects. *J. Insect Physiol.* 44:157–164.

Strand, M.R., and T. Noda. 1991. Alterations in the haemocytes of *Pseudoplusia includens* after parasitism by *Microplitis demolitor. J. Insect Physiol.* 37:839–850.

Strand, M.R., and E.A. Wong. 1991. The growth and role of *Microplitis demolitor* teratocytes in parasitism of *Pseudoplusia includens. J. Insect Physiol.* 37:503–515.

Su, X.-D., L.N. Gastinel, D.E. Vaughn, I. Faye, P. Poon, and P.J. Bjorkman. 1998. Crystal structure of hemolin: A horseshoe shape with implications for homophilic adhesion. *Science* 281:991–995.

Sun, S.C., I. Lindström, H.G. Boman, I. Faye, and O. Schmidt. 1990. Hemolin: An insect-immune protein belonging to the immunoglobulin superfamily. *Science* 250:1729–1732.

Theopold, U., O. Schmidt, K. Söderhäll, and M.S. Dushay. 2004. Coagulation in arthropods: Defence, wound closure and healing. *Trends Immunol.* 25:289–294.

Todres, E., J.B. Nardi, and H.M. Robertson. 2000. The tetraspanin superfamily in insects. *Insect Mol. Biol.* 9:581–590.

Tong, Y.R., H.B. Jiang, and M.R. Kanost. 2005. Identification of plasma proteases inhibited by *Manduca sexta* serpin-4 and serpin-5 and their association with components of the prophenol oxidase activation pathway. *J. Biol. Chem.* 280:14932–14942.

Tong, Y.R., and M.R. Kanost. 2005. *Manduca sexta* serpin-4 and serpin-5 inhibit the prophenol oxidase activation pathway. *J. Biol. Chem.* 280:14923–14931.

Van Kooyk, Y., and C.G. Figdor. 2000. Avidity regulation of integrins: The driving force in leukocyte adhesion. *Curr. Opinion Cell Biol.* 12:542–547.

Wang, X., J.F. Fuchs, L.C. Infanger, et al. 2005. Mosquito innate immunity: Involvement of ß 1,3-glucan recognition protein in melanotic encapsulation immune responses in *Armigeres subalbatus. Mol. Biochem. Parasitol.* 139:65–73.

Wang, Y., and H.B. Jiang. 2004a. Prophenoloxidase (proPO) activation in Manduca sexta: An analysis of molecular interactions among proPO, proPO-activating proteinase-3, and a cofactor. *Insect Biochem. Mol. Biol.* 34:731–742.

Wang, Y., and H.B. Jiang. 2004b. Purification and characterization of *Manduca sexta* serpin-6: A serine proteinase inhibitor that selectively inhibits prophenoloxidase-activating proteinase-3. *Insect Biochem. Mol. Biol.* 34:387–395.

Wang, Y., and H. Jiang. 2006. Interaction of beta-1,3-glucan with its recognition protein activates hemolymph proteinase 14, an initiation enzyme of the prophenoloxidase activation system in *Manduca sexta. J. Biol. Chem.* 281:9271–9278.

Wang, Y., and H. Jiang. 2007. Reconstitution of a branch of the *Manduca sexta* prophenoloxidase activation cascade in vitro: Snake-like hemolymph proteinase 21 (HP21) cleaved by HP14 activates prophenol oxidase- activating proteinase-2 precursor. *Insect Biochem. Mol. Biol.* 37:1015–1025.

Wang, Y., H. Jiang, and M.R. Kanost. 1999. Biological activity of *Manduca sexta* paralytic and plasmatocyte spreading peptide and primary structure of its hemolymph precursor. *Insect Biochem. Molec. Biol.* 29:1075–1086.

Wang, C., and R.J. St. Leger. 2006. A collagenous protective coat enables *Metarhizium anisopliae* to evade insect immune responses. *Proc. Natl. Acad. Sci. U.S.A.* 103:6647–6652.

Wang, Y., Z. Zou, and H.B. Jiang. 2006. An expansion of the dual clip-domain serine proteinase family in *Manduca sexta*: Gene organization, expression, and evolution of prophenoloxidase-activating proteinase-2, hemolymph proteinase 12, and other related proteinases. *Genomics* 87:399–409.

Watanabe, A., S. Miyazawa, M. Kitami, H. Tabunoki, K. Ueda, and R. Sato. 2006. Characterization of a novel C-type lectin, *Bombyx mori* multibinding protein, from the *B. mori* hemolymph: Mechanism of wide-range microorganism recognition and role in immunity. *J. Immunol.* 177:4594–4604.

Waterhouse, R.M., Z.Y. Xi, E. Kriventseva, et al. 2007. Evolutionary dynamics of immune-related genes and pathways in disease vector mosquitoes. *Science* 316:1738–1743.

Wiegand, C., D. Levin, J.P. Gillespie, E. Willott, M.R. Kanost, and T. Trenczek. 2000. Monoclonal antibody MS13 identifies a plasmatocyte membrane protein and inhibits encapsulation and spreading reactions of *Manduca sexta* hemocytes. *Arch. Insect Biochem. Physiol.* 45:95–108.

Willott, E., T. Trenczek, L.W. Thrower, and M.R. Kanost. 1994. Immunochemical identification of insect hemocyte populations: Monoclonal antibodies distinguish four major hemocyte types in *Manduca sexta. Eur. J. Cell Biol.* 65:417–423.

Xavier, R., T. Brennan, Q. Li, C. McCormack, and B. Seed. 1998. Membrane compartmentation is required for efficient T cell activation. *Immunity* 8:723–732.

Ye, S., A.L. Cech, R. Belmares, et al. 2001. The structure of a Michaelis serpin-protease complex. *Nature Struct. Biol.* 8:979–983.

Yoshida, H., K. Kinoshita, and M. Ashida. 1996. Purification of a peptidoglycan recognition protein from hemolymph of the silkworm, *Bombyx mori. J. Biol. Chem.* 271:13854–13860.

Yu, X.Q., H. Gan, and M.R. Kanost. 1999. Immulectin, an inducible C-type lectin from an insect, *Manduca sexta*, stimulates activation of plasma prophenol oxidase. *Insect Biochem. Mol. Biol.* 29:585–597.

Yu, X.Q., H.B. Jiang, Y. Wang, and M.R. Kanost. 2003. Nonproteolytic serine proteinase homologs are involved in prophenoloxidase activation in the tobacco hornworm, *Manduca sexta. Insect Biochem. Mol. Biol.* 33:197–208.

Yu, X.Q., and M. R. Kanost. 1999. Developmental expression of *Manduca sexta* hemolin. *Arch. Insect Biochem. Physiol.* 42:198–212.

Yu, X.-Q., and M.R. Kanost. 2000. Immulectin-2, a lipopolysaccharide-specific lectin from an insect, *Manduca sexta*, is induced in response to Gram-negative bacteria. *J. Biol. Chem.* 275:37373–37381.

Yu, X.-Q., and M.R. Kanost. 2002. Binding of hemolin to bacterial lipopolysaccharide and lipoteichoic acid—An immunoglobulin superfamily member from insects as a pattern-recognition receptor. *Eur. J. Biochem.* 269:1827–1834.

Yu, X.-Q., and M.R. Kanost. 2003. *Manduca sexta* lipopolysaccharide-specific immulectin-2 protects larvae from bacterial infection. *Dev. Comp. Immunol.* 27:189–196.

Yu, X.-Q., and M.R. Kanost. 2004. Immulectin-2, a pattern recognition receptor that stimulates hemocyte encapsulation and melanization in the tobacco hornworm, *Manduca sexta. Dev. Comp. Immunol.* 9:891–900.

Yu, X.-Q., and M.R. Kanost. 2008. Activation of lepidopteran insect innate immune responses by C-type immulectins. In *Animal lectins: A functional view*, ed. G.R. Vasta and H.A. Ahmed. Boca Raton: CRC Press. pp. 383–396.

Yu, X.-Q., E. Ling, M.E. Tracy, and Y. Zhu. 2006. Immulectin-4 from the tobacco hornworm *Manduca sexta* binds to lipopolysaccharide and lipoteichoic acid. *Insect Mol. Biol.* 15:119–128.

Yu, X.-Q., O. Prakash, and M.R. Kanost. 1999. Structure of a paralytic peptide from an insect, *Manduca sexta. J. Peptide Res.* 54:256–261.

Yu, X.-Q., Y. Zhu, C. Ma, J.A. Fabrick, and M.R. Kanost. 2002. Pattern recognition proteins in *Manduca sexta* plasma. *Insect Biochem. Molec. Biol.* 32:1287–1293.

Zhao, L., and M.R. Kanost. 1996. In search of a function for hemolin, a hemolymph protein from the immunoglobulin superfamily. *J. Insect Physiol.* 42:73–79.

Zhao, P.C., J.J. Li, Y. Wang, and H.B. Jiang. 2007. Broad-spectrum antimicrobial activity of the reactive compounds generated in vitro by *Manduca sexta* phenoloxidase. *Insect Biochem. Mol. Biol.* 37:952–959.

Zhu, Y. 2001. Identification of immune-related genes from the tobacco hornworm, *Manduca sexta*, and characterization of two immune-inducible proteins, serpin-3 and leureptin. Ph.D. dissertation, Kansas State University, Manhattan, KS.

Zhu, Y., T.J. Johnson, A.A. Myers, and M.R. Kanost. 2003a. Identification by subtractive suppression hybridization of bacteria-induced genes expressed in *Manduca sexta* fat body. *Insect Biochem. Mol. Biol.* 33:541–559.

Zhu, Y.F., Y. Wang, M.J. Gorman, H.B. Jiang, and M.R. Kanost. 2003b. *Manduca sexta* serpin-3 regulates prophenoloxidase activation in response to infection by inhibiting prophenoloxidase-activating proteinases. *J. Biol. Chem.* 278:46556–46564.

Zhuang, S., L. Kelo, J.B. Nardi, and M.R. Kanost. 2007a. Neuroglian on hemocyte surfaces is involved in homophilic and heterophilic interactions of the innate immune system of *Manduca sexta. Dev. Comp. Immunol.* 31:1159–1167.

Zhuang, S., L. Kelo, J.B. Nardi, and M.R. Kanost. 2007b. An integrin-tetraspanin interaction required for cellular innate immune responses of an insect, *Manduca sexta. J. Biol. Chem.* 282:22563–22572.

Zhuang, S., L. Kelo, J.B. Nardi, and M.R. Kanost. 2008. Multiple α subunits of integrin are involved in cell-mediated responses of the *Manduca sexta* immune system. *Dev. Comp. Immunol.* 32:365–79.

Zou, Z., J.D. Evans, Z. Lu, et al. 2007. Comparative genomic analysis of the *Tribolium* immune system. *Genome Biol.* 8:R177.

Zou, Z. and H. Jiang. 2005a. Gene structure and expression profile of *Manduca sexta* prophenoloxidase-activating proteinase-3 (PAP-3), an immune protein containing two clip domains. *Insect Mol. Biol.* 14:433–442.

Zou, Z., and H.B. Jiang. 2005b. *Manduca sexta* serpin-6 regulates immune serine proteinases PAP-3 and HP8—cDNA cloning, protein expression, inhibition kinetics, and function elucidation. *J. Biol. Chem.* 280:14341–14348.

Zou, Z., Y. Wang, and H.B. Jiang. 2005. *Manduca sexta* prophenoloxidase activating proteinase-1 (PAP-1) gene: Organization, expression, and regulation by immune and hormonal signals. *Insect Biochem. Mol. Biol.* 35:627–636.

15 Lepidopterans as Model Mini-Hosts for Human Pathogens and as a Resource for Peptide Antibiotics

Andreas Vilcinskas

CONTENTS

INTRODUCTION

Infectious diseases have plagued humanity throughout history and still represent a major threat for human beings. Driven by the continuous and serious demand for novel therapeutic strategies, the research on the interactions of human pathogens and their virulence factors with the host defense system has been elaborated predominantly using mammalian models such as mice, rats, and rabbits, which provide powerful experimental systems to reproduce human infection. Apart from ethical concerns associated with the exploitation of mammalian host models, the high overall costs for large numbers of individuals, logistic problems, and complex cross talk between the adaptive and the innate immune systems argue for the exploration of alternative host models (Mylonakis, Casadevall, and Ausubel 2007).

If the goal is to study interactions between pathogens and the host innate immune system, the choices include invertebrate model systems lacking adaptive immunity and antibodies that arose during evolution of vertebrates. The latter share with invertebrates the evolutionarily conserved innate immune system. Our knowledge about its genetic architecture and function has been remarkably expanded by investigation of genetically tractable invertebrate models with short reproductive cycles such as the fruit fly *Drosophila melanogaster* (Lemaitre and Hoffmann 2007) and the nematode *Caenorhabitis elegans* (Sifiri et al. 2003; Mylonakis and Aballay 2005). The advantages of *C. elegans* and *D. melanogaster* are that their genomes have been completely sequenced and that microarrays, RNAi libraries, and mutant strains are available, which permit analysis of host-pathogen interactions at the molecular level. On the other hand, the larger size of lepidopteran caterpillars facilitates precise injection of antibiotics or number of pathogens, easy manipulation, and collection of tissue and hemolymph samples to study pathophysiology, for example, with proteomic approaches. A number of lepidopteran species such as the tobacco hornworm, *Manduca sexta* (Silva et al. 2002; Kanost, Jiang, and Yu 2004); the silkworm *Bombyx mori* (Cheng et al. 2006); and the greater wax moth, *Galleria mellonella* (Vilcinskas and Götz 1999) are classic and widely used models for exploring interactions between entomopathogens and their hosts. Whereas current knowledge about lepidopteran innate immunity is comprehensively reviewed in Chapter 14 of this volume, this chapter highlights the use of lepidopterans as mini-host models for human pathogens and as a reservoir for peptide antibiotics with therapeutic potential in medicine and plant protection.

THE GREATER WAX MOTH, *GALLERIA MELLONELLA*, AS A MINI-HOST FOR HUMAN PATHOGENS

In recent years, the larvae of *G. mellonella* have emerged as favorite mini-hosts for pathogens causing severe diseases in humans and as experimental systems to study efficacy of therapeutic drugs (Kavanagh and Reeves 2004; Scully and Bidochka 2006). *G. mellonella* provides a number of advantages in comparison with other lepidopteran models such as *B. mori* or *M. sexta*. First, the small size of *G. mellonella* larvae and the low overall costs of breeding large numbers facilitate the use of this lepidopteran model as an inexpensive whole-animal high-throughput infection assay (Mylonakis, Casadevall, and Ausubel 2007). Second, *G. melonella* larvae are commercially available worldwide; for example, they are sold as bait for fishermen or as food for pets (reptiles). Third, the major advantage of *G. mellonella* that has been outlined in the recent literature is that this heterologous host can be adapted in the laboratory to human physiological temperature (37°C). It is essential to mimic the physiological conditions in mammals because human pathogens are adapted to the physiological temperature of their host, which is often required for the synthesis and the release of their pathogenic or virulence factors (Fuchs and Mylonakis 2006). Therefore, in order to study infection and host response operative at human temperatures, one must select a thermotolerant model system such as *G. mellonella* (Mylonakis, Casadevall, and Ausubel 2007). Fourth, a positive correlation exists between the pathogenicity of bacteria and fungi when evaluated in *G. mellonella* and mice (Jander et al. 2000; Brennan et al. 2002). These advantages have convinced an increasing number of researchers to favor *G. mellonella* as a mini-host model for prominent pathogenic bacteria and fungi that are responsible for severe human diseases: *Bacillus cereus* (Fedhila et al. 2006), *Enterococcus faecalis* (Park et al. 2007), *Francisella tularensis* (Aperis et al. 2007), *Pseudomonas aeruginosa* (Miyata et al. 2003), *Staphylococcus aureus* (Garcia-Lara, Needham, and Foster 2005), *Candida albicans* (Bergin et al. 2006), and *Cryptococcus neoformans* (Mylonakis et al. 2005). In essence, these studies show that *G. mellonella* enables rapid screening of pathogen mutant libraries and can serve to investigate evolutionarily conserved aspects of microbial virulence and host response.

G. MELLONELLA AS A MODEL HOST FOR PATHOGENIC BACTERIA

Many entomopathogenic bacteria are closely related to human pathogens, or can infect hosts as diverse as insects and mammals. For example, *B. thuringiensis*, which is used to control pest and vector insects worldwide, is highly related to *B. anthracis*, the infective agent of anthrax. Both bacteria have been reported to use a similar set of pathogenic/virulence factors, among which metalloproteinases play a prominent predominant role (Silva et al. 2002; Chung et al. 2006). *B. cereus*, an opportunistic bacterium frequently associated with food-borne infections causing gasteroenteritis, is closely related to *B. thuringinsis* and *B. anthracis* and shares common features such as the presence of a capsule and production of toxins. *G. mellonella* is also susceptible to *B. cereus* and has been used to identify its virulence factors (Fedhila et al. 2006). Another recent study illustrating the use of *G. mellonella* for functional analysis of virulence factors from human pathogenic bacteria focused on proteolytic enzymes produced by *E. faecalis*. Enterococci belong to the most common hospital-acquired pathogens causing a wide variety of diseases in humans. Among its virulence factors *E. faecalis* was shown to release metalloproteinases that destroy defense molecules both in *G. mellonella* hemolymph and in human serum (Park et al. 2007). *Burkholderia* is an important bacterial genus with a complex taxonomy encompassing nine species of both ecological and pathogenic importance, which are collectively termed the *Burkholderia cepacia* complex (Bcc). Very recently, *G. mellonella* has been established as an alternative infection model for Bcc. Genetically mutated Bcc were tested in comparison to a wild-type Bcc strain in order to show concomitant reduction of Bcc virulence and increased survival of *G. mellonella* larvae (Seed and Dennis 2008).

Despite the obvious similarities that have been reported between closely related insect and human pathogens, we should take into account that there are also differences. For example, *G. mellonella* has been exploited to study the effects of toxins that are released via the type III secretion system by *P. aeruginosa*, a versatile pathogen that is capable of causing disease in plants, nematodes, insects, mice, and humans (Miyata et al. 2003), whereas the entomopathogenic soil bacterium, *P. entomophila*, lacks a type III secretion system (Vodovar et al. 2006). But the latter is obviously present in other insect pathogens such as *Photorhabdus luminescens*, which secretes toxins similar to those of *Yersinia pestis* (Joyce, Watson, and Clarke 2006). However, the high level of correlation between effects caused by *P. aeruginosa* type III secretions in *G. mellonella* and in a mammalian tissue culture system validated the suitability of this model (Miyata et al. 2003). Mylonakis and co-workers used *G. mellonella* larvae as a suitable host system to study the efficacy of antibiotics against *Francisella tularensis*, the causative agent of tularemia, which has been considered a category A bioterrorism agent by the Centers for Disease Control and Prevention in the United States (Aperis et al. 2007). These studies illustrate the potential of *G. melonella* as a high-throughput whole-animal system to evaluate antibiotics before they are tested in mammalian models.

G. mellonella as a Model Host for Pathogenic Fungi

In addition to its value for testing bacterial pathogens, *G. mellonella* has attracted great attention as a mini-host model in the study of pathogenesis and virulence factors of medically important fungi (Chamilos et al. 2007). For example, it has been used to establish the role of gliotoxin in the virulence of the human pathogenic fungus *Aspergillus fumigatus* (Reeves et al. 2004). Its virulence against *G. mellonella* larvae depends upon the stage of conidial germination. Nongerminated conidia are phagocytosed by immune-competent larval hemocytes and appear avirulent; by contrast, germinated conidia are not engulfed and are capable of killing the host (Renwick et al. 2006). *G. mellonella* larvae have also been established as a powerful tool to study cryptococcal host-pathogen interactions (London, Orozco, and Mylonakis 2006). *Cryptocuccus neoformans* causes morbidity and mortality in immune-deficient patients such as transplant recipients and HIV-infected individuals. Several of its virulence-related genes previously determined to contribute to mammalian infection such as CAP59, GRA1, RAS1, and PKA1 have also been shown to play a role in its interaction with the *G. mellonella* immune system. For instance, the *MFα1* gene is induced during the proliferative stage of the infection, similar to the findings in mammals (London, Orozco, and Mylonakis 2006).

This lepidopteran mini-host has also been used to study pathogenicity of *C. albicans*. The virulence of this pathogenic yeast in mice correlates with that in *G. mellonella* (Brennan et al. 2002). Furthermore, preexposure to yeast protects *G. mellonella* larvae from a subsequent lethal infection by *C. albicans*, and this effect is mediated by increased expression of antimicrobial peptides (Bergin et al. 2006). In agreement, it was shown that prechallenge with yeast or bacterial cells mediates survival after subsequent challenge with a normally lethal dose of yeast or prolongs survival of larvae exposed to entomopathogenic fungi such as *Beauveria bassiana* and *Metarhizium anisopliae*. These protective effects of preexposure to bacterial and fungal elicitors of innate immune responses have also been attributed to antimicrobial peptides and inhibitors of microbial proteinases that are induced and secreted within the hemolymph (Vilcinskas and Matha 1997a; Vilcinskas and Wedde 1997). Taken together, these studies have shown/demonstrated that *G. mellonella* is particularly promising as a readily available mini-host for pathogenic fungi.

QUALITATIVE AND QUANTITATIVE ANALYSIS OF INNATE IMMUNE RESPONSES IN *G. MELLONELLA*

Analysis of Cellular Defense Mechanisms in *G. mellonella*

A number of immune-related effector molecules have been isolated, cloned, and characterized from *G. mellonella* during the past decade. Before highlighting their therapeutic potential in medicine and plant protection, methods that have been established to examine its cellular and humoral immune responses qualitatively and quantitatively will be addressed.

The cellular immune response in *G. mellonella*, as in other lepidopterans (see Chapter 14), encompasses phagocytosis and encapsulation of intruding microbes in the hemocoel. These defense mechanisms have been attributed to plasmatocytes and granular cells that represent the predominant cell types circulating within the hemolymph. Both immune-competent hemocyte types are capable of discriminating microbial and host surfaces and become adhesive upon contact with pathogen-associated molecular patterns, resulting in removal of hemocytes from the hemolymph stream. The relatively large size of *G. mellonella* facilitates collection of hemolymph samples, for example, by piercing an abdominal foot. The hemocyte density in isolated hemolymph samples can easily be calculated with hemocytometers, which are commonly used to quantify cells in mammalian blood samples. Fluctuations in the total hemocyte count in *G. mellonella* hemolymph samples have proved to be a valuable parameter to quantify cellular defense reactions during infection with both entomopathogenic and human pathogenic fungi (Vilcinskas, Matha, and Götz 1997a; Bergin, Brennan, and Kavanagh 2003).

If the number of bacterial or fungal cells that enter the hemocoel via injection or natural infection is too high to be engulfed by immune-competent hemocytes, then they are entrapped within multilayered sheets of plasmatocytes and granular cells. This multicellular encapsulation is a complex process that separates entrapped pathogens or parasites from host tissues and is accomplished by melanization of the aggregates. The formation of melanized microbial traps in *G. mellonella* is easy to quantify by dissecting larvae and counting the dark nodules attached to tissues. Phagocytosis or multicellular entrapping of microbes within the *G. mellonella* host can be determined by using fluorescently labeled cells (e.g., with fluorescein isothiocyanate, FITC) and fluorescence microscopy (Rohloff, Wiesner, and Götz 1994). Besides these approaches to quantify cellular defense reactions in vivo, *G. mellonella* also offers the opportunity to investigate interactions of insect and human pathogens with immune-competent hemocytes in vitro.

A simple method has been introduced to separate plasmatocytes responsible for phagocytosis of microbes from other hemocytes and to cultivate them in primary cell cultures (Wiesner and Götz 1993). The ability of plasmatocytes to engulf microbes is dependent on their ability to attach and spread on foreign surfaces. A number of combined assays have been elaborated to enable quantification of phagocytic activity, attachment, and spreading of isolated and cultured *G. mellonella* plasmatocytes in vitro. These assays allowed investigation of the impact of entomopathogenic fungi and their virulence factors on the cellular defense (Vilcinskas, Matha, and Götz 1997a,b; Griesch and Vilcinskas 1998). Combined in vivo and in vitro approaches demonstrated that entomopathogenic fungi that are capable of directly infecting the insect host through its cuticle produce so-called

G. MELLONELLA AS A SOURCE OF TRANSGENES TO RENDER DISEASE-RESISTANT PLANTS

Phytopathogens account for severe and increasing worldwide crop losses ($30–$50 billion annually) and, therefore, considerably threaten human nutrition. Similar to the dilemma of human pathogens that have become resistant to clinically applied antibiotics from their inappropriate use, one of the hottest problems in agriculture is the rapidly propagating resistance of phytopathogens against pesticides that once held them under control. The consequences reach from significantly increased agricultural production costs to growing public concerns about the adverse impact of agrochemicals on human health and the environment (Osusky et al. 2000). Consequently, there is a pressing demand to explore novel approaches in modern plant protection measures, and tuning up the crops' own defense mechanisms to render them disease resistant has emerged as a superior strategy in sustainable agriculture (Kogel and Langen 2005). This can be achieved, for example, by transferring a single gene encoding an antimicrobial peptide without any alteration of valuable traits of the plant genome. Since expression of antifungal genes derived from plants provides only limited protection against pathogens because of their acquired tolerance to plant antifungal peptides during coevolution, transgenic expression of insect-derived antimicrobial peptides in particular has emerged as a powerful tool to create disease-resistant crops (Jaynes et al. 1987; Vilcinskas and Gross 2005; Yevtushenko et al. 2005; Coca et al. 2006).

Not surprisingly, the first insect-derived genes used to engineer disease-resistant crops were those encoding attacin and cecropin. Attacin binds to bacterial lipopolysaccharide and inhibits synthesis of the bacterial outer-membrane proteins (Carlsson et al. 1998). Its transgenic expression was shown to confer resistance to fire blight in apple (Ko et al. 2000) and pear (Reynoird et al. 1999). Besides its potent antibacterial activity, cecropin was also found to exert antifungal activity in vitro (Ekengren and Hultmark 1999) and, therefore, has been used as a transgene to render fungal disease–resistant crops. First attempts failed because, as a consequence of posttranslational degradation, cecropin persists only a short time in crop plants, which was an impediment to its utilization (Mills et al. 1994). To compensate for this disadvantage, modified cecropin (Owens and Heutte 1997) and cecropin-mellitin hybrids with increased stability have been constructed (Huang et al. 1997; Yevtushenko et al. 2005) and used successfully to engineer crops exhibiting enhanced resistance against bacterial and fungal phytopathogens (Coca et al. 2006).

In comparison to the diversity of antibacterial peptides that have been identified in insects, the number of peptides that have been discovered to inhibit fungi exclusively is rather limited. Drosomycin from *D. melanogaster* (Fehlbaum et al. 1994) and heliomicin from *H. virescens* (Lamberty et al. 1999) share sequence similarity with plant defensins and target the same binding site on sensitive fungi (Thevissen et al. 2004). Since fungal pathogens plausibly acquired tolerance to plant defensins during coevolution with their plant hosts, it is not surprising that transgenic expression of drosomycin or heliomicin conferred only moderately enhanced resistance against fungal pathogens in tobacco due to their similarity with plant defensins (Banzet et al. 2002). In contrast, gallerimycin, a novel antifungal peptide from *G. mellonella* (Schuhmann et al. 2003), has recently been introduced as a powerful tool to transfer resistance against fungal pathogens from insects to plants. Gallerimycin has been proven to be particularly suitable as a transgene because (1) it belongs to the limited number of known peptides that exclusively inhibit mycelium-forming fungi, (2) it confers higher resistance to pathogenic fungi than other antifungal peptides from insect origin tested in tobacco, and (3) it has a signal sequence that is recognized by the plant and directs the synthesized peptide into intercellular spaces, where it is not subjected to cleavage by endogenous proteases (Langen et al. 2006). In addition, a recent phylogenetic analysis demonstrates that gallerimycin is more related to fungal than insect defensins such as drosomycin (Altincicek and Vilcinskas 2007).

In order to prevent adverse impacts that may accompany constitutive transgenic expression of antifungal peptides from insects in plants such as impaired growth and reduced yield, we designed a plant transformation vector containing the gallerimycin coding sequence under control of the

Agrobacterium tumefaciens mannopine synthase *mas* P2′ promoter. Because this promoter has been shown to be responsive to both wounding and fungal infection in tobacco, it provides the desired on-demand expression of insect transgenes, which rules out the possibility that antifungal peptides constitutively present in plants can promote selection for resistant strains among fungal pathogens (Langen et al. 2006). In order to explore further the potential of antimicrobial peptides to transfer disease resistance from insects to plants, we are currently elaborating technologies in our institute that enable combined pathogen-induced expression of several lepidopteran antimicrobial peptides in crops. On-demand expression of gallerimycin along with other antimicrobial peptides with distinct modes of action or with insect-derived proteinase inhibitors may prevent selection of phytopathogen resistance against transferred genes conferring protection. Development of transgenic expression of insect antimicrobial peptides has just started and may provide tools to keep ahead of the evolutionary adaptability of plant pathogens (Vilcinskas and Gross 2005).

REFERENCES

Altincicek, B., M. Linder, D. Linder, K. Preissner, and A. Vilcinskas. 2007. Microbial metalloproteinases mediate sensing of invading pathogens and activate innate immune responses in the lepidopteran model host *Galleria mellonella*. *Infect. Immun.* 75:175–83.

Altincicek, B., S. Stötzel, M. Wygrecka, K. Preissner, and A. Vilcinskas. 2008. Host-derived extracellular nucleic acids enhance innate immune responses, induce coagulation, and prolong survival upon infection in insects. *J. Immunol.* 181:2705–12.

Altincicek, B., and A. Vilcinskas. 2007. Identification of immune-related genes from an apterygote insect, the firebrat *Thermobia domestica*. *Insect Biochem. Mol. Biol.* 37:726–31.

Aperis, G., B. Fuchs, C. Anderson, J. Warner, S. Calderwood, and E. Mylonakis. 2007. *Galleria mellonella* as a model host to study infection by the *Francisella tularensis* live vaccine strain. *Microbes Infect.* 9:729–34.

Banzet, N., M. Latorse, P. Bulet, E. Francois, C. Derpierre, and M. Dubal. 2002. Expression of insect cystein-rich antifungal peptides in transgenic tobacco enhances resistance to a fungal disease. *Plant Sci.* 162:995–1006.

Bergin, D., M. Brennan, and K. Kavanagh. 2003. Fluctuations in haemocyte density and microbial load may be used as indicators of fungal pathogenicity in larvae of *Galleria mellonella*. *Microbes Infect.* 5:1389–95.

Bergin, D., L. Murphy, J. Keenan, M. Clynes, and K. Kavanagh. 2006. Pre-exposure to yeast protects larvae of *Galleria mellonella* from a subsequent lethal infection by *Candida albicans* and is mediated by the increased expression of antimicrobial peptides. *Microbes Infect.* 8:2105–12.

Bergin, D., E. Reeves, J. Renwick, F. Wientjes, and K. Kavanagh. 2005. Superoxide production in *Galleria mellonella* hemocytes: Identification of proteins homologous to the NADPH oxidase comples of human neutrophils. *Infect. Immun.* 73:4161–70.

Bettencourt, R., O. Terenius, and I. Faye. 2002. *Hemolin* silencing by ds-RNA injected into Cecropia pupae is lethal to next generation embryos. *Insect Mol. Biol.* 11:267–71.

Breithaupt, M. 1999. The new antibiotics. *Nat. Biotechnol.* 17:1165–69.

Brennan, M., D.Y. Thomas, M. Whiteway, and K. Kavanagh. 2002. Correlation between virulence of *Candida albicans* mutants in mice and in *Galleria mellonella* larvae. *FEMS Immunol. Med. Microbiol.* 34:153–57.

Brown, S., A. Howard, A. Kasprzak, K. Gordon, and P. East. 2008. The discovery and analysis of a diverged family of novel antifungal moricin-like peptides in the wax moth *Galleria mellonella*. *Insect Biochem. Mol. Biol.* 38:201–12.

Bulet, P., and R. Stocklin. 2005. Insect antimicrobial peptides: Structures, properties and gene regulation. *Protein Pept. Lett.* 12:3–11.

Carlsson, A., T. Nystrom, H. de Cock, and H. Bennich. 1998. Attacin—an insect immune protein—binds LPS and triggers the specific inhibition of bacterial outer membrane protein synthesis. *Microbiology* 144:2179–88.

Chamilos, G., M. Lionakis, R. Lewis, and D. Kontojiannis. 2007. Role of mini-host models in the study of medically important fungi. *Lancet Infect. Dis.* 7:42–55.

Cheng, T., P. Zhao, C. Liu, et al. 2006. Structures, regulatory regions, and inductive expression patterns of antimicrobial peptide genes in the silk worm *Bombyx mori*. *Genomics* 87:356–65.

Chung, M.-C., T. Popova, B. Millis, et al. 2006. Secreted neutral metalloproteases of *Bacillus anthracis* as candidate pathogenic factors. *J. Biol. Chem.* 281:31408–18.

Clermont, A., M. Wedde, V. Seitz, L. Podsiadlowski, M. Hummel, and A. Vilcinskas. 2004. Cloning and expression of an inhibitor against microbial metalloproteinases from insects (IMPI) contributing to innate immunity. *Biochem. J.* 382:315–22.

Coca, M., G. Penas, J. Gomez, et al. 2006. Enhanced resistance to the rice blast fungus *Magnaporthe grisea* conferred by expression of a cecropin A gene in transgenic rice. *Planta* 233:392–406.

Collins, M.S., and D. Papagianis. 1974. Inhibition by lysozyome on the growth of the sperule phase of *Coccidoides immitis in vitro*. *Infect. Immun.* 10:616–23.

Cytrynska, M., P. Mak, A. Zdybicka-Barabas, P. Suder, and T. Jakubowicz. 2006. Purification and characterization of eight peptides from *Galleria mellonella* immune hemolymph. *Peptides* 28:533–46.

Ekengren, S., and D. Hultmark. 1999. *Drosophila* cecropin as an antifungal agent. *Insect Biochem. Mol. Biol.* 29:965–72.

Fedhila, S., N. Daou, D. Lereclus, and C. Nielsen-LeRoux. 2006. Identification of *Bacillus cereus* internalin and other candidate virulence genes specifically induced during oral infection in insects. *Mol. Microbiol.* 62:339–55.

Fehlbaum, P., P. Bulet, L. Michaut, et al. 1994. Insect Immunity. Septic injury of *Drosophila* induces the synthesis of a potent antifungal peptide with sequence homology to plant antifungal peptides. *J. Biol. Chem.* 269:33159–63.

Fröbius, A., M. Kanost, P. Götz, and A. Vilcinskas. 2000. Isolation and characterization of novel inducible serine protease inhibitors from larval hemolymph of the greater wax moth, *Galleria mellonella*. *Eur. J. Biochem.* 267:2046–53.

Fuchs, B., and E. Mylonakis. 2006. Using non-mammalian hosts to study fungal virulence and host defense. *Curr. Opin. Microbiol.* 9:346–51.

Garcia-Lara, J., A. Needham, and S. Foster. 2005. Invertebrates as animal models for *Staphylococcus aureus* pathogenesis: A window into host-pathogen interaction. *FEMS Immunol. Med. Microbiol.* 43:311–23.

Gillespie, J., A. Bailey, B. Cobb, and A. Vilcinskas. 2000. Fungal elicitors of insect immune responses. *Arch. Insect Biochem. Physiol.* 44:49–68.

Girard, P.-A., Y. Boublik, C. Wheat, et al. 2008. X-tox: An atypical defensin derived family of immune-related proteins specific to Lepidoptera. *Dev. Comp. Immunol.* 32:575–84.

Griesch, J., and A. Vilcinskas. 1998. Proteases released by entomopathogenic fungi impair phagocytic activity, attachment and spreading of plasmatocytes isolated from hemolymph of the greater wax moth *Galleria mellonella*. *Biocontrol Sci. Technol.* 8:517–31.

Hara, S., A. Asaoka, and M. Yamakava. 1996. Effect of moricin, a novel antibacterial peptide of *Bombyx mori* (Lepidoptera, Bombycidae) on the growth of methicillin-resistant *Staphylococcus aureus*. *Appl. Entomol. Zool.* 31:465–66.

Hara, S., and M. A. Yamakawa. 1995. Moricin, a novel type of antibacterial peptide isolated from the silkworm, *Bombyx mori*. *J. Biol. Chem.* 270:29923–27.

Huang, Y., R. Nordeen, M. Di, L. Owens, and J. McBeath. 1997. Expression of engineered cecropin gene cassette in transgenic tobacco plants confers resistance to *Pseudomonas syringae pv. tabaci*. *Phytopathology* 87:494–99.

Jander, G., L. Rahme, F. Ausubel, and E. Drenkard. 2000. Positive correlation between virulence of *Pseudomonas aeroginosa* mutants in mice and insects. *J. Bacteriol.* 182:3843–45.

Jaynes, J., K. Xanthopoulos, L. Destefano-Beltran, and J. Dodds. 1987. Increasing bacterial disease resistance in plants utilizing antibacterial genes from insects. *BioEssays* 6:263–70.

Jolles, J., F. Schoentgen, G. Croizier, L. Croizier, and P. Jolles. 1979. Insect lysozymes from three species of Lepidoptera: Their structural relatedness to the C (chicken) type lysozyme. *J. Mol. Evol.* 14:267–71.

Joyce, S., R. Watson, and D. Clarke. 2006. The regulation of pathogenicity and mutualism in *Photorhabdus*. *Curr. Opin. Microbiol.* 9:127–32.

Kanost, M., H. Jiang, and X.-Q. Yu. 2004. Innate immune response of a lepidopteran insect, *Manduca sexta*. *Immunol. Rev.* 198:97–105.

Kamaya, T. 1970. Lytic action of lysozyme on *Candida albicans*. *Mycopathol. Mycologia Appl.* 37:320–30.

Kavanagh, K., and E.P. Reeves. 2004. Exploiting the potential of insects for the *in vivo* pathogenicity testing of microbial pathogens. *FEMS Microbiol.* 28:101–12.

Kim, C.H., J.H. Lee, I. Kim, et al. 2004. Purification and cDNA cloning of a cecropin-like peptide from the greater wax moth *Galleria mellonella*. *Mol. Cells* 17:262–66.

Ko, K., J.L. Norelli, J.P. Reynoird, W.W. Boresjza, S.K. Brown, and H.S. Aldwinckle. 2000. Effect of untranslated leader sequence of AMV RNA 4 and signal peptide of pathogenesis related protein 1b on attacin gene expression and resistance to fire blight in transgenic apple. *Biotechnol. Lett.* 22:373–81.

Kogel, K.H., and G. Langen. 2005. Induced disease resistance and gene expression in cereals. *Cell. Microbiol.* 7:1555–64.

Lamberty, M., S. Ades, J.S. Uttenweiler, et al. 1999. Insect immunity. Isolation from the lepidopteran *Heliothis virescens* of a novel insect defensin with potent antifungal activity. *J. Biol. Chem.* 274:9320–26.

Langen, G., J. Imani, B. Altincicek, G. Kieseritzky, K.-H. Kogel, and A. Vilcinskas. 2006. Transgenic expression of gallerimycin, a novel antifungal insect defensin from the greater wax moth *Galleria mellonella*, confers resistance against pathogenic fungi in tobacco. *Biol. Chem.* 387:549–57.

Lee, Y.S., E.K. Yun, W.S. Jang, et al. 2004. Purification, cDNA cloning and expression of an insect defensin from the great wax moth, *Galleria mellonella. Insect Mol. Biol.* 13:65–72.

Lemaitre, B., and J. Hoffmann. 2007. The host defence of *Drosophila melanogaster. Annu. Rev. Immunol.* 25:697–743.

London, R., B. Orozco, and E. Mylonakis. 2006. The pursuit of cryptococcal pathogenesis: Heterologous hosts and the study of cryptococcal host-pathogen interactions. *FEMS Yeast Res.* 6:567–73.

Maeda, H. 1996. Role of microbial proteases in pathogenesis. *Microbiol. Immunol.* 40:685–99.

Mills, D., F.A. Hammerschlag, R.O. Nordeen, and L.D. Owens. 1994. Evidence for the breakdown of cecropin B by proteinases in the intercellular fluid of peach leaves. *Plant Sci.* 104:17–22.

Miyata, S., M. Casey, D. Frank, F. Ausubel, and E. Drenkard. 2003. Use of the *Galleria mellonella* caterpillar as a model host to study the role of type III secretion system in Pseudomonas aeroginosa pathogenesis. *Infect. Immun.* 71:2404–13.

Mohrig, W., and B. Messner. 1968. Immunreaktionen bei Insekten. I. Lysozyme als grundlegender antibakterieller Faktor im humoralen Abwehrgeschehen. *Biol. Zentralbl.* 87:439–47.

Mylonakis, E., and A. Aballay. 2005. Worms and flies as genetically tractable animal models to study host-pathogen interactions. *Infect. Immun.* 73:3833–41.

Mylonakis, E., A. Casadevall, and F. Ausubel. 2007. Exploiting amoeboid and non-vertebrate animal model systems to study the virulence of human pathogen fungi. *PLoS Pathogens* 3:0859–0865.

Mylonakis, E., R. Moreno, J. El Khoury, et al. 2005. *Galleria mellonella* as a model system to study *Cryptococcus neoformans* pathogenesis. *Infect. Immun.* 73:3842–50.

Okada, M., and S. Natori. 1983. Purification and characterization of an antibacterial protein from haemolymph of *Sarcophaga peregrina* (flesh-fly) larvae. *Biochem. J.* 211:727–34.

Osusky, M., G. Zhou, L. Osuska, R.E. Hancock, W.W. Kay, and S. Misra. 2000. Transgenic plants expressing cationic peptide chimeras exhibit broad-spectrum resistance to phytopathogens. *Nat. Biotechnol.* 18:1162–66.

Owens, L.D., and T.M. Heutte. 1997. A single amino acid substitution in the antimicrobial defense protein cecropin B is associated with diminished degradation by leaf intercellular fluid. *Mol. Plant Microbe Interact.* 10:525–28.

Park, S., K.M. Kim, J.H. Lee, S.J. Seo, and I.H. Lee. 2007. Extracellular gelatinase of *Enterococcus faecalis* destroys a defense system in insect hemolymph and human serum. *Infect. Immun.* 75:1861–69.

Powning, R.F., and W.J. Davidson. 1976. Studies on the insect bacteriolytic enzymes-II. Some physical and enzymatic properties of lysozme from haemolymph of *Galleria mellonella. Comp. Biochem. Physiol.* 55:221–28.

Reeves, E.P., C.G. Messina, S. Doyle, and K. Kavanagh. 2004. Correlation between gliotoxin production and virulence of *Aspergillus fumigatus* in *Galleria mellonella. Mycopathologia* 158:73–79.

Renwick, J., P. Daly, E. Reeves, and K. Kavanagh. 2006. Susceptibility of larvae of *Galleria mellonella* to infection by *Aspergillus fumigatus* is dependent upon stage of conidial germination. *Mycopathologia* 161:377–84.

Renwick, J., E. Reeves, F. Wientjes, and K. Kavanagh. 2007. Translocation of proteins homologous to human neutrophil p47phox and p67phox to cell membrane in activated hemocytes of *Galleria mellonella. Dev. Comp. Immunol.* 31:347–59.

Reynoird, J., F. Mourgues, J. Norelli, H.S. Aldwinckle, M. Brisset, and E. Chevreau. 1999. First evidence for improved resistance to fire blight in transgenic pear expressing the *attacin E* gene from *Hyalophora cecropia. Plant Sci.* 149:23–31.

Rohloff, L., A. Wiesner, and P. Götz. 1994. A fluorescence assay demonstrating stimulation of phagocytosis by haemolymph molecules of *Galleria mellonella*. *J. Insect Physiol.* 40:1045–49.

Saido-Sakanaka, H., J. Ishibashi, E. Momotani, and M. Yamakawa. 2005. Protective effects of synthetic antibacterial oligopeptides based on the insect defensins on Methicillin-resistant *Staphylococcus aureus* in mice. *Dev. Comp. Immunol.* 29:469–77.

Schuhmann, B., V. Seitz, A. Vilcinskas, and L. Podsiadlowski. 2003. Cloning and expression of gallerimycin, an antifungal peptide expressed in immune response by the greater wax moth, *Galleria mellonella*. *Arch. Insect Biochem. Physiol.* 53:125–33.

Scully, L., and M. Bidochka. 2006. Developing insects as models for current and emerging human pathogens. *FEMS Microbiol. Lett.* 263:1–9.

Seed, K., and J. Dennis. 2008. Development of *Galleria mellonella* as an alternative infection model for the *Burkholderia cepacia* complex. *Infect. Immun.* 76:1267–75.

Seitz, V., A. Clermont, M. Wedde, et al. 2003. Identification of immunorelevant genes from greater wax moth (*Galleria mellonella*) by a subtractive hybridization approach. *Dev. Comp. Immunol.* 27:207–15.

Sifiri, C.D., J. Begun, F.M. Ausubel, and S. Calderwood. 2003. *Caenorhabditis elegans* as a model host for *Staphylococcus aureus* pathogenesis. *Infect. Immun.* 71:2208–17.

Silva, C.P., N. Waterfield, P. Daborn, et al. 2002. Bacterial infection of a model insect: *Photorhabdus luminescens* and *Manduca sexta*. *Cell. Microbiol.* 4:329–39.

Steiner, H., D. Hultmark, A. Engstom, H. Bennich, and H.G. Boman. 1981. Sequence and specificity of two antibacterial proteins involved in insect immunity. *Nature* 292:246–48.

Terenius, O., R. Bettencourt, S.Y. Lee, W. Li, K. Söderhäll, and I. Faye. 2007. RNA interference of hemolin causes depletion of phenoloxidase activity in *Hyalophora cecropia*. *Devel. Comp. Immunol.* 31:571–75.

Thevissen, K., D.C. Warnecke, I.E. Francois, et al. 2004. Defensins from insects and plants interact with fungal glucosylceramides. *J. Biol. Chem.* 279:3900–05.

Travis, J., and J. Potempa. 2000. Bacterial proteinases as targets for development of second generation antibiotics. *Biochim. Biophys. Acta* 1477:35–50.

Vilcinskas, A., and P. Götz. 1999. Parasitic fungi and their interactions with the insect immune system. *Adv. Parasitol.* 43:267–13.

Vilcinskas, A., and J. Gross. 2005. Drugs from bugs: The use of insects as a valuable source of transgenes with potential in modern plant protection strategies. *J. Pest Sci.* 78:187–91.

Vilcinskas, A., and V. Matha. 1997a. Effect of the entomopathogenic fungus *Beauveria bassiana* on humoral immune response of *Galleria mellonella* larvae (Lepidoptera: Pyralidae). *Eur. J. Entomol.* 94:461–72.

Vilcinskas, A., and V. Matha. 1997b. Antimycotic activity of lysozyme and its contribution to antifungal humoral defence reactions in *Galleria mellonella*. *Animal Biol.* 6:13–23.

Vilcinskas, A., V. Matha, and P. Götz. 1997a. Effects of the entomopathogenic fungus *Metarhizium anisopliae* and its secondary metabolites on morphology and cytoskeleton of plasmatocytes isolated from *Galleria mellonella*. *J. Insect Physiol.* 43:1149–59.

Vilcinskas, A., V. Matha, and P. Götz. 1997b. Inhibition of phagocytic activity of plasmatocytes isolated from *Galleria mellonella* by entomogenous fungi and their secondary metabolites. *J. Insect Physiol.* 43:475–83.

Vilcinskas, A., and M. Wedde. 1997. Inhibition of *Beauveria bassiana* proteases and fungal development by inducible protease inhibitors in the haemolymph of *Galleria mellonella* larvae. *Biocontrol Sci. Technol.* 7:591–601.

Vodovar, N., D. Vallenet, S. Cruveiller, et al. 2006. Complete genome sequence of the entomopathogenic and metabolically versatile soil bacterium *Pseudomonas entomophila*. *Nat. Biotechnol.* 24:673–79.

Wedde, M., C. Weise, P. Kopacek, P. Franke, and A. Vilcinskas. 1998. Purification and characterization of an inducible metalloprotease inhibitor from the hemolymph of greater wax moth larvae, *Galleria mellonella*. *Eur. J. Biochem.* 255:534–43.

Wedde, M., C. Weise, C. Nuck, B. Altincicek, and A. Vilcinskas. 2007. The insect metalloproteinase inhibitor gene of the lepidopteran *Galleria mellonella* encodes two distinct inhibitors. *Biol. Chem.* 388:119–27.

Wiesner, A., and P. Götz. 1993. Silica beads induce cellular and humoral immune responses in *Galleria mellonella* larvae and in isolated plasmatocytes, obtained by a newly adapted nylon wool separation. *J. Insect Physiol.* 39:865–76.

Yevtushenko, D.P., R. Romero, B.S. Forward, R.E. Hancock, W.W. Kay, and S. Misra. 2005. Pathogen-induced expression of a cecropin A-melittin antimicrobial peptide gene confers antifungal resistance in transgenic tobacco. *J. Exp. Bot.* 56:1685–95.

Yu, K.H., K.N. Kim, J.H. Lee, et al. 2002. Comparative study on characteristics of lysozymes from the hemoymph of three lepidopteran larvae, *Galleria mellonella, Bombyx mori, Agrius convolvuli. Dev. Comp. Immunol.* 26:707–13.

Zasloff, M. 2002. Antimicrobial peptides of multicellular organisms. *Nature* 415:389–95.

Zhu, Y., T. Johnson, A. Meyers, and M. Kanost. 2003. Identification by subtractive suppression hybridization of bacteria-induced genes expressed in *Manduca sexta* fat body. *Insect Biochem. Mol. Biol.* 33:541–59.

16 Intrahemocoelic Toxins for Lepidopteran Pest Management

Nina Richtman Schmidt and Bryony C. Bonning

CONTENTS

INTRODUCTION

Lepidopteran pests are among the most economically important insect pests severely impacting agricultural productivity and contributing to an estimated 25 percent loss of global annual crop production (Oerke 1994). The most common approaches for management of lepidopteran pests are application of classical chemical insecticides or the use of preprotected transgenic crops that express *Bacillus thuringiensis* (Bt)-derived toxins that target the midgut epithelial cells (Huang et al. 2002; Pray et al. 2002; Toenniessen, O'Toole, and DeVries 2003; Lawrence 2005). Although Bt transgenic crops are increasingly being adopted by growers, there is continued concern over the potential for development of resistance to Bt toxins, and the potential nontarget and environmental impacts (Bates et al. 2005; Heckel et al. 2007; Rosi-Marshall et al. 2007; Sisterson et al. 2007). Alternative lepidopteran-active toxin transgenes would be beneficial for a number of reasons including combining toxins (so-called stacking or pyramiding) to delay the development of resistance (Cao et al. 2002), or use of new transgenes once resistance to Bt toxins has developed for maintenance of the genetically modified (GM) crop industry. The marketing of competing cultivars that express alternative toxins could also reduce the cost of seed to growers.

Whereas the focus for transgenic technologies has been on gut-active effectors such as Bt and protease inhibitors, toxins that act within the hemocoel rather than in the gut provide an untapped resource for management of lepidopteran pests. Research conducted over the last two decades on the genetic optimization of baculovirus insecticides for increased speed of kill has effectively screened numerous intrahemocoelic effectors for toxicity when delivered to lepidopteran larvae by a virus (Kamita et al. 2005). The use of an entomopathogenic fungus delivery system for intrahemocoelic toxins has also been demonstrated (Wang and Leger 2007). Based on their target site, intrahemocoelic toxins are not expected to be orally active against Lepidoptera. Nevertheless, there are exceptions, the most notable being the orally active atracotoxins derived from spider venom (King 2007). For toxins that require entry into the hemocoel for toxic action, lectins that enter the insect hemocoel from the gut by an unknown mechanism can be used for delivery of intrahemocoelic toxins to their target sites (Fitches et al. 2002, 2004; Trung, Fitches, and Gatehouse 2006). Here, we provide an overview of the delivery of intrahemocoelic toxins to lepidopteran pests by recombinant baculovirus insecticides, by entomopathogenic fungi, and from transgenic plants (Figure 16.1).

BACULOVIRUS DELIVERY OF INTRAHEMOCOELIC TOXINS

Baculoviruses are insect-specific viruses that primarily infect species within the order Lepidoptera (Adams and McClintock 1991). The two genera, nucleopolyhedrovirus (NPV) and granulovirus (GV), differ in occlusion body morphology with single (GV) and multiple (NPV) virions occluded in granules or polyhedra respectively. Baculoviruses are considered to be safe and selective insecticides and have been used for management of a variety of agricultural and forestry pests (Federici 1999; Moscardi 1999). More extensive use of baculovirus insecticides for pest management has been restricted by the slow speed of kill of the targeted pest relative to the speed of kill resulting from application of classical chemical insecticides, and by the narrow host range of the virus,

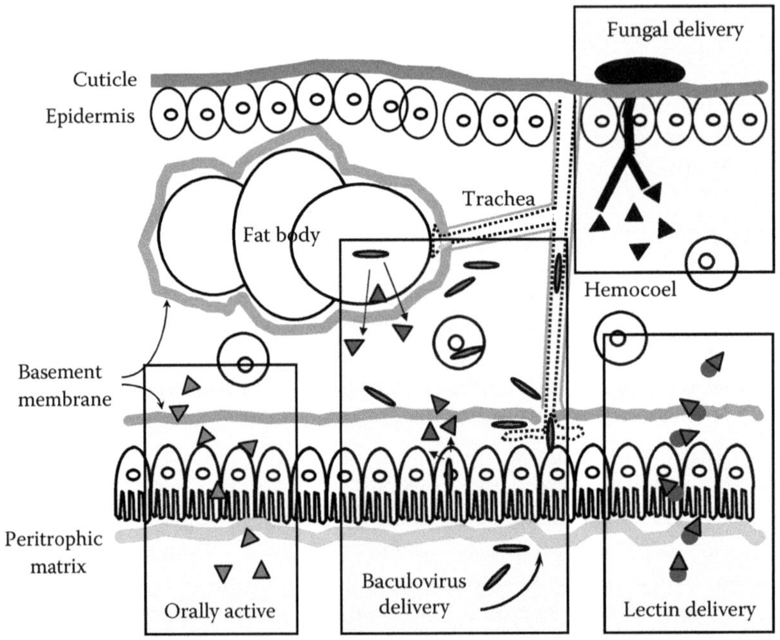

FIGURE 16.1 ***A color version of this figure follows page 176.*** Delivery of intrahemocoelic toxins for management of lepidopteran pests. In a few cases such as the spider venom–derived atracotoxin Hvla (Khan et al. 2006), intrahemocoelic toxins (depicted as triangles) are orally active and can be delivered directly via transgenic plants. For intrahemocoelic toxins that are not orally active, lectins such as the snowdrop lectin, *Galanthus nivalis* agglutin (GNA, depicted by circles), can be used for delivery of toxins into the hemocoel (Fitches et al. 2004; Trung, Fitches, and Gatehouse 2006). Alternatively, insect pathogens can be used as vectors for expression of the toxin and secretion into the hemocoel of the infected insect (Harrison and Bonning 2000b). Baculoviruses (depicted as rods) infect the midgut epithelium and other tissues such as the fat body, and virus-expressed toxin secreted into the hemocoel results in death of the host insect (Kamita et al. 2005). Entomopathogenic fungi (depicted at top right) such as *Metarhizium anisopliae* can also be engineered for release of intrahemocoelic toxins into the hemocoel (Wang and Leger 2007).

which may not include all pest species within a given cropping system. For the past two decades, baculovirus insecticides have been genetically optimized to enhance the speed of kill (Kamita et al. 2005; Inceoglu, Kamita, and Hammock 2006). It can take from several days to several weeks for a wild-type baculovirus to kill the host insect, according to the virus-host combination. The primary approach has been to incorporate a gene into the baculovirus genome that encodes a toxin or other physiological effecter. On replication of the virus in the host insect, the toxin is produced and disrupts insect well-being sufficiently to reduce feeding damage caused by the infected insect or to result in death of the infected insect (Figure 16.2). Hence the baculovirus serves as a toxin delivery system. Some thirty intrahemocoelic toxins have been screened for relative insecticidal efficacy in this way (Figure 16.3; Kamita et al. 2005), and some of these recombinant baculovirus insecticides are competitive with classical chemical insecticides in maintaining pest insect populations below the economic threshold level.

Hormones and Enzymes

Early studies focused on baculovirus expression of insect hormones or enzymes to disrupt insect physiology by overexpressing these agents or expressing them at inappropriate times during development. Examples include baculovirus expression of diuretic hormone for disruption of water balance (Maeda 1989), eclosion hormone for premature molting and eclosion (Eldridge et al. 1991), prothoracicotrophic hormone (PTTH) for potential overproduction of ecdysteroids (O'Reilly et al. 1995), and juvenile hormone esterase for premature reduction of juvenile hormone titers and potential premature metamorphosis (Hammock et al. 1990; Bonning et al. 1997). These transgenes had relatively low impact on reducing survival time of the host insect, in some cases because of efficient regulatory systems for maintaining appropriate levels of the hormone or enzyme, and in other cases because of the impact of transgene expression on the virus itself. For example, baculovirus expression of PTTH inhibited the pathogenicity of the baculovirus (O'Reilly et al. 1995).

Pheromone biosynthesis activating neuropeptide (PBAN) is secreted from the subesophageal ganglion and is required for pheromone synthesis and release in some lepidopterans (pheromone biosynthesis is described in Chapter 10). A gene that encodes multiple neuropeptides including PBAN, derived from the corn earworm, *Helicoverpa zea*, when expressed by a recombinant baculovirus, resulted in a 26 percent reduction in the survival time of neonates compared with larvae infected with the wild-type virus (Ma et al. 1998). Which neuropeptide of those expressed by the recombinant baculovirus resulted in the insecticidal action is unclear.

In contrast to baculovirus expression of the hormones and enzymes listed above that target hormonal regulation of insect development, one of the fastest recombinant baculovirus insecticides is one that expresses a cathepsin L-like protease that targets the basement membrane (Harrison and Bonning 2001). This protease, derived from the flesh fly, *Sarcophaga peregrina*, selectively degrades two substrate proteins in the basement membrane (Homma et al. 1994; Homma and Natori 1996; Fujii-Taira et al. 2000). The recombinant virus was constructed to address whether disruption of the basement membrane barrier to virus movement within the host insect would facilitate virus dissemination, thereby enhancing the speed of kill by the virus (Harrison and Bonning 2001). Rather than facilitating virus dissemination, however, baculovirus expression of the protease resulted in sufficient damage to the basement membrane and underlying tissues to cause death of the host insect (Li et al. 2007; Philip et al. 2007; Tang et al. 2007).

Neurotoxins

Insect-selective neurotoxins have been isolated from a variety of venomous creatures including parasitic wasps, spiders, scorpions, sea anemones, and mites (De Lima et al. 2007; Gurevitz et al. 2007; King 2007). Such venoms provide a tremendous resource for isolation of insect-specific

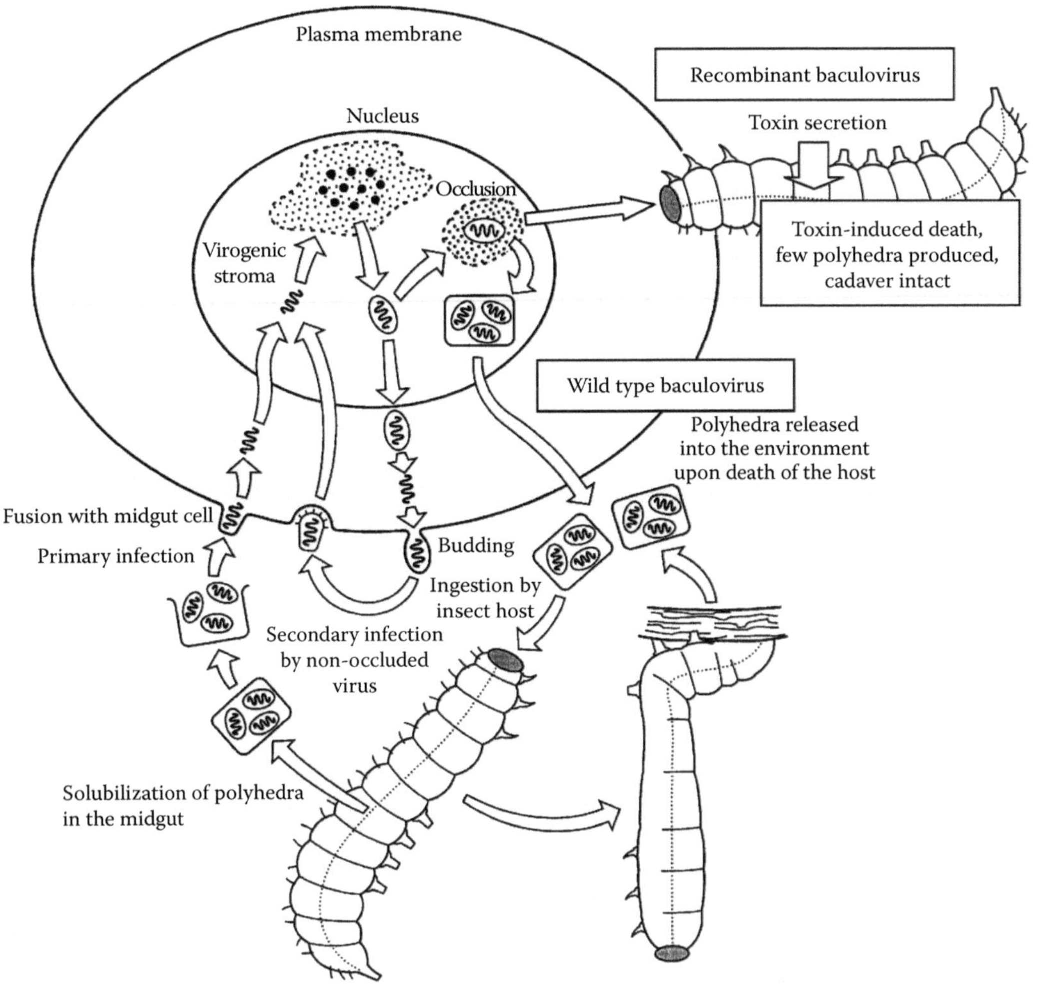

FIGURE 16.2 Life cycle of nucleopolyhedrovirus. Polyhedra ingested by a lepidopteran larva dissociate in the alkaline environment of the midgut, releasing the infectious occlusion-derived virus (ODV). ODV infect midgut cells and release rod-shaped nucleocapsids (NC) that move into the nucleus. Virus replication ensues within the virogenic stroma in the nucleus. Initially, budded viruses (BV) are produced by the infected cell. BVs disseminate the virus within the host insect and initiate infection of other cells and tissues. Subsequently, newly synthesized nucleocapsids are retained within the nucleus and become embedded in a polyhedrin matrix to produce polyhedra. For wild-type virus infections, the host insect becomes overwhelmed by the virus infection and climbs prior to death, hanging in a characteristic position by the prolegs. Subsequent disturbance of the fragile cuticle by the elements releases up to 10^{10} polyhedra per cadaver. In contrast, for recombinant baculoviruses that express intrahemocoelic toxins, the larva dies prematurely as a result of toxin action rather than from the virus infection itself. For viruses expressing paralytic neurotoxins, the larva may fall from the plant. Significantly fewer polyhedra are present in the cadaver compared with wild-type virus infections, and the cadaver retains its integrity.

toxins, with as many as a thousand different peptide toxins in spider venom glands, for example (Escoubas 2006; Escoubas, Sollod, and King 2006). These toxins typically cause rapid paralysis by acting on the major ion channels such as the Na^+, K^+, Ca^{2+}, and Cl^- channels. Baculovirus expression of such neurotoxins resulted in significant reductions in feeding damage caused by infected larvae relative to that caused by larvae infected with wild-type virus (Figure 16.3). The paralytic neurotoxins have the added bonus that larvae may fall from the plant, thereby decreasing the likelihood of

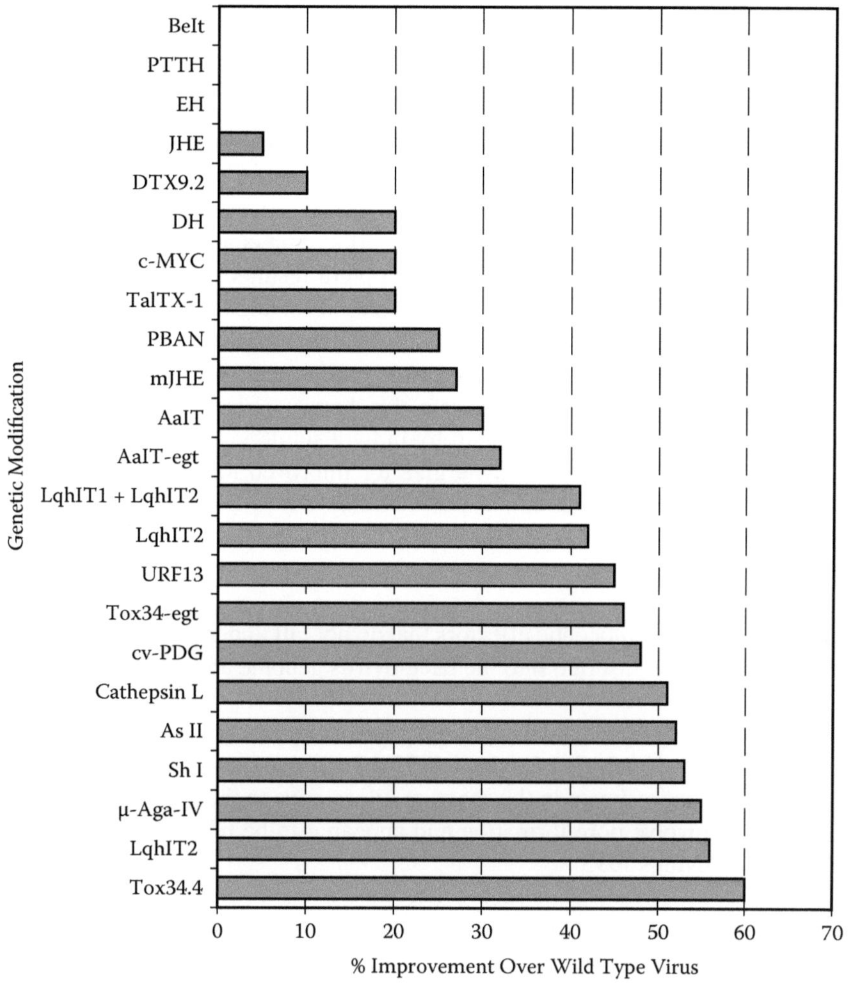

FIGURE 16.3 Improvement in insecticidal efficacy of recombinant baculoviruses expressing intrahemocoelic toxins relative to wild-type virus. The improvement in speed of kill relative to that of wild-type virus is shown for insertion of genes encoding intrahemocoelic effectors into baculovirus genomes. Direct comparison between the different recombinant viruses is not appropriate because of the different parent viruses and strategies used in construction. Abbreviations and references: Tox34.4, mite toxin (Carbonell et al. 1988; Burden et al. 2000); LqhIT2, scorpion-derived toxin under the iel promoter (Harrison and Bonning 2000a); μ-Aga-IV, spider toxin (Prikhod'ko et al. 1996); Sh I, sea anemone toxin (Prikhod'ko et al. 1996); As II, sea anemone toxin (Prikhod'ko et al. 1996); cathepsin L, basement membrane–degrading cathepsin L (Harrison and Bonning 2001); cv-PDG, glycosylase (Petrik et al. 2003); Tox34-egt, expression of Tox34 by the early DA26 promoter at the *egt* locus (Popham, Li, and Miller 1997); URF13, maize pore-forming protein (Korth and Levings 1993); LqhIT2, scorpion *Leiurus quinquestriatus* insect toxin 2 (Froy et al. 2000); LqhIT1 + LqhIT2, coexpression of LqhIT1 and LqhIT2 toxins derived from the scorpion *L. quinquestriatus* (Regev et al. 2003); AaIT-egt, insertion of AaIT at the *egt* locus (Chen et al. 2000); AaIT, AaIT toxin derived from the scorpion *Androctonus australis* (McCutchen et al. 1991; Stewart et al. 1991); mJHE, modified JHE (Bonning et al. 1997); PBAN, pheromone biosynthesis activating neuropeptide (Ma et al. 1998); TalTX-1, spider toxin (Hughes et al. 1997); c-MYC, expression of antisense transcription factor gene, *c-myc* (Lee et al. 1997); DH, diuretic hormone (Maeda 1989); DTX9.2, spider toxin (Hughes et al. 1997); JHE, juvenile hormone esterase (Hammock et al. 1990); EH, eclosion hormone (Eldridge, O'Reilly, and Miller 1992); PTTH, prothoracicotropic hormone (O'Reilly et al. 1995); Belt, mite insectotoxin-1 (Carbonell et al. 1988). After Kamita et al. (2005).

further damage to the plant and reduced foliar contamination (Cory et al. 1994; Hoover et al. 1995). The excitatory neurotoxin AaIT, derived from the buthid North African scorpion, *Androctonus australis* (Hector), and the excitatory and depressant toxins from the yellow Israeli scorpion *Leiurus* spp. have been particularly well studied (Zlotkin et al. 1991; Zlotkin, Fishman, and Elazar 2000). Recombinant baculoviruses expressing these toxins have been field tested in the United States, in China, and in the United Kingdom (Maeda et al. 1991; McCutchen et al. 1991; Stewart et al. 1991; Zlotkin et al. 1995; Black et al. 1997; Chen et al. 2000; Harrison and Bonning 2000a; Treacy, Rensner, and All 2000; Zlotkin, Fishman, and Elazar 2000; Sun et al. 2002; Sun et al. 2004).

Baculovirus expression of AaIT within the Lepidoptera resulted in greater-than-expected efficacy based on assays with injected toxin. The explanation for this is that baculovirus replication in the tracheal epithelia allows the toxin to bypass the neural lamella that protects lepidopteran neurons. Baculovirus delivery provides direct access to the central nervous system. In contrast, the injected neurotoxin was unable to penetrate this barrier (Elazar, Levi, and Zlotkin 2001). Baculovirus expression of more than one neurotoxin by a single virus has also been investigated, and in some cases, toxins were found to synergize for improved toxicity (Herrmann et al. 1995; Prikhod'ko et al. 1998; Regev et al. 2003).

Other Gene Products

Several other genes for intrahemocoelic effectors that do not fall under the toxin, insect hormone, or enzyme categories have been inserted into the baculovirus genome and resulted in increased speed of kill.

URF13 is a mitochondrially encoded protein derived from maize that forms pores in the inner mitochondrial membrane (Korth and Levings 1993). Baculoviruses expressing URF13 caused a significant (45 percent) improvement in the speed of kill relative to the wild-type virus. The mechanism of action did not involve pore formation and appeared to be related to disruption of cellular functions (Korth and Levings 1993).

A recombinant baculovirus was constructed to express antisense *c-myc* (Lee et al. 1997). c-MYC is a transcription factor involved in a variety of physiological processes including cell growth, proliferation, and apoptosis. Expression of antisense *c-myc* resulted in a reduction in survival time of infected larvae and in feeding damage caused by infected larvae. As pointed out by the authors, use of the antisense technology avoids problems associated with appropriate posttranslational processing or secretion of recombinant proteins, and also decreases the likelihood of resistance because no foreign protein is produced.

In an attempt to construct baculoviruses resistant to inactivation by UV light, Petrik et al. (2003) produced a recombinant baculovirus that expressed an algal virus pyrimidine dimer-specific glycolase, cv-PDG, which is involved in repair of DNA damage. The budded virus, but not polyhedra, was indeed more resistant to UV inactivation. Unexpectedly, the survival time of neonate *Spodoptera frugiperda* was reduced up to 48 percent relative to neonates infected with wild-type virus (Petrik et al. 2003).

FUNGAL DELIVERY OF INTRAHEMOCOELIC TOXINS

Pr1 is a protease produced by the entomopathogenic fungus *Metarhizium anisopliae*, which facilitates penetration of the host cuticle by the fungus. *M. anisopliae* was transformed to overexpress the cuticle-degrading protease (Leger 1995). Overexpression of Pr1 in the hemolymph caused activation of the prophenoloxidase system of the insect immune system, in addition to increasing the speed of kill of the fungus relative to the wild type and decreasing feeding damage by the tobacco hornworm, *Manduca sexta* (Leger et al. 1996; Hu and St. Leger 2002). In contrast, overexpression of an endochitinase to facilitate cuticle penetration did not enhance the virulence of the fungus relative to the wild type, suggesting that chitinase activity was not a limiting factor in cuticle penetration (Screen, Hu, and St. Leger 2001).

The potential use of *M. anisopliae* for delivery of intrahemocoelic toxins such as those expressed by recombinant baculoviruses has been demonstrated (Wang and Leger 2007). Fungal expression of the insect-selective neurotoxin AaIT using the MCL1 promoter, which provides rapid and high-level expression in the hemolymph, increased infectivity 22-fold against fifth instar *M. sexta*. In addition to reducing the number of conidia required to kill *M. sexta*, AaIT expression also reduced the survival time of larvae by 28 percent compared with those infected with wild-type *M. anisopliae*. Contractive paralysis typical of AaIT neurotoxicity was also observed. Genetically enhanced *M. anisopliae* will provide a particularly useful tool for management of insects that are not susceptible to Bt toxins (Wang and Leger 2007).

DELIVERY OF INTRAHEMOCOELIC TOXINS FROM TRANSGENIC PLANTS

On the basis that intrahemocoelic toxins by definition have target sites within the insect, rather than in the insect gut, such toxins are not expected to be orally active. Possibly as a consequence of this assumption, there are relatively few reports on testing of the oral toxicity of intrahemocoelic toxins against Lepidoptera. Exceptions include the scorpion-derived neurotoxins SFI1 and AaIT, which were not orally active (Zlotkin, Fishman, and Shapiro 1992; Fitches et al. 2004), and the spider-derived atracotoxin ω-ACTX-Hv1a, which was orally active (Khan et al. 2006). For the neurotoxins, the absence of oral bioassay data may also result from the amount of material required for such bioassays being prohibitive for purification of sufficient toxin from venom, and the difficulties associated with production of some recombinant toxins in an active form (Taniai, Inceoglu, and Hammock 2002).

Toxins such as ω-ACTX-Hv1a that are orally active must be sufficiently stable to withstand the harsh environment of the insect midgut. These orally active toxins must have some resistance to degradation by digestive proteases, and cross the midgut epithelium in sufficient quantities to exert their toxic effect within the insect hemocoel. The mechanism for movement of such toxins into the lepidopteran hemocoel is unknown.

Although it has been shown that very small amounts (0.3 percent) of the scorpion venom–derived neurotoxin AaIT administered orally can enter the hemocoel of the blowfly *Sarcophaga falculata*, AaIT was not orally active against *M. sexta* or *Helicoverpa armigera* (Zlotkin, Fishman, and Shapiro 1992). Using radioiodinated toxin, Zlotkin et al. showed that degradation products but no intact toxin were present in the hemolymph of the lepidopteran larvae (Zlotkin, Fishman, and Shapiro 1992). In apparent contradiction to these results, there have been several reports documenting the successful production of insect-resistant transgenic plants expressing AaIT: in poplar against the gypsy moth, *Lymantria dispar* (Wu et al. 2000), and in tobacco and cotton against the cotton bollworm, *Helicoverpa armigera* (Yao et al. 1996; Wu et al. 2008). However, insufficient information was provided in these reports to confirm that the insecticidal effects resulted from the action of AaIT. For example, the toxicity of the plant-expressed AaIT was not confirmed, and evidence for AaIT toxicity such as larval paralysis was not presented. Given the relatively low toxicity of AaIT on injection of lepidopteran larvae, very high levels of toxin would be required to see an effect (Elazar, Levi, and Zlotkin 2001). Given that AaIT has been shown to lack oral toxicity against two lepidopteran species (Zlotkin, Fishman, and Shapiro 1992), the data presented on plant expression of AaIT should be interpreted with caution.

There are two additional reports of transgenic rice expressing a spider venom gene (Huang et al. 2001) and transgenic rapeseed expressing an *M. sexta* chitinase gene and a scorpion toxin gene, *Bmk*, from the Manchurian scorpion, *Buthus martensii* (Wang et al. 2005) that claim resistance against larvae of the striped stem borer, *Chino suppressalis*, and the diamondback moth, *Plutella xylostella* (L.) syn. *P. maculipennis* (Curtis), respectively. These papers also fail to demonstrate symptoms associated with neurotoxin-mediated effects.

Orally Active Toxins

In recent years, the genes encoding insect-selective venom toxins derived from spiders, scorpions, and braconid wasps have been incorporated into plant genomes to target phytophagous insects such as those in the order Lepidoptera. Tobacco plants were transformed to express the orally active atracotoxin ω-ACTX-Hv1a from the Australian funnel-web spider, *Hadronyche versuta*. Transgenic lines provided superior protection from second instar *H. armigera* and *Spodoptera littoralis* at a rate of 100 percent mortality within a period of 72 hours (Khan et al. 2006). Toxins in this ω-ACTX-1 peptide group are 36–37 residues in length, have a stable cysteine-knot structure, and are sufficiently small that they can penetrate the blood-brain barrier to access their target sites within the central nervous system (Fletcher et al. 1997). Interestingly, larvae also succumbed to topically applied ω-ACTX-Hv1a (Khan et al. 2006).

TSP14

Hymenopteran parasitoids release teratocytes at egg hatch into the hemocoel of lepidopteran larvae (lepidopteran parasitoids are discussed in Chapter 17). The teratocytes produce teratocyte secretory protein, TSP14, which inhibits protein synthesis, growth, and development of the parasitized lepidopteran host to support endoparasite development (Dahlman et al. 2003). The gene encoding TSP14 was stably engineered into tobacco plants (Maiti et al. 2003). When larvae of *M. sexta* and the tobacco budworm, *Heliothis virescens*, were fed on the transgenic plants, mortality was increased and growth rates reduced when compared with larvae fed on control plants. Although mortality resulted from TSP14 inhibition of protein synthesis in the host (Maiti et al. 2003), it was unclear from this study whether TSP14 was acting in the gut or in the hemocoel of the larvae as expected.

Lectin Delivery of Intrahemocoelic Toxins

Lectins are proteins that bind carbohydrates and are resistant to proteolysis in the guts of herbivores. In plants, lectins serve as defensive compounds against herbivores including lepidopteran larvae and have been shown to negatively impact lepidopteran pests such as the tomato moth, *Lacanobia oleracea*, when delivered via transgenic plants (Gatehouse et al. 1999). Lectins bind to glycan receptors along the intestinal tract, resulting in discomfort or toxicity. Lectins vary widely in binding specificity, and some plant lectins such as the mannose-specific snowdrop lectin, *Galanthus nivalis* agglutinin (GNA), have the ability to enter the insect hemocoel through the midgut, most likely by endocytosis (Fitches et al. 2001). This characteristic has allowed for use of lectins for delivery of toxins into the insect hemocoel. Following uptake, GNA has been detected in the gut, Malpighian tubules, and hemolymph (Fitches et al. 2001).

GNA is of particular interest for insect pest management because it shows high insect toxicity, but low toxicity to nontarget organisms, including mammals (Hilder et al. 1995; Down et al. 1996; Gatehouse et al. 1996; Sauvion et al. 1996). GNA also exhibits low toxicity to Lepidoptera (Gatehouse et al. 1995; Fitches, Gatehouse, and Gatehouse 1997; Gatehouse et al. 1997; Rao et al. 1998; Stoger et al. 1998); however, when fused to the neuropeptide *M. sexta* allatostatin, the combination suppressed feeding and decreased growth of fifth instar *L. oleracea*. Further, allostatin was detected in the hemolymph, indicating that GNA served to transport this peptide through the gut epithelium and into the hemocoel (Fitches et al. 2002). GNA has also been used as a carrier protein to transport the insecticidal spider venom–derived toxin *Segestria florentina* toxin 1 (SFI1) into the hemocoel. Oral ingestion of SFI1 or GNA separately resulted in no mortality in *L. oleracea*, but fusion of the two resulted in 100 percent mortality of larvae within six days when incorporated into diet (Fitches et al. 2004). The presence of the GNA-SFI1 fusion in the hemocoel was demonstrated by immunoblotting. Fusion of the toxin ButaIT derived from the red scorpion *Mesobuthus tamulus* to GNA resulted in increased levels of mortality and decreased growth rates following ingestion by larvae of *L. oleracea* when compared with either the toxin or GNA alone. Again the fused ButalT-GNA was detected intact in the insect hemolymph (Trung, Fitches, and Gatehouse 2006). These studies highlight the potential for transgenic plant expression of lectin-intrahemocoelic toxin fusions for management of lepidopteran pests.

FUTURE PROSPECTS

1. Venoms will continue to provide an excellent resource for isolation of insect-selective toxins that can be employed for management of pest Lepidoptera. Bacteria associated with entomopathogenic nematodes are also receiving attention to that end (ffrench-Constant, Dowling, and Waterfield 2007).
2. Given the specificity and more consistent efficacy of insect-selective neurotoxins relative to other intrahemocoelic effectors, it is likely that these toxins will be particularly useful for lepidopteran pest management in the future. Lepidoptera-resistant transgenic plants expressing TSP14 or atracotoxins are likely candidates for commercialization.
3. Genetically optimized baculovirus insecticides have particularly good prospects for management of *H. armigera* in China. Genetically optimized entomopathogenic fungi may also find a niche in the insecticide market against pests that are not susceptible to toxins derived from Bt.
4. The mechanism by which orally active neurotoxins such as ω-ACTX-Hv1a (and indeed the lectins) enter the hemocoel of lepidopteran larvae is unknown but could provide leads for novel control strategies for lepidopteran pests.

Because of the importance of finding alternatives to Bt-derived transgenes for transgenic plants resistant to lepidopteran pests, research into the use of intrahemocoelic toxins for pest management is likely to increase in scope.

ACKNOWLEDGMENTS

This material is based in part upon work supported by USDA NRI 2003-35302-13558 as well as Hatch Act and State of Iowa funds. NRS was funded by a Henry and Sylvia Richardson Research Incentive grant.

REFERENCES

Adams, J.R., and J.T. McClintock. 1991. Baculoviridae. Nuclear polyhedrosis viruses, part 1: Nuclear polyhedrosis viruses of insects. In *Atlas of Invertebrate Viruses*, ed. J.R. Adams and J.R. Bonami, 87–204. Boca Raton, FL: CRC Press.

Bates, S.L., J.Z. Zhao, R.T. Roush, and A.M. Shelton. 2005. Insect resistance management in GM crops: Past, present and future. *Nat. Biotechnol.* 23:57–62.

Black, B.C., L.A. Brennan, P.M. Dierks, and I.E. Gard. 1997. Commercialization of baculoviral insecticides. In *The Baculoviruses*, ed. L.K. Miller, 341–88. New York: Plenum Press.

Bonning, B.C., V.K. Ward, M. van Meer, T.F. Booth, and B.D. Hammock. 1997. Disruption of lysosomal targeting is associated with insecticidal potency of juvenile hormone esterase. *Proc. Natl. Acad. Sci. U.S.A.* 94:6007–12.

Burden, J.P., R.S. Hails, J.D. Windass, M.-M. Suner, and J.S. Cory. 2000. Infectivity, speed of kill, and productivity of a baculovirus expressing the itch mite toxin Txp-1 in second and fourth instar larvae of *Trichoplusia ni*. *J. Invertebr. Pathol.* 75:226–36.

Cao, J., J.-Z. Zhao, J.D. Tang, A.M. Shelton, and E.D. Earle. 2002. Broccoli plants with pyramided *cry1Ac* and *cry1C* Bt genes control diamondback moths resistant to Cry1A and Cry1C proteins. *Theor. Appl. Genet.* 105:258–64.

Carbonell, L.F., M.R. Hodge, M.D. Tomalski, and L.K. Miller. 1988. Synthesis of a gene coding for an insect-specific scorpion neurotoxin and attempts to express it using baculovirus vectors. *Gene* 73:409–18.

Chen, X., X. Sun, Z. Hu, M. Li, et al. 2000. Genetic engineering of *Helicoverpa armigera* single-nucleopolyhedrovirus as an improved pesticide. *J. Invertebr. Pathol.* 76:140–46.

Cory, J.S., M.L. Hirst, T. Williams, et al. 1994. Field trial of a genetically improved baculovirus insecticide. *Nature* 370:138–40.

Dahlman, D.L., R.L. Rana, E.J. Schepers, T. Schepers, F.A. DiLuna, and B.A. Webb. 2003. A teratocyte gene from a parasitic wasp that is associated with inhibition of insect growth and development inhibits host protein synthesis. *Insect Mol. Biol.* 12:527–34.

De Lima, M.E., S.G. Figueiredo, A.M. Pimenta, et al. 2007. Peptides of arachnid venoms with insecticidal activity targeting sodium channels. *Comp. Biochem. Physiol. C Toxicol. Pharmacol.* 146:264–79.

Down, R.E., A.M.R. Gatehouse, G.M. Davison, et al. 1996. Snowdrop lectin inhibits development and decreases fecundity of the glasshouse potato aphid (*Aulacorthum solani*) when administered in vitro and via transgenic plants both in laboratory and glasshouse trials. *J. Insect Physiol.* 42:1035–45.

Elazar, M., R. Levi, and E. Zlotkin. 2001. Targeting of an expressed neurotoxin by its recombinant baculovirus. *J. Exp. Biol.* 204:2637–45.

Eldridge, R., F.M. Horodyski, D.B. Morton, et al. 1991. Expression of an eclosion hormone gene in insect cells using baculovirus vectors. *Insect Biochem.* 21:341–51.

Eldridge, R., D.R. O'Reilly, and L.K. Miller. 1992. Efficacy of a baculovirus pesticide expressing an eclosion hormone gene. *Biol. Control* 2:104–10.

Escoubas, P. 2006. Molecular diversification in spider venoms: A web of combinatorial peptide libraries. *Mol. Divers.* 10:545–54.

Escoubas, P., B. Sollod, and G.F. King. 2006. Venom landscapes: Mining the complexity of spider venoms via a combined cDNA and mass spectrometric approach. *Toxicon* 47:650–63.

Federici, B.A. 1999. Naturally occurring baculoviruses for insect pest control. *Methods Biotechnol.* 5:301–20.

ffrench-Constant, R.H., A. Dowling, and N.R. Waterfield. 2007. Insecticidal toxins from *Photorhabdus* bacteria and their potential use in agriculture. *Toxicon* 49:436–51.

Fitches, E., N. Audsley, J.A. Gatehouse, and J.P. Edwards. 2002. Fusion proteins containing neuropeptides as novel insect control agents: Snowdrop lectin delivers fused allatostatin to insect haemolymph following oral ingestion. *Insect Biochem. Molec. Biol.* 32:1653–61.

Fitches, E., M.G. Edwards, C. Mee, et al. 2004. Fusion proteins containing insect-specific toxins as pest control agents: Snowdrop lectin delivers fused insecticidal spider venom toxin to insect haemolymph following oral ingestion. *J. Insect Physiol.* 50:61–71.

Fitches, E., A.M.R. Gatehouse, and J.A. Gatehouse. 1997. Effects of snowdrop lectin (GNA) delivered via artificial diet and transgenic plants on the development of the tomato moth (*Lacanobia oleracea*) larvae in the laboratory and glasshouse trials. *J. Insect Physiol.* 43:727–39.

Fitches, E., S.D. Woodhouse, J.P. Edwards, and J.A. Gatehouse. 2001. In vitro and in vivo binding of snowdrop (*Galanthus nivalis* agglutinin; GNA) and jackbean (*Canavalia ensiformis*; Con A) lectins within tomato moth (*Lacanobia oleracea*) larvae; mechanisms of insecticidal action. *J. Insect Physiol.* 47:777–87.

Fletcher, J.I., R. Smith, S.I. O'Donoghue, et al. 1997. The structure of a novel insecticidal neurotoxin, omega-atracotoxin-HV1, from the venom of an Australian funnel web spider. *Nat. Struct. Biol.* 4:559–66.

Froy, O., N. Zilberberg, N. Chejanovsky, J. Anglister, and E. Loret. 2000. Scorpion neurotoxins: Structure/function relationships and application in agriculture. *Pest Manag. Sci.* 56:472–74.

Fujii-Taira, I., Y. Tanaka, K.J. Homma, and S. Natori. 2000. Hydrolysis and synthesis of substrate proteins for cathepsin L in the brain basement membranes of *Sarcophaga* during metamorphosis. *J. Biochem.* 128:539–42.

Gatehouse, A.M.R., G.M. Davison, C.A. Newell, et al. 1997. Transgenic potato plants with enhanced resistance to the tomato moth (*Lacanobia oleracea*) larvae; mechanisms of insecticidal action. *Mol. Breeding* 3:49–63.

Gatehouse, A.M.R., G.M. Davidson, J.N. Stewart, et al. 1999. Concanavalin A inhibits development of tomato moth (*Lacanobia oleracea*) and peach-potato aphid (*Myzus persicae*) when expressed in transgenic potato. *Mol. Breeding* 5:153–65.

Gatehouse, A.M.R., R.E. Down, K.S. Powell, et al. 1996. Transgenic potato plants with enhanced resistance to the peach-potato aphid *Myzus persicae*. *Ent. Exp. Appl.* 79:295–307.

Gatehouse, A.M.R., K.S. Powell, E.J.M.V. Damme, and J.A. Gatehouse. 1995. Insecticidal properties of plant lectins. In *Lectins, biomedical perspectives*, ed. A. Pusztai and S. Bardocz, 35–57. London: Taylor and Francis.

Gurevitz, M., I. Karbat, L. Cohen, et al. 2007. The insecticidal potential of scorpion beta-toxins. *Toxicon* 49:473–89.

Hammock, B.D., B.C. Bonning, R.D. Possee, T.N. Hanzlik, and S. Maeda. 1990. Expression and effects of the juvenile hormone esterase in a baculovirus vector. *Nature* 344:458–61.

Harrison, R.L., and B.C. Bonning. 2000a. Use of scorpion neurotoxins to improve the insecticidal activity of *Rachiplusia ou* multicapsid nucleopolyhedrovirus. *Biol. Control* 17:191–201.

Harrison, R.L., and B.C. Bonning. 2000b. Genetic engineering of biocontrol agents for insects. In *Lectins, biomedical perspectives*, ed. A. Pusztai and S. Bardocz, 35–57. London: Taylor and Francis.

Harrison, R.L., and B.C. Bonning. 2001. Use of proteases to improve the insecticidal activity of baculoviruses. *Biol. Control* 20:199–209.

Heckel, D.G., L.J. Gahan, S.W. Baxter, et al. 2007. The diversity of Bt resistance genes in species of Lepidoptera. *J. Invertebr. Pathol.* 95:192–97.

Herrmann, R., H. Moskowitz, E. Zlotkin, and B.D. Hammock. 1995. Positive cooperativity among insecticidal scorpion neurotoxins. *Toxicon* 33:1099–1102.

Hilder, V.A., K.S. Powell, A.M.R. Gatehouse, et al. 1995. Expression of snowdrop lectin in transgenic tobacco plants results in added protection against aphids. *Transgenic Res.* 4:18–25.

Homma, K., S. Jurata, and S. Natori. 1994. Purification, characterization, and cDNA cloning of procathepsin L from the culture medium of NIH-Sape-4, an embryonic cell line of *Sarcophaga peregrina* (flesh fly), and its involvement in the differentiation of imaginal discs. *J Biol. Chem.* 269:15258–64.

Homma, K., and S. Natori. 1996. Identification of substrate proteins for cathepsin L that are selectively hydrolyzed during the differentiation of imaginal discs of *Sarcophaga peregrina. Eur. J. Biochem.* 240:443–47.

Hoover, K., C.M. Schultz, S.S. Lane, B.C. Bonning, S.S. Duffey, and B.D. Hammock. 1995. Reduction in damage to cotton plants by a recombinant baculovirus that causes moribund larvae of *Heliothis virescens* to fall off the plant. *Biol. Control* 5:419–26.

Hu, G., and R.J. St. Leger. 2002. Field studies using a recombinant mycoinsecticide (*Metarhizium anisopliae*) reveal that it is rhizosphere competent. *Appl. Environ. Microbiol.* 68:6383–87.

Huang, J., S. Rozelle, C. Pray, and Q. Wang. 2002. Plant biotechnology in China. *Science* 295:674–76.

Huang, J.Q., Z.M. Wel, H.L. An, and Y.X. Zhu. 2001. *Agrobacterium tumefaciens*-mediated transformation of rice with the spider insecticidal gene conferring resistance to leaffolder and striped stem borer. *Cell Res.* 11:149–55.

Hughes, P.R., H.A. Wood, J.P. Breen, S.F. Simpson, A.J. Duggan, and J.A. Dybas. 1997. Enhanced bioactivity of recombinant baculoviruses expressing insect-specific spider toxins in lepidopteran crop pests. *J. Invertebr. Pathol.* 69:112–18.

Inceoglu, A.B., S.G. Kamita, and B.D. Hammock. 2006. Genetically modified baculoviruses: A historical overview and future outlook. *Adv. Virus Res.* 68:323–60.

Kamita, S.G., K.-D. Kang, B.D. Hammock, and A.B. Inceoglu. 2005. Genetically modified baculoviruses for pest insect control. In *Comprehensive molecular insect science*, ed. L.I. Gilbert et al., Vol. 6, 271–322. Oxford: Elsevier.

Khan, S.A., Y. Zafar, R.W. Briddon, K.A. Malik, and Z. Mukhtar. 2006. Spider venom toxin protects plants from insect attack. *Transgenic Res.* 15:349–57.

King, G.F. 2007. Modulation of insect Ca(v) channels by peptidic spider toxins. *Toxicon* 49:513–30.

Korth, K.L., and C.S. Levings. 1993. Baculovirus expression of the maize mitochondrial protein URF13 confers insecticidal activity in cell cultures and larvae. *Proc. Natl. Acad. Sci. U.S.A.* 90:3388–92.

Lawrence, S. 2005. Agbio keeps on growing. *Nat. Biotechnol.* 23:281.

Lee, S.-Y., X. Qu, W. Chen, et al. 1997. Insecticidal activity of a recombinant baculovirus containing an antisense *c-myc* fragment. *J. Gen. Virol.* 78:273–81.

Leger, R.J.S. 1995. The role of cuticle-degrading proteases in fungal pathogenesis of insects. *Can. J. Bot.* (Suppl. 1) 73:S1119–25.

Leger, R.J.S., L. Joshi, M.J. Bidochka, and D.W. Roberts. 1996. Construction of an improved mycoinsecticide overexpressing a toxic protease. *Proc. Natl. Acad. Sci. U.S.A.* 93:6349–54.

Li, H., H. Tang, R.L. Harrison, and B.C. Bonning. 2007. Impact of a basement membrane-degrading protease on dissemination and secondary infection of *Autographa californica* multiple nucleopolyhedrovirus in Heliothis virescens (Fabricus). *J. Gen. Virol.* 88:1109–19.

Ma, P.W.K., T.R. Davis, H.A. Wood, D.C. Knipple, and W.L. Roelofs. 1998. Baculovirus expression of an insect gene that encodes multiple neuropeptides. *Insect Biochem. Mol. Biol.* 28:239–49.

Maeda, S. 1989. Increased insecticidal effect by a recombinant baculovirus carrying a synthetic diuretic hormone gene. *Biochem. Biophys. Res. Commun.* 165:1177–83.

Maeda, S., S.L. Volrath, T.N. Hanzlik, et al. 1991. Insecticidal effects of an insect-specific neurotoxin expressed by a recombinant baculovirus. *Virology* 184:777–80.

Maiti, I.B., N. Oey, D.L. Dahlman, and B.A. Webb. 2003. Antibiosis-type resistance in transgenic plants expressing a teratocyte secretory peptide (TSP) gene from a hymenopteran endoparasite (Microplitis croceipes). *Plant Biotech. J.* 1:209–19.

McCutchen, B.F., P.V. Choudary, R. Crenshaw, et al. 1991. Development of a recombinant baculovirus expressing an insect-selective neurotoxin: Potential for pest control. *Bio/Technol.* 9:848–52.

Moscardi, F. 1999. Assessment of the application of baculoviruses for control of Lepidoptera. *Annu. Rev. Entomol.* 44:257–89.

Oerke, E.-C. 1994. Estimated crop losses due to pathogens, animal pests and weeds. In *Crop production and crop protection: Estimated losses in major food and cash crops*, ed. E.-C. Oerke et al., 72–78. Amsterdam: Elsevier.

O'Reilly, D.R., T.J. Kelly, E.P. Masler, et al. 1995. Overexpression of *Bombyx mori* prothoracicotropic hormone using baculovirus vectors. *Insect Biochem. Mol. Biol.* 25:475–85.

Petrik, D.T., A. Iseli, B.A. Montelone, J.L. Van Etten, and R.J. Clem. 2003. Improving baculovirus resistance to UV inactivation: Increased virulence resulting from expression of a DNA repair enzyme. *J. Invertebr. Pathol.* 82:50–56.

Philip, J.M.D., E. Fitches, R.L. Harrison, B.C. Bonning, and J.A. Gatehouse. 2007. Characterisation of functional and insecticidal properties of a recombinant cathepsin L-like proteinase from flesh fly (*Sarcophaga peregrina*), which plays a role in differentiation of imaginal discs. *Insect Biochem. Molec. Biol.* 37:589–600.

Popham, H.J.R., Y. Li, and L.K. Miller. 1997. Genetic improvement of *Helicoverpa zea* nuclear polyhedrosis virus as a biopesticide. *Biol. Control* 10:83–91.

Pray, C.E., J. Huang, R. Hu, and S. Rozelle. 2002. Five years of Bt cotton in China—the benefits continue. *Plant J.* 31:423–30.

Prikhod'ko, G.G., H.J.R. Popham, T.J. Felcetto, et al. 1998. Effects of simultaneous expression of two sodium channel toxin genes on the properties of baculoviruses as biopesticides. *Biol. Control* 12:66–78.

Prikhod'ko, G.G., M. Robson, J.W. Warmke, et al. 1996. Properties of three baculovirus-expressing genes that encode insect-selective toxins: μ-Aga-IV, As II, and Sh I. *Biol. Control* 7:236–44.

Rao, K.V., K.S. Rathore, T.K. Hodges, et al. 1998. Expression of snowdrop lectin (GNA) in the phloem of transgenic rice plants confers resistance to rice brown planthopper. *Plant J.* 15:469–77.

Regev, A., H. Rivkin, B. Inceoglu, et al. 2003. Further enhancement of baculovirus insecticidal efficacy with scorpion toxins that interact cooperatively. *FEBS Lett.* 537:106–10.

Rosi-Marshall, E.J., J.L. Tank, T.V. Royer, et al. 2007. Toxins in transgenic crop byproducts may affect headwater stream ecosystems. *Proc. Natl. Acad. Sci. U.S.A.* 104:16204–08.

Sauvion, N., Y. Rahbe, W.J. Peumans, E.J.V. Damme, J.A. Gatehouse, and A.M.R. Gatehouse. 1996. Effects of GNA and other binding lectins on development and fecundity of the peach-potato aphid *Myzus persicae*. *Ent. Exp. Appl.* 79:285–93.

Screen, S.E., G. Hu, and R.J. St. Leger. 2001. Transformants of *Metarhizium anisopliae* sf. *anisopliae* overexpressing chitinase from *Metarhizium anisopliae* sf. *acridum* show early induction of native chitinase but are not altered in pathogenicity to *Manduca sexta*. *J. Invertebr. Pathol.* 78:260–66.

Sisterson, M.S., Y. Carriere, T.J. Dennehy, and B.E. Tabashnik. 2007. Nontarget effects of transgenic insecticidal crops: Implications of source-sink population dynamics. *Environ. Entomol.* 36:121–27.

Stewart, L.M.D., M. Hirst, M.L. Ferber, A.T. Merryweather, P.J. Cayley, and R.D. Possee. 1991. Construction of an improved baculovirus insecticide containing an insect-specific toxin gene. *Nature* 352:85–88.

Stoger, E., S. Williams, P. Christou, R.E. Down, and J.A. Gatehouse. 1998. Expression of the insecticidal lectin from snowdrop (*Galanthus nivalis* aggluitin; GNA) in transgenic wheat plants: Effects on predation by the grain aphid *Sitobion avenae*. *Mol. Breeding* 5:65–73.

Sun, X., X. Chen, Z. Zhang, et al. 2002. Bollworm responses to release of genetically modified *Helicoverpa armigera* nucleopolyhedroviruses in cotton. *J. Invertebr. Pathol.* 81:63–69.

Sun, X., H. Wang, X. Sun, et al. 2004. Biological activity and field efficacy of a genetically modified *Helicoverpa armigera* single-nucleocapsid nucleopolyhedrovirus expressing an insect-selective toxin from a chimeric promoter. *Biol. Control* 29:124–37.

Tang, H., H. Li, S.M. Lei, R.L. Harrison, and B.C. Bonning. 2007. Tissue specificity of a baculovirus-expressed, basement membrane-degrading protease in larvae of *Heliothis virescens*. *Tissue Cell* 39:431–43.

Taniai, K., A.B. Inceoglu, and B.D. Hammock. 2002. Expression efficiency of a scorpion neurotoxin, AaHIT, using baculovirus in insect cells. *Appl. Entomol. Zool.* 37:225–32.

Toenniessen, G.H., J.C. O'Toole, and J. DeVries. 2003. Advances in plant biotechnology and its adoption in developing countries. *Curr. Opin. Plant Biol.* 6:191–98.

Treacy, M.F., P.E. Rensner, and J.N. All. 2000. Comparative insecticidal properties of two nucleopolyhedrovirus vectors encoding a similar toxin gene chimer. *J. Econ. Entomol.* 93:1096–1104.

Trung, N.P., E. Fitches, and J.A. Gatehouse. 2006. A fusion protein containing a lepidopteran-specific toxin from the South Indian red scorpion (*Mesobuthus tamulus*) and snowdrop lectin shows oral toxicity to target insects. *BMC Biotechnol.* 6:18–30.

Wang, J.X., Z.L. Chen, J.Z. Du, Y. Sun, and A.H. Liang. 2005. Novel insect resistance in *Brassica napus* developed by transformation of chitinase and scorpion toxin genes. *Plant Cell Rep.* 24:549–55.

Wang, C., and R.J.S. Leger. 2007. Expressing a scorpion neurotoxin makes a fungus hyperinfectious to insects. *Nat. Biotechnol.* 25:1455–56.

Wu, J., X. Luo, Z. Wang, Y. Tian, A. Liang, and Y. Sun. 2008. Transgenic cotton expressing synthesized scorpion insect toxin AaHIT gene confers enhanced resistance to cotton bollworm (*Heliothis armigera*) larvae. *Biotechnol Lett.* 30:547–54.

Wu, N.F., Q. Sun, B. Yao, et al. 2000. Insect-resistant transgenic poplar expressing AaIT gene. *Sheng Wu Gong Cheng Xue Bao* 16:129–33.

Yao, B., Y. Fan, Q. Zheng, and R. Zhao. 1996. Insect-resistant tobacco plants expressing insect-specific neurotoxin AaIT. *Chin. J. Biotechnol.* 12:67–72.

Zlotkin, E., M. Eitan, V.P. Bindokas, et al. 1991. Functional duality and structural uniqueness of depressant insect-selective neurotoxins. *Biochemistry* 30:4814–20.

Zlotkin, E., Y. Fishman, and M. Elazar. 2000. AaIT: From neurotoxin to insecticide. *Biochimie* 82:869–81.

Zlotkin, E., L. Fishman, and J.P. Shapiro. 1992. Oral toxicity to flesh flies of a neurotoxic polypeptide. *Arch. Insect Biochem. Physiol.* 21:41–52.

Zlotkin, E., H. Moskowitz, R. Herrmann, M. Pelhate, and D. Gordon. 1995. Insect sodium channel as the target for insect-selective neurotoxins from scorpion venom. *ACS Symp. Ser.* 591:56–85.

17 The Interactions between Polydnavirus-Carrying Parasitoids and Their Lepidopteran Hosts

Michael R. Strand

CONTENTS

INTRODUCTION

As discussed elsewhere in this volume (see Chapter 1), the Lepidoptera comprise one of the mega-diverse orders of holometabolous insects with approximately 160,000 named species including many of the most economically important pests in agriculture and forestry (Powell 2003). Most moths and butterflies are herbivores as larvae, and the order as a whole represents the single largest evolutionary lineage of animals specialized to consume living plants. Commensurate with this diversity is a similarly large community of natural enemies that prey upon Lepidoptera. The most species-rich of these natural enemies are parasitoid wasps (Hymenoptera). Almost all Lepidoptera are attacked by one or more species of parasitoid wasps, but the vast majority of parasitoids are specialists that parasitize a particular life stage of one or a few species of hosts (Whitfield 1998; Pennacchio and Strand 2006).

Parasitoids are divided into two broad categories on the basis of how they develop and interact with their hosts (Askew and Shaw 1986; Pennacchio and Strand 2006). Idiobionts are either ectoparasitoids that paralyze their hosts or endoparasitoids that attack sessile host stages like eggs or pupae. Hosts attacked by idiobionts also cease development following parasitism. Most koinobionts in contrast are endoparasitoids of larval-stage insects, and their hosts continue to develop as the parasitoid's offspring mature. Not surprisingly, larval stage Lepidoptera attacked by koinobionts possess formidable defenses to resist endoparasitoid attack, and endoparasitoids have reciprocally evolved a diversity of counterstrategies for overcoming host defenses and developing in the internal environment of the host. Among these counterstrategies is the evolution of symbiotic associations with different types of microorganisms that wasps inject into hosts when ovipositing (Pennacchio and Strand 2006). The activity of these symbionts then causes physiological alterations in hosts that allow the parasitoid's offspring to develop successfully. The best studied and most diverse of these parasitoid-microbial associations occur between thousands of wasp species in the families Braconidae and Ichneumonidae and viruses in the family Polydnaviridae. The structure and organization of polydnavirus (PDV) genomes (Kroemer and Webb 2004a; Webb and Strand 2005; Dupuy, Huguet, and Drezen 2006) and the role of PDVs in parasitism (Schmidt, Theopold, and Strand 2001; Webb and Strand 2005; Pennacchio and Strand 2006; Gill et al. 2006) have been discussed in several reviews. In this chapter, I update this literature, focusing on the functional activities of PDVs from the perspective of lepidopteran biology, and the potential use of PDV gene products in insect control and biotechnology.

THE POLYDNAVIRUS LIFE CYCLE

The PDVs associated with braconid wasps are currently classified into one genus called bracoviruses (BVs), and the PDVs associated with ichneumonids are classified into a second genus called ichnoviruses (IVs). All species of braconids in seven subfamilies (Cardiochilinae, Cheloninae, Dirrhoponae, Mendesellinae, Khoikhoiinae, Miricinae, Microgastrinae; 17,000 species) are thought to carry BVs, whereas all ichneumonids, which are classified into two subfamilies (Campopleginae, Banchinae; 13,000 species), are thought to carry IVs (Stoltz and Vinson 1979; Whitfield and Asgari 2003; Webb and Strand 2005). Recent studies strongly suggest BVs arose from a nudivirus ancestor but the origins of IVs remain unknown (Bezier et al., 2009). Comparative genomic analyses indicate that each PDV carried by a given parasitoid species is genetically unique with relatedness of PDV isolates paralleling wasp phylogeny (Whitfield 2002; Espagne et al. 2004; Webb et al. 2006; Tanaka et al. 2007; Murphy et al. 2008). BVs and IVs also share a similar life cycle (Figure 17.1). In wasps, BVs and IVs persist as proviruses that are stably integrated into the genome of all cells of both sexes including the germ line. Transmission from one wasp generation to the next occurs vertically via inheritance of proviral DNA in eggs and sperm. Replication in contrast is restricted to female wasps in specialized cells that form a region of the reproductive tract called the calyx. Replication is first detected in the late pupal stage and usually continues after the female emerges as an adult (see Webb and Strand 2005). BV virions consist of a single unit membrane enveloping one or more cylindrical nucleocapsids of variable length that are released from calyx cells by cell lysis (Stoltz, Vinson, and MacKinnon 1976). IV virions from campoplegine ichneumonids consist of two unit membranes that envelop a single fusiform nucleocapsid of uniform size that are released from calyx cells by budding (Volkoff et al. 1995). Virions from only one banchine ichneumonid have been examined to date; they too consist of two unit membranes but envelop multiple fusiform-shaped nucleocapsids (Lapointe et al. 2007). Following replication and release from calyx cells, virions accumulate to high concentrations in the lumen of the lateral oviducts to form a suspension of virus called calyx fluid (Figure 17.1). When a wasp lays an egg into its host, she injects a quantity of virus that infects various host cell types and organs. PDVs do not replicate in the parasitized host insect, but expression of viral gene products prevents the host's immune system from killing the parasitoid's offspring and also induces a variety of developmental alterations that likely facilitate parasitoid growth (summarized by Webb and Strand 2005; Dupuy, Huguet, and Drezen 2006; Pennacchio and Strand 2006). Thus, a true mutualism exists between PDVs and wasps, because viral transmission depends on survival of the parasitoid, and parasitoid survival depends on infection and expression of gene products encoded by the virus.

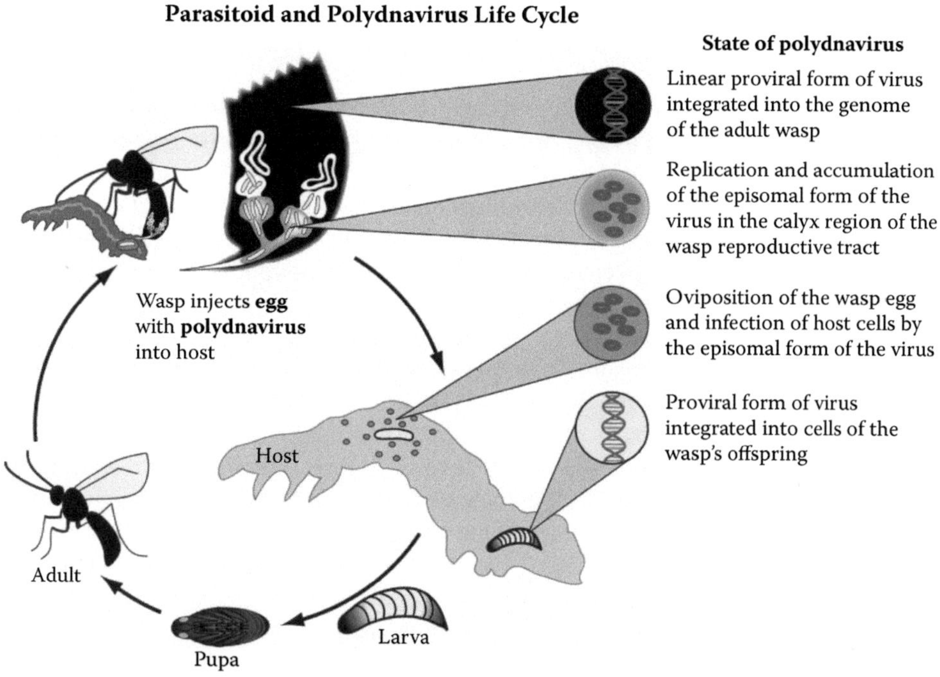

FIGURE 17.1 The polydnavirus life cycle. Polydnaviruses exist as integrated proviruses in the genomes of wasps, and wasps infect their lepidopteran hosts by injecting the episomal form of the virus packaged in virions. The parasitoid develops by ovipositing one or more eggs into the host, which hatch into wasp larvae that feed on hemolymph and/or host tissues. Upon completing development, the wasp larva emerges from the host, pupates, and develops into an adult wasp.

POLYDNAVIRUS GENOMES AND EXPRESSION ACTIVITY IN PARASITIZED LEPIDOPTERA

General Organizational Features of PDV Genomes

As previously noted, PDV genomes exist in two states: an episomal form that is packaged into nucleocapsids during replication, and a linear proviral form that is integrated into the wasp genome. The episomal form of BV and IV genomes consists of multiple, circular, double-stranded DNAs that range in size from 2 to over 30 kilobases (kb; Kroemer and Webb 2004a; Webb and Strand 2005). The number of DNA segments varies among PDV isolates from as few as 6 in some BVs to well over 100 in the case of viruses from banchine ichneumonids (Stoltz et al. 1995; Tanaka et al. 2007). Generally, BV genomes consist of fewer but larger DNA segments, but IVs genomes consist of a larger number of smaller segments. Aggregate genome sizes for PDVs vary from 180 kb to more than 500 kb (Espagne et al. 2004; Webb et al. 2006; Lapointe et al. 2007; Tanaka et al. 2007). It is also well known that the genomic segments of BVs and IVs exist in nonequimolar concentrations in calyx fluid, with some segments consistently being in greater abundance than others. In *Microplitis demolitor* bracovirus (MdBV), for example, the total genome consists of 15 segments, but 5 segments and their associated nucleocapsids account for more than 60 percent of the viral DNAs in calyx fluid (Beck, Inman, and Strand 2007). Studies with BVs indicate that each nucleocapsid contains a single genomic segment with capsid size correlating with the size of the genomic segment it contains (Albrecht et al. 1994; Beck, Inman, and Strand 2007). In contrast, it is currently unknown whether IV nucleocapsids contain single or multiple genomic segments. Unlike BVs, the nonequimolar production of genomic segments by IVs is also associated with a phenomenon called segment nesting (see Kroemer and Webb 2004a; Webb and Strand 2005). In effect, some genomic segments in IVs are unique, whereas others give rise to additional segments after excision from the wasp genome via intramolecular recombination (Xu and Stoltz 1993; Cui and Webb 1997; Webb and Cui 1998).

Sequence analysis indicates that in the proviral form, the episomal segments of BVs are integrated into the genome of wasps as one or more tandem arrays that form macroloci (Belle et al. 2002; Desjardins et al. 2007). The proviral segments of IVs on the other hand appear to be dispersed within the genomes of ichneumonids (Fleming and Summers 1991). Studies with BVs suggest macroloci are amplified before excision of individual viral segments and packaging into virions (Pasquier-Barre et al. 2002; Drezen et al. 2003; Marti et al. 2003). Studies with *Chelonus inanitus* bracovirus (CiBV) also suggest the more abundant episomal segments are present in greater copy number in the nonexcised proviral form (Annaheim and Lanzrein 2007). Webb (1998) in contrast suggests that IV replication may occur in a rolling-circle-type mechanism with excision of proviral DNA being followed by amplification of circular episomes.

To date, the encapsidated genomes of two BVs from microgastrine braconids (*Cotesia congregata* BV [CcBV] and *Microplitis demolitor* BV [MdBV]), three IVs from campoplegine ichneumonids (*Campoletis sonorensis* IV [CsIV], *Hyposoter fugitives* IV [HfIV], and *Tranosema rostrale* IV), and one isolate from a banchine ichneumonid (*Glypta fumiferanae* [GfV]) have been fully or nearly fully sequenced (Espagne et al. 2004; Webb et al. 2006; Tanaka et al. 2007; Lapointe et al. 2007). Database searches further indicate that approximately six other BVs or IVs have been partially sequenced. Together, these results identify several shared features in addition to the general genomic properties discussed above. These include very low coding densities, strong A + T biases, and the finding that the majority of predicted genes encoded by PDVs form families of related gene variants of which a majority are expressed in parasitized host insects (see below). The encapsidated genomes of PDVs in contrast lack genes coding for polymerases or other proteins required for replication, which accounts for why PDVs do not replicate in parasitized host insects. PDVs from individual species of wasps also encode several novel single-copy genes or gene families not found in other PDV species.

PDV Gene Families

Comparative studies to date identify only one gene family encoded by both BVs and IVs, ankyrin genes (*ank*; Figure 17.2). However, phylogenetic studies indicate that BV and IV *ank* genes form distinct clusters that suggest they do not derive from a common ancestor (Webb et al. 2006). Excluding chelonines, two gene families, *ank* and protein tyrosine phosphatase genes (*ptp*), appear to be shared among all BVs carried by microgastrine and cardiochiline braconids studied to date; whereas six gene families are shared among IVs from campoplegine ichneumonids (*ank*, innexin genes [*inex*], cysteine-motif genes [*cys-motif*], repeat element genes [*rep*], N-family genes [*N-family*], and polar-residue rich genes [*pol-res*]; Figure 17.2). The lone banchine-associated virus sequenced to date, GfV, exhibits a degree of genome segmentation that is far greater than that reported for IVs from campoplegine ichneumonids. Gene families identified in the GfV genome include predicted *anks* and *ptps*, but phylogenetic analysis also indicates that the *ank* genes are not embedded within those of campoplegine ichneumonids; whereas the *ptp* family members are distinct from the PTPs encoded by BVs. GfV also encodes a novel family of NTPase-like genes absent from the genomes of other PDVs. Notably, sequence data generated to date for BVs from chelonine braconids like CiBV (see above) thus far reveal few if any genes shared with other PDVs (Annaheim and Lanzrein 2007). Comparisons within the same or closely related wasp genera reveal that associated PDVs encode similar gene families, whereas viruses associated with more distantly related wasps exhibit greater differences. For example, CcBV encodes cystatin (*cys*), *crv1*-like, and *C-lectin* related genes for which homologues have been identified in BVs from other wasp species in the genus *Cotesia* (Figure 17.2). In contrast, BVs from parasitoids in other genera of Microgastrinae like MdBV lack

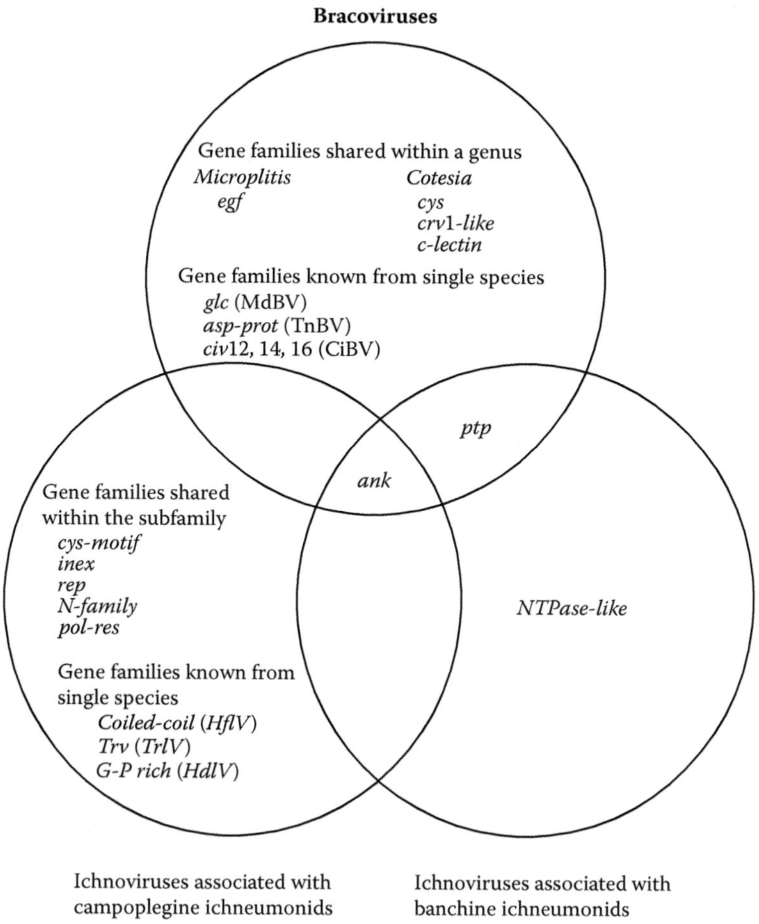

FIGURE 17.2 Venn diagrams of shared and unique gene families from braco- and ichnoviruses. *ank* genes have been identified in nearly all PDVs studied to date, whereas other gene families are restricted to only BVs and IVs from banchine ichneumonids (*ptp*), IVs from campoplegine ichneumonids, BVs from wasps in a given genus, or a single species.

these genes but encode other gene families (*egf*) for which homologues exist in BVs associated with other wasps in the genus *Microplitis* (Figure 17.2).

PDV gene families appear to arise through a combination of gene duplication and inter-genomic segment recombination events (Espagne et al. 2004; Webb et al. 2006; Friedman and Hughes 2006). The size of gene families also varies considerably between species. Among campoplegine ichneumonids, for example, the cysteine-motif gene family of CsIV consists of 10 members, whereas only five and one *cys-motif* gene have been identified in the genomes of HfIV and TrIV, respectively (Tanaka et al. 2007). Why gene family size varies among species is currently unknown, although it likely reflects differences in the host ranges of the associated parasitoids or the function of a given virus in parasitism (see below). Expression studies reveal that members of various gene families encoded by BVs and IVs are differentially expressed in tissues of virus-infected hosts (see Kroemer and Webb 2004b; Pruijssers and Strand 2007). Host tissues with especially important roles in immunity and development, like hemocytes, the fat body, and nervous system, are the primary sites of expression for most gene family members, whereas other major organs like the gut and Malpighian tubules exhibit little or no PDV gene activity. Functional studies also indicate that in some cases PDV gene family members have specialized activities that differ from other family members (see below). Novel genes and gene families restricted to viruses associated with a particular wasp species or genus also likely reflect traits acquired during the radiation of different PDV-carrying lineages driven by positive selective pressure for functions unique to particular host species parasitized by a given species or taxon of wasps. Far less clear is whether the virulence genes encoded by PDVs derive from an ancestral virus, their associated wasp or its host(s), or other sources (Webb and Strand 2005). For example, sequence analysis of CcBV reveals that many genes are highly diverged and no more closely related to hymenopteran or other insect homologues than they are to homologues from humans and other mammals (Bezier et al. 2008).

That BVs share more similarities with one another than with IVs, and vice versa, is fully consistent with the phylogeny of their associated wasps under conditions of Mendelian inheritance as proviruses. Sequence data combined with differences in morphology also suggest, however, that BVs, IVs from campoplegine ichneumonids, and IVs from banchine ichneumonids have distinct origins. If true, similarities in the life cycle and general organization of their genomes reflect convergence driven by the similar roles each of these viral groups play in parasitism (Webb and Strand 2005; Pennacchio and Strand 2006).

IMMUNOLOGICAL INTERACTIONS BETWEEN PDV-CARRYING PARASITOIDS AND LEPIDOPTERA

The most important function of PDVs is the protection of the developing parasitoid from the host's immune system. Like other insects, the innate immune system of Lepidoptera consists of both humoral and cellular effector responses (Chapter 14 discusses in detail the immune system in the tobacco hornworm, *Manduca sexta*). Humoral effector responses refer to soluble molecules with defense activities such as antimicrobial peptides (AMPs), complement-like proteins, and products generated by the phenoloxidase (PO) cascade like melanin (Cerenius and Soderhall 2004; Theopold et al. 2004; Imler and Bulet 2005; Kanost and Gorman 2008). Cellular responses refer to defenses directly mediated by insect immune cells (hemocytes) such as phagocytosis, encapsulation, and clotting (Gillespie, Kanost, and Trenczek 1997; Strand 2008a). Circulating hemocytes in Lepidoptera consist of four subpopulations called granulocytes (= granular cells), plasmatocytes, spherule cells, and oenocytoids (Lavine and Strand 2002; Strand 2008a). Granulocytes are the most abundant hemocyte type in circulation and function as the professional phagocytes. Plasmatocytes are usually larger than granulocytes and are the main capsule-forming hemocytes. Nonadhesive hemocytes include oenocytoids that contain PO and spherule cells, which are potential sources of cuticular components. Progenitor cells called prohemocytes reside in hematopoietic organs as well as at low frequency in circulation (Gardiner and Strand 2000; Nakahara et al. 2003; Nardi et al. 2003, 2006; Ling et al. 2005). Taken together, lepidopteran larvae possess a well-developed innate immune response against parasitoids and a variety of other parasites and pathogens.

LEPIDOPTERAN DEFENSE AGAINST PARASITOID ATTACK

The primary defense against parasitoids is encapsulation that begins (Phase 1) when the host recognizes the parasitoid egg or larva as foreign (Strand 2008a). Recognition of parasitoids and other invaders involves both receptors on the surface of hemocytes as well as humoral pattern-recognition molecules that enhance the recognition of foreign targets by binding to their surface (i.e., opsonization). Cell surface receptors with roles in phagocytosis or encapsulation include scavenger receptors (Ramet et al. 2001; Kocks et al. 2005; Philips, Rubin, and Perrimon 2005), a transmembrane form of the Down's syndrome cell adhesion protein (Dscam; Ramet et al. 2002; Moita et al. 2005; Dong, Taylor, and Dimopoulos 2006), and integrins (Lavine and Strand 2003; Irving et al. 2005; Levin et al. 2005; Wertheim et al. 2005; Moita et al. 2005). Humoral pattern-recognition receptors include hemolin, LPS-binding protein, gram-negative bacteria recognition protein (GNBPs), soluble PGRPs (PGRP-SA and PGRP-SD), GRPs, complement-like thioester-containing proteins (TEPs), and immunolectins (Levashina et al. 2001; Irving et al. 2005; Moita et al. 2005; Dong, Taylor, and Dimopoulos 2006; Ling and Yu 2006; Terenius et al. 2007). Recognition results in a small number of hemocytes, usually granulocytes, binding the surface of the parasitoid and also stimulates proliferation and differentiation of additional hemocytes (Phase 2; Pech and Strand 1996; Irving et al. 2005). Bound granulocytes then release cytokines that recruit additional hemocytes, primarily plasmatocytes, which form the bulk of the capsule enveloping the parasitoid. Tetraspanin proteins have recently been shown to serve as ligands for integrins in the sphingid moth *Manduca sexta*, which mediate hemocyte binding within a capsule (Zhuang et al. 2007). Granulocyte-stimulated activation and binding of plasmatocytes also involves the immunoglobulin superfamily member neuroglian (Nardi et al. 2006). In Phase 3, capsules often melanize due to activation of the PO cascade. The parasitoid is then killed by asphyxiation and/or compounds generated as a consequence of producing melanin. Capsule formation in Lepidoptera usually begins 2–6 hours after parasitism and is completed by 48 hours under standard conditions.

PDV-Mediated Responses against the Lepidopteran Immune System

Most studies indicate that PDVs prevent their associated parasitoid from being recognized and encapsulated by expressing immunosuppressive gene products in the parasitized host (Davies, Strand, and Vinson 1987; Lavine and Beckage 1995; Strand and Pech 1995; Schmidt, Theopold, and Strand 2001; Webb and Strand 2005; Strand 2008b). In some parasitoid-host systems, viral infection appears only to suppress encapsulation of the wasp (Doucet and Cusson 1996), whereas in others infection results in broader immunosuppressive effects that prevent encapsulation of a variety of foreign targets as well as disabling other cellular and humoral immune defenses (summarized by Webb and Strand 2005; Strand 2008b).

Relatively few PDV genes have been experimentally linked with causing these immunosuppressive effects, and less still is known about the host molecules or signaling pathways with which PDV genes interact. Among BVs, MdBV carried by the microgastrine braconid *M. demolitor* broadly immunosuppresses host insects by inhibiting hemocyte binding to foreign targets, blocking phagocytosis, inducing apoptosis of granulocytes, and inhibiting inducible humoral defenses (Strand 2008b). All of these effects manifest themselves within 12 hours of MdBV infecting a host and involve gene products that are expressed and function within hemocytes, or that are expressed in hemocytes and other tissues like the fat body and are secreted into the hemolymph. Functional assays implicate members of four gene families in causing these alterations. Inhibition of encapsulation and phagocytosis are primarily mediated by the interaction between Glc1.8, a surface mucin, and two PTPs, PTP-H2 and PTP-H3, that localize to focal adhesion complexes in infected immune cells (Beck and Strand 2003, 2005; Pruijssers and Strand 2007). Recent studies also indicate that PTP-H2 is an apoptosis inducer (Suderman, Pruijssers, and Strand 2008). The Toll and immune deficiency (imd) signaling pathways regulate expression of a diversity of immune genes following parasitoid attack or infection by other organisms. Both pathways involve specific NF-κB transcription factors that are normally regulated by endogenous inhibitor κB (IκB) proteins. The twelve-member *ank* gene family of MdBV shares significant homology with insect IκBs and two family members, ANK-H4 and ANK-N5, suppress Toll and imd signaling by binding insect NF-κBs (Thoetkiattikul, Beck, and Strand 2005). As previously noted, melanin formation is regulated by the PO cascade, which consists of multiple serine proteinases that terminate with the zymogen proPO (Kanost and Gorman 2008). The number of proteolytic steps in the PO cascade is unknown for any insect, but the serine proteinases that activate proPO, called proPAPs, are themselves activated by multiple upstream proteinases. One member of the MdBV *egf* gene family, EGF1.0, inhibits melanization by functioning as a dual activity inhibitor that blocks both proPAP processing and the enzymatic activity of processed PAPs (Beck and Strand 2007; Lu et al. 2008). In contrast, other EGF protein family members, like EGF0.4, are also likely serine proteinase inhibitors, but their unidentified target enzymes have functions unrelated to the PO cascade (Beck and Strand 2007).

Most other studies on BV-mediated immunosuppression involve viruses from wasps in the genus *Cotesia*. Recent studies with *C. plutellae* BV (CpBV) implicate a virally encoded PTP, lectin, histone, and an EP1-like gene in disrupting hemocyte adhesion to foreign surfaces and hemocyte proliferation (Gad and Kim 2008; Ibrahim and Kim 2008; Kwon and Kim 2008; Lee, Nalini, and Kim 2008). Infection of *Pieris rapae* by CrBV from *C. rubecula* results in transient cytoskeletal alterations in hemocytes and suppression of melanization. The *CrV1* gene has been implicated in causing the former, while the latter involves a nonviral serine proteinase homologue present in the venom gland secretion of *C. rubecula* (Asgari, Schmidt, and Theopold 1997; Zhang et al. 2004). Disruption of encapsulation by CcBV from *C. congregata* is associated with alterations to the cytoskeleton of host hemocytes and formation of aggregations rather than loss of adhesion. CcBV and other *Cotesia* BVs also reduce melanization. *Toxoneuron nigriceps* BV (TnBV) encodes a gene (*TnBV1*) that appears to activate caspases in the absence of apoptosis (Lapointe et al. 2005) as well as *ank* genes that interact with insect NF-kBs (Falabella et al. 2007). Unlike other BV-carrying braconids, wasps in the genus *Chelonus* are all egg-larval parasitoids. Little is known about the role of cheloonine BVs in immune suppression, but studies with CiBV indicate that viral infection of hosts does selectively protect *C. inanitus* larvae from encapsulation (Lanzrein et al. 1998).

IV genes implicated in immunosuppression include the Cys-motif proteins VHv1.1 and VHv1.4 of CsIV from *C. sonorensis* that are expressed in host fat body and that are secreted into the hemolymph. Both proteins bind to the surface of hemocytes. Expression of VHv1.1 also reduces encapsulation of *C. sonorensis* eggs in hosts infected with a VHv1.1-expressing recombinant baculovirus (Li and Webb 1994; Cui, Soldevila, and Webb 1997). Other IV genes have been more indirectly implicated in immunosuppression, because of homology to previously described BV genes or because they are expressed in key immune tissues like hemocytes or the fat body. IV encoded *ank* genes, for example, may function as NF-κB inhibitors, but structural differences with BV *ank* genes as well as recent expression studies suggest the possibility they may have other functions (Kroemer and Webb 2004b; Webb et al. 2006; see below). Products of the CsIV *innex* genes form functional gap junctions that have diverse functions related to intercellular communication (Turnbull et al. 2005). Given observations that gap junctions form between many hemocytes in capsules, it is possible that IV *innex* gene products may play a role in altering immune cell functions. The P30 ORF from HdIV is likewise predicted to encode a mucinlike protein with weak similarity to Glc1.8 from MdBV, but its function in immunity is unclear (Galibert et al. 2003).

Although viral gene expression is usually required for blocking host immune defenses toward the parasitoid, some BVs and IVs appear to passively protect newly laid parasitoid eggs by coating their surface, which prevents the host from either recognizing the parasitoid egg as foreign or interferes with binding by host hemocytes. Eggs of *Cotesia kariyai*, for example, are coated by CkBV virions as well as other proteins secreted from ovarial cells. Two ovarial immunoevasive proteins, IEP1 and 2, have been identified that are not encoded within the CkBV genome but are detected on the surface of CkBV virions. IEP proteins also confer protection to eggs from encapsulation and elimination of CkBV by hemocytes (Hayakawa and Yazaki 1997). IEP ovarial proteins also coat the eggs of other braconids in the genera *Cotesia* and *Toxoneuron* (Davies and Vinson 1986; Asgari et al. 1998). Among ichneumonids, *Venturia canescens* produces viruslike particles (VLPs) morphologically similar to the IVs associated with other members of the *Campopleginae*, but these particles appear to lack any nucleic acid. Hosts parasitized by *V. canescens* also remain capable of mounting an encapsulation response against numerous foreign targets but are unable to encapsulate *V. canescens* because of VLPs and a hemomucin that coat the surface of their eggs (Theopold et al. 1996). TrIV similarly coats the eggs of *Tranosema rosele* in a manner that also likely confers passive protection from encapsulation (Cusson et al. 1998).

Developmental Interactions between Lepidoptera and PDV-Carrying Parasitoids

The second alteration in lepidopteran physiology that is strongly linked to PDV gene expression is changes in the growth and development of parasitized hosts. All PDV-carrying ichneumonids are solitary (one offspring produced per host), whereas PDV-carrrying braconids are either solitary or gregarious (multiple offspring per host). Chelonine braconids (e.g., *Chelonus*, *Ascogaster*) oviposit into the egg stage of Lepidoptera, but almost all other PDV-carrying wasps parasitize larvae during multiple instars. The progeny of PDV-carrying parasitoids either complete their immature development in final instar of the host or develop rapidly and complete development in an earlier instar (Pennacchio and Strand 2006). Lepidoptera parasitized by solitary species usually exhibit dramatic reductions in weight gain, delays in molting, and the inability to pupate. Hosts parasitized by gregarious species are also inhibited from pupating but usually exhibit less severe alterations in weight gain. The exceptions to these trends are, again, chelonine braconids, which cause hosts to initiate precocious metamorphosis one instar earlier than normal but prevent the host from actually pupating (Lanzrein, Pfister-Wilhelm, and von Niederhäusern 2001).

Alterations in Lepidopteran Endocrine and Metabolic Physiology following Parasitism or Infection by PDVs

The alterations described above are associated with changes in both endocrine and metabolic physiology (Thompson and Dahlman 1998; Beckage and Gelman 2004; Pennacchio and Strand 2006). Endocrine alterations include increases in hemolymph juvenile hormone (JH) titers and a failure of ecdysteroid titers to rise to levels that normally occur in nonparasitized hosts during a larval-larval or larval-pupal molt. Most studies indicate that elevated JH titers correlate with reductions in the activity of host metabolic enzymes like JH esterase (Balgopal et al. 1996; Dong, Zang, and Dahlman 1996; Schafellner, Marktl, and Schopf 2007) or the secretion of JH from the parasitoid larva (Cole et al. 2002). Only a few studies report decreases in the synthesis and secretion of JH by the corpora allata of the host (Cole et al. 2002; Li et al. 2003). In contrast, a diversity of alterations has been implicated in suppression of ecdysteroid titers including reduced synthesis and release of PTTH (Tanaka, Agui, and Hiruma 1987; Hayakawa 1995), insensitivity of prothoracic glands (PTGs) to PTTH stimulation (Kelly et al. 1998), reduced biosynthetic activity in PTGs (Tanaka, Agui, and Hiruma 1987; Pennacchio, Digilio, and Tremblay 1995), and premature death of PTG cells (Dover, Tanaka, and Vinson 1995). Experiments with different *Chelonus* species indicate that precocious metamorphosis occurs because JH titers decrease earlier than normal through a combination of corpora allata (CA) inactivation, increased juvenile hormone esterase (JHE) activity, and an altered ecdysteroid titer (Reed and Brown 1998; Lanzrein, Pfister-Wilhelm, and von Niederhäusern 2001). Metabolic alterations include changes in the abundance of hemolymph proteins, free amino acids, and carbohydrates (Thompson 1993; Thompson and Dahlman 1998). Further, alterations in protein abundance appear to be associated with selective alterations in protein synthesis. For example, parasitism by *C. sonorensis* and CsIV infection greatly reduces the abundance of arylphorin, a major hemolymph storage protein, but does not affect other major hemolymph proteins like lipophorin and transferrin (Shelby and Webb 1994, 1997). Triglycerides and glycogen deposits in fat body are also greatly reduced in hosts parasitized by *C. sonorensis*, while levels of free sugars in hemolymph increase (Vinson 1990). Similar alterations have also been reported in hosts parasitized by braconids (see Thompson and Dahlman 1998; Pennacchio and Strand 2006).

The Role of PDV Gene Products in Altering Host Growth and Development

The alterations in endocrine and metabolic physiology described are almost certainly linked, given the essential role that the insect endocrine system plays in regulating nutrient storage and mobilization (Vinson 1990; Britton et al. 2002; Strand and Casas 2007). Elevating host JH titers and blocking the rise in 20E levels could also be partially responsible for altering titers of specific storage proteins that are normally upregulated late in larval development of Lepidoptera, whereas carbohydrate and lipid metabolism in insects is under regulation by the insulin-signaling pathway (Brown et al. 2008). Pennacchio and Strand (2006) emphasized that a central theme in parasitoid life history is the disabling of metabolic sinks like metamorphosis and/or reproduction, and the mobilization of nutrient stores that together redirect energetic resources away from host tissues and toward the parasitoid's progeny. Thus, the primary adaptive significance of altering host endocrine physiology and reproduction is more likely metabolic, whereas associated phenomena like arrested development and inhibition of pupation are indirect consequences of redirecting host nutritional resources.

Less clear is the identity of the PDV gene products responsible for causing these alterations. Expression of PTP genes from TnBV in host PTGs (Falabella et al. 2006) has been suggested to disrupt PTTH signaling and/or ecdysteroid biosynthesis via altered phosphorylation of pathway components. Several PTP genes encoded by MdBV from *M. demolitor* are also differentially expressed in the nervous system and prothoracic glands of infected hosts (Pruijssers and Strand 2007), while two novel genes encoded by CiBV appear involved in regulating developmental arrest following the onset of precocious metamorphosis (Bonvin et al. 2005). Two members of the cys-motif gene family, VHv1.1 and VHv1.4, encoded by CsIV, disable protein translation in certain host tissues (fat body, hemocytes, testis) as well as in lepidopteran-derived cell lines like TN368 cells from *Trichoplusia ni* (Shelby and Webb 1997; Kim 2005). Injection or feeding of recombinant VHv1.1 and VHv1.4 from CsIV causes host larvae to gain weight more slowly and to pupate abnormally more frequently than seen in control larvae (Fath-Goodin et al. 2006). Interestingly, a gene product (TSP14) secreted by teratocytes from the braconid *M. croceipes* shares some structural similarities with IV cys-motif genes and in bioassays exhibits similar activity to VH1.1 and VHv1.4 (Dahlman et al. 2003).

PROSPECTIVE USES FOR PDV GENES IN PEST MANAGEMENT

Given the essential roles of PDVs in altering the immune response and development of host insects, studies are beginning to explore the potential use of PDV gene products for control of lepidopteran and other insect pests, and as tools for use in biotechnology applications. It is important to reemphasize that, in addition to the specific PDV genes and gene families identified to date from fewer than a dozen species, an estimated thirty thousand species of parasitoids in total carry these viruses, and many of these wasps attack a diversity of important agricultural pest insects. Given evidence that each PDV-parasitoid association is genetically unique, these viruses collectively represent a vast repository of virulence molecules with potential for development as environmentally compatible, new-generation pesticides or as pharmacological agents with diverse applications.

Three strategies for using PDV genes are currently being pursued. The first is to use PDV gene products with immunosuppressive activity to enhance the efficacy of other insecticidal agents like baculoviruses. The idea is that suppressing host immune defenses could enhance the pathogenicity and/or host range of other control agent of interest. This was explored by Washburn et al. (2000) who coinfected *M. sexta* larvae with the baculovirus *Autographa californica* Multicapsid Nucleopolyhedrosis virus (AcMNPV) and the bracovirus CcBV. It has also been examined using AcMNPV and the ichnovirus CsIV with other lepidopterans (Rivkin et al. 2006). In both cases, coinfection with PDVs synergized baculovirus infection. Parasitism by PDV-carrying parasitoids or infection of hosts with PDVs alone is also well known to increase the susceptibility of lepidopteran larvae to other natural enemies including bacteria, other viruses, and parasitoids (Shelby and Webb 1999; Kadash, Harvey, and Strand 2003; Stoltz and Makkay 2003). Prior studies have also demonstrated a correlation between the expression patterns of PDV genes and host range (Cui, Soldevila, and Webb 2000), as well as establishing that several immunosuppressive PDV genes have biological activity in lepidopteran and nonlepidopteran insects outside the normal host range of the associated parasitoid (Thoetkiattikul, Beck, and Strand 2005; Beck and Strand 2007). Taken together, these results provide proof of principle that PDV immunosuppressive genes could be used to inhibit components of the immune system in pest species and, in the process, enhance the efficacy of several potential biological control agents.

The second approach focuses on using PDV genes with growth-inhibiting activity directly as candidate insecticidal agents delivered via expression in transgenic crops or by conventional application. As previously discussed, although cys-motif–related gene products are not naturally delivered to the gut of host insects during parasitism, feeding trials using recombinant VHv1.1 and 1.4 from CsIV and the related teratocyte protein TSP14 from *M. croceipes* exhibit growth-reducing activity similar to what occurs during parasitism (Maiti et al. 2003; Fath-Goodin et al. 2006). Growth-reducing activity and reduced feeding damage are also found when these proteins are constituatively expressed in transgenic tobacco (Gill et al. 2006), indicating that at least some PDV gene products have potential to reduce crop plant damage by Lepidoptera.

The third strategy under consideration involves using PDV-derived proteins as reagents for biotechnology applications. One example that has recently been commercially developed involves the use of PDV genes to enhance the productivity of baculovirus expression sytems (BEVs) that are extensively used to produce recombinant proteins. During studies to characterize expression patterns and cellular localization of *ank* gene family members encoded by CsIV, Kroemer and Webb (2004b) observed that recombinant baculoviruses expressing two *ank* gene family members exhibited enhanced longevity compared with control viruses or recombinant viruses expressing other family members. Given that longevity of insect cells is a key rate-limiting factor in protein production, enhancing host cell survival has the potential to increase recombinant protein production. Although mode of action remains unclear, stably transformed cell lines that express ANK proteins or dual BEVS that coexpress *ank* genes in combination with a novel gene of interest yield 4- to 15-fold increases in recombinant protein production.

REFERENCES

Albrecht, U., T. Wyler, R. Pfister-Wilhelm, et al. 1994. PDV of the parasitic wasp *Chelonus inanitus* (Braconidae): Characterization, genome organization and time point of replication. *J. Gen. Virol.* 75:3353–63.

Annaheim, M., and B. Lanzrein. 2007. Genome organization of the *Chelonus inanitus* polydnavirus: Excision sites, spacers, and abundance of proviral and excised segments. *J. Gen. Virol.* 88:450–57.

Asgari, S., O. Schmidt, and U. Theopold. 1997. A polydnavirus-encoded protein of an endoparasitoid wasp is an immune suppressor. *J. Gen. Virol.* 78:3061–70.

Asgari, S., U. Theopold, C. Wellby, and O. Schmidt. 1998. A protein with protective properties against the cellular defense reactions in insects. *Proc. Natl. Acad. Sci. U.S.A.* 95:3690–95.

Askew, R.R., and M.R. Shaw. 1986. Parasitoid communities: Their size, structure and development. In *Insect parasitoids*, ed. J.K. Waage and D. Greathead, 225–63. London: Academic Press.

Balgopal, M.M., B.A. Dover, W.G. Goodman, and M.R. Strand. 1996. Parasitism by *Microplitis demolitor* induces alterations in the juvenile hormone titers and juvenile hormone esterase activity of its host, *Pseudoplusia includens*. *J. Insect Physiol.* 42:337–45.

Beck, M.H., R.B. Inman, and M.R. Strand. 2007. *Microplitis demolitor* bracovirus genome segments vary in abundance and are individually packaged in virions. *Virology* 359:179–89.

Beck, M., and M.R. Strand. 2003. RNA interference silences *Microplitis demolitor* bracovirus genes and implicates glc1.8 in disruption of adhesion in infected host cells. *Virology* 314:521–35.

Beck, M., and M.R. Strand. 2005. Glc1.8 from *Microplitis demolitor* bracovirus induces a loss of adhesion and phagocytosis in insect high five and S2 cells. *J. Virol.* 79:1861–70.

Beck, M., and M.R. Strand. 2007. A novel polydnavirus protein inhibits the insect prophenoloxidase activation pathway. *Proc. Natl. Acad. Sci. U.S.A.* 104:19267–72.

Beckage, N.E., and D.B. Gelman. 2004. Wasp parasitoid disruption of host development: Implications for new biologically based strategies for insect control. *Annu. Rev. Entomol.* 49:299–330.

Belle, E., N.E. Beckage, J. Rousselet, M. Poirie, F. Lemeunier, and J.-M. Drezen. 2002. Visualization of polydnavirus sequences in a parasitoid wasp chromosome. *J. Virol.* 76:5793–96.

Bézier, A., M. Annaheim, J. Herbinière, et al. 2009. Polydnaviruses of braconid wasps derive from an ancestral nudivirus. *Science* 323:926–30.

Bézier, A., J. Herbinière, C. Serbielle, et al. 2008. Bracovirus gene products are highly divergent from insect proteins. *Arch. Insect Biochem. Physiol.* 17:172–87.

Bonvin, M., D. Marti, S. Wyder, D. Kojic, M. Annaheim, and B. Lanzrein. 2005. Cloning, characterization and analysis by RNA interference of various genes of the *Chelonus inantius* polydnavirus. *J. Gen. Virol.* 86:973–83.

Britton, J.S., W.K. Lockwood, L. Li, S.M. Cohen, and B.A. Edgar. 2002. *Drosophila*'s insulin/PI3-kinase pathway coordinates cellular metabolism with nutritional conditions. *Dev. Cell* 2:239–49.

Brown, M.R., K.D. Clark, M. Gulia, et al. 2008. An insulin-like peptide regulates egg maturation and metabolism in the mosquito *Aedes aegypti*. *Proc. Natl. Acad. Sci. U.S.A.* 105:5716–21.

Cerenius, L., and K. Soderhall. 2004. The prophenoloxidse-activating system in invertebrates. *Immunol. Rev.* 198:116–26.

Cole, T. J., N.E. Beckage, F.F. Tan, A. Srinivasan, and S.B. Ramaswamy. 2002. Parasitoid-host endocrine relations: Self-reliance or co-optation? *Insect Biochem. Mol. Biol.* 32:1673–79.

Cui, L., A.I. Soldevila, and B.A. Webb. 1997. Expression and hemocyte-targeting of a *Campoletis sonorensis* polydnavirus cysteine-rich gene in *Heliothis virescens* larvae. *Arch. Insect Biochem. Physiol.* 36:251–71.

Cui, L., A.I. Soldevila, and B.A. Webb. 2000. Relationships between polydnavirus gene expression and host range of the parasitoid wasp *Campoletis sonorensis*. *J. Insect Physiol.* 46:1397–1407.

Cui, L., and B.A. Webb. 1997. Homologous sequences in the *Campolitis sonorensis* polydnavirus genome are implicated in replication and nesting of the W segment family. *J. Virol.* 71:8504–13.

Cusson, M., C. Lucarotti, D. Stoltz, P. Krell, and D. Doucet. 1998. A polydnavirus from the spruce budworm parasitoid *Tranosema rostrale* (Ichneumonidae). *J. Invertebr. Pathol.* 72:50–56.

Dahlman, D.L., E.L. Scheppers, T. Scheppers, R.L. Rana, F.A. Diluna, and B.A. Webb. 2003. A gene from a parasitic wasp expressed in teratocytes inhibits host protein synthesis, growth and development. *Insect Mol. Biol.* 12:527–34.

Davies, D.H., M.R. Strand, and S.B. Vinson. 1987. Changes in differential haemocyte count and *in vitro* behaviour of plasmatocytes from host *Heliothis virescens* caused by *Campoletis sonorensis* PDV. *J. Insect Physiol.* 33:143–53.

Davies, D.H., and S.B. Vinson. 1986. Passive evasion by eggs of braconid parasitoid *Cardiochiles nigriceps* of encapsulation in vitro by haemocytes of host *Heliothis virescens*. Possible role for fibrous layer in immunity. *J. Insect Physiol.* 32:1003–10.

Desjardins, C.A., D.E. Gundersen-Rindal, J.B. Hostetler, et al. 2007. Structure and evolution of a proviral locus of *Glyptapanteles indiensis* bracovirus. *BMC Microbiol.* 7:61.

Dong, Y.M., H.E. Taylor, and G. Dimopoulos. 2006. AgDscam, a hypervariable immunoglobulin domain-containing receptor of the *Anopheles gambiae* immune system. *PLoS Biol.* 4:1137–46.

Dong, K., D. Zang, and D.L. Dahlman. 1996. Down-regulation of juvenile hormone esterase and arylphorin production in *Heliothis virescens* larvae parasitized by *Microplitis croceipes*. *Arch. Insect Biochem. Physiol.* 32:237–48.

Doucet, D., and M. Cusson. 1996. Role of calyx fluid in alterations of immunity in *Choristoneura fumiferana* larvae parasitized by *Tranosema rostrale*. *Entomol. Exp. Appl.* 81:21–30.

Dover, B.A., T. Tanaka, and S.B. Vinson. 1995. Stadium-specific degeneration of host prothoracic glands by *Campoletis sonorensis* caly fluid and its association with host ecdysteroid titers. *J. Insect Physiol.* 41:947–55.

Drezen, J.-M., B. Provost, E. Espagne, et al. 2003. Polydnavirus genome: Integrated vs. free virus. *J. Insect Physiol.* 49:407–17.

Dupuy, C., E. Huguet, and J.-M. Drezen. 2006. Unfolding the evolutionary history of polydnaviruses. *Virus Res.* 117:81–89.

Espagne, E., C. Dupuy, E. Huguet, et al. 2004. Genome sequence of a polydnavirus: Insights into symbiotic virus evolution. *Science* 306:286–89.

Falabella, P., P. Cacciaupi, P. Varricchio, C. Malva, and P. Pennacchio. 2006. Protein tyrosine phosphatases of *Toxoneuron nigriceps* bracovirus as potential disrupters of host prothoracic gland function. *Arch. Insect Biochem. Physiol.* 61:157–69.

Falabella, P., P. Varricchio, B. Provost, et al. 2007. Characterization of the IkappaB-like gene family in polydnaviruses associated with wasps belonging to different braconid subfamilies. *J. Gen. Virol.* 88:92–104.

Fath-Goodin, A., T.A. Gill, S.B. Martin, and B.A. Webb. 2006. Effect of *Campoletis sonorensis* cys-motif proteins on *Heliothis virescens* development. *J. Insect Physiol.* 52:576–85.

Fleming, J.G., and M.D. Summers. 1991. PDV DNA is integrated in the DNA of its parasitoid wasp host. *Proc. Natl. Acad. Sci. U.S.A.* 88:9770–74.

Friedman, R., and A.L. Hughes. 2006. Pattern of gene duplication in the *Cotesia congregata* bracovirus. *Infect. Genet. Evol.* 6:315–22.

Gad, W., and Y. Kim. 2008. A viral histone H4 encoded by *Cotesia plutellae* bracovirus inhibits haemocyte-spreading behaviour of the diamondback moth, *Plutellae xylostella. J. Gen. Virol.* 89:931–38.

Galibert, L., J. Rocher, M. Ravallec, M. Duonor-Cerutti, B.A. Webb, and A.N. Volkoff. 2003. Two *Hyposoter didymator* ichnovirus genes expressed in the lepidopteran host encode secreted or membrane-associated serine and threonine rich proteins in segments that may be nested. *J. Insect Physiol.* 49:441–51.

Gardiner, E.M.M., and M.R. Strand. 2000. Hematopoiesis in larval *Pseudoplusia includens* and *Spodoptera frugiperda. Arch. Insect Bich. Physiol.* 43:147–64.

Gill, T.A., A. Fath-Goodin, I.I. Maiti, and B.A. Webb. 2006. Potential uses of Cys-motif and other polydnavirus genes in biotechnology. *Adv. Virus Res.* 68:393–425.

Gillespie, J.P., M.R. Kanost, and T. Trenczek. 1997. Biological mediators of insect immunity. *Annu. Rev. Entomol.* 42:611–43.

Hayakawa, Y. 1995. Growth-blocking peptide: An insect biogenic peptide that prevents the onset of metamorphosis. *J. Insect Physiol.* 41:1–6.

Hayakawa, Y., and K. Yazaki. 1997. Envelope proteins of parasitic wasp symbiont virus, protects the wasp eggs from the cellular immune reactions by the host insect. *Eur. J. Biochem.* 246:820–26.

Ibrahim, A.M., and Y. Kim. 2008. Transient expression of protein tyrosine phosphatases encoded by *Cotesia plutellae* bracovirus inhibits insect cellular immune responses. *Naturwissenschaften* 95:25–32.

Imler, J.-L., and P. Bulet. 2005. Antimicrobial peptides in *Drosophila*, structures, activities and gene regulation. In *Mechanisms of epithelial defense*, ed. D. Kabelitz and J.M. Schroder, Vol. 86, 1–21. Basil: Karger.

Irving, P., J. Ubeda, D. Doucet, et al. 2005. New insights into *Drosophila* larval haemocyte functions through genome-wide analysis. *Cell Microbiol.* 7:335–50.

Kadash, K., J.A. Harvey, and M.R. Strand 2003. Cross-protection experiments with parasitoids in the genus *Microplitis* (Hymenoptera: Braconidae) suggest a high level of specificity in their associated bracoviruses. *J. Insect Physiol.* 49:473–82.

Kanost, M.R., and M.J. Gorman. 2008. Phenoloxidases in insect immunity. In *Insect immunity*, ed. N.E. Beckage, 69–96. San Diego: Academic Press.

Kelly, T.J., D.B. Gelman, D.A. Reed, and N.E. Beckage. 1998. Effects of parasitization by *Cotesia congregata* on the brain-prothoracic gland axis of its host, *Manduca sexta. J. Insect Physiol.* 44:323–32.

Kim, Y. 2005. Identification of host translation inhibitory factor of *Campoletis sonorensis* ichnovirus on the tobacco budworm, *Heliothis virescens. Arch. Insect Biochem. Physiol.* 59:230–44.

Kocks, C., J.-H. Cho, N. Nehme, et al. 2005. Eater, a transmembrane protein mediating phagocytosis of bacterial pathogens in *Drosophila. Cell* 123:335–46.

Kroemer, J.A., and B.A. Webb. 2004a. Polydnavirus genes and genomes: Emerging gene families and new insights into polydnavirus replication. *Annu. Rev. Entomol.* 49:431–56.

Kroemer, J.A., and B.A. Webb. 2004b. Divergences in protein activity and cellular localization within the *Campoletis sonorensis* ichnovirus vankyrin family. *J. Virol.* 80:12219–28.

Kwon, B., and Y. Kim. 2008. Transient expression of an EP1-like gene encoded in *Cotesia plutellae* bracovirus suppresses the hemocyte population in the diamondback moth, *Plutella xylostella. Dev. Comp. Immunol.* 32:932–42.

Lanzrein, B., R. Pfister-Wilhelm, and F. von Niederhäusern. 2001. Effects of an egg-larval parasitoid and its polydnavirus on development and the endocrine system of the host. In *Endocrine interactions of insect parasites and pathogens*, ed. J.P. Edwards and R.J. Weaver, 95–109. Oxford: BIOS Sci. Publishers.

Lanzrein, B., R. Pfister-Wilhelm, T. Wyler, T. Trenczek, and P. Stettler. 1998. Overview of parasitism associated effects on host haemocytes in larval parasitoids and comparison with effects of the egg-larval parasitoid *Chelonus inanitus* on its host *Spodoptera littoralis. J. Insect Physiol.* 44:817–31.

Lapointe, R., K. Tanaka, W.E. Barney, et al. 2007. Genomic and morphological features of a banchine polydnavirus: Comparison with bracoviruses and ichnoviruses. *J. Virol.* 81:6491–6501.

Lapointe, R., R. Wilson, L. Vilaplana, et al. 2005. Expression of a *Toxoneuron nigriceps* polydnavirus (TnBV) encoded protein, causes apoptosis-like programmed cell death in lepidopteran insect cells. *J. Gen. Virol.* 86:963–71.

Lavine, M.D., and N.B. Beckage. 1995. Polydnaviruses: Potent mediators of host insect immune dysfunction. *Parasitol. Today* 11:368–78.

Lavine, M.D., and M.R. Strand. 2002. Insect hemocytes and their role in cellular immune responses. *Insect Biochem. Mol. Biol.* 32:1237–42.

Lavine, M.D., and M.R. Strand. 2003. Hemocytes from *Pseudoplusia includens* express multiple alpha and beta integrin subunits. *Insect Mol. Biol.* 12:441–52.

Lee, S., M. Nalini, and Y. Kim. 2008. A viral lectin encoded in *Cotesia plutellae* bracovirus and its immunosuppressive effect on host hemocytes. *Comp. Biochem. Physiol. A* 149:351–61.

Levashina, E.A., L.F. Moita, S. Blandin, G. Vriend, M. Lagueux, and F.C. Kafatos. 2001. Conserved role of a complement-like protein in phagocytosis revealed by dsRNA knockout in cultured cells of the mosquito, *Anopheles gambiae*. *Cell* 104:709–18.

Levin, D.M., L.N. Breuer, S.F. Zhuang, S.A. Anderson, J.B. Nardi, and M.R. Kanost. 2005. A hemocyte-specific integrin requered for hemocytic encapsulation in the tobacco hornworm, *Manduca sexta*. *Insect Biochem. Mol. Biol.* 35:369–80.

Li, S., P. Falabella, I. Kuriachan, et al. 2003. Juvenile hormone synthesis, metabolism, and resulting haemolymph titre in *Heliothis virescens* larvae parasitized by *Toxoneuron nigriceps*. *J. Insect Physiol.* 49:1023–30.

Li, X., and B.A. Webb. 1994. Apparent functional role for a cysteine-rich polydnavirus protein in suppression of the insect cellular immune response. *J. Virol.* 68:7482–89.

Ling, E., K. Shirai, R. Kanekatsu, and K. Kiguchi. 2005. Hemocyte differentiation in the hematopoietic organs of the silkworm, *Bombyx mori*: Prohemocytes have the function of phagocytosis. *Cell Tiss. Res.* 320:535–43.

Ling, E.J., and X.Q. Yu. 2006. Cellular encapsulation and melanization are enhanced by immulectins, pattern recognition receptors from the tobacco hornworm *Manduca sexta*. *Dev. Comp. Immunol.* 30:289–99.

Lu, Z., M.H. Beck, H. Jiang, Y. Wang, and M.R. Strand. 2008. The viral protein Egf1.0 is a dual activity inhibitor of prophenoloxidase activating proteinases 1 and 3 from *Manduca sexta*. *J. Biol. Chem.* 283: 21325–233.

Maiti, I.B., N. Dey, D.L. Dahlman, and B.A. Webb. 2003. Antibiosis-type resistance in transgenic plants expressing teratocyte secretory peptide (TSP) gene from a hymenopteran endoparasite (*Microplitis croceipes*). *Plant Biotechnol.* 1:209–19.

Marti, D., C. Grossniklaus-Burgin, S. Wyder, T. Wyler, and B. Lanzrein. 2003. Ovary development and polydnavirus morphogenesis in the parasitic wasp *Chelonus inanitus*. I. Ovary morphogenesis, amplification of viral DNA and ecdysteroid titres. *J. Gen. Virol.* 84:1141–50.

Moita, L.F., R. Wang-Sattler, K. Michel, et al. 2005. In vivo identification of novel regulators and conserved pathways of phagocytosis in *A. gambiae*. *Immunity* 23:65–73.

Murphy, N., J.C. Banks, J.B. Whitfield, and A.D. Austin. 2008. Phylogeny of the parasitic microgastroid subfamilies (Hymenoptera: Braconidae) based on sequence data from seven genes, with an improved time estimate of the origin of the lineage. *Mol. Phylogenet. Evol.* 47:378–95.

Nakahara, Y., Y. Kanamori, M. Kiuchi, and M. Kamimura. 2003. In vitro studies of hematopoiesis in the silkworm, cell proliferation in and hemocyte discharge from the hematopoietic organ. *J. Insect Physiol.* 49:907–16.

Nardi, J.B., B. Pilas, C.M. Bee, S. Zhuang, K. Garsha, and M.R. Kanost. 2006. Neuroglian-positive plasmatocytes of *Manduca sexta* and the initiation of hemocyte attachment to foreign surfaces. *Dev. Comp. Immunol.* 30:447–62.

Nardi, J.B., B. Pilas, E. Ujhelyi, K. Garsha, and M.R. Kanost. 2003. Hematopoietic organs of *Manduca sexta* and hemocyte lineages. *Dev. Genes Evol.* 213:477–91.

Pasquier-Barre, F., C. Dupuy, E. Huguet, et al. 2002. Polydnavirus replication: The EP1 segment of the parasitoid wasp *Cotesia congregata* is amplified within a larger precursor molecule. *J. Gen. Virol.* 83:2035–45.

Pech, L.L., and M.R. Strand. 1996. Granular cells are required for encapsulation of foreign targets by insect haemocytes. *J. Cell Sci.* 109:2053–60.

Pennacchio, F., M.C. Digilio, and E. Tremblay. 1995. Biochemical and metabolic alterations in *Acyrthosiphon pisum* parasitized by *Aphidius ervi*. *Arch. Insect Biochem. Physiol.* 30:351–67.

Pennacchio, F., and M.R. Strand. 2006. Evolution of developmental strategies in parasitic Hymenoptera. *Annu. Rev. Entomol.* 51:233–58.

Philips, J.A., E.J. Rubin, and N. Perrimon. 2005. *Drosophila* RNAi screen reveals C35 family member required for mycobacterial infection. *Science* 309:1248–51.

Powell, J.A. 2003. Lepidoptera (moths, butterflies). In *Encyclopedia of insects*, ed. V.H. Resh and R.T. Carde, 631–64. San Diego: Academic Press.

Pruijssers, A.J., and M.R. Strand. 2007. PTP-H2 and PTP-H3 from *Microplitis demolitor* bracovirus localize to focal adhesions and are antiphagocytic in insect immune cells. *J. Virol.* 81:1209–19.

Ramet, M., P. Manfruelli, A. Pearson, B. Mathey-Prevot, and R.A.B. Ezekowitz. 2002. Functional genomic analysis and identification of a *Drosophila* receptor for *E. coli*. *Nature* 416:644–48.

Ramet, M., A. Pearson, P. Manfruelli, et al. 2001. *Drosophila* scavenger receptor CI is a pattern recognition receptor for bacteria. *Immunity* 15:1027–38.

Reed, D.A., and J.J. Brown. 1998. Host/parasitoid interactions: Critical timing of parasitoid-derived products. *J. Insect Physiol.* 44:721–32.

Rivkin, H., J.A. Kroemer, A. Bronshtein, E. Belausov, B.A. Webb, and N. Chejanovsky. 2006. Response of immunocompetent and immunocompromised *Spodoptera littoralis* larvae to baculovirus infection. *J. Gen. Virol.* 87:2217–25.

Schafellner, C., R.C. Marktl, and A. Schopf. 2007. Inhibition of juvenile hormone esterase in *Lymantria dispar* (Lepidoptera: Lymantriidae) larvae parasitized by *Glyptapanteles liparidis* (Hymenoptera: Braconidae). *J. Insect Physiol.* 53:858–68.

Shelby, K.S., and B.A. Webb. 1994. Polydnavirus infection inhibits synthesis of an insect plasma protein, arylphorin. *J. Gen. Virol.* 75:2285–94.

Shelby, K.S., and B.A. Webb. 1997. Polydnavirus infection inhibits translation of specific growth-associated host proteins. *Insect Biochem. Mol. Biol.* 27:263–70.

Shelby, K.S., and B.A. Webb. 1999. Polydnavirus-mediated suppression of insect immunity. *J. Insect Physiol.* 45:507–14.

Schmidt, O., U. Theopold, and M. Strand. 2001. Innate immunity and its evasion and suppression by hymenopteran endoparasitoids. *Bioessays* 23:344–51.

Stoltz, D.B., N.E. Beckage, G.W. Blissard, et al. 1995. Polydnaviridae. In *Virus taxonomy*, ed. F.A. Murphy, C.M. Fauquet, D.H.L. Bishop, et al., 143–47. New York: Springer-Verlag.

Stoltz, D., and A. Makkay. 2003. Overt viral diseases induced from apparent latency following parasitization by the ichneumonid wasp, *Hyposoter exiguae. J. Insect Physiol.* 49:483–90.

Stoltz, D.B., and S.B. Vinson. 1979. Viruses and parasitism in insects. *Adv. Virus Res.* 24:125–71.

Stoltz, D.B., S.B. Vinson, and E.A. MacKinnon. 1976. Baculovirus-like particles in the reproductive tracts of female parasitoid wasps. *Can. J. Microbiol.* 22:1013–23.

Strand, M.R. 2008a. Insect hemocytes and their role in immunity. In *Insect immunity*, ed. N.E. Beckage, 25–47. San Diego: Academic Press.

Strand, M.R. 2008b. Polydnavirus abrogation of the insect immune system. In *Encyclopedia of virology*, ed. B.W.J. Mahy and M.H.V. van Regenmortel, Vol. 4, 250–256. London: Elsevier.

Strand, M.R., and J. Casas. 2007. Parasitoid and host nutritional physiology in behavioral ecology. In *Parasitoid behavioral ecology*, ed. E. Wajnberg, J. van Alphen, and C. Bernstein, 113–28. Oxford: Blackwell Press.

Strand, M.R., and L.L. Pech. 1995. Immunological basis for compatibility in parasitoid-host relationships. *Annu. Rev. Entomol.* 40:31–56.

Suderman, R.J., A.J. Pruijssers, and M.R. Strand. 2008. Protein tyrosine phosphatase-H2 from a polydnavirus induces apoptosis of insect cells. *J. Gen. Virol.* 89:1411–20.

Tanaka, T., N. Agui, and K. Hiruma. 1987. The parasitoid *Apanteles kariyai* inhibits pupation of its host, *Pseudaletia separata*, via disruption of prothoracicotropic hormone release. *Gen. Comp. Endocrinol.* 67:364–74.

Tanaka, K., R. Lapointe, W.E. Barney, et al. 2007. Shared and species-specific features among ichnovirus genomes. *Virology* 363:26–35.

Terenius, O., R. Bettencourt, S.Y. Lee, W. Li, K. Soderhall, and I. Faye. 2007. RNA interference of hemolin causes depletion of phenoloxidase activity in *Hyalophora cecropia. Dev. Comp. Immunol.* 31:571–75.

Theopold, U., C. Samakovlis, H. Erdjumentbromage, et al. 1996. *Helix pomatia* lectin, an inducer of *Drosophila* immune response, binds to hemomucin, a novel surface mucin. *J. Biol. Chem.* 271:12708–15.

Theopold, U., O. Schmidt, K. Soderhall, and M.S. Dushay. 2004. Coagulation in arthropods, defence, wound closure and healing. *Trends Immunol.* 25:289–94.

Thoetkiattikul, H., M.H. Beck, and M.R. Strand. 2005. Inhibitor kappaB-like proteins from a polydnavirus inhibit NF-kappaB activation and suppress the insect immune response. *Proc. Natl. Acad. Sci. U.S.A.* 102:11426–31.

Thompson, S.N. 1993. Redirection of host metabolism and effects on parasite nutrition. In *Parasites and pathogens of insects*, ed. N.E. Beckage, S.N. Thompson, and B.A. Federici, Vol. 1, 125–44, New York: Academic Press.

Thompson, S.N., and D.L. Dahlman. 1998. Aberrant nutritional regulation of carbohydrate synthesis by parasitized *Manduca sexta. J. Insect Physiol.* 44:745–54.

Turnbull, M.W., A.N. Volkoff, B.A. Webb, and P. Phelan. 2005. Functional gap junction genes are encoded by insect viruses. *Curr. Biol.* 15:R491–92.

Vinson, S.B. 1990. Physiological interactions between the host genus *Heliothis* and its guild of parasitoids. *Arch. Insect Biochem. Physiol.* 13:63–81.

Volkoff, A.-N., M. Ravallec, J. Bossy, et al. 1995. The replication of *Hyposoter didymator* PDV: Cytopathology of the calyx cells in the parasitoid. *Biol. Cell* 83:1–13.

Washburn, J.O., E.J. Haas-Stapleton, F.F. Tan, N.E. Beckage, and L.E. Volkman. 2000. Co-infection of *Manduca sexta* larvae with PDV from *Cotesia congregata* increases susceptibility to fatal infection by *Autographa californica* M Nucleopolyhedrovirus. *J. Insect Physiol.* 46:179–90.

Webb, B.A. 1998. Polydnavirus biology, genome structure, and evolution. In *The insect viruses*, ed. L.K. Miller and L.A. Balls, 105–39. New York: Plenum Press.

Webb, B.A., and L. Cui. 1998. Relationships between polydnavirus genomes and viral gene expression. *J. Insect Physiol.* 44:785–93.

Webb, B.A., and M.R. Strand. 2005. The biology and genomics of polydnaviruses. In *Comprehensive molecular insect science*, Vol. 6, ed. L.I. Gilbert, K. Iatrou, and S.S. Gill, 323–60. San Diego: Elsevier.

Webb, B.A., M.R. Strand, S.E. Dickey, et al. 2006. Polydnavirus genomes reflect their dual roles as mutualists and pathogens. *Virology* 347:160–74.

Wertheim, B., A.R. Kraaijeveld, E. Schuster, et al. 2005. Genome wide expression in response to parasitoid attack in *Drosophila*. *Genome Biol.* 6: R94.

Whitfield, J.B. 1998. Phylogeny and evolution of host-parasitoid interactions in Hymenoptera. *Annu. Rev. Entomol.* 43:129–51.

Whitfield, J.B. 2002. Estimating the age of the polydnavirus/braconid wasp symbiosis. *Proc. Natl. Acad. Sci. U.S.A.* 99:7508–13.

Whitfield, J.B., and S. Asgari. 2003. Virus or not? Phylogenetics of polydnaviruses and their wasp carriers. *J. Insect Physiol.* 49:397–405.

Xu, D., and D. Stoltz. 1993. Polydnavirus genome segment families in the ichneumonid parasitoid *Hyposoter fugitivus*. *J. Virol.* 67:1340–49.

Zhang, G., Z.-Q. Lu, H. Jiang, and S. Asgari. 2004. Negative regulation of prophenoloxidase (proPO) activation by a clip-domain serine proteinase homolog (SPH) from endoparasitoid venom. *Insect Biochem. Mol. Biol.* 34:477–83.

Zhuang, S., L. Kelo, J.B. Nardi, and M.R. Kanost. 2007. An integrin-tetraspanin interaction required for cellular innate immune responses of an inset, *Manduca sexta*. *J. Biol. Chem.* 282:22563–72.

18 Densovirus Resistance in *Bombyx mori*

Keiko Kadono-Okuda

CONTENTS

BOMBYX DENSOVIRUS

Bombyx mori densovirus (*Bm*DNV), a virus that impacts the sericultural industry by destructive damage throughout the rearing of larvae in many farms, has long been seen as a serious problem. The midgut columnar cells infected by this virus contain hypertrophied nuclei where the virus multiplies. Consequently, the nucleoplasm of infected cells is densely stained with DNA-chromophilic methyl green or Feulgen reagent (Watanabe et al. 1976). The infected susceptible larvae become flaccid and develop diarrhea as midgut tissue is destroyed, a condition known as flacherie, and finally die.

*Bm*DNV belongs to Parvoviridae. The Parvoviridae is divided into two subfamilies, Parvovirinae, whose hosts are vertebrates, and Densovirinae, whose hosts are arthropods. These viruses have single-stranded DNA and, with a genome size of about 4 to 6 kb, belong to the smallest of all viruses.

The subfamily Densovirinae was first found and classified in *Galleria mellonella* (Meynadier et al. 1964; Kurstak and Cote 1969). Currently it is classified into four genera: (1) *Densovirus*, which occurs in lepidopteran hosts such as *Junonia coenia* and *G. mellonella*, and dipteran hosts such as *Toxorhynchites splendens*; (2) *Iteravirus*, which is found in lepidopteran hosts such as *B. mori*, *Casphalia extranea*, and *Sibine fusca*; (3) *Pefudensovirus*, known from *Periplaneta fuliginosa* densovirus; and (4) the genus of smallest genome size, *Brevidensovirus*, found in mosquitoes such as *Aedes aegypti* and *A. albopictus* (Diptera). In the past, viruses considered to be densovirus have been found in insects of six orders including Orthoptera, Hemiptera, and Odonata, and also in crustaceans, shrimps, and crabs. However, their genus classification has not been verified yet (see 8th ICTV Report, http://www.ncbi.nlm.nih.gov/ICTVdb/Ictv/index.htm). Recently, mosquito densovirus has been studied to control mosquitoes that transmit pathogens such as malaria, dengue fever, and West Nile fever. The study takes advantage of its host specificity and stability in nature (Carlson, Suchman, and Buchatsky 2006). On the other hand, there are reports that densovirus is chronically infected in mosquito-cultured cell lines (Jousset et al. 1993; Chen et al. 2004), suggesting densovirus is widely distributed in mosquitoes in nature.

*Bm*DNV was first found as a flacherie disease of an infectious flacherie virus (IFV, *Iflavirus*), the so-called Picorna virus, in a Japanese sericultural farm in 1968, but it was characterized as different from *Bm*IFV (Shimizu 1975). The same virus was isolated from an infection solution of the Sakashiro strain of *Bm*IFV and later named *B. mori* densovirus type 1 (*Bm*DNV-1, *Iteravirus*). Subsequently, two new types of *Bm*DNV were found independently in Japan and China. The one from Japan was classified as *B. mori* densovirus type 2 (*Bm*DNV-2), and a Chinese isolate was classified as *Bm*DNV-3 or Z (Kawase and Kurstak 1991). *Bm*DNV-2 and -3/Z showed very similar symptoms (chronic infection compared with *Bm*DNV-1 as acute) and serological homology. In addition, it was found that *Bm*DNV-2 and *Bm*DNV-3/Z have bipartite genomes (Bando et al. 1995), including a gene that encodes DNA polymerase (Hayakawa et al. 2000; Wang et al. 2007), and so are considered to be very closely related. These characteristics are not seen in *Bm*DNV-1 or other Parvoviridae; therefore, Tijssen and Bergoin (1995) proposed that they should be reclassified as the new genus *Bidensovirus*. However, by those discriminating characters the latest ICTV 8th report excluded *Bm*DNV-2 and -3/Z from Parvoviridae (Bergoin and Tijssen 2000; Fauquet et al. 2005), leaving their higher classification undetermined. Now they are tentatively referred to as Parvo-like viruses, and will be treated as a single group here.

In the silkworm *B. mori*, resistance to *Bm*DNV-1, *Bm*DNV-2 is controlled by different genes. Therefore, these genes will be referred to as *Bm*DNV-1 and *Bm*DNV-2 resistance. In this review, the resistance genes will be briefly described and their genetic nature characterized. Then, positional cloning of the *Bm*DNV-2 resistance gene, *nsd-2*, will be outlined.

DENSOVIRUS RESISTANCE GENES IN *BOMBYX MORI*

Interestingly, some silkworm strains show absolute resistance (nonsusceptibility) to *Bm*DNV. These strains cannot be infected with *Bm*DNV, no matter how much virus is inoculated. Crosses between resistant and susceptible silkworms revealed that the resistance is controlled by both dominant and recessive genes. Four resistance genes have been reported so far. These are the *Nid-1* and *nsd-1* genes against *Bm*DNV-1 and *nsd-2* and *nsd-Z* genes against *Bm*DNV-2. Details on each gene from current studies are reported below.

RESISTANCE GENES AGAINST *BM*DNV-1

Nid-1 (No infection with DNV-1) is one of two known genes controlling susceptibility/nonsusceptibility to *Bm*DNV-1, and like the other densovirus resistance genes (*nsd-1, nsd-2*, and *nsd-Z*), it acts as a single major gene. Moreover, it is the only gene in which a dominant allele is responsible for *Bm*DNV-resistance (Eguchi, Furuta, and Ninaki 1986). *Nid-1* was detected quite recently. So far only five strains have been identified as *Nid-1*-carriers, while there are many *nsd-1* and/or *nsd-2*-carrying strains in Japan (Furuta 1995; K. Kadono-Okuda, unpubl. obs.). It has been speculated that this mutation occurred recently in a Japanese noncommercial strain (K. Kadono-Okuda, unpubl.), accounting for the fact that it has not spread as extensively as *nsd-1* and *nsd-2* resistant mutations.

Single-pair backcrosses (BC_1) between resistant and susceptible strains revealed that *Nid-1* is linked to the *Bm* (Black moth) and *bts* (brown tail and head spot) mutations of LG 17. Its location was determined at 31.1 cM by a three-point cross with triple-dominant ($+^{bts}$, *Nid-1*, $+^{ow}$) and triple-recessive (*bts*, $+^{Nid-1}$, *ow*) mutant strains, where *ow* is one of many oily mutations whose larval epidermis is translucent because of abnormalities of uric acid metabolism (Eguchi et al. 2007).

Hara et al. (2008) performed the linkage and mapping analysis of *Nid-1* by means of restriction fragment length polymorphisms (RFLPs) that used expressed sequence tag (EST) clones as probes. By taking advantage of this dominant resistance gene, Eguchi et al. (1998) succeeded in breeding a *Bm*DNV-1 resistant hybrid race between a near-isogenic Japanese race containing *Nid-1* and a wild-type Chinese race. On the basis of their results, we proceeded to carry out fine mapping of *Nid-1* and succeeded in narrowing down the *Nid-1* candidate region to within about 34 kb (K. Kadono-Okuda, unpubl. data). Future work will be aimed at isolating and verifying a candidate gene using available genomic resources in the silkworm (see Chapter 2 for a review on silkworm genomics).

By contrast to *Nid-1*, *nsd-1* (nonsusceptibility to DNV-1) ensures resistance against *Bm*DNV-1 by a recessive mutation, expressed only in a homozygous state. Its locus was determined at 8.3 cM on LG 21 by Eguchi, Ninaki, and Hara (1991) and later confirmed by mapping with five linked random amplification of polymorphic DNA (RAPD) markers (Abe et al. 1998). Little is known about its mode of action except that upon infection of cultured embryos the resistant or susceptible phenotype is expressed after the embryonic reversal stage when the midgut is formed (Furuta 1983).

It is interesting to note that in crosses between $+^{nsd\text{-}1}$ and *Nid-1*, the dominant susceptible allele of the *nsd-1* gene versus the dominant resistant allele of the *Nid-1* gene, the latter is epistatic (Abe, Watanabe, and Eguchi 1987). Even if a silkworm has$+^{nsd\text{-}1}/+^{nsd\text{-}1}$ on LG 21, it becomes resistant in the presence of a heterozygous *Nid-1* gene on LG 17. This indicates that *nsd-1* and *Nid-1* are independent genes involved in different aspects of the process of virus invasion and proliferation.

Resistance Genes against *Bm*DNV-2

The *nsd-2* (nonsusceptibility to DNV-2) gene is a recessive mutation underlying the resistance against *Bm*DNV-2. Abe et al. (2000) reported two RAPD markers linked to $+^{nsd\text{-}2}$, a wild-type susceptible allele of the *nsd-2* locus. Later, Ogoyi et al. (2003) discovered three RFLP markers on LG 17 closely linked to *nsd-2*, with zero crossover values. Following these reports, our group performed positional cloning of *nsd-2* and succeeded to identify a candidate gene. In addition, we functionally restored the mutation by transforming silkworms with a $+^{nsd\text{-}2}$ gene (Ito et al. 2008). Details of this study are described in the next section.

Mapping with mutant strains representing each linkage group Qin and Yi (1996) revealed that the gene *nsd-Z* (nonsusceptibility to DNV-3, also called DNV-Z, for the Zhenjiang, Chinese, strain of DNV) is linked to the *Se* (White-sided egg) marker, located on LG 15. In addition, simple sequence repeat (SSR) mapping detected seven markers encompassing *nsd-Z* (Li et al. 2006); however, the linkage group was not confirmed directly. Nakagaki et al. (1999) compared the amount of *Bm*DNV-Z DNA produced between susceptible and resistant strains by quantitative PCR after inoculation at the fourth instar. The virus DNAs increased logarithmically and reached a maximum at 48 hours postinfection (hpi) in the susceptible larvae, but the resistant larvae allowed only a slight increase until 12 hpi and then viral DNA decreased rapidly, leading the authors to propose that some of the viral DNAs replicated and accumulated in midgut of the resistant race but were then degraded. The mechanism of resistance-expression by *nsd-Z* was further studied by Han et al. (2007), who compared the replication levels of *Bm*DNV-3 DNAs in the fifth instar larvae of susceptible and resistant Chinese strains. A similar growth mode was observed as previously described (Nakagaki et al. 1999): In the resistant strain virus replication increased from 6–10 copies per cell at 2 hpi to only 150–200 copies per cell at 96 hpi. The authors proposed that the resistance to *Bm*DNV-3 in some of the silkworm strains was a kind of chronic viral replication carried by the host without causing flacherie. Breeding of near-isogenic lines for *nsd-Z* was attempted, and the progress of successive backcrosses was observed with SSR markers including *nsd-Z*-linked markers. For example, a marker belonging to LG 11 (not linked with *nsd-Z*) showed rapid replacement and homozygosity at the fourth backcross generation, whereas another marker closely linked to *nsd-Z* was not replaced (Li et al. 2007). By subjecting such near isogenic strains to fluorescent differential display, Chen et al. (2007) detected a protein kinase C inhibitor gene that was highly expressed in resistant but not susceptible silkworms. It is still unclear how this gene relates to densovirus resistance and whether this phenomenon is common to the other resistant, *nsd-Z*-carrying strains, and further analysis will be required to clarify the underlying mechanisms.

Possible Expression Sites of the Resistance Genes

In mosaic silkworms produced by the use of a hereditary mosaic mutation, *mo*, in which double fertilization yielded cells that were heterozygous resistant (*Nid-1*/+) or homozygous susceptible (+/+), the susceptibility of midgut cells differed from each other (Abe, Kobayashi, and Watanabe 1990). This suggested that the dominant resistance factor encoded by *Nid-1* does not function by being secreted into the midgut lumen but functions within each cell or on the midgut cell surface. Similar results were obtained with *nsd-2* in another mosaic silkworm strain, in which individual cells were homozygous resistant (*nsd-2*/*nsd-2*) or heterozygous susceptible (*nsd-2*/+): Each midgut cell discretely exerted resistance/susceptibility in the same way as *Nid-1* (Abe et al. 1993). These findings, in which independent dominant and recessive mutations on different linkage groups act cell-autonomously, indicate that the two genes controlling resistance act at different steps in viral infection.

ISOLATION AND ANALYSIS OF *NSD-2*

Recent advances in silkworm genome analysis have made many tools available to enable positional or map-based cloning such as linkage maps based on SSRs, RFLPs, RAPDs, and single-nucleotide polymorphisms (SNPs) (Miao et al. 2005; Nguu et al. 2005; Yasukochi et al. 2006; Yamamoto et al. 2006, 2008). In addition, positional information has been reported for large sets of EST clones, bacterial artificial chromosome (BAC) clone-end sequences, and predicted expressed genes on extended physical-genetic maps and scaffolds (Yamamoto et al. 2008; see Chapter 2 for the current status of silkworm genetics and genomics). The requirements for positional cloning are (1) markers located at regular intervals in each linkage group and (2) parental strains in which the target phenotype can be clearly identified in the segregants in F_2 or BC_1 (backcross) generations. Even if the linkage group(s) of the gene(s) responsible for the objective trait to be identified is unknown, initial linkage analysis is made easy in Lepidoptera, including silkworms, by tracking the inheritance of one or two representative markers for each linkage group in a small number of BC_1 offspring (approximately 15 individuals, depending on the genetic complexity of the trait) of F_1 females. This takes advantage of complete linkage due to the lack of crossing over in the female sex (the absence of meiotic recombination in lepidopteran females is documented in Chapter 3). After identification of the linkage group(s) of interest, a rough linkage map can be made using about 50 BC_1 segregants of reciprocal backcrosses between parental females (the homozygous recessive trait–containing parent) and F_1 males, in which crossover occurs during meiosis, and using markers scattered on the target linkage group (Figure 18.1A; Ogoyi et al. 2003; for an example of this mapping strategy, see Chapter 6).

If the location of the target trait on the linkage map is known, markers corresponding to this region can be used preferentially. When the "candidate region" is included within the nearest upstream and downstream markers but some segregants are still not linked to the target trait, it can be narrowed down using inside markers from segregants showing recombination with either of the distal flanking markers. Since all individuals become linked with the trait when the target region is reduced to some extent, narrowing will be started again with new BC_1 segregants by screening with the closest current upstream and downstream markers. For regions with no polymorphic markers, it is now possible to retrieve the sequences around the target from the silkworm genome database (http://sgp.dna.affrc.go.jp/KAIKObase/) for designing PCR primers. In some cases, the target area can be narrowed down adequately with hundreds of silkworms, but in other cases, thousands of segregants are necessary to define a region small enough to predict a candidate gene. It is also possible that some candidate genes can be identified before the region is narrowed down sufficiently to define a single gene.

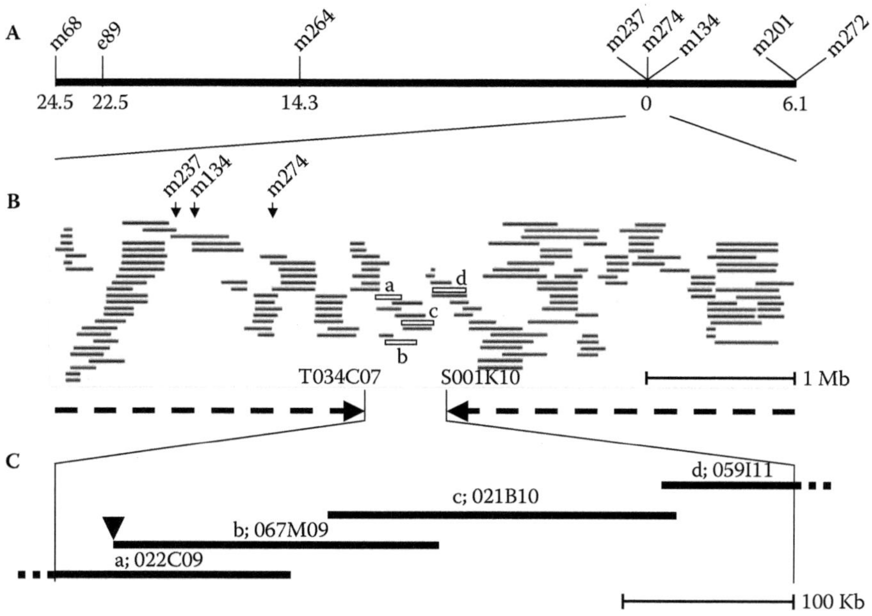

FIGURE 18.1 Strategy for positional cloning of *nsd-2*. (A) RFLP map of *nsd-2*, modified from the previous report (Ogoyi et al. 2003). EST markers are shown above the map; the distance between the markers and *nsd-2* is shown in centiMorgan units (cM) below the map. EST markers m237, m274, and m134, which showed no recombination with *nsd-2* in the initial study involving 49 male informative backcross (BC_1) progeny, served as the starting point for the walk (Ogoyi et al. 2003). (B) BAC contig covering three EST markers. Each line shows a BAC clone. BACs with open lines (a–d) are illustrated in C. The dotted lines indicate the result of linkage analysis using 206 BC_1 progeny surviving virus treatment to narrow the region linked to *nsd-2*. (C) Minimum BAC tiling path for the region closely linked to *nsd-2*. The arrowhead indicates the locus where the deletion was found in resistant strain J150. (Adapted from Ito et al., 2008).

Recently, the unified physical-genetic linkage map (Yamamoto et al. 2008) and combined data from genome projects of China and Japan produced extremely long scaffolds assembled from WGS sequences (International Silkworm Genome Consortium 2008). In addition, linkage group assignments for most sequences are emerging, so that it will be relatively easy to obtain sequence information for a target region. Early positional cloning experiments involved successive rounds of screening BAC libraries with the nearest flanking markers from both ends of the candidate region (Koike et al. 2003), constructing contigs by DNA fingerprinting, and then screening recombinant BC_1 segregants with primers designed from BAC end sequences on both ends of the contig to narrow down the region on the chromosome. Repeating these procedures, the BAC contig was extended to the target gene. At present, such troublesome procedures are not needed; for example, it is quite possible to identify the responsible gene only by examining gene predictions from the region of the genome sequence specified by the linkage analysis. Following is a brief description of how the densovirus type 2 resistance gene, *nsd-2*, was identified and its function demonstrated by restoration of the densovirus susceptibility using transgenesis.

Narrowing Down the *nsd-2* Candidate Region by Chromosome Walking and Linkage Analysis

First, chromosome walking was performed to find markers to narrow down the *nsd-2* candidate region. Three EST markers, ml34, m237, and m274, on LG 17, which were closely linked to *nsd-2* as verified by Ogoyi et al. (2003), were used as probes to start the walk. By screening high-density replica filters of a silkworm BAC library with these probes and then repeated screening with the end sequences, a BAC contig about 5-Mb long covering the upstream and downstream regions from the starting points of the three markers was constructed (Figure 18.1B). Then, based on the established physical map, linkage analysis was performed with BC_1 offspring surviving virus selection, and the region linked to *nsd-2* was narrowed down to about 400 kb (Figure 18.1C). PCR primers designed for closely spaced loci in this region were used to amplify genomic DNA from the parental strains of the BC_1 segregants. A 6.3 kb long deletion accompanied by a 34 bp insertion was found in the resistant (J150) but not in the susceptible (No. 908) strain. Using this sequence information, specific primers were designed and used to compare three resistant strains and seven susceptible strains using PCR. The deletion and 34 bp insertion were common characteristics only in the resistant strains (Ito et al. 2008).

Full-Length cDNA Sequence of the *nsd-2* Candidate Gene

By conducting a search with the 6-kb deletion using the KAIKOGAAS automated annotation system (http://kaikogaas.dna.affrc.go.jp/), one candidate gene was predicted, and full-length cDNA sequences of the candidate gene from resistant and susceptible strains were determined with RACE (rapid amplification of cDNA end). The susceptible type was composed of 14 exons, whereas the resistant type lacked exons 5 to 13. Additionally, as a result of a frame shift caused by the deletion, a new stop codon appeared in the beginning of exon 14 (Figure 18.2). To examine the products of the candidate gene, primers were designed to the 5' UTR and 3' UTR regions and RT-PCR was performed with RNA prepared from the larval midgut of resistant and susceptible strains. It was found that the size of the candidate gene products of all resistant types was smaller than that of susceptible types, reflecting the size difference in the PCR products amplified from the genomic DNA (Ito et al. 2008).

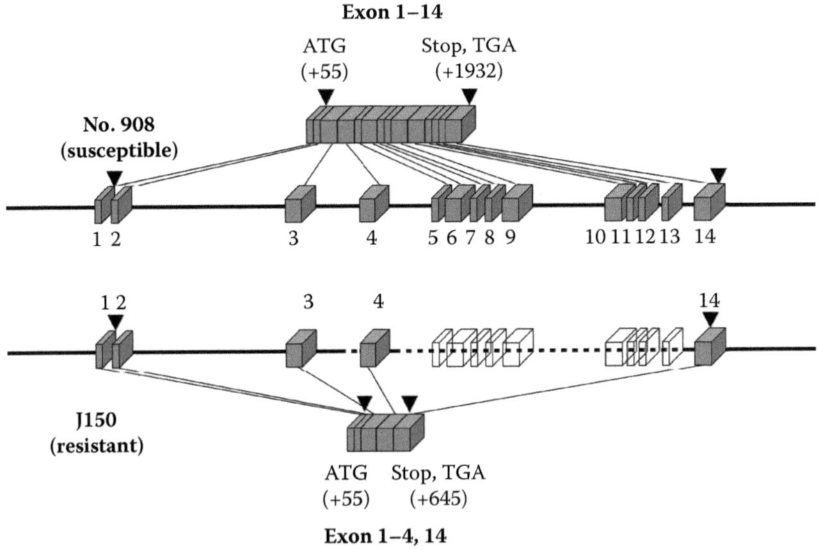

FIGURE 18.2 Schematic structure of candidate genes for *nsd-2* and $+^{nsd\text{-}2}$. Relative position and size of exons and introns in susceptible strain No. 908 (upper) and resistant strain J150 (lower) are shown. The dotted line indicates the deletion in J150. The arrowheads show the start and stop codons.

Expression Analysis of the *nsd-2* Candidate Gene

RT-PCR was performed with the RNA from the silk gland, foregut, midgut, hindgut, Malpighian tubule, fat body, testis plus ovary and central nervous system in susceptible and resistant fourth instar, day 1 larvae. The RT-PCR product was detected only in the midgut (Ito et al. 2008). This agrees with findings by Seki and Iwashita (1983) that the infection of this densovirus is midgut specific. The developmental expression pattern of this gene was examined in developing eggs on days 1, 4, and 10, and in larvae, pupae, and adults of No. 908 and J150; the RNA appeared only during the late embryonic and larval feeding stages. Importantly, no expression of the candidate gene was detected in the early and middle embryonic stages, in the molting stage at each instar, in the pupa, or in adult silk moths (Ito et al. 2008). These results indicate that this gene is expressed only in the period when the midgut carries out digestion and absorption of food.

Structural Analysis of the *nsd-2* Candidate Gene

Translated amino acid sequences of *nsd-2* from No. 908 had a high homology of 63 percent with a *Manduca sexta* amino acid transporter (Castagna et al. 1998; Feldman, Harvey, and Stevens 2000). Amino acid transporters are membrane proteins located on the lumen side of epidermal cells, consistent with the finding that the translated amino acid sequence secondary structure of No. 908 was a 12-pass transmembrane protein as defined by SOSUI (http://bp.nuap.nagoya-u.ac.jp/sosui/), a protein structure prediction program that predicts transmembrane regions from the hydrophobicity of constituent amino acids. On the other hand, because of the large deletion it was inferred that the resistant strain, J150, had only 3-pass transmembrane structures for the lead portion of the protein (Figure 18.3). Since the protein functions as a monomer to transport specific amino acids in *M. sexta* (Castagna et al. 1998; Feldman, Harvey, and Stevens 2000), it is assumed that the silkworm *nsd-2* candidate gene encodes a functional protein involved in amino acid transport. On the one hand, because resistant strains lacking the internal exons encoded by this gene grow normally, it is thought to be nonessential for the silkworm. However, during *nsd-2* isolation we found another gene nearby that has extremely high homology (69 percent) with *nsd-2*. Currently, this gene, which may function as an amino acid transporter instead of the *nsd-2* candidate, is being analyzed in detail (K. Kadono-Okuda, unpubl.). On the other hand, infection cannot occur without a functional *nsd-2*, suggesting that the product of this gene is an essential factor for the virus.

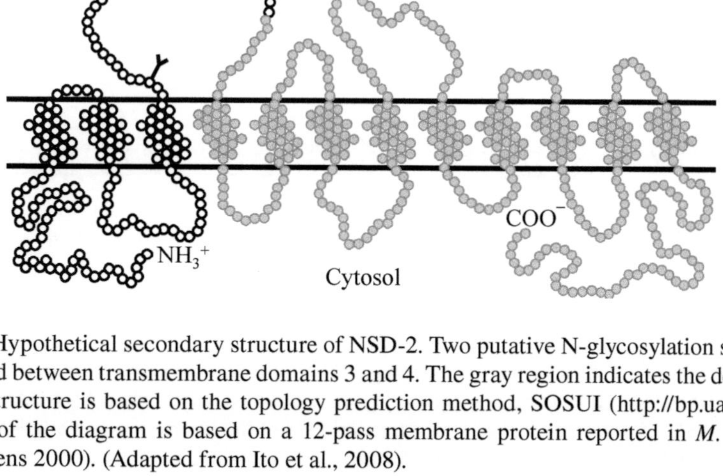

FIGURE 18.3 Hypothetical secondary structure of NSD-2. Two putative N-glycosylation sites (Y-like symbols) are indicated between transmembrane domains 3 and 4. The gray region indicates the deletion in NSD-2. The secondary structure is based on the topology prediction method, SOSUI (http://bp.uap.nagoya-u.ac.jp/sosui); the style of the diagram is based on a 12-pass membrane protein reported in *M. sexta* (Feldman, Harvey, and Stevens 2000). (Adapted from Ito et al., 2008).

Functional Proof of the *nsd-2* Candidate Gene

Complementation tests are the most direct and essential proof for the function of isolated candidate genes. Currently, no system of in vitro or cultured cell lines is available to study infection of this virus. Moreover, successful examples of RNAi to knock down the function of a wild-type gene are rare in silkworms (Ohnishi, Hull, and Matsumoto 2006), and efficiency is low (Tabunoki et al. 2004), making this an uncertain approach for testing the *nsd-2* candidate gene. Therefore, to determine whether the isolated candidate gene is responsible for virus resistance, resistant silkworms were transformed with the $+^{nsd-2}$ candidate sequence using the GAL4-UAS system (Tamura et al. 2000). A GAL4 line that strongly expresses GAL4 in the midgut was chosen. Upon inoculation with the virus only the transformed GAL4/UAS silkworms carrying and expressing the wild-type (susceptibility) candidate gene showed a notable susceptibility phenotype (Figure 18.4; Ito et al. 2008). From these results, it was concluded that the isolated candidate gene is *nsd-2* itself, the *Bm*DNV-2 resistance gene, and the wild-type membrane protein expressed by $+^{nsd-2}$ is required for infection by the virus.

FUTURE APPLICATIONS OF *nsd-2* ISOLATION

The achievement of isolating the *Bm*DNV-2 resistance gene, *nsd-2*, is the first successful example of isolating a mutant gene in the silkworm using the method of map-based or positional cloning. It is significant not only in clarifying a key aspect of the resistance mechanism of a host insect against an entomopathogenic microorganism, but also in elucidating the cause of a form of absolute virus resistance.

Host-specific or tissue-specific infection mechanisms in insect viruses have been studied in NPV (Bonning 2005); however, those studies focused on virus genes, not on the response mechanism of the host. Therefore, verifying the interaction between *Bm*DNV-2 and the midgut membrane protein identified in the present study will contribute greatly to understanding a virus infection mechanism from the perspective of the host insect. It is important to compare these viral infection

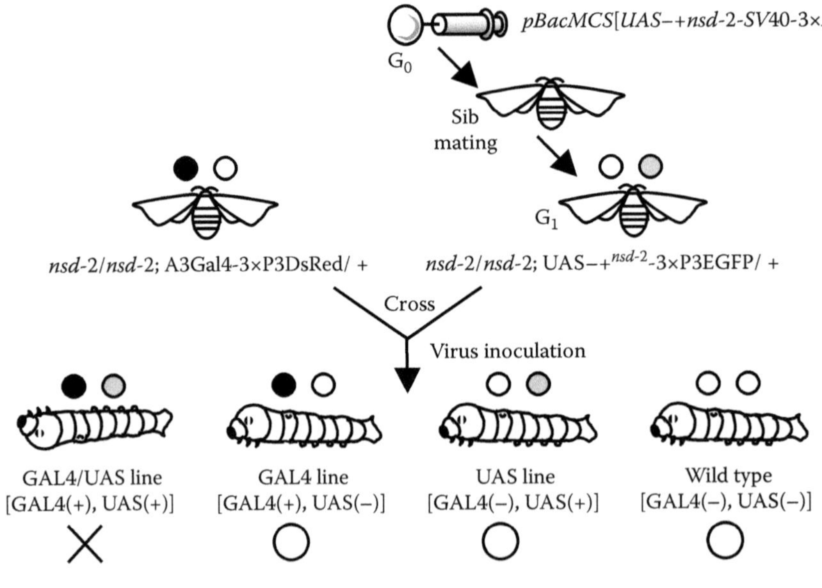

FIGURE 18.4 Virus susceptibility of progeny of the transformed silkworm. Four lines of progeny (●◉/●○/○◉/○○) were generated by crossing a GAL4 strain (●○) carrying a DsRed marker with a UAS strain (○◉) injected with the UAS target gene marked with GFP. The transgene ($+^{nsd-2}$) was expressed only in the GAL4/UAS line (●◉) in which GAL4 recognized the UAS sequence and activated transcription of the inserted gene. Only larvae of the GAL4/UAS line showed a notably susceptible phenotype (X); the others remained resistant (○).

mechanisms with those of human, canine, or feline Parvovirus. In Parvovirinae, it is well known that human Parvovirus B19 infects only humans, and canine and feline Parvovirus infect only dog and cat family members, respectively. This suggests that the relationship between each virus and its receptor is highly specific. In the first clarification of the infection mechanism of Parvovirus, Brown, Anderson, and Young (1993) found that the receptor for the B19 virus is the P antigen on the erythroblast surface. On the other hand, canine and feline Parvovirus bind to host-specific transferrin receptors on the surface of intestinal canal cells, enter the cell by endocytosis, and then move through the cytoplasm to the nucleus. That each virus specifically infects host canines or felines relates to differences in the virus surface structure and the amino acid sequence of each transferrin receptor (Truyen, Agbandje, and Parrish 1994; Truyen et al. 1996; Parker and Parrish 2000; Palermo, Hueffer, and Parrish 2003). Vertebrate and invertebrate parvoviruses are very small DNA viruses with genome sizes of about 4 to 6 kb. It is remarkable that the variation present in such small genome structures can determine host and tissue specificities. Studying these phenomena between the silkworm and *Bm*DNV will bring new information on infection mechanisms occurring in Parvoviridae, including densoviruses. Even if *Bm*DNV-2 is a new virus that does not belong to Parvoviridae, it can be used to investigate the action of a complete resistance gene of the silkworm against a new form of major insect pathogen.

By introducing the virus-resistance gene into strains that have the virus-susceptible allele, for example, by substituting the resistance allele or by disrupting it, it will be possible to carry out more accurate marker-assisted selection or breeding for the silkworm industry, in which the gene itself is used directly as a marker. It is far more rapid and accurate to select the recessive gene-containing resistant progeny directly by genotyping than to carry out the selection by virus-inoculation because additional side-crosses and testing by virus inoculation are required in every step to discriminate the heterozygotes for a recessive mutation from wild-type homozygotes, which have the same susceptible phenotype (see production of near-isogenic or congenic lines; Abe et al. 1995, 2000; Chen et al. 2007). In addition, if a highly susceptible cell line is established by transducing susceptible genes such as $+^{nsd\text{-}2}$, it is expected that the molecular and cellular biological study of this virus, which now lags far behind because no cultured cell lines susceptible to infection by this virus are available, will be rapidly developed.

The methods that we used for isolating *nsd-2* by classical positional cloning have been shown to be highly effective. In the silkworm, at least three more resistance genes against other forms of *Bm*DNV have yet to be studied, namely, the dominant *Nid-1* and recessive *nsd-1* mutations against *Bm*DNV-1, and the recessive *nsd-Z* mutation against *Bm*DNV-Z/3. The relationship between these genes and *nsd-2* will be of interest, as will their potential to provide new tools for silkworm protection in sericulture. Although as a prerequisite for the isolation of phenotypic mutations by positional cloning it is essential that the target characters are clearly segregated among large numbers of BC_1 or F_2 populations, this is feasible for these virus resistance genes. Additionally, with the ongoing improvement of the silkworm genome database (Xia et al. 2004; International Silkworm Genome Consortium 2008; Yamamoto et al. 2008; see Chapter 2 for details on silkworm genomics) positional cloning methods are now rapidly becoming streamlined, and labor-intensive chromosome walking as shown in Figure 18.1 will no be longer necessary. Presently, the isolation of the *Bm*DNV resistance genes and many other silkworm mutant genes is in progress.

EVOLUTIONARY NOTES ON THE ORIGIN OF DNV-RESISTANCE GENES

Can we trace the evolution of the silkworm by examining the three DNV-resistance genes found in Japan? *B. mandarina*, which shares a common ancestor and is the closest living relative of *B. mori* (Arunkumar, Metta, and Nagaraju 2006), carries the $+^{nsd\text{-}2}$ (susceptible) gene (Ito et al. 2008). In general, Indian and Chinese local varieties are susceptible to *Bm*DNV-2, whereas modern Japanese varieties and improved races tend to be resistant (Furuta 1994, 1995). Does this mean that silkworms were domesticated in China and later brought to Japan, where the *nsd-2* deletion mutation arose in the course of breed improvement? In fact, most of the Japanese ancestral varieties such as Seihaku,

Dainyorai, or Matamukashi are still susceptible to *Bm*DNV-2, and it is assumed that *Bm*DNV-2 was a pathogenic virus against the common ancestors of the silkworm and *B. mandarina*. Hence, it will be interesting to know whether a few resistant strains that exist among Chinese indigenous varieties contain the same *nsd-2* mutation. By contrast, the resistance/susceptibility expressed by strains to *Bm*DNV-1 tends to show a mirror-image relationship to that of *Bm*DNV-2 (Furuta 1994, 1995), as the Japanese *B. mandarina* is resistant to *Bm*DNV-1 in a recessive inheritance manner (K. Kadono-Okuda, unpubl. obs.). Therefore, it will be of interest to investigate resistance to *Bm*DNV-1 in the Chinese *B. mandarina*. If it is also resistant, we can speculate that *Bm*DNV-1 was not a virus native to the *Bombyx* genus and that a susceptible mutation arose during the breeding process in Japan.

Watanabe, Kurihara, and Wang (1988) reported that almost half of the larvae of the mulberry pyralid, *Glyphodes pyloalis* Walker, in mulberry farms potentially contain antigens that react to three antibodies against *Bm*DNV-1, *Bm*DNV-2, and infectious flacherie virus. In addition, the homogenates of those *G. pyloalis* were infectious to the silkworm when administered *per os*. If these were exactly the same as *Bm*DNVs, inapparently or subclinically infected *G. pyloalis* may have been the original host. In contrast to the time when a serological diagnosis was the only method to detect the virus, this question can now be answered. As for *Nid-1*, so far only six strains have been discovered as dominantly resistant to *Bm*DNV-1; these were found by three independent research organizations (R. Eguchi and H. Abe, unpubl. obs.). Hence, it is hypothesized that *Nid-1* was generated by a recent mutation. In the present period when most commercial races have already been established, this gene could hardly have spread to other strains by breeding. If the pedigree of the six strains was recorded, it can be determined in which strain and when the mutation occurred.

REFERENCES

Abe, H., T. Harada, M. Kanehara, T. Shimada, F. Ohbayashi, and T. Oshiki. 1998. Genetic mapping of RAPD markers linked to the densonucleosis refractoriness gene, *nsd-1*, in the silkworm, *Bombyx mori*. *Genes Genet. Syst.* 73:237–42.

Abe, H., M. Kobayashi, and H. Watanabe. 1990. Mosaic infection with a densonucleosis virus in the midgut epithelium of the silkworm, *Bombyx mori*. *J. Invertebr. Pathol.* 55:112–17.

Abe, H., K. Kobayashi, T. Shimada, et al. 1993. Infection of a susceptible/nonsusceptible mosaic silkworm, *Bombyx mori*, with densonucleosis virus type-2 is not lethal. *J. Seric. Sci. Jpn.* 62:367–75.

Abe, H., T. Shimada, G. Tsuji, T. Yokoyama, T. Oshiki, and M. Kobayashi. 1995. Identification of random amplified polymorphic DNA linked to the densonucleosis virus type-1 susceptibility gene of the silkworm, *Bombyx mori*. *J. Seric. Sci. Jpn.* 64:262–64.

Abe, H., T. Sugasaki, M. Kanehara, et al. 2000. Identification and genetic mapping of RAPD markers linked to the densonucleosis refractoriness gene, *nsd-2*, in the silkworm, *Bombyx mori*. *Genes Genet. Syst.* 75:93–96.

Abe, H., H. Watanabe, and R. Eguchi. 1987. Genetical relationship between nonsusceptibilities of the silkworm, *Bombyx mori*, to two densonucleosis viruses. *J. Seric. Sci. Jpn.* 56:443–44.

Arunkumar, K.P., M. Metta, and J. Nagaraju. 2006. Molecular phylogeny of silkmoths reveals the origin of domesticated silkmoth, *Bombyx mori* from Chinese *Bombyx mandarina* and paternal inheritance of *Antheraea proylei* mitochondrial DNA. *Mol. Phylogenet. Evol.* 40:419–27.

Bando, H., T. Hayakawa, S. Asano, K. Sahara, M. Nakagaki, and T. Iizuka. 1995. Analysis of the genetic information of a DNA segment of a new virus from silkworm. *Arch Virol.* 140:1147–55.

Bergoin, M., and P. Tijssen. 2000. Molecular biology of Densovirinae. *Contrib. Microbiol.* 4:12–32.

Bonning, B.C. 2005. Baculoviruses: Biology, biochemistry, molecular biology. In *Comprehensive molecular insect science*, Vol. 6: *Control*, ed. S. Gill, K. Iatrou, and L. Gilbert, 233–70. Oxford: Elsevier.

Brown, K.E., S.M. Anderson, and N.S. Young. 1993. Erythrocyte P antigen: Cellular receptor for B19 parvovirus. *Science* 262:114–17.

Carlson, J., E. Suchman, and L. Buchatsky. 2006. Densoviruses for control and genetic manipulation of mosquitoes. *Adv. Virus. Res.* 68:361–92.

Castagna, M., C. Shayakul, D. Trotti, V.F. Sacchi, W.R. Harvey, and M.A. Hediger. 1998. Cloning and characterization of a potassium-coupled amino acid transporter. *Proc. Natl. Acad. Sci. U.S.A.* 95:5395–400.

Chen, K.-P., H.-Q. Chen, X.-D. Tang, Q. Yao, L.-L. Wang, and X. Han. 2007. *bmpkci* is highly expressed in a resistant strain of silkworm (Lepidoptera: Bombycidae): Implication of its role in resistance to BmDNV-Z. *Eur. J. Entomol.* 104:369–76.

Chen, S., L. Cheng, Q. Zhang, et al. 2004. Genetic, biochemical, and structural characterization of a new densovirus isolated from a chronically infected *Aedes albopictus* C6/36 cell line. *Virology* 318:123–33.

Eguchi, R., Y. Furuta, and O. Ninaki. 1986. Dominant nonsusceptibility to densonucleosis virus in the silkworm, *Bombyx mori. J. Seric. Sci. Jpn.* 55:177–78.

Eguchi, R., W. Hara, A. Shimazaki, et al. 1998. Breeding of the silkworm race "Taisei" non-susceptible to a densonucleosis virus type 1. *J. Seric. Sci. Jpn.* 67:361–66.

Eguchi, R., K. Nagayasu, O. Ninagi, and W. Hara. 2007. Genetic analysis on the dominant non-susceptibility to densonucleosis virus type 1 in the silkworm, *Bombyx mori. Sanshi-Konchu Biotec.* 76:159–63.

Eguchi, R., O. Ninaki, and W. Hara. 1991. Genetical analysis on the nonsusceptibility to densonucleosis virus in the silkworm, *Bombyx mori. J. Seric. Sci. Jpn.* 60:384–89.

Fauquet, C.M., M.A. Mayo, J. Maniloff, U. Desselberger, and L.A. Ball, eds. 2005. *Virus taxonomy. VIIIth report of the international committee on taxonomy of viruses*. London: Elsevier/Academic Press.

Feldman, D.H., W.R. Harvey, and B.R. Stevens. 2000. A novel electrogenic amino acid transporter is activated by K^+ or Na^+, is alkaline pH-dependent, and is Cl^--independent. *J. Biol. Chem.* 275:24518–26.

Furuta, Y. 1983. Multiplication of the infectious flacherie virus and the densonucleosis virus in cultured silkworm embryos. *J. Seric. Sci. Jpn.* 52:245–46.

Furuta, Y. 1994. Susceptibility of Indian races of the silkworm, *Bombyx mori*, to the nuclear polyhedrosis virus and densonucleosis viruses. *Acta Seric. Entomol.* 8:29–36.

Furuta, Y. 1995. Susceptibility of the races of the silkworm, *Bombyx mori*, preserved in NISES to the nuclear polyhedrosis virus and densonucleosis viruses. *Bull. Natl. Inst. Seric. Entomol. Sci.* 15:119–45.

Han, X., Q. Yao, L. Gao, Y.J. Wang, F. Bao, and K.P. Chen. 2007. Replication of *Bombyx mori* Densonucleosis Virus (Zhenjiang isolate) in different silkworm strains. *Sheng Wu Gong Cheng Xue Bao*. 23:145–51.

Hara, W., Y. An, R. Eguchi, et al. 2008. Mapping of a novel virus resistant gene, *Nid-1*, in the silkworm, *Bombyx mori*, based on the restriction fragment length polymorphism (RFLP). *J. Insect Biotech. Sericol.* 77:59–66.

Hayakawa, T., K. Kojima, K. Nonaka, et al. 2000. Analysis of proteins encoded in the bipartite genome of a new type of parvo-like virus isolated from silkworm—structural protein with DNA polymerase motif. *Virus Res.* 66:101–08.

International Silkworm Genome Consortium. 2008. The genome of a lepidopteran model insect, the silkworm *Bombyx mori. Insect Biochem. Mol. Biol.* 38:1036–45.

Ito, K., K. Kidokoro, H. Sezutsu, et al. 2008. Deletion of a gene encoding an amino acid transporter in the midgut membrane causes resistance to a *Bombyx* parvo-like virus. *Proc. Natl. Acad. Sci. U.S.A.* 105:7523–27.

Jousset, F.X., C. Barreau, Y. Boublik, and M. Cornet. 1993. A parvo-like virus persistently infecting a C6/36 clone of *Aedes albopictus* mosquito cell line and pathogenic for *Aedes aegypti* larvae. *Virus Res.* 29:99–114.

Kawase, S., and E. Kurstak. 1991. Parvoviridae of invertebrates: Densonucleosis viruses. In *Viruses of invertebrates*, ed. E. Kurstak, 315–43. New York: Marcel Dekker.

Koike, Y., K. Mita, M.G. Suzuki, et al. 2003. Genomic sequence of a 320-kb segment of the Z chromosome of *Bombyx mori* containing a *kettin* ortholog. *Mol. Genet. Genomics* 269:137–49.

Kurstak, E., and J.-R. Cote. 1969. Proposition de classification du virus de la densonucleose (VDV) basee sur l'etude de la structure moleculaire et des proprietes physicochimiques. *C. R. Acad. Sci. Paris.* 268:616–19.

Li, M., Q. Guo, C. Hou, et al. 2006. Linkage and mapping analyses of the densonucleosis non-susceptible gene *nsd-Z* in the silkworm *Bombyx mori* using SSR markers. *Genome* 49:397–402.

Li, M., C. Hou, Y. Zhao, A. Xu, X. Guo, and Y. Huang. 2007. Detection of homozygosity in near isogenic lines of non-susceptible to Zhenjiang strain of densonucleosis virus in silkworm. *Afr. J. Biotechnol.* 6:1629–33.

Meynadier, G., C. Vago, G. Plantevin, and P. Atger. 1964. Virose de un type inhabituel chez le Lépidoptère. "*Galleria. mellonella.*" *Rev. Zool. Agric. Appl.* 63:207–08.

Miao, X.X., S.J. Xub, M.H. Li, et al. 2005. Simple sequence repeat-based consensus linkage map of *Bombyx mori. Proc. Natl. Acad. Sci. U.S.A.* 102:16303–08.

Nakagaki, M., T. Morinaga, C. Zhow, Z. Kajiura, and R. Takei. 1999. Increasing curves of two virus DNAs in the midgut epithelium of silkworm infected with *Bombyx mori* densonucleosis virus type 2 (*Bm*DNV-2). *J. Seric. Sci. Jpn.* 68:173–80.

Nguu, E.K., K. Kadono-Okuda, K. Mase, E. Kosegawa, and W. Hara. 2005. Molecular linkage map for the silkworm, *Bombyx mori*, based on restriction fragment length polymorphism of cDNA clones. *J. Insect Biotechnol. Seriocol.* 74:5–13.

Ogoyi, D.O., K. Kadono-Okuda, R. Eguchi, et al. 2003. Linkage and mapping analysis of a non-susceptibility gene to densovirus (*nsd-2*) in the silkworm, *Bombyx mori*. *Insect Mol. Biol.* 12:117–24.

Ohnishi, A., J.J. Hull, and S. Matsumoto. 2006. Targeted disruption of genes in the *Bombyx mori* sex pheromone biosynthetic pathway. *Proc. Natl. Acad. Sci. U.S.A.* 103:4398–403.

Palermo, L.M., K. Hueffer, and C.R. Parrish. 2003. Residues in the apical domain of the feline and canine transferrin receptors control host-specific binding and cell infection of canine and feline parvoviruses. *J. Virol.* 77:8915–23.

Parker, J.S., and C.R. Parrish. 2000 Cellular uptake and infection by canine parvovirus involves rapid dynamin-regulated clathrin-mediated endocytosis, followed by slower intracellular trafficking. *J. Virol.* 74:1919–30.

Qin, J., and W.Z. Yi. 1996. Genetic linkage analysis of *nsd-Z*, the nonsusceptibility gene of *Bombyx mori* to the Zhenjiang (China) strain densonucleosis virus. *Sericologia* 36:241–44.

Seki, H., and Y. Iwashita. 1983. Histopathological features and pathogenicity of a densonucleosis virus of the silkworm, *Bombyx mori*, isolated from sericultural farms in Yamanashi prefecture. *J. Seric. Sci. Jpn.* 52:400–05.

Shimizu, T. 1975. Pathogenicity of an infection flacherie virus of the silkworm *Bombyx mori*, obtained from sericultural farms in the suburbs of Ina city. *J. Seric. Sci. Japan.* 44:45–48.

Tabunoki, H., S. Higurashi, O. Ninagi, et al. 2004. A carotenoid-binding protein (CBP) plays a crucial role in cocoon pigmentation of silkworm (*Bombyx mori*) larvae. *FEBS Lett.* 567:175–78.

Tamura T., C. Thibert, C. Royer, et al. 2000. Germline transformation of the silkworm *Bombyx mori* L. using a *piggyBac* transposon-derived vector. *Nat. Biotechnol.* 18:81–84.

Tijssen, P., and M. Bergoin. 1995. Densonucleosis viruses constitute an increasingly diversified subfamily among the parvoviruses. *Semin. Virol.* 6:347–55.

Truyen, U., M. Agbandje, and C.R. Parrish. 1994. Characterization of the feline host range and a specific epitope of feline panleukopenia virus. *Virology* 200:494–503.

Truyen, U., J.F. Evermann, E. Vieler, and C.R. Parrish. 1996. Evolution of canine parvovirus involved loss and gain of feline host range. *Virology* 215:186–89.

Wang, Y.J., Q. Yao, K.P. Chen, Y. Wang, J. Lu, and X. Han. 2007. Characterization of the genome structure of *Bombyx mori* densovirus (China isolate). *Virus Genes* 35:103–08.

Watanabe, H., Y. Kurihara, and Y.-X. Wang. 1988. Mulberry pyralid, *Glyphodes pyloalis*: Habitual host of non-occluded viruses pathogenic to the silkworm, *Bombyx mori*. *J. Invertebr. Pathol.* 52:401–08.

Watanabe, H., S. Maeda, M. Matsui, and T. Shimizu. 1976. Histopathology of the midgut epithelium of the silkworm, *Bombyx mori*, infected with a newly-isolated virus from the flacherie-diseased larvae. *J. Seric. Sci. Jpn.* 45:29–34.

Xia, Q., Z. Zhou, C. Lu, et al. 2004. A draft sequence for the genome of the domesticated silkworm (*Bombyx mori*). *Science* 306:1937–40.

Yamamoto, K., J. Narukawa, K. Kadono-Okuda, et al. 2006. Construction of a single nucleotide polymorphism linkage map for the silkworm, *Bombyx mori*, based on bacterial artificial chromosome end sequences. *Genetics* 173:151–61.

Yamamoto, K., J. Nohata, K. Kadono-Okuda, et al. 2008. A BAC-based integrated linkage map of the silkworm *Bombyx mori*. *Genome Biol.* 9:R21.1–14.

Yasukochi, Y., L.A. Ashakumary, K. Baba, A. Yoshido, and K. Sahara. 2006. A second-generation integrated map of the silkworm reveals synteny and conserved gene order between lepidopteran insects. *Genetics* 173:1319–28. Erratum 2008 178:1837.

Index

C

D

E

F

I

M

N

T

Y

Z

FIGURE 1.1 Representatives of superfamilies containing model systems. A: Bombycoidea, *Anthela oressarcha* (A. Zwick); B: Bombycoidea, *Antheraea larissa* (A. Kawahara); C: Noctuoidea, *Trichoplusia ni* (M. Dreiling); D: Noctuoidea, *Tyria jacobaeae* (D. Dictchburn); E: Papilionoidea, *Bicyclus anynana* (A. Monteiro and W. Piel); F: Papilionoidea, *Heliconius erato* (K. Garwood); G: Pyraloidea, *Ostrinia nubilalis* (S. Nanz); H: Tortricoidea, *Cydia pomonella* (N. Schneider).

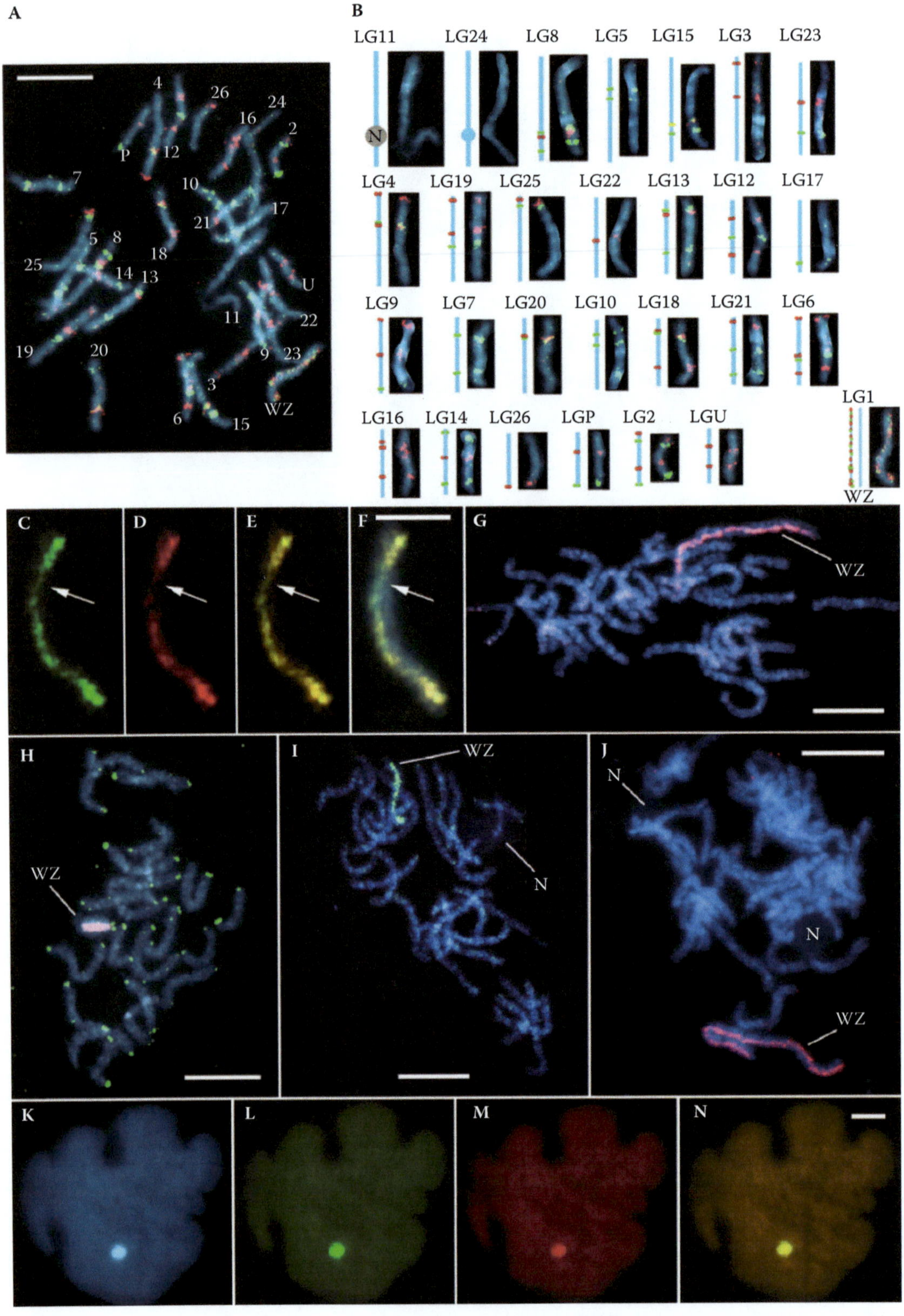

FIGURE 3.1 Fluorescence *in situ* hybridization (FISH) images. ***(See text for full caption.)***

FIGURE 5.1 Phenotypic plasticity in *Bicyclus anynana* wing pattern development. Two distinct seasonal forms occur in the wild and can be mimicked in the laboratory by rearing larvae at 27°C or 20°C. The emerging adults will then resemble the natural wet-season (left) and dry-season (right) forms, respectively. The distal section of the ventral surface of both forewings and hindwings is shown. The dorsal surface (not visible) does not show plasticity in relation to rearing temperature. The adaptive significance and underlying physiological basis of the alternative phenotypes is discussed in the text.

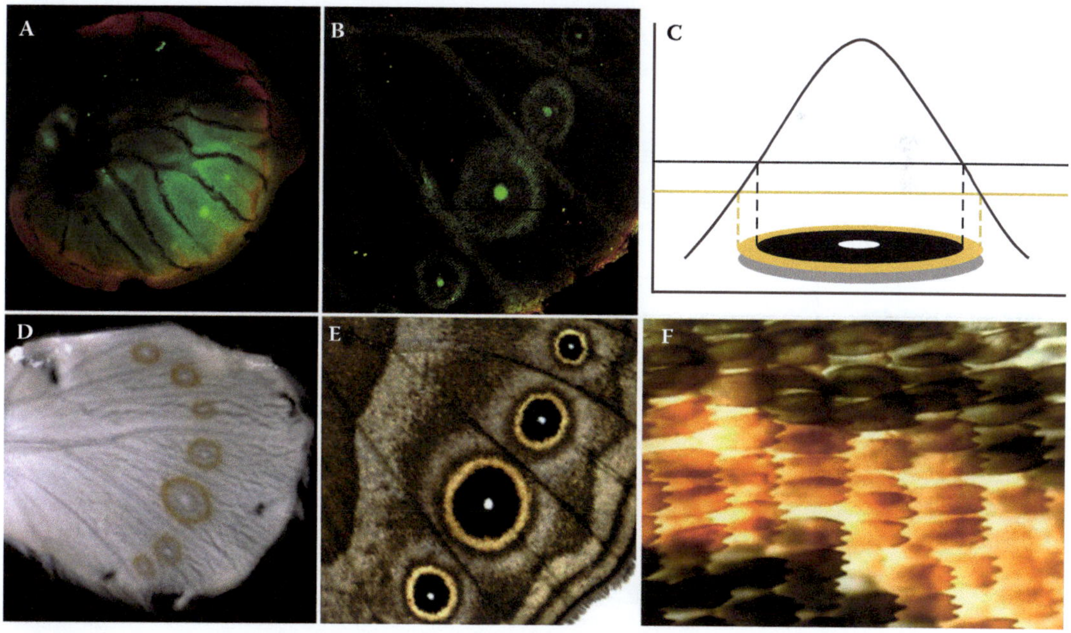

FIGURE 5.2 Eyespot development in *Bicyclus anynana*. ***(See text for full caption.)***

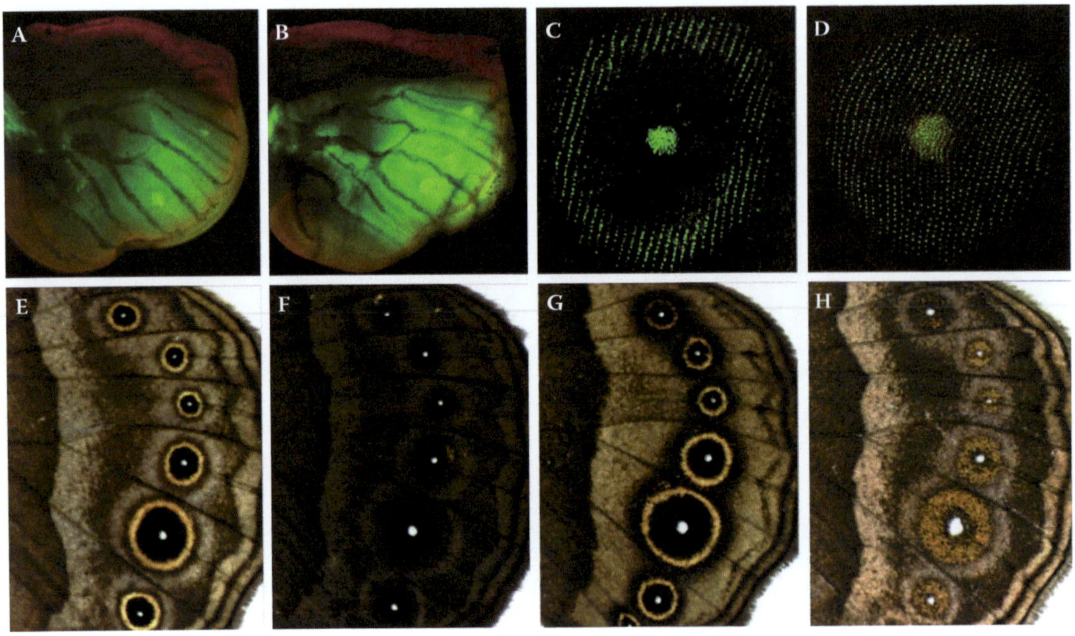

FIGURE 5.4 Mutations of large effect on *Bicyclus anynana* wing patterns. ***(See text for full caption.)***

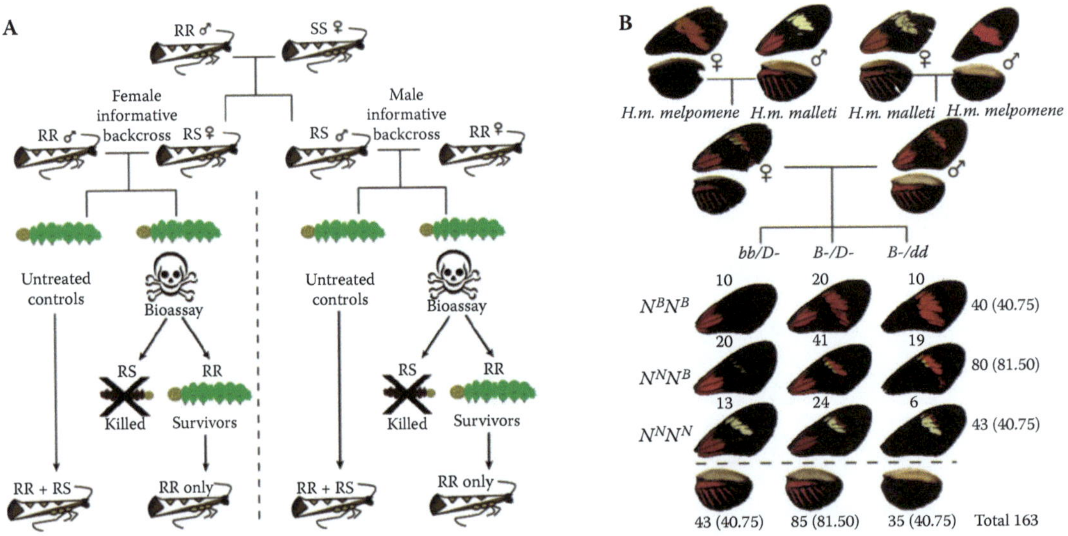

FIGURE 6.1 Cross designs for genetic mapping. ***(See text for full caption.)***

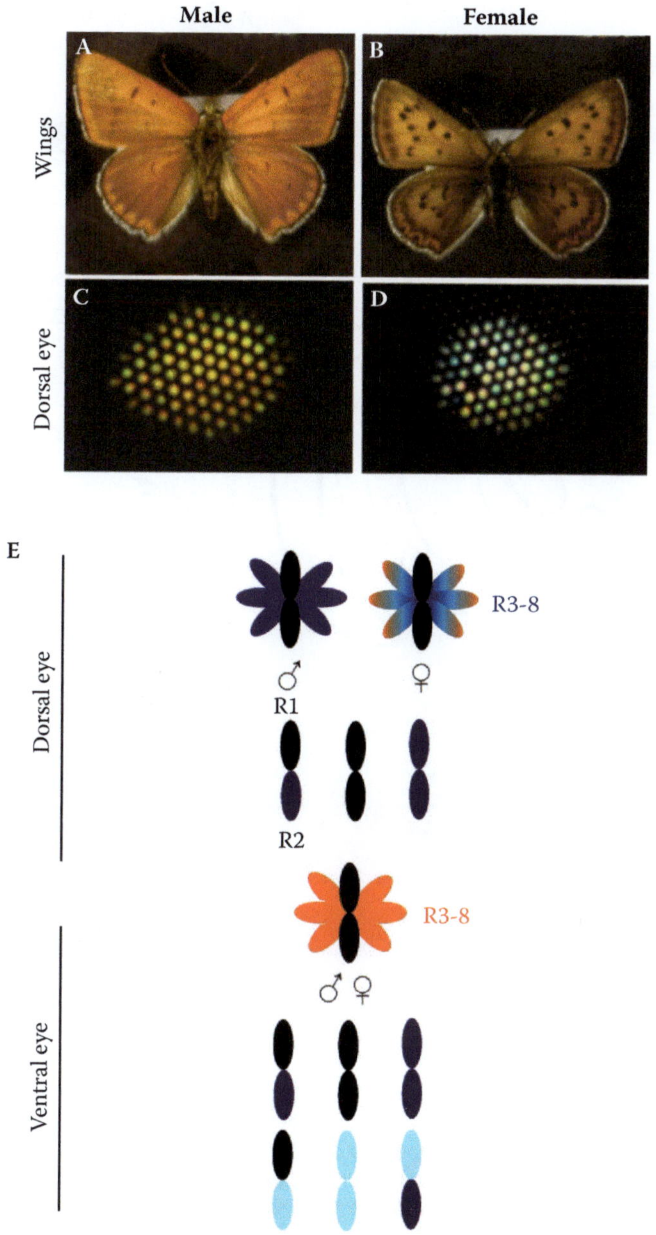

FIGURE 7.5 Sex differences in wing color pattern, eyeshine, and opsin expression patterns in *Lycaena rubidus*. (A) UV-reflecting scales (iridescent purple) on the lower forewing and outer hind wing margins of males. (B) Non-UV-reflecting scales on wings of females. Eyeshine from the dorsal eye of a male (C) and a female (D) showing strongly sexually dimorphic coloration. (E) Diagram summarizing the pattern of opsin expression in the *L. rubidus* eye. Dark blue indicates *BRh1* opsin mRNA expression. Orange indicates *LWRh* opsin mRNA expression. Dark blue and orange indicate coexpression of *BRh1* and *LWRh* opsin mRNAs. Black indicates *UVRh* opsin mRNA expression. Light blue indicates *BRh2* opsin mRNA expression. Adapted from Sison-Mangus et al. (2006).

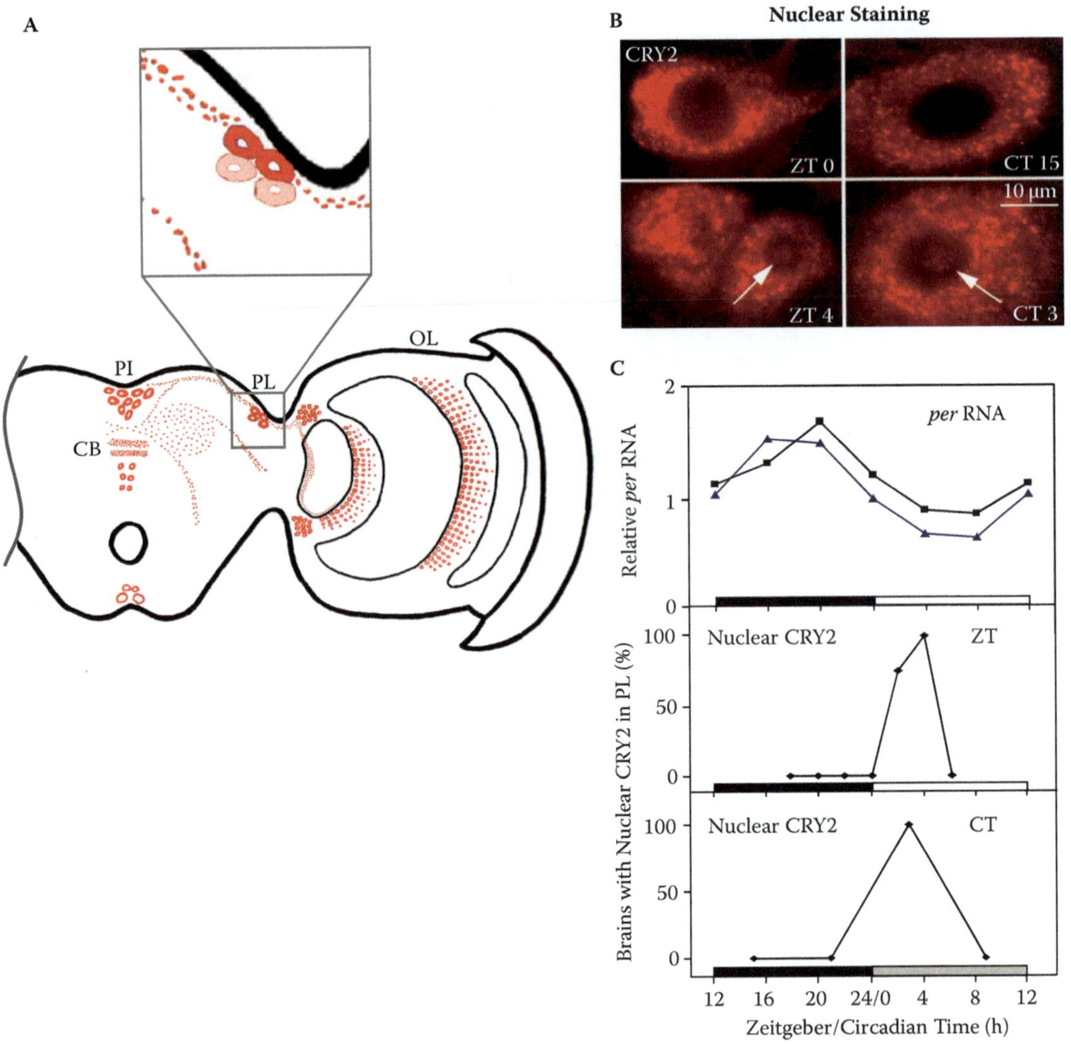

FIGURE 8.3 Cellular location of clock cells in monarch butterfly brain. (A) Schematic diagram of a partial frontal brain section illustrating the topography of CRY2-positive cells and neuronal projections, revealed by using monarch-specific antibodies. (OL) Optic lobe; (PL) pars lateralis; (PI) pars intercerebralis; (CB) central body. Top, enlarged PL regions showing the four clock protein-positive cells; the two red cells coexpress PER, TIM, CRY1, and CRY2, while the two pink cells coexpress TIM and CRY2. Modified from Zhu et al. (2008). (B) CRY2 nuclear staining in PL cells. Top left, zeitgeber time [ZT] 0; bottom left, ZT4; top right, circadian time [CT] 15; bottom right, CT3. CRY2 staining was not found in the nucleus at ZT0 or CT15, but it was found in the nucleus in PL at ZT4 and CT3 (arrows). From Zhu et al. (2008). (C) Comparison of *per* RNA levels in the brain with temporal patterns of CRY2 nuclear staining in PL. *Per* RNA levels for two sets of dissected brains without photoreceptors (black and blue lines) collected at 4-hour intervals over 24 hours in LD (upper). Semiquantitative assessment of nuclear CRY2 immunostaining in PL at seven ZT (middle) and CT (lower) times plotted as a percentage of brains examined (n = 4–5 brains for each time point). From Zhu et al. (2008).

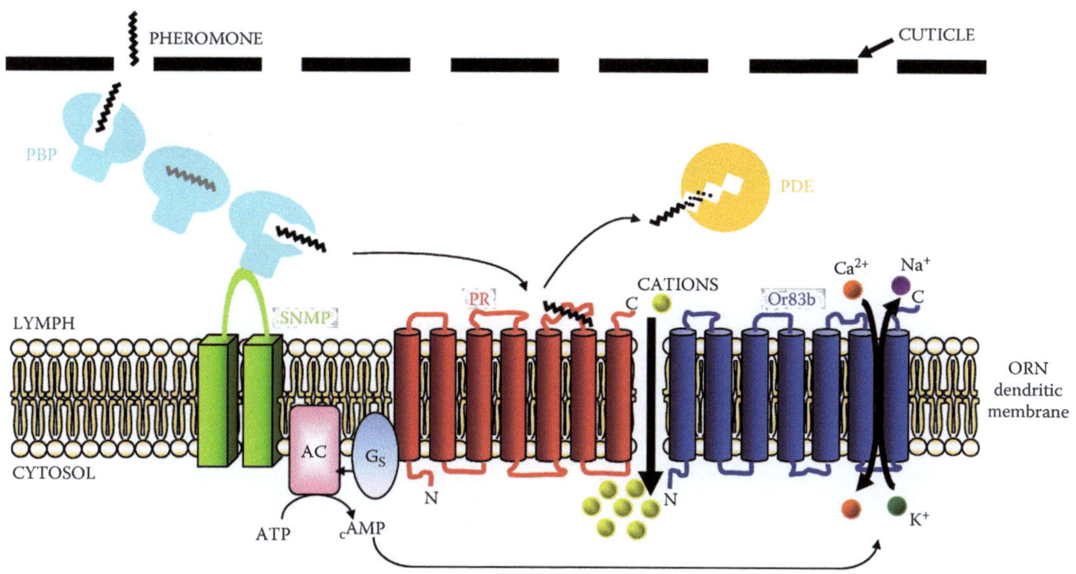

FIGURE 10.3 Schematic model of pheromone reception in moths. (Modified from Rützler and Zwiebel 2005.) ***(See text for full caption.)***

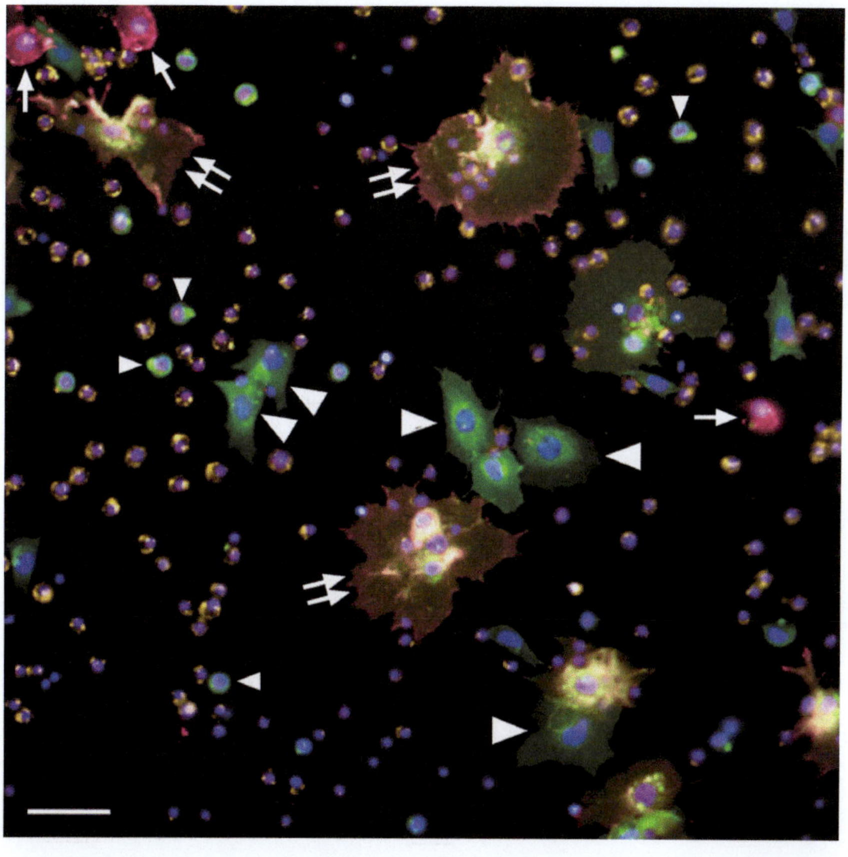

FIGURE 14.2 *Manduca sexta* hemocytes cultured for 60 minutes in Grace's medium. ***(See text for full caption.)***

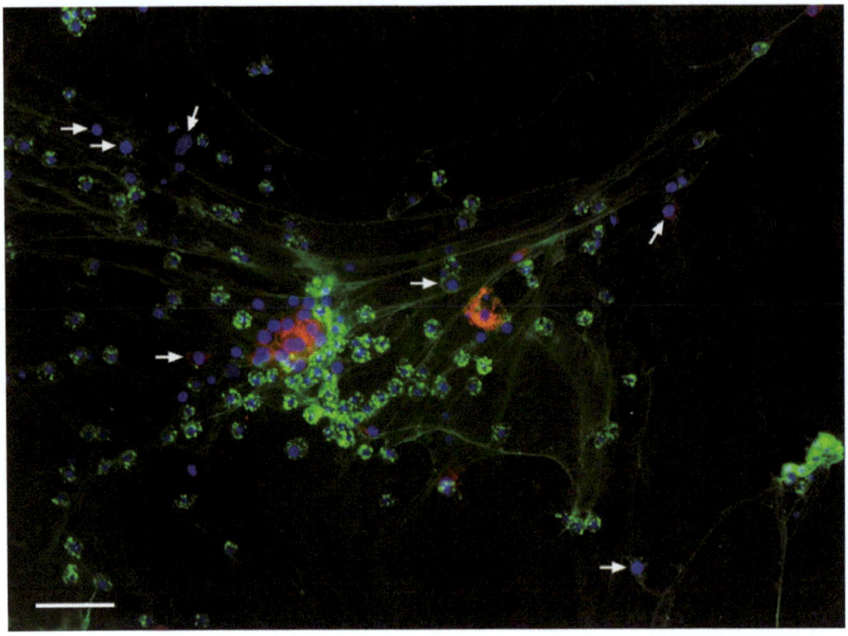

FIGURE 14.4 Labeling of adhesion molecules in aggregation of *Manduca sexta* hemocytes. ***(See text for full caption.)***

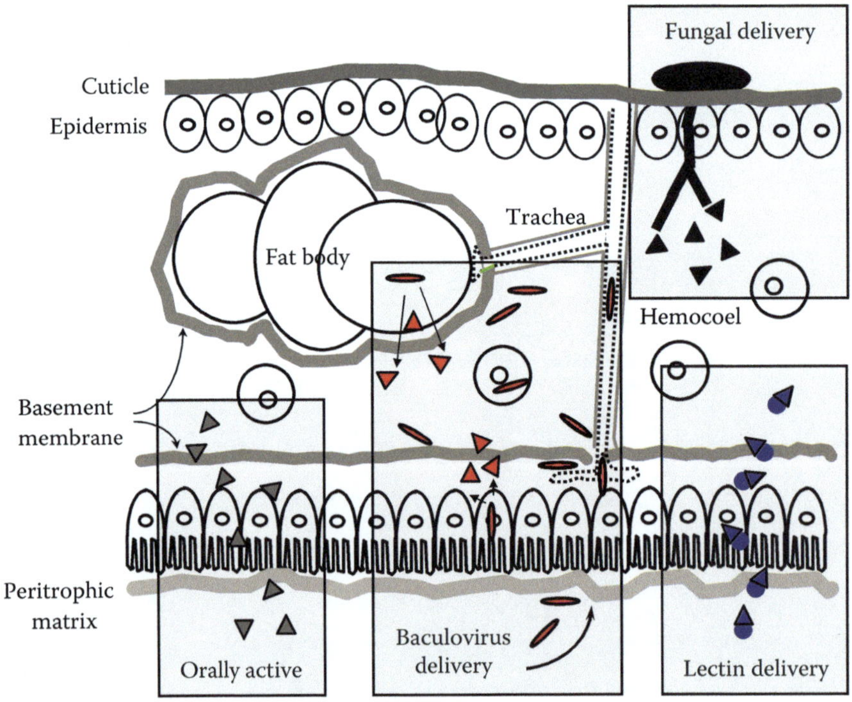

FIGURE 16.1 Delivery of intrahemocoelic toxins for management of lepidopteran pests. ***(See text for full caption.)***